Biological Science and Innovation for Achieving the Sustainable Development Goals

Karen Ghazaryan
https://orcid.org/0000-0002-8973-2595
Yerevan State University, Armenia

Karen Trchounian
Yerevan State University, Armenia

Abhishek Singh
https://orcid.org/0000-0002-0049-6910
Yerevan State University, Armenia

IGI Global
Scientific Publishing
Publishing Tomorrow's Research Today

Vice President of Editorial Melissa Wagner
Director of Acquisitions Mikaela Felty
Director of Book Development Jocelynn Hessler
Production Manager Mike Brehm
Cover Design Jose Rosado

Published in the United States of America by
IGI Global Scientific Publishing
701 East Chocolate Avenue
Hershey, PA, 17033, USA
Tel: 717-533-8845 | Fax: 717-533-7115
Website: https://www.igi-global.com E-mail: cust@igi-global.com

Library of Congress Cataloging-in-Publication Data

Names: Ghazaryan, Karen editor | Trchounian, Karen editor | Singh, Abhishek editor
Title: Biological science and innovation for achieving the sustainable development goals / edited by Karen Ghazaryan, Karen Trchounian, Abhishek Singh.
Description: Hershey, PA : IGI Global Scientific Publishing, [2026] | Includes bibliographical references and index. | Summary: "This book highlights how biological sciences drive progress toward the UN Sustainable Development Goals by addressing challenges such as climate change, biodiversity loss, and public health. Through interdisciplinary research and practical insights, it bridges science, policy, and innovation to promote sustainable development and environmental resilience worldwide"-- Provided by publisher.
Identifiers: LCCN 2025053774 (print) | LCCN 2025053775 (ebook) | ISBN 9798337315249 hardcover | ISBN 9798337315256 paperback | ISBN 9798337315263 ebook
Subjects: LCSH: Sustainable agriculture--Research | Environmental sciences--Research | Biotechnology--Environmental aspects | Sustainable development
Classification: LCC S494.5.S86 B555 2026 (print) | LCC S494.5.S86 (ebook)
LC record available at https://lccn.loc.gov/2025053774
LC ebook record available at https://lccn.loc.gov/2025053775

British Cataloguing in Publication Data
A Cataloguing in Publication record for this book is available from the British Library.

This volume, published as part of the BSES-SDGs conference, is dedicated with respect and pride to the 90th anniversary of the Faculty of Biology, the 10th anniversary of the Research Institute of Biology, and the great scientists whose hard work and contributions have greatly advanced biological science at Yerevan State University.

Prof. Armen Takhtajian (1910-2009): An eminent botanist whose classification scheme for flowering plants reshaped global plant systematics. He authored about 400 works, including major monographs such as *Higher Plants* (1956), *Die Evolution der Angiospermen* (1959), and *Flowering Plants: Origin and Dispersal* (1969). His "Takhtajian system" remains influential in botanical institutes worldwide.

Prof. Mikhail Chailakhyan (1902-1991): A pioneering Armenian plant physiologist who discovered the universal flowering hormone "florigen" and established the hormonal theory of flowering, fundamentally linking light, hormones, and plant development. His pioneering studies on photoperiodism and phytohormones transformed modern plant physiology and laid the groundwork for improving global crop productivity.

Prof. Hrachya Buniatyan (1907-1981): Founder of biochemistry and neurochemistry in Armenia. He investigated brain biochemistry, vitamins, phosphatides, and biogenic amines, discovered a new nicotinamide coenzyme, highlighted the role of γ-aminobutyric acid in metabolism, and uncovered copper-containing proteins in the brain.

Prof. Ezras Hasratyan (1903-1981): A distinguished Armenian physiologist renowned for his pioneering research on higher nervous activity and the adaptive plasticity of the nervous system. He revealed the protective and therapeutic roles of neural inhibition and advanced understanding of the body's response to stress, trauma, and heat shock. His studies provided crucial insights into neurophysiological adaptation and recovery, forming the basis for modern medical physiology.

Prof. Norair Sissakian (1907-1966): A biochemist and one of the founders of cosmic biology. His research spanned plant biochemistry, technical biochemistry, radiation biology, and the preparation of humans for space flight. He led the development of over 10 research laboratories, chaired international bioastronautics committees, and the far side lunar crater "Sisakyan" was named after him.

Their collective legacy continues to guide the Yerevan State University, motivating its professors and researchers to pursue excellence, innovation, and sustainability in the fields of biological and environmental sciences.

Table of Contents

Foreword xxiii

Preface xxv

Chapter 1
FocA Maintains Constant Intracellular Formate Levels During Fermentative Growth of *Escherichia coli* 1
Christopher Erdmann, Institute of Microbiology, Germany
Gary Sawers, Institute of Microbiology, Germany

Chapter 2
Hydrogenase Activity and Polyhydroxyalkanoate Accumulation in *Cupriavidus necator H16* Cultivated on Dairy and Biodiesel Industry By-Products 21
Anna Poladyan, Yerevan State University, Armenia
Ofelya Karapetyan, Yerevan State University, Armenia
Ela Minasyan, Yerevan State University, Armenia
Karen Trchounian, Yerevan State University, Armenia

Chapter 3
Exploring the Resilience of Traditional Barley (*Hordeum vulgare* L.) Genotype in Mitigating Salinity Stress for Sustainable Agriculture and Food Security 39
Anna Harutyunyan, Faculty of Biology, Yerevan State University, Armenia
Abhishek Singh, Faculty of Biology, Yerevan State University, Armenia
Hrant Khachatryan, Faculty of Biology, Yerevan State University, Armenia
Hasmik Movsesyan, Faculty of Biology, Yerevan State University, Armenia
Karen Ghazaryan, Faculty of Biology, Yerevan State University, Armenia

Chapter 4
Evaluation of Wheat Straw for Biofuel Production Using Various Microbial Strains 61
Anna Galstyan, Research Institute of Biology, Armenia
Nare Patvakanyan, Department of Biochemistry, Microbiology, and Biotechnology, Armenia
Anahit Shirvanyan, Research Institute of Biology, Armenia
Liana Vanyan, Research Institute of Biology, Armenia
Ayaulym Daniyarova, Research School of Food Engineering, Kazakhstan
Kamila Baichiyeva, Research School of Food Engineering, Kazakhstan
Kairat Bekbayev, Research School of Food Engineering, Kazakhstan
Karen Trchounian, Research Institute of Biology, Armenia

Chapter 5
Effect of Nanoparticles on Germination and Seedling Tolerance Traits of Armenian Genotypes of Wheat (*Triticum aestivum* L.) Under Salinity Stress ... 85
Armine Chakhmakhchyan, Faculty of Biology, Yerevan State University, Armenia
Abhishek Singh, Faculty of Biology, Yerevan State University, Armenia
Nare Darbinyan, Faculty of Biology, Yerevan State University, Armenia
Sakshi Singh, Faculty of Biology, Yerevan State University, Armenia
Karen Ghazaryan, Faculty of Biology, Yerevan State University, Armenia

Chapter 6
Physiological and Biochemical Responses of Salt-Tolerant Cultivated Amaranthus "Ultra" to Salinity Stress 103
Gohar Margaryan, Faculty of Biology, Yerevan State University, Armenia
Abhishek Singh, Faculty of Biology, Yerevan State University, Armenia
Hrant Khachatryan, Faculty of Biology, Yerevan State University, Armenia
Karen Ghazaryan, Faculty of Biology, Yerevan State University, Armenia

Chapter 7
The Influence of *Dicranum scoparium* Moss Extract on the Growth of Yeasts *Candida guilliermondii* NP-4 131
Gayane Henrik Semerjyan, Yerevan State Unisversity, Armenia
Inessa Henrik Semerjyan, Yerevan State University, Armenia
Syuzan Ashot Marutyan, Yerevan State University, Armenia
Hripsime Masis Ter-Sahakyan, Yerevan State University, Armenia
Hasmik Maenlest Karapetyan, Yerevan State University, Armenia
Seda Victor Marutyan, Yerevan State University, Armenia

Chapter 8
Recellularized Rat Liver Scaffold: An In Vitro Model for Biosafety Assessments 151
Olga Kalinovskaia, L.A. Orbeli Institute of Physiology, Armenia
Mikhail Poliakov, L.A. Orbeli Institute of Physiology, Armenia
Zaruhi Karabekian, L.A. Orbeli Institute of Physiology, Armenia

Chapter 9
Isolation, Characterization, and Anti-Cancer Properties of Microalgal Mono-galactosyldiacylglycerol (MGDG) 175
Baghish Harutyunyan, Scientific and Production Center "Armbiotechnology" of the National Academy of Sciences of Armenia, Armenia
Anna Zakoyan, Scientific and Production Center "Armbiotechnology" of the National Academy of Sciences of Armenia, Armenia
Ani Aghabekyan, Scientific and Production Center "Armbiotechnology" of the National Academy of Sciences of Armenia, Armenia
Angelina Sayadyan, Scientific and Production Center "Armbiotechnology" of the National Academy of Sciences of Armenia, Armenia
Elena Kalita, CANDLE Synchrotron Research Institute, Institute of Molecular Biology NAS RA, Armenia
Nelly Babayan, Faculty of Biology, Institute of Molecular Biology NAS RA, Yerevan State University, Armenia
Iren Shahazizyan, Faculty of Biology, Research Institute of Biology, Yerevan State University, Armenia
Vigen B. Goginyan, Scientific and Production Center "Armbiotechnology" of the National Academy of Sciences of Armenia, Armenia

Chapter 10
Ecological Resources and Environmental Solutions for Sustainable Development in Togo and Sierra Leone: Challenges of Innovative Management in a Multicultural and Globalized African Context...................... 197
Sèmèvo Magloire Assogba, Baikal State University, Russia
Malame Yakinambe, Baikal State University, Russia
Agossou Elie Adona, Baikal State University, Russia
Victor Yakovlevich Kuzevanov, Baikal State University, Russia
Shirley Koroma, Baikal State University, Russia

Chapter 11
Assessment of Heavy Metals Contamination in Soils of Mining Area of Tumanyan Region, Armenia: Distribution and Environmental Risks 217
Hasmik Vardumyan, Faculty of Biology, Yerevan State University, Armenia
Abhishek Singh, Faculty of Biology, Yerevan State University, Armenia
Rupesh Kumar Singh, Centre for the Research and Technology of Agro-Environmental and Biological Sciences (CITAB), Inov4Agro, University of Trás-os-Montes and Alto Douro, Portugal
João Ricardo Sousa, Centre for the Research and Technology of Agro-Environmental and Biological Sciences (CITAB), Inov4Agro, University of Trás-os-Montes and Alto Douro, Portugal
Karen Ghazaryan, Faculty of Biology, Yerevan State University, Armenia

Chapter 12
Assessing the Environmental Impacts of Pesticides: A Case Study From Armenia 241
Meri Arzumanyan, Faculty of Biology, Research Institute of Biology, Armenia
Lusine Margaryan, Faculty of Biology, Research Institute of Biology, Armenia
Narine Zakaryan, Faculty of Biology, Research Institute of Biology, Armenia
Ruzanna Matevosyan, Faculty of Biology, Research Institute of Biology, Armenia
Inessa Eloyan, Faculty of Biology, Research Institute of Biology, Armenia
Iren Shahazizyan, Faculty of Biology, Research Institute of Biology, Armenia
Anzhela Sargsyan, Faculty of Biology, Research Institute of Biology, Armenia
Anahit Atoyants, Research Institute of Biology, Armenia
Siranush Nanagulyan, Faculty of Biology, Research Institute of Biology, Armenia

Chapter 13
Antioxidant Enzyme Activities in Yeast *Candida guilliermondii* NP-4 Exposed to Millimeter Waves 263
Anna Muradyan, Yerevan State University, Armenia
Syuzan Marutyan, Yerevan State University, Armenia
Hasmik Karapetyan, Yerevan State University, Armenia
Meri Hayrapetyan, Yerevan State University, Armenia
Seda Marutyan, Yerevan State University, Armenia

Chapter 14
Antibiotic Activity of Plant Metabolites and Possibilities of Their Application 283
Aram Aramyan, Yerevan State University, Armenia
Silvard Tadevosyan, Research Institute of Biology, Armenia
Alvard Minasyan, Research Institute of Biology, Armenia
Anush Babayan, Research Institute of Biology, Armenia
Naira Sahakyan, Research Institute of Biology, Armenia

Chapter 15
Adaptation of Ixodid Ticks (Ixodidae) to Parasitism on Cattle in Northeastern Armenia 303
Martin Yakov Rukhkyan, Scientific Center of Zoology and Hydroecology, Armenia
Oleg Valery Shcherbakov, Institute of Zoology, Armenia
Ruzan Levon Hovhannisyan, Institute of Zoology, Armenia

Chapter 16
Global Scientific Trends in Biochar Applications and Research for Salinity and Drought Stresses Management: A Bibliometric Perspective 315
Sakshi Singh, Faculty of Biology, Yerevan State University, Armenia
Abhishek Singh, Faculty of Biology, Yerevan State University, Armenia
Rupesh Kumar Singh, Centro de Investigação e Tecnologias Agroambientais e Biológicas, Universidade de Trás-os-Montes e Alto Douro, Vila Real, Portugal
João Ricardo Sousa, Centro de Investigação e Tecnologias Agroambientais e Biológicas, Universidade de Trás-os-Montes e Alto Douro, Vila Real, Portugal
Karen Ghazaryan, Faculty of Biology, Yerevan State University, Armenia

Chapter 17
Prospects of Algae-Based Biofuel Production for the Development of a Carbon-Neutral Economy 347
Maria Vetrova, Saint Petersburg State University, Russia

Compilation of References 363

About the Contributors 431

Index 435

Detailed Table of Contents

Foreword xxiii

Preface xxv

Chapter 1
FocA Maintains Constant Intracellular Formate Levels During Fermentative Growth of *Escherichia coli* 1
Christopher Erdmann, Institute of Microbiology, Germany
Gary Sawers, Institute of Microbiology, Germany

Formic acid is a key molecule in enterobacterial mixed-acid fermentation. Formate is either converted to H_2 and CO_2 by formate hydrogenlyase (FHL-1), or translocated from the cell by the formate channel, FocA. FocA also imports formate into non-growing, stationary-phase cells. *Escherichia coli* has a second formate channel, FocB, but it is unclear whether it also translocates formate. The authors show that formate levels remain constant during growth, but decrease 10-fold in stationary phase; most of the formate accumulates in the culture medium. Reducing the growth rate of fermentating cells by adding the toxic formate analogue hypophosphite does not negatively impact formate homeostasis. Hypophosphite sensitivity was lost in a focA mutant, but not in a focB mutant. Intracellular formate accumulated significantly more in the focA mutant compared with the parental strain, and was mainly converted to H_2 and CO_2 by FHL-1. Together, these data indicate that FocA is mainly responsible for maintenance of formate homeostasis, probably functioning with FHL-1, during pH-neutral glucose fermentation.

Chapter 2
Hydrogenase Activity and Polyhydroxyalkanoate Accumulation in *Cupriavidus necator H16* Cultivated on Dairy and Biodiesel Industry By-Products 21
Anna Poladyan, Yerevan State University, Armenia
Ofelya Karapetyan, Yerevan State University, Armenia
Ela Minasyan, Yerevan State University, Armenia
Karen Trchounian, Yerevan State University, Armenia

Cupriavidus necator H16 expresses O_2-tolerant [NiFe]-hydrogenases (Hyds), which present promising anodic biocatalysts in enzymatic fuel cells (EFC). Bacteria also produce polyhydroxyalkanates (PHA). Depolymerization of this polymer leads to the production of crotonic acid (CA), which was estimated by the production of CA. Cheese and curd whey (SW and AW) with a combination of glycerol (G) assessed for the growth of *C. necator H16*, PHA accumulation, and Hyds activity. Bacterial growth, PHA accumulation, and Hyd activity were observed under microaerophilic conditions for 7 days. Prolonged bacterial growth was observed in glycerol-supplemented samples. Maximum biomass formation was observed on the 5th day in AW (OD_{600}3.1) and on the 6th day in SW+G (OD_{600} 3.3). pH and ORP decrease and Hyd activity were detected under all conditions; however, SW+G sample showed enhanced Hyd activity (0.17 ± 0.03 $U \cdot min^{-1} \cdot g^{-1}$CDW) compared with G-free samples. A clear correlation was shown between PHA accumulation and Hyd activity: with the Hyd activity decline, CA concentration increased, reaching up to 2.4 ± 0.05 mg/mL. The results will advance hydrogen-based EFC systems as well as the sustainable synthesis of biodegradable PHAs.

Chapter 3
Exploring the Resilience of Traditional Barley (*Hordeum vulgare* L.) Genotype in Mitigating Salinity Stress for Sustainable Agriculture and Food Security 39
Anna Harutyunyan, Faculty of Biology, Yerevan State University, Armenia
Abhishek Singh, Faculty of Biology, Yerevan State University, Armenia
Hrant Khachatryan, Faculty of Biology, Yerevan State University, Armenia
Hasmik Movsesyan, Faculty of Biology, Yerevan State University, Armenia
Karen Ghazaryan, Faculty of Biology, Yerevan State University, Armenia

Understanding the salinity stress processes is crucial for crop development and sustainable agriculture. Salinity stress has the highest level of adverse impact on crop growth and development, agricultural yield, and food security compared to other abiotic stressors. Therefore, it is important to protect, investigate, and document traditional plants that can reduce salt damage and increase salt tolerance in newly grown crops for food security and sustainable agriculture. This study aimed to examine the response of the cereal crop barley (*Hordeum vulgare* L.) genotypes to salinity stress. The seeds of the local genotype of barley (Alashkert) were grown for 3 months in the greenhouse of Yerevan State University (Republic of Armenia) in perlite pots while applying various concentrations of NaCl (0, 100, 200, 300, 400, and 500 mM). In response to increased salt stress, barley stem length and diameter, root, stem, leaf biomass and water content, chlorophyll content, and gas exchange were all reduced. However, the data also demonstrate that the barley cultivar absorbs more Na^+ ions and increases the Na^+/K^+ ratio within its leaves, shoots, and roots under high-salinity conditions. The findings suggest that the "Alashkert" local barley genotype has potential as a crop for use in saline agriculture in Armenia and other countries. Therefore, our in-depth primary analysis of barley genotypes at the physiological, morphological, and biochemical levels provides comprehensive insights into the potential to improve agricultural practices for crop improvement and management programs.

Chapter 4
Evaluation of Wheat Straw for Biofuel Production Using Various Microbial Strains 61
Anna Galstyan, Research Institute of Biology, Armenia
Nare Patvakanyan, Department of Biochemistry, Microbiology, and Biotechnology, Armenia
Anahit Shirvanyan, Research Institute of Biology, Armenia
Liana Vanyan, Research Institute of Biology, Armenia
Ayaulym Daniyarova, Research School of Food Engineering, Kazakhstan
Kamila Baichiyeva, Research School of Food Engineering, Kazakhstan
Kairat Bekbayev, Research School of Food Engineering, Kazakhstan
Karen Trchounian, Research Institute of Biology, Armenia

Wheat straw (WS) is one of the most widely generated agro-industrial residues, which has high-carbohydrate content; however, its complex structure requires optimized processing and efficient microbial strains to achieve high conversion yields. Thermal-treated (TT) and acid-treated (AT) WS were used as substrates for *S. cerevisiae* to assess bioethanol production, while untreated WS as substrate for hydrogen, ethanol, and butanol production by *Clostridium* species. The highest ethanol yield (80 ± 2 g L^{-1}) for *S. cerevisiae* ATCC 13007 strain was produced using non-diluted AT hydrolysates during oxygen-limited, while the highest ethanol yield (92 ± 3 g L^{-1}) for *S. cerevisiae* ATCC 9804 observed in non-diluted AT hydrolysates under aerobic conditions. Six percent WS with glucose supplementation maximizes and sustains hydrogen yields in *C. beijerinckii* producing up to 30 mmol L^{-1} butanol and 72 mmol L^{-1} ethanol. The findings contribute to the valorization of agro-industrial residues and the advancement of renewable biofuel production strategies.

Chapter 5
Effect of Nanoparticles on Germination and Seedling Tolerance Traits of Armenian Genotypes of Wheat (*Triticum aestivum* L.) Under Salinity Stress ... 85
Armine Chakhmakhchyan, Faculty of Biology, Yerevan State University, Armenia
Abhishek Singh, Faculty of Biology, Yerevan State University, Armenia
Nare Darbinyan, Faculty of Biology, Yerevan State University, Armenia
Sakshi Singh, Faculty of Biology, Yerevan State University, Armenia
Karen Ghazaryan, Faculty of Biology, Yerevan State University, Armenia

Soil salinity is considered to be one of the major abiotic stresses, worldwide which disrupts water uptake, nutrient balance, and physiological processes and limits wheat growth and productivity leading to poor germination, stunting, and loss in yields. Nanoparticles (NPs) have great potential in improving plant salt stress tolerance. The effects of zinc oxide and silicon dioxide nanoparticles (ZnO-NPs and SiO_2 NPs) and (50 mg/L) on seed germination in two Armenian wheat genotypes, Vars and Rima, at various levels of salt stress, as well as their early growth and physiological responses were studied.

Chapter 6
Physiological and Biochemical Responses of Salt-Tolerant Cultivated Amaranthus "Ultra" to Salinity Stress 103
Gohar Margaryan, Faculty of Biology, Yerevan State University, Armenia
Abhishek Singh, Faculty of Biology, Yerevan State University, Armenia
Hrant Khachatryan, Faculty of Biology, Yerevan State University, Armenia
Karen Ghazaryan, Faculty of Biology, Yerevan State University, Armenia

In arid and semiarid areas of the world, drought and soil salinity are the main reasons for low production. The Ararat Plain, an important agricultural zone in Armenia, suffers strongly from salinity stress and the potential for agriculture in this region is largely reduced. Hence, it is important to select salt-tolerant crops and plant them in these areas where they can grow while helping in land reclamation to cope with food security challenges. Due to the increasingly serious soil salinity worldwide, in this part of the study the authors tested salinity tolerance ability of specific cultivar of Amaranthus "Ultra". This study therefore attempted to characterize the salt stress resistance of A. "Ultra" and examine to what extent these responses are associated with salt tolerance mechanisms.

Chapter 7
The Influence of *Dicranum scoparium* Moss Extract on the Growth of Yeasts *Candida guilliermondii* NP-4 131
Gayane Henrik Semerjyan, Yerevan State Unisversity, Armenia
Inessa Henrik Semerjyan, Yerevan State University, Armenia
Syuzan Ashot Marutyan, Yerevan State University, Armenia
Hripsime Masis Ter-Sahakyan, Yerevan State University, Armenia
Hasmik Maenlest Karapetyan, Yerevan State University, Armenia
Seda Victor Marutyan, Yerevan State University, Armenia

This study aimed to evaluate the effect of D. scoparium moss extract on biomass accumulation, colony formation ability, and growth kinetics of yeasts *Candida guilliermondii* NP-4, as well as its effect on the activity of ATPases, antioxidant enzymes catalase, and superoxid dismutase. The antiradical activity of the moss extract was determined using the free radical method. The total content of phenolic substances and total content of flavonoids of moss was determined. *D. scoparium* moss extracts contained 17.53 µg GAE g-1 of total phenolic compounds and 7.88 µg QE mL-1 of total flavonoids, with an IC_{50} value of 51.487 µg mL-1. Treatment with the moss extract inhibited yeast colony formation and mitotic activity. The moss extract led to a 64% reduction in malondialdehyde levels and enhanced superoxide dismutase activity by 43%. There was a significant increase in total and H+-ATPase activity in the cell homogenates and mitochondria. The data obtained prove the potential application of *D. scoparium* extract as a natural antifungal agent with additional antioxidative properties.

Chapter 8
Recellularized Rat Liver Scaffold: An In Vitro Model for Biosafety Assessments 151
Olga Kalinovskaia, L.A. Orbeli Institute of Physiology, Armenia
Mikhail Poliakov, L.A. Orbeli Institute of Physiology, Armenia
Zaruhi Karabekian, L.A. Orbeli Institute of Physiology, Armenia

At present, fully decellularized animal organ scaffolds are widely used in research for cell biology, tissue engineering, and regenerative medicine. These scaffolds serve as enhanced 3D models, providing natural microenvironments that support and guide cell interactions. In this study, the authors aimed to generate a reproducible 3D in vitro liver model for drug metabolism and toxicity testing that is based on decellularized rat liver scaffolds and to validate the model's suitability for toxicity tests. To restore liver-specific functions, various rat and human cell lines were used, including hepatocytes and macrophages. To date, the authors have achieved high reproducibility in the preparation of the scaffolds and demonstrated low cytotoxicity of liver scaffold material. Immunogenicity assays and experiments on recellularization of rat liver scaffolds are ongoing. The model's metabolic functionality is to be confirmed by assessing liver-specific markers. The development of such a model should enable efficient assessment of drug candidates while reducing the need for animal testing.

Chapter 9
Isolation, Characterization, and Anti-Cancer Properties of Microalgal Monogalactosyldiacylglycerol (MGDG) 175
Baghish Harutyunyan, Scientific and Production Center "Armbiotechnology" of the National Academy of Sciences of Armenia, Armenia
Anna Zakoyan, Scientific and Production Center "Armbiotechnology" of the National Academy of Sciences of Armenia, Armenia
Ani Aghabekyan, Scientific and Production Center "Armbiotechnology" of the National Academy of Sciences of Armenia, Armenia
Angelina Sayadyan, Scientific and Production Center "Armbiotechnology" of the National Academy of Sciences of Armenia, Armenia
Elena Kalita, CANDLE Synchrotron Research Institute, Institute of Molecular Biology NAS RA, Armenia
Nelly Babayan, Faculty of Biology, Institute of Molecular Biology NAS RA, Yerevan State University, Armenia
Iren Shahazizyan, Faculty of Biology, Research Institute of Biology, Yerevan State University, Armenia
Vigen B. Goginyan, Scientific and Production Center "Armbiotechnology" of the National Academy of Sciences of Armenia, Armenia

Glycoglycerolipids are amphiphilic compounds found in thylakoid membranes of cyanobacteria, algae, and plants, attracting interest for their therapeutic potential, especially anticancer applications. They interact with biological molecules and show anticancer, antiviral, and anti-inflammatory properties. Among them, monogalactosyldiacylglycerol (MGDG) inhibits cancer cell proliferation through DNA polymerase inhibition, kinesin repression, and T-cell activation. While plant sources are well studied, microalgal MGDGs remain understudied. This work isolated, purified, and characterized MGDGs from the green microalgae Ch. vulgaris and N. oleoabundans and evaluated their anticancer activity. Microalgae were cultured in a photobioreactor, lipids were extracted and MGDGs purified by chromatography. HPLC and NMR confirmed their profiles. In MTT assays, microalgal MGDGs markedly reduced MCF-7 cell viability, with stronger effects when combined with 5-aminolevulinic acid, while parsley MGDG showed moderate activity. These findings highlight the strong anticancer potential of microalgal MGDGs.

Chapter 10
Ecological Resources and Environmental Solutions for Sustainable Development in Togo and Sierra Leone: Challenges of Innovative Management in a Multicultural and Globalized African Context...................... 197
Sèmèvo Magloire Assogba, Baikal State University, Russia
Malame Yakinambe, Baikal State University, Russia
Agossou Elie Adona, Baikal State University, Russia
Victor Yakovlevich Kuzevanov, Baikal State University, Russia
Shirley Koroma, Baikal State University, Russia

The nations of West Africa, notably Togo and Sierra Leone, face complex ecological and managerial challenges in a globalized and culturally diverse environment. This study focuses on issues related to small and medium-sized enterprises (SMEs), logistics, e-commerce, and intercultural management, which are key factors in achieving the sustainable development of these countries. SMEs in these two nations face limited access to finance, poor infrastructure, and insufficient managerial skills. Simultaneously, the rapid-growing e-commerce sector encounters significant challenges such as last-mile delivery problems, lack of adequate addressing systems, and high transport costs. Integrating sustainable development principles reveals a direct link to adapting management practices to local cultural contexts. The difficulties in adapting management practices, particularly in commercial interactions and international cooperation with partners such as Russia, China, the United States, and the European Union, have a directly affect the effectiveness of collaborations. This chapter proposes integrated strategies for sustainable development, including digitalization, change management approaches, promotion of green technologies, and the integration of local and regional supply chains. These strategies align the Sustainable Development Goals (SDGs) of the 2030 Agenda, particularly SDG 8 (Decent Work), SDG 9 (Industry, Innovation and Infrastructure), SDG 11 (Sustainable Cities and Communities, Eco-Cities), and SDG 17 (Partnerships to Achieve the Goals). Geospatial solutions will significantly advance e-commerce development.

Chapter 11
Assessment of Heavy Metals Contamination in Soils of Mining Area of Tumanyan Region, Armenia: Distribution and Environmental Risks 217
Hasmik Vardumyan, Faculty of Biology, Yerevan State University, Armenia
Abhishek Singh, Faculty of Biology, Yerevan State University, Armenia
Rupesh Kumar Singh, Centre for the Research and Technology of Agro-Environmental and Biological Sciences (CITAB), Inov4Agro, University of Trás-os-Montes and Alto Douro, Portugal
João Ricardo Sousa, Centre for the Research and Technology of Agro-Environmental and Biological Sciences (CITAB), Inov4Agro, University of Trás-os-Montes and Alto Douro, Portugal
Karen Ghazaryan, Faculty of Biology, Yerevan State University, Armenia

Heavy metal pollution of soils has brought potential risk to the environment and human health. The chapter is a long-term study evaluating the extent of heavy metal pollution in soils of one of Armenia's most important mining regions. Soil samples were taken from various distances (0–20 cm) of the surface of both active and abandoned mining sites. The sampling sites were selected based on accessibility and their distance from contaminant sources. Three other baseline soil properties, pH, texture, and organic matter content, were also taken to aid interpretation. Analytical methods for ecological risk assessment indices like potential ecological risk index (PERI), pollution index (PI), Nemerow integrated pollution index (NIPI) were useful for this study. The findings revealed that the concentration levels of some heavy metals were considerably higher than the local and world set standards, in close proximity to tailings and ore processing areas. Spatial distribution maps indicate a spatial association of heavy metals uptake with the distance from mining activities. According to the ecological risk assessment, there is high potential of phytotoxicity and wide contamination of food chain in treated sites. These results highlight the necessity for frequent monitoring and the application of reparation techniques such as phytoremediation to reduce heavy-metal risks in mining-affected areas of Armenia. The work adds to the limited literature on post-mining land use planning and environmental health in South Caucasus.

Chapter 12
Assessing the Environmental Impacts of Pesticides: A Case Study From Armenia 241
Meri Arzumanyan, Faculty of Biology, Research Institute of Biology, Armenia
Lusine Margaryan, Faculty of Biology, Research Institute of Biology, Armenia
Narine Zakaryan, Faculty of Biology, Research Institute of Biology, Armenia
Ruzanna Matevosyan, Faculty of Biology, Research Institute of Biology, Armenia
Inessa Eloyan, Faculty of Biology, Research Institute of Biology, Armenia
Iren Shahazizyan, Faculty of Biology, Research Institute of Biology, Armenia
Anzhela Sargsyan, Faculty of Biology, Research Institute of Biology, Armenia
Anahit Atoyants, Research Institute of Biology, Armenia
Siranush Nanagulyan, Faculty of Biology, Research Institute of Biology, Armenia

This research investigates the impacts of pesticides use on the soil-plant-fungi-animal chain in the mountain regions of Armenia, using Tradescantia plants for mutation detections, micromycetes, molluscs richness for secondary impact of pesticides on the non-target biodiversity. Data sampling has been conducted from 2022 to 2023 in the potato fields, during spring, summer, and autumn seasons on three different locations: Gogaran, Arevshat, and Lichk mountain-steppe regions of Armenia. Soil chemical analysis revealed that experimental areas contained four dangerous active components of pesticides including Arsenous. Using two model tests of Tradescantia, Trad-SHM (stamen hair mutations of Tradescantia) and Trad-MN (micronuclei in tetrads of plant microspores), a significant increase in the level of mutation events was shown in all three experimental soil variants compared to their background ones. A total 95 species of micromycetes was recorded, out of which the diversity noticeably declined. As a result, some beneficial species of Mucoromycetes were destroyed, which led to the emergence of a resistant pathogenic fungal species *Agroathelia rolfsii*. As for molluscs species richness remained constant throughout the experiment. However, molluscs species abundance and mutation levels appeared varying in experimental and control locations. The authors observed genotoxic effects of pesticides on *Georginapaeous hohenackeri* with high levels of DNA damage in pesticide used plots. These findings highlight the importance of adopting sustainable agricultural practices to mitigate negative impacts on non-target species and preserve biodiversity, thereby ensuring the integrity of ecosystems and underscoring the urgent need for stringent pesticide regulation and monitoring the usage in Armenia.

Chapter 13
Antioxidant Enzyme Activities in Yeast *Candida guilliermondii* NP-4
Exposed to Millimeter Waves 263
Anna Muradyan, Yerevan State University, Armenia
Syuzan Marutyan, Yerevan State University, Armenia
Hasmik Karapetyan, Yerevan State University, Armenia
Meri Hayrapetyan, Yerevan State University, Armenia
Seda Marutyan, Yerevan State University, Armenia

Modern telecommunication technologies use the high frequency microwave radiation, the application of which continues to expand rapidly. Exposure to non-ionizing radiation can induce oxidative stress in cells. The biochemical processes involved in overcoming this stress are highly energy-consuming and require additional energy expenditure which consequently enhances the ATP hydrolysis. The indicator of oxidative stress is the intensification of lipid peroxidation, with malondialdehyde (MDA) serving as a key biomarker. In eukaryotic cells, various bioactive dietary supplements, including royal jelly (RJ), may exert beneficial effects on mitigating oxidative stress due to their antioxidant and other biologically active components. Yeasts represent a convenient model for studying the development of oxidative stress and the mechanisms of its overcoming in eukaryotic cells, because they are similar to higher eukaryotes by their structural, reproductive, and functional peculiarities. This study aimed to investigate the effect of 51.8 GHz microwave radiation on ATPase activity, lipid peroxidation, and antioxidant enzyme activity in homogenates and mitochondria of *Candida guilliermondii* NP-4, as well as to assess the potential role of RJ in migtigating of oxidative stress in yeast cells. The results demonstrated that the exposure to millimeter waves of 51.8 GHz and the following repair period led to intensification of lipid peroxidation, evidenced by an increase in MDA level by 75%. In repaired cells, the MDA content was reduced by ≈27%. Microwave radiation also stimulated catalase and superoxide dismutase (SOD) activity by 65% and 30%, respectively, in repaired cells the SOD activity increased by 13%, while catalase activity - by 25%.

Chapter 14
Antibiotic Activity of Plant Metabolites and Possibilities of Their
Application 283
Aram Aramyan, Yerevan State University, Armenia
Silvard Tadevosyan, Research Institute of Biology, Armenia
Alvard Minasyan, Research Institute of Biology, Armenia
Anush Babayan, Research Institute of Biology, Armenia
Naira Sahakyan, Research Institute of Biology, Armenia

The growing threat of antibiotic resistance demands the search for novel alternatives to existing antimicrobial agents. This study aimed to investigate the antibacterial potential of four natural flavonoids—catechin, gallic acid, rutin, and quercetin—and one sugar alcohol, myo-inositol, all abundant in plants used in Armenian folk medicine. Tested bacterial strains included both Gram-negative and Gram-positive species. Myo-inositol demonstrated growth inhibition zones of 7.0–9.0 mm for E. coli K12, 11.0–18.0 mm for B. subtilis, and 12.0–21.0 mm for S. aureus. Rutin showed 12.0–15.0 mm zones for B. subtilis and 15.0–16.0 mm for S. aureus. MTT assays revealed that quercetin (0.25 mg/mL) reduced the MIC of kanamycin fourfold for E. coli pARG25 and halved the MIC of ampicillin for E. coli DH5α-pUC18. Gallic acid and rutin reduced the kanamycin MIC twofold, while catechin halved the ampicillin MIC. These findings indicate the potential of flavonoid-based agents in combating antibiotic resistance, encouraging further investigation into their mechanisms of action and clinical relevance.

Chapter 15
Adaptation of Ixodid Ticks (Ixodidae) to Parasitism on Cattle in Northeastern Armenia 303
Martin Yakov Rukhkyan, Scientific Center of Zoology and Hydroecology, Armenia
Oleg Valery Shcherbakov, Institute of Zoology, Armenia
Ruzan Levon Hovhannisyan, Institute of Zoology, Armenia

The ecology, abundance, and behavioral characteristics of ixodid ticks, parasitizing on the body of farm animals and having epidemiological significance, were studied. A total of 175 cows and 130 sheep were examined. Manual collection of ticks from animals did not cause them any discomfort or stress. Species diversity of ixodid ticks in the mountain-forest zone of Armenia is rather high. The following tick species were detected: *Ixodes ricinus, Ixodes redikorzevi, Ixodes trianguliceps, Rhipicephslus sanguineus, Rhipicephslus annulatus, Rhipicephslus bursa, Hyalomma marginatum, H. asiaticum*, and *Dermacentor marginatus*. Typically, ticks will first cling to the animal forelimbs, head, and neck that is facilitated by the grazing posture of the animal. The distribution of ticks over the animal's body also depends on the host's ability to self-clean and “self-defense”. The research data on seasonal dynamics of ticks in the region can be used by agricultural enterprises in planning the timing and conditions of tick control measures in Dilijan National Park.

Chapter 16
Global Scientific Trends in Biochar Applications and Research for Salinity and Drought Stresses Management: A Bibliometric Perspective 315
Sakshi Singh, Faculty of Biology, Yerevan State University, Armenia
Abhishek Singh, Faculty of Biology, Yerevan State University, Armenia
Rupesh Kumar Singh, Centro de Investigação e Tecnologias Agroambientais e Biológicas, Universidade de Trás-os-Montes e Alto Douro, Vila Real, Portugal
João Ricardo Sousa, Centro de Investigação e Tecnologias Agroambientais e Biológicas, Universidade de Trás-os-Montes e Alto Douro, Vila Real, Portugal
Karen Ghazaryan, Faculty of Biology, Yerevan State University, Armenia

This bibliometric analysis offers a global overview of published research on use of biochar for combating soil salinity and drought in agriculture during 2013-2026. A total of 110 related articles were reviewed, with 4243 citations indicating high academic attention. Journal articles were the most represented, followed by reviews and book chapters, in line with the trend of synthesizing knowledge and conceptual frameworks. Keyword mapping, word clouds, treemaps, and BERT topic modelling found biochar-mediated soil amendment, crop stress tolerance, sustainable agriculture, and emerging cross-sections with nanotechnology as the core topics. Most of the leading contributions were from China, Pakistan, and India, assisted mainly by large publishers and funding agencies. Open access publications have increased in visibility, and network analysis has revealed solid international collaborations and under-explored research frontiers. The results provide insights into research patterns, knowledge gaps, and future prospects for biochar-based approaches targeting improvement of soil health and crop resilience against abiotic stresses.

Chapter 17
Prospects of Algae-Based Biofuel Production for the Development of a Carbon-Neutral Economy 347
Maria Vetrova, Saint Petersburg State University, Russia

In the context of aggravation of the climate agenda and continuing instability in the global energy market, the diversification of the energy structure in favour of innovative and low-carbon technologies and resources, among which bioenergy occupies a special place, is becoming an urgent area of the countries' strategies. According to the Global Bioenergy Statistics Report 2024, biofuels, including bioethanol, biodiesel, biochar, and biogas, account for about 9% of global energy production, which is characterised by low greenhouse gas emissions and high energy efficiency. Taking into account the current trends in the development of bioenergy technologies and its potential to address the climate agenda, this study analyses the possibilities of implementing biocarbon projects and provides an economic assessment of alternative options for biomass production with a focus on the production of third-generation biofuels based on seaweed. The object of special attention is the systematisation of technologies and infrastructural conditions for scaling up third-generation bioenergy, including for diversification of the global energy balance and development of a carbon-neutral economy. As one of the key results of the study, recommendations on the formation of institutional and infrastructural environment for achieving commercial feasibility and effective scaling of biocarbon projects have been developed.

Compilation of References 363

About the Contributors 431

Index 435

Foreword

It is with immense pleasure that I present the edited volume *Biological Science and Innovation for Achieving the Sustainable Development Goals*, published by IGI Global, USA. This volume stands as a significant intellectual outcome of the International Conference on "Global Sciences and Environmental Solutions for the Achievement of Sustainable Development Goals (SDGs)," held from September 24 to 26, 2025, at Yerevan State University, Armenia. The conference was organized within the framework of two remarkable milestones: the 90th Anniversary of the Faculty of Biology and the 10th Anniversary of the Research Institute of Biology, both symbolizing YSU's enduring legacy of excellence in scientific education, innovation, and research.

This publication also honors the rich scientific heritage of eminent biologists of Yerevan State University such as Prof. Armen Takhtajan, Prof. Mikhail Chalikyan, Prof. Hrachya Buniatyan, Prof. Ezras Hasratyan, and Prof. Norair Sissakian whose outstanding contributions continue to inspire future generations of professors and scientists. This edited volume embodies the collaborative spirit and evidence-based synergy that characterized the conference, uniting distinguished researchers, academics, and innovators from around the globe. It highlights the pivotal role of biological and environmental sciences in addressing the United Nations Sustainable Development Goals (SDGs) through scientific creativity, technological advancement and ecological stewardship.

The chapters encompass a broad spectrum of themes, from molecular biology, biotechnology, and environmental management to nanobiotechnology and sustainable agriculture, reflecting the dynamic, integrative, and interdisciplinary nature of modern biological research. Through rigorous analysis and evidence-based insights, this volume underscores how biological innovations can drive sustainable food systems, biodiversity conservation, climate resilience, and environmental restoration, thereby contributing meaningfully to a greener and more sustainable planet.

Today, the Faculty of Biology at Yerevan State University continues to play a vital role in the global scientific landscape, fostering international collaborations, advancing frontier research in biological and environmental sciences, and training new generations of scholars dedicated to sustainable development. Through its vibrant research culture and academic excellence, the faculty remains a beacon of scientific progress and innovation, contributing to the advancement of global biosciences.

I extend my deepest appreciation to the editors Dr. Karen Ghazaryan, Dr. Karen Trehounian, and Dr. Abhishek Singh for their visionary leadership, scholarly dedication, and organizational excellence in shaping this volume. Their tireless efforts, combined with contributions from renowned scientists across continents, have resulted in a comprehensive and thought-provoking reference for academia, policymakers, and innovators worldwide.

I am confident that this edited volume will serve as a valuable academic contribution and inspire future generations to continue fostering international collaboration, scientific innovation, and sustainable progress in alignment with the global goals of humanity.

Rafayel Barkhudaryan

Yerevan State University, Armenia

Preface

Biological sciences are central to addressing the complex and interconnected challenges of the twenty-first century, including climate change, environmental degradation, food insecurity, public health risks, and sustainable resource management, which are key priorities of the United Nations Sustainable Development Goals (SDGs). When integrated with innovation, emerging technologies, and sustainability principles, biological science offers powerful pathways for enhancing ecosystem resilience, promoting sustainable agriculture, and supporting inclusive socioeconomic development. This volume of the *Biological Science and Innovation for Achieving the Sustainable Development Goals* book represents a significant intellectual outcome of the International Conference on *"Biological Sciences and Environmental Solutions for the Achievement of Sustainable Development Goals (SDGs)"*, held from *September 24 to 26, 2025*, at *Yerevan State University (YSU), Armenia*. The conference coincided with two important institutional milestones the *90th Anniversary of the Faculty of Biology* and the *10th Anniversary of the Research Institute of Biology* reflecting YSU long-standing commitment to excellence in scientific education, innovation, and research. The chapters compiled in this book capture the high scientific quality, interdisciplinary exchange, and global perspectives presented at this landmark academic event. Recent advances in biotechnology, environmental biology, molecular sciences, and ecological research have substantially improved our understanding of biological systems at the molecular and ecosystem levels. These advances have enabled innovative solutions, such as climate-resilient crops, sustainable biofuels, environmentally safe nanoparticles, microbial and algal technologies, biochar-based strategies, and robust biosafety frameworks. Translating these scientific developments into practical applications is essential for accelerating progress toward SDGs. The chapters in this volume cover an interdisciplinary range of topics, including microbial physiology and intracellular regulation, biofuel and biodiesel-related bioprocesses, sustainable agriculture under drought and salinity stress, and crop resilience supported by biochemical, nanotechnological, and biochar-based approaches, including meta-analytical assessments of the same. Additional contributions addressed antioxidant and stress response mechanisms, biosafety assessments, microalgal bioactive compounds, and environmental health challenges, such as heavy metal contamination and pesticide impacts. Collectively, this volume highlights biological innovations and ecosystem-based solutions for environmental protection, climate resilience, and sustainable development.

This book serves researchers, educators, students, policymakers, and professionals in the fields of biological sciences and environmental sustainability. We hope that this will inspire interdisciplinary research supporting Sustainable Development Goals and building a resilient future.

Chapter 1
FocA Maintains Constant Intracellular Formate Levels During Fermentative Growth of Escherichia coli

Christopher Erdmann
Institute of Microbiology, Germany

Gary Sawers
http://orcid.org/0000-0003-0862-2683
Institute of Microbiology, Germany

ABSTRACT

Formic acid is a key molecule in enterobacterial mixed-acid fermentation. Formate is either converted to H_2 and CO_2 by formate hydrogenlyase (FHL-1), or translocated from the cell by the formate channel, FocA. FocA also imports formate into non-growing, stationary-phase cells. Escherichia coli has a second formate channel, FocB, but it is unclear whether it also translocates formate. The authors show that formate levels remain constant during growth, but decrease 10-fold in stationary phase; most of the formate accumulates in the culture medium. Reducing the growth rate of fermentating cells by adding the toxic formate analogue hypophosphite does not negatively impact formate homeostasis. Hypophosphite sensitivity was lost in a focA mutant, but not in a focB mutant. Intracellular formate accumulated significantly more in the focA mutant compared with the parental strain, and was

DOI: 10.4018/979-8-3373-1524-9.ch001

mainly converted to H_2 and CO_2 by FHL-1. Together, these data indicate that FocA is mainly responsible for maintenance of formate homeostasis, probably functioning with FHL-1, during pH-neutral glucose fermentation.

INTRODUCTION

Formic acid is a valuable commodity chemical for both the industrial and sustainable energy sectors (Kottenhahn et al., 2018; Tyagi et al., 2025). During mixed-acid fermentation, formate is generated from pyruvate through the activity of the anaerobically inducible glycyl-radical enzyme, pyruvate formate-lyase (PflB) (Knappe et al., 1984). PflB is encoded by the *focA-pflB* operon, which also encodes the pentameric transmembrane protein, FocA (Sawers & Böck, 1989; Suppmann & Sawers, 1994). FocA belongs to the superfamily of formate-nitrite transporters (FNT) and functions in the bidirectional translocation of formic acid across the cytoplasmic membrane (Kammel, Pinske, & Sawers, 2022; Lü et al., 2011; Waight et al., 2010; Wang et al., 2009). During fermentative growth on glucose, the intracellular formate level increases and this induces synthesis of the formate hydrogenlyase (FHL-1) complex (McDowall et al., 2014; Rossmann et al., 1991; Steinhilper et al., 2022). This membrane-associated complex has two cytoplasmically-oriented catalytic components, formate dehydrogenase-H and hydrogenase-3 (McDowall et al., 2014; Steinhilper et al., 2022), which together function to disproportionate formic acid to H_2 and CO_2 (Peters & Sargent, 2023). Alternatively, formic acid ($pK_a = 3.75$) can be translocated out of the cell by FocA (Vanyan et al., 2024). However, once the external pH of the culture medium begins to decrease below 6.5, due to extracellular accumulation of mixed-acid fermentation products, formic acid (Metcalfe et al., 2022; Suppmann & Sawers, 1994) is re-imported into the cell and disproportionated by FHL-1. As formic acid uptake by FocA does not occur in mutants unable to generate an active FHL-1 complex (Beyer et al., 2013), this suggests direct or indirect coupling between FocA-dependent formate uptake and FHL-1 activity. It is proposed that the actions of PflB, FHL-1 and FocA combine to maintain intracellular formate homeostasis during fermentative growth at neutral pH (Beyer et al., 2013; Kammel et al., 2024; Kammel, Pinske, & Sawers, 2022; Peters & Sargent, 2023; Sawers, 2025). Notably, *E. coli* synthesizes a homologue of FocA, called FocB, which is encoded by an operon (*hyf*) that also encodes a second FHL-2 complex (Andrews et al., 1997). FocB also translocates formate, but in late-stationary phase cells and under particular conditions of pH (Babayan et al., 2024; Hakobyan et al., 2018).

The *E. coli focA* gene was identified in mutants showing resistance to the toxic formate analogue, hypophosphite (Suppmann & Sawers, 1994). Hypophosphite's toxicity is due to its irreversible inhibition of PflB activity, which impairs anaerobic growth on glucose and, consequently, formate and acetyl CoA generation (Plaga et al., 1988). Hypophosphite is imported into *E. coli* mainly via FocA (Hunger et al., 2017; Kammel et al., 2021; Suppmann & Sawers, 1994). Thus, combining measurement of changes in intracellular and extracellular formate levels with use of hypophosphite to manipulate growth, the importance of FocA and FocB in controlling intracellular formate levels during fermentation can be assessed. Therefore, the aim of this study was to analyze *focA* and *focB* mutants to determine whether FocA and FocB control formate homeostasis during fermentative growth at neutral pH. Addressing this question is relevant to improving the efficiency of microbial formic acid production.

MATERIALS AND METHODS

Bacterial Strains and Growth Conditions

Four isogenic strains of *E. coli* were used in this study, all of which carry a chromosomally integrated *λ(fdhF$_P$::lacZ)* gene fusion (Falke et al., 2010): DH4100 (Hunger et al., 2014) is a derivative of MC4100 (F$^-$ *araD* Δ(*argF lac*) *U169 ptsF25 deoC1 relA1 fblB530 rpsL 150 λ*$^-$) (Casadaban, 1976); DH701 is derived from DH4100, but has additional mutations in the *focA* gene (codons 114 and 115 are changed to UAG and UAA) preventing synthesis of full-length FocA (Hunger et al., 2014); DH1003 is like DH4100, but has a deletion in the *focB* gene, which was introduced by P1$_{kc}$-mediated transduction (Miller, 1972) after the phage was grown on the Keio-collection mutant, BW25113 Δ*focB* (Baba et al., 2006; Maeda et al., 2008); and DH1004 is like DH701, but has an additional deletion in the *focB* gene. DH1003 and DH1004 were constructed as part of this study. Deletion of the Kan^R-cassette from the mutated *focB* gene was done using the method of Cherepanov and Wackernagel (Cherepanov & Wackernagel, 1995).

All four strains were grown anaerobically in liquid M9-minimal medium (Sambrook et al., 1989) with 0.8% (w/v) glucose as a carbon source, and at 37 °C. Anaerobic cultures were inoculated at a dilution of 1:100 from aerobically grown overnight cultures, cultivated in the same M9-glucose minimal medium, exactly as described (Kammel et al., 2021; Kammel, Trebbin, Pinske et al, 2022). Where indicated, sodium hypophosphite was added to a final concentration of 0.5 mM from the beginning of growth (Hunger et al., 2017; Kammel et al., 2021). Kanamycin was used at a final concentration of 50 μg ml^{-1} for strain selection.

Hypophosphite-Sensitivity Test and Analysis of Intra- and Extracellular Formate Levels

To compare anaerobic growth and to generate growth curves, strains were cultivated in the absence and presence of 0.5 mM sodium hypophosphite. These growth studies using the four strains described were carried out using microtiter plates and under an atmosphere comprising 80% N_2 and 20% CO_2, as previously described (Kammel et al., 2021).

The analysis of changes in intracellular formate levels via measurement of ß-galactosidase enzyme activity and the determination of the formate concentration in the growth medium using high performance liquid chromatography (HPLC) were done exactly as described (Kammel et al., 2021; Kammel, Trebbin, Pinske et al, 2022). For these analyses, the strains were grown in sealed Hungate tubes, as described (Kammel et al., 2021; Kammel, Trebbin, Pinske et al, 2022). Unless otherwise noted, five samples were taken to determine β-galactosidase enzyme activity and extracellular formate concentration. These included the timepoints when cells reached the early-, middle- (two timepoints were taken) and the late-exponential phases ($OD_{600\ nm}$ ~ 0.3, 0.4, 0.6, and 0.8, respectively) and the early-stationary phase ($OD_{600\ nm}$ ~ 0.9-1.0) of cultivation. The experiments were performed with triplicate technical replicates and with minimally three biological replicates. Data are presented as standard deviation of the mean.

RESULTS

The Effect of Hypophosphite on Growth

Hypophosphite is known to reduce both the growth rate and the final cell density attained by fermenting cultures of *E. coli* (Hunger et al., 2017; Kammel et al., 2021; Suppmann & Sawers, 1994); however, detailed comparative growth analyses between the parental wild-type and *focA* and *focB* mutant strains have not been undertaken. In particular, any role of FocB in conferring sensitivity towards hypophosphite during glucose fermentation at neutral pH has yet to be assessed. Therefore, the four strains at the focus of the current study, DH4100 (parental), DH701 (*focA* mutant), DH1003 (Δ*focB* mutant), and DH1004 (*focA*, Δ*focB* mutant), were grown under fermentative conditions in defined, liquid M9-glucose minimal medium, pH 7 (see Methods), either without addition of further supplements, or in the presence of 0.5 mM sodium hypophosphite (Figure 1). The growth curves are depicted as linear, and not as semi-logarithmic, graphs to highlight any differences between the final optical densities of the cultures. A typical anaerobic growth curve of the parental strain, DH4100, is shown in panel A of Figure 1, with entry into stationary phase beginning around 9 h after the start of anaerobic cultivation; the cultures reached a final OD_{600} of approximately 0.9 at around 11 h. Growth-restriction of the same strain occurred after 5 h when 0.5 mM hypophosphite was supplemented, and this was accompanied by an approximate 20% decrease in the anaerobic growth-rate. The cells entered the stationary phase after approximately 9 h of cultivation (Figure 1A). The final OD_{600} attained by the culture was 0.6 (Figure 1A). Similar reductions in anaerobic growth rate and final stationary-phase OD_{600} due to cultivation in the presence of hypophosphite have been noted previously (Kammel et al., 2021; Kammel, Trebbin, Pinske et al, 2022).

Figure 1. Sodium hypophosphite restricts anaerobic growth of E. coli strains in a FocA-dependent manner. Strains were grown in M9-glucose minimal medium in the absence (upper curve in each graph) or presence (lower curve) of 0.5 mM hypophosphite. Panel A, DH4100 (parental); panel B, DH701 (focA); panel C, DH1003 (ΔfocB); panel D, DH1004 (focA, ΔfocB). Growth curves are depicted linearly on both axes and the vertical arrows indicate the points at which samples were taken for analysis of formate levels (see Figure 2 below). Experiments were performed with three biological replicates and each replicate was done with three technical replicates. The data are presented as the standard deviation of the mean.

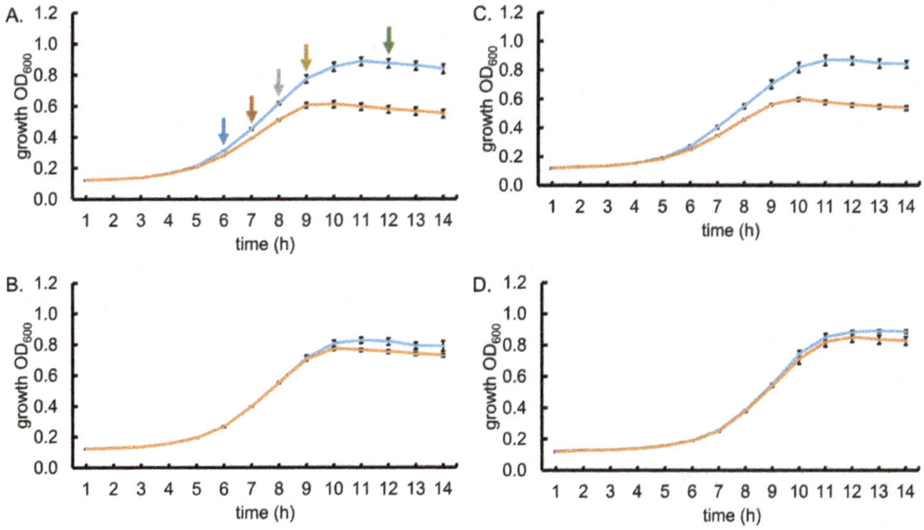

Sensitivity to Hypophosphite During Growth at Neutral pH Is FocA-Dependent

Growth of the *focA* mutant, DH701 with glucose revealed a similar profile to that obtained for the parental strain, with the exception that the final OD_{600} attained upon entry into the stationary phase was 0.8 (Figure 1B). This finding suggests a minor negative influence on growth when FocA was not synthesized by the fermenting cells. However, in contrast to what was observed for the parental strain, DH4100, addition of sodium hypophosphite to the culture of strain DH701 (*focA*) failed to have a significant impact on either growth-rate or the final density attained by the culture (Figure 1B). This result confirms that hypophosphite must be imported into the cells to cause growth restriction and that during glucose fermentation at neutral pH, uptake is mainly via the formate channel, FocA (Hunger et al., 2017; Kammel et al., 2021).

Hypophosphite-Sensitive Fermentative Growth at Neutral pH Is Not Dependent on FocB

To determine whether a deletion in the *focB* gene had any impact on the growth of the respective strain to hypophosphite, the same experiment performed with DH4100 and DH701 was repeated with strains DH1003 (Δ*focB*) and DH1004 (*focA*, Δ*focB*), (Figure 1C, D). The results show that the growth profiles of DH1003 in the absence and presence of hypophosphite were very similar to those observed for the parental strain, DH4100 (compare with Figure 1A). This result suggests that during glucose fermentation at pH 7, FocB does not contribute significantly to the import of hypophosphite.

Very similar profiles for strain DH1004 (*focA*, Δ*focB*) to those determined for strain DH701 (*focA*) during growth in the absence or presence of sodium hypophosphite were observed (Figure 1D). Together, these experiments confirm that under the growth conditions used in these experiments, FocA is solely responsible for uptake of hypophosphite, while FocB is not involved in uptake of hypophosphite during fermentation with glucose at pH 7.

Role of FocA in Controlling Intracellular Formate Levels During pH-Neutral Fermentation

It was shown recently that intracellular formate levels remain relatively constant in wild-type *E. coli* cells during fermentative growth, but are altered in a *focA* mutant (Kammel & Sawers, 2022). It is unclear, however, whether FocB plays any role in this homeostatic mechanism. It is also unclear to what extent a reduced growth-rate and, for example, an earlier entry (at a lower final OD) into stationary phase affects intracellular formate levels. To address these questions, we first determined whether hypophosphite had any effect on intracellular formate levels during exponential growth and upon entry into the stationary phase. This was done by monitoring gene expression using a formate-responsive $\lambda fdhF_p$::*lacZ* reporter construct (Falke et al., 2010). Use of this reporter system allows changes in β-galactosidase enzyme activity to be monitored and these changes signify corresponding changes, either through an increase or a decrease, in intracellular formate levels (Hunger et al., 2014). Thus, this reporter system acts as a proxy *in lieu* of being able to measure intracellular formic acid concentration, which due to its small size and rapid diffusion is extremely difficult to quantify (Kammel et al., 2024; Metcalfe et al., 2022). Nevertheless, the reporter system allows direct monitoring of the effects of mutations or growth conditions on changes in intracellular levels of formate.

Samples of cultures were removed at the five indicated timepoints (see Figure 1A) throughout the exponential growth phase and upon entry into the early stationary phase, and β-galactosidase enzyme activity was determined (Figure 2A). Analysis of enzyme activity in cells of the parental strain, DH4100, revealed that activity remained relatively constant (between 200 and 250 U) from the early- to the late-exponential phase of growth. This result indicates that a constant level of formate is present in these cells and confirms earlier findings (Kammel & Sawers, 2022). Because of its small size and rapid equilibrium across the cytoplasmic membrane, it is experimentally difficult to ascertain the precise formate/formic acid concentration within these cells. However, formate levels must be minimally in the low mM concentration, because the dissociation constant of the FhlA transcription factor for formate has been determined to be approximately 5 mM (Hopper & Böck, 1995; Leonhartsberger et al., 2002). Activation of FhlA by formate is essential for synthesis of the FHL-1 complex under fermentative growth conditions (Rossmann et al., 1991; Schlensog et al., 1994).

Figure 2. FocA-dependent formate homeostasis during anaerobic, exponential growth of E. coli. Strains were grown in the absence or presence of 0.5 mM hypophosphite (HP), as described (see Methods). Panel A shows changes in intracellular formate levels as determined by measuring β-galactosidase enzyme activity (represented in Miller units (Miller, 1972)); panel B shows the extracellular formate concentration determined for the same samples. The shades of the histograms correlate with the shades of the vertical arrows shown in Figure 1, and indicate when samples were taken for analysis. Experiments were performed with three biological replicates and each replicate was done with three technical replicates. The data are presented as the standard deviation of the mean.

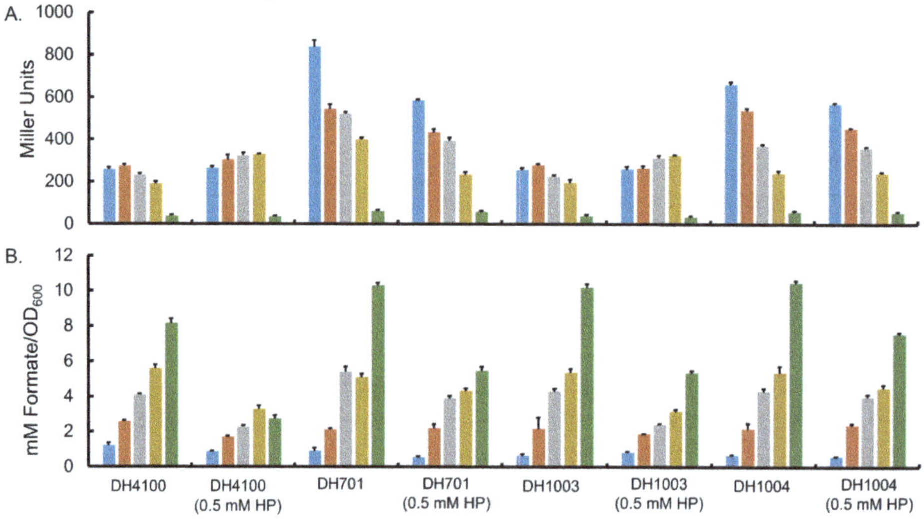

Notably, intracellular formate levels decreased by roughly 10-fold upon entry of all tested strains into the early stationary phase (Figure 2A). This indicates that once cells enter the stationary phase, formic acid (formate plus H^+) is rapidly removed from the cytoplasm by the action of the FHL-1 complex (Falke et al., 2010; Peters & Sargent, 2023). A contribution is also made by FocA-dependent formic acid translocation out of the cell, if the gene encoding the transmembrane protein is present in the strain.

Effect of Hypophosphite on Intracellular Formate Levels

When the parental strain, DH4100, was cultivated in the presence of 0.5 mM hypophosphite, β-galactosidase enzyme activity increased only marginally, to between 250 and 300 U, but the decrease in intracellular formate levels upon entry into the early stationary phase was also observed (Figure 2A). These results indicate that, despite an approximate 20% reduction in growth rate in the presence of hypophosphite, and the 35% lower final cell density attained upon entry into stationary phase (compare Figure 1A), intracellular formate levels were maintained at a relatively constant level. Thus, despite hypophosphite inhibiting PflB enzyme activity, the fermenting cell manages to maintain a relatively constant level of the intracellular formate pool.

Importance of FocA in Maintaining Intracellular Formate Control

The absence of the FocA protein (strain DH701 in Figure 2) significantly disrupted intracellular formate levels during the exponential growth phase. The β-galactosidase enzyme activity increased between 2- and 3-fold compared to the same point during exponential phase growth of the parental strain, DH4100. A similar profile was measured during cultivation in the presence of hypophosphite, but with an approximate 15-20% reduction in total activity (Figure 2A). These findings indicate that in the absence of FocA, formate homeostasis cannot be maintained.

Effect of a *focA* Mutation and Hypophosphite on Extracellular Formate Levels

Next, the extracellular formate concentrations were determined for DH4100 and for DH701 (*focA*) in the same samples used to analyze changes in intracellular formate levels. The results for the parental strain, DH4100, showed a steady accumulation of formate in the culture medium, reaching a concentration of 5 mM OD_{600}^{-1} at the end of the exponential phase of growth, increasing to 8 mM OD_{600}^{-1} in the early stationary phase (Figure 2B). This suggests that the sharp decrease in intracellular formate levels in early stationary phase cells is due to formic acid being exported into the culture medium by the parental strain, DH4100 (Vanyan et al., 2024).

Hypophosphite treatment of strain DH4100 reduced the concentration of formate that accumulated in the growth medium by approximately 40%, which presumably reflects the lower level of formate produced due to inhibition of the PflB enzyme reaction by hypophosphite (Plaga et al., 1988). A 2- to 3-fold reduction in the formate level was determined in early stationary phase cultures (Figure 2B). Clearly, the efflux of formic acid (Vanyan et al., 2024) is adjusted in rate and amount to ensure that the intracellular level of formate remains at a relatively constant level, presumably to ensure that synthesis of the FHL-1 complex remains optimal (Metcalfe et al., 2022; Rossmann et al., 1991).

Surprisingly, the concentration of formate excreted into the culture medium by the *focA* mutant, DH701, during fermentative growth on glucose was only marginally lower than that excreted by the parental strain, DH4100, although a higher concentration of formate accumulated in the early stationary phase culture medium (Figure 2B). Indeed, during the exponential growth phase, the concentration of formate excreted by both strains was nearly identical. This indicates that while the absence of FocA caused a 3- to 4-fold accumulation of formate intracellularly, a steady proportion of this was excreted from the cell by diffusion across the cytoplasmic membrane as neutral formic acid (p*Ka* = 3.75), while a significant, currently undefined, proportion was likely disproportionated into H_2 and CO_2 (Peters & Sargent, 2023). Growth of strain DH701 (*focA*) in the presence of 0.5 mM hypophosphite did not cause such a significant decrease in extracellular formate levels as was determined for the parental strain (Figure 2B). This is because hypophosphite uptake into growing cells requires FocA and without efficient hypophosphite uptake, PflB enzyme activity cannot be inhibited. This is also consistent with the improved fermentative growth of strain DH701 (*focA*) compared with DH4100 (parental) when hypophosphite was added to the medium (see Figure 1A and B).

Effect of a *focB* Deletion on Extracellular Formate Levels

To determine whether FocB plays any significant role in formate homeostasis under the growth conditions tested in this study, we next analyzed β-galactosidase enzyme activity in strain DH1003 (Δ*focB*) during glucose fermentation at neutral pH (Figure 2A). The β-galactosidase enzyme activity profile was nearly identical to that of the parental strain, DH4100, indicating minimal impact of the *focB* gene deletion on formate homeostasis under the studied growth condition.

Growth in both the absence and presence of hypophosphite resulted in activity profiles that were nearly identical to those observed for the parental strain, DH4100 (compare Figure 1A and C). Similarly, the concentrations of formate excreted by DH1003 (Δ*focB*), both in the absence and presence of hypophosphite, were very similar to those determined for DH4100 (Figure 2B). This finding reflects minimal impact of the *focB* gene deletion on formic acid efflux by the strain. Moreover, this also indicates that, while FocA is essential to maintain controlled intracellular formate levels, FocB does not play a significant role in controlling formate levels under the conditions tested in these experiments. This conclusion was further supported by the analysis of the β-galactosidase enzyme activity determined during growth of the *focA*-Δ*focB* double-null mutant, DH1004 (Figure 2A), which revealed an activity profile similar to that determined for the *focA* mutant, DH701. The activity was, however, reduced by approximately 20% in late-exponential phase cells of DH1004 compared with the activity measured in the same phase for DH701. This finding might indicate a minor impact on formate levels when both formate channels are absent. However, when strain DH1004 was grown in the presence of 0.5 mM hypophosphite, the enzyme activity profile was very similar to that of DH701 (*focA*) grown with hypophosphite (Figure 2A). Moreover, the profile of the excreted formate levels in the culture medium of DH1004 was very similar to that of the *focA* mutant DH701 (Figure 2B). This result again strongly suggests that FocB does not significantly influence intracellular and extracellular formate levels during fermentative growth at near-neutral pH.

DISCUSSION

The findings of the current study show that intracellular formate levels remain constant when *E. coli* cells grow by fermentation at high initial glucose concentration and at an initial pH of 7. A strain unable to synthesize the formate channel, FocA, loses the ability to exert control over the level of intracellular formate and the anion accumulates intracellularly from the early stages of growth. These data therefore are consistent with the findings of an earlier study (Kammel & Sawers, 2022). Using hypophosphite to manipulate the growth rate and reduce the final optical density at which fermenting cells enter the stationary phase had no apparent negative impact on formate levels in cells synthesizing FocA, as long as the cells were growing. As soon as cells entered the stationary phase, however, intracellular formate levels decreased significantly. It is proposed that this occurs through a combination of the activity of the formate-induced FHL-1 complex, which disproportionates formate plus a proton to H_2 and CO_2, and efflux of formic acid through FocA (Figure 3). Although the *focA* mutant also excretes formic acid, it does so with presumably slower kinetics, because unless a further unknown formate transporter is present in these cells, only neutral formic acid can diffuse across the cytoplasmic membrane to escape from the cytoplasm and this is determined by the pK_a of formic acid/formate (3.75) and the prevailing intracellular pH.

Figure 3. Control of formate homeostasis during fermentation at neutral pH. Pyruvate formate-lyase (PflB), formate hydrogenlyase (FHL-1) and the formate channel, FocA, balance the intracellular formate level, keeping it constant during growth. An increase in cytoplasmic H^+ concentration (acidification) in the stationary phase may be the signal to increase formic acid efflux through FocA and its disproportionation to H_2 and CO_2 by FHL-1. FHL-2 and FocB have a limited role under these conditions and are therefore not shown. The dashed arrows indicate diffusion across the membrane.

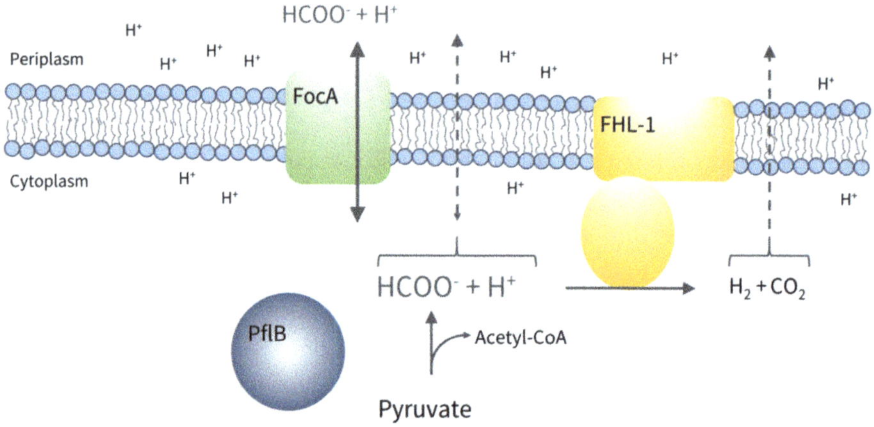

The second formate channel, FocB, encoded by *focB* on the *E. coli* genome (Andrews et al., 1997), appears not to have a major role in excreting formic acid when *E. coli* is cultivated at neutral pH and with high glucose concentrations. Although it remains to be determined under precisely which conditions FocB unfolds its physiological function, results from the Trchounian lab suggest that FocB, along with the FHL-2 complex (Babayan et al., 2024; Grigoryan et al., 2025; Hakobyan et al., 2018), might have a significant role during growth at higher pH, or when fermenting cells deal with osmotic stress (Babayan et al., 2024). Those studies suggest that FocB, together with FHL-2, is particularly important in regulating formate metabolism under energy-limiting conditions, including in the late stationary phase, and when cells are exposed to slightly alkaline pH (Babayan et al., 2024; Hakobyan et al., 2018). At pH 5.5, both FocA and FocB appear to share the role of regulating intracellular formate levels in the stationary phase (Babayan et al., 2024; Grigoryan et al., 2025). Clearly, however, both FocA and FocB help to control the intracellular pH by influencing the activities of both F_oF_1-ATPase and H^+/monovalent cation antiporters (Babayan et al., 2024; Grigoryan et al., 2025; Hakobyan et al., 2018).

During fermentation at neutral pH, the question arises as to whether FocA itself might sense and directly control the intracellular formate levels and, consequently, either directly or indirectly sense pH. The protonation status of the highly conserved histidine residue 209 within the pore of FocA potentially has a role in sensing the accumulation of H^+ in the cytoplasm and periplasm, which would suggest that possibly intracellular (and possibly also periplasmic) pH is what it monitors. The possible involvement of this key amino acid residue within the FocA and FocB channels has been discussed extensively (Kammel et al., 2024; Kammel, Trebbin, Pinske et al, 2022; Sawers, 2025). Importantly, it has been recently shown that in the efflux direction, formic acid is translocated through the FocA pore (Vanyan et al., 2024) and if metabolism slows down, *e.g.* at the latter stages of exponential-phase growth, then formate and H^+ begin to accumulate and must be removed from the cytoplasm. The H209 residue in FocA (and the equivalent residue in FocB) is important in controlling formic acid efflux and in preventing it happening in an uncontrolled manner (Kammel et al., 2024). In both the efflux and uptake directions, H209 appears to exert its control through the protonation of the formate anion, which is the reason we suggest that H209 may be the important in pH-sensing.

Nevertheless, the fact that a *focA* mutant does not show dramatically impaired growth during fermentation strongly suggests that another protein or protein complex, such as FHL-1, is also likely to be involved in this process. The finding that mutants of *E. coli* incapable of synthesizing an active FHL-1 complex accumulate formate in the growth medium in stationary-phase cultures (Beyer et al., 2013) suggests that FHL-1, together with FocA and PflB determine and control intracellular formate levels during fermentation at neutral pH (Figure 3), as has been recently proposed (Kammel et al., 2024; Peters & Sargent, 2023; Sawers, 2025). The observation that the *focA* mutant, DH701, also exhibits strongly reduced intracellular formate levels in early-stationary phase cells supports the proposal that both FocA and FHL-1, probably working together, impose homeostatic control on formate levels; in the absence of FocA, the cell deals with excess formic acid via the FHL-1 complex, which ensures that the cytoplasm maintains pH balance by removing formic acid as gaseous products, CO_2 and H_2 (Metcalfe et al., 2022; Peters & Sargent, 2023). It is conceivable that FHL-2 and FocB perform a similar function at pH extremes and under energy-limitating or stressful conditions (Babayan et al., 2024).

FUTURE DIRECTIONS

This current study raises important questions regarding how intracellular formate homeostasis is maintained in growing cells and what signal(s) initiates formate efflux and disproportionation upon entry into the stationary phase. Moreover, it will be important to determine whether FocB is functional exclusively together with the FHL-2 complex in late-stationary phase cells, or whether FocB can also function together with FHL-1 under certain conditions (Andrews et al., 1997; Trchounian et al., 2012). It appears clear that FocB also translocates formic acid, but this still needs to be demonstrated definitively. A further important aim will be to determine to what extent changes in intracellular pH, as reflected by alterations in endogenous and exogenous formate levels, are sensed and act as a signal in changing the activities of the F_0F_1-ATPase and the cation/H^+ antiporters to help the cell maintain pH control. It is likely that the combined actions of PflB, FocA and FocB, as well as both FHL complexes are involved in this process.

ACKNOWLEDGMENT

This work was supported by the Deutsche Forschungsgemeinschaft (grant SA 494 11/1), which is gratefully acknowleged. Lydia Beyer is thanked for help in the construction of strains.

REFERENCES

Andrews, S. C., Berks, B. C., McClay, J., Ambler, A., Quail, M. A., Golby, P., & Guest, J. R. (1997). A 12-cistron *Escherichia coli* operon (*hyf*) encoding a putative proton-translocating formate hydrogenlyase system. *Microbiology*, *143*(11), 3633–3647. DOI: 10.1099/00221287-143-11-3633 PMID: 9387241

. Baba, T., Ara, T., Hasegawa, M., Takai, Y., Okumura, Y., Baba, M., Datsenko, K. A., Tomita, M., Wanner, B. L., & Mori, H. (2006). Construction of *Escherichia coli* K-12 in-frame, single-gene knockout mutants: the Keio collection. *Molecular Systems Biology*, 2, 8. DOI: 10.1038/msb4100050

Babayan, A., Vassilian, A., Poladyan, A., & Trchounian, K. (2024). Role of the *Escherichia coli* FocA and FocB formate channels in controlling proton/potassium fluxes and hydrogen production during osmotic stress in energy-limited, stationary phase fermenting cells. *Biochimie*, *221*, 91–98. DOI: 10.1016/j.biochi.2024.01.017 PMID: 38307245

Beyer, L., Doberenz, C., Falke, D., Hunger, D., Suppmann, B., & Sawers, R. G. (2013). Coordinating FocA and pyruvate formate-lyase synthesis in *Escherichia coli*: Preferential translocation of formate over other mixed-acid fermentation products. *Journal of Bacteriology*, *195*(7), 1428–1435. DOI: 10.1128/JB.02166-12 PMID: 23335413

Casadaban, M. J. (1976). Transposition and fusion of the *lac* genes to selected promoters in *Escherichia coli* using bacteriophage lambda and Mu. *Journal of Molecular Biology*, *104*(3), 541–555. DOI: 10.1016/0022-2836(76)90119-4 PMID: 781293

Cherepanov, P. P., & Wackernagel, W. (1995). Gene disruption in *Escherichia coli*: TcR and KmR cassettes with the option of Flp-catalyzed excision of the antibiotic-resistance determinant. *Gene*, *158*(1), 9–14. DOI: 10.1016/0378-1119(95)00193-A PMID: 7789817

Falke, D., Schulz, K., Doberenz, C., Beyer, L., Lilie, H., Thiemer, B., & Sawers, R. G. (2010). Unexpected oligomeric structure of the FocA formate channel of *Escherichia coli*: A paradigm for the formate-nitrite transporter family of integral membrane proteins. *FEMS Microbiology Letters*, *303*(1), 69–75. DOI: 10.1111/j.1574-6968.2009.01862.x PMID: 20041954

Grigoryan, L., Babayan, A., Vassilian, A., Poladyan, A., Sawers, G., & Trchounian, K. (2025). *Escherichia coli* FocA/B-dependent H^+ and K^+ fluxes: Influence of exogenous versus endogenous formate. *Biophysical Reports*, *5*(3), 100225. DOI: 10.1016/j.bpr.2025.100225 PMID: 40784583

Hakobyan, B., Pinske, C., Sawers, G., Trchounian, A., & Trchounian, K. (2018). pH and a mixed carbon-substrate spectrum influence FocA- and FocB-dependent, formate-driven H_2 production in *Escherichia coli*. *FEMS Microbiology Letters*, *365*, 21. DOI: 10.1093/femsle/fny233 PMID: 30247527

Hopper, S., & Böck, A. (1995). Effector-mediated stimulation of ATPase activity by the sigma 54-dependent transcriptional activator FHLA from *Escherichia coli*. *Journal of Bacteriology*, *177*(10), 2798–2803. DOI: 10.1128/jb.177.10.2798-2803.1995 PMID: 7751289

Hunger, D., Doberenz, C., & Sawers, R. G. (2014). Identification of key residues in the formate channel FocA that control import and export of formate. *Biological Chemistry*, *395*(7-8), 813–825. DOI: 10.1515/hsz-2014-0154 PMID: 24659605

Hunger, D., Röcker, M., Falke, D., Lilie, H., & Sawers, R. G. (2017). The *C*-terminal six amino acids of the FNT channel FocA are required for formate translocation but not homopentamer integrity. *Frontiers in Microbiology*, *8*, 1616. DOI: 10.3389/fmicb.2017.01616 PMID: 28878762

Kammel, M., Erdmann, C., & Sawers, R. G. (2024). The formate-hydrogen axis and its impact on the physiology of enterobacterial fermentation. *Advances in Microbial Physiology*, *84*, 51–82. DOI: 10.1016/bs.ampbs.2024.02.002 PMID: 38821634

Kammel, M., Hunger, D., & Sawers, R. G. (2021). The soluble cytoplasmic *N*-terminal domain of the FocA channel gates bidirectional formate translocation. *Molecular Microbiology*, *115*(4), 758–773. DOI: 10.1111/mmi.14641 PMID: 33169422

Kammel, M., Pinske, C., & Sawers, R. G. (2022). FocA and its central role in fine-tuning pH homeostasis of enterobacterial formate metabolism. *Microbiology*, *168*(10), 10. DOI: 10.1099/mic.0.001253 PMID: 36197793

Kammel, M., & Sawers, R. G. (2022). The FocA channel functions to maintain intracellular formate homeostasis during *Escherichia coli* fermentation. *Microbiology, 168*(4), 4. DOI: 10.1099/mic.0.001168 PMID: 35377837

Kammel, M., Trebbin, O., Pinske, C., & Sawers, R. G. (2022). A single amino acid-exchange converts FocA into a unidirectional efflux channel for formate. *Microbiology, 168*(1), 001132. DOI: 10.1099/mic.0.001132 PMID: 35084298

Knappe, J., Neugebauer, F. A., Blaschkowski, H. P., & Gänzler, M. (1984). Post-translational activation introduces a free radical into pyruvate formate-lyase. *Proceedings of the National Academy of Sciences of the United States of America, 81*(5), 1332–1335. DOI: 10.1073/pnas.81.5.1332 PMID: 6369325

Kottenhahn, P., Schuchmann, K., & Müller, V. (2018). Efficient whole cell biocatalyst for formate-based hydrogen production. *Biotechnology for Biofuels and Bioproducts, 11*(1), 93. DOI: 10.1186/s13068-018-1082-3 PMID: 29619089

Leonhartsberger, S., Korsa, I., & Böck, A. (2002). The molecular biology of formate metabolism in enterobacteria. *Journal of Molecular Microbiology and Biotechnology, 4*, 269–276. https://pubmed.ncbi.nlm.nih.gov/11931558/ PMID: 11931558

Lü, W., Du, J., Wacker, T., Gerbig-Smentek, E., Andrade, S. L. A., & Einsle, O. (2011). pH-dependent gating in a FocA formate channel. *Science, 332*(6027), 352–354. DOI: 10.1126/science.1199098 PMID: 21493860

Maeda, T., Sanchez-Torres, V., & Wood, T. K. (2008). Metabolic engineering to enhance bacterial hydrogen production. *Microbial Biotechnology, 1*(1), 30–39. DOI: 10.1111/j.1751-7915.2007.00003.x PMID: 21261819

McDowall, J. S., Murphy, B. J., Haumann, M., Palmer, T., Armstrong, F. A., & Sargent, F. (2014). Bacterial formate hydrogenlyase complex. *Proceedings of the National Academy of Sciences of the United States of America, 111*(38), E3948–E3956. DOI: 10.1073/pnas.1407927111 PMID: 25157147

Metcalfe, G., Sargent, F., & Hippler, M. (2022). Hydrogen production in the presence of oxygen by *Escherichia coli* K-12. *Microbiology, 168*(3), 001167. DOI: 10.1099/mic.0.001167 PMID: 35343886

Miller, J. (1972). *Experiments in Molecular Genetics*. Cold Spring Harbor Laboratory.

Peters, K., & Sargent, F. (2023). Formate hydrogenlyase, formic acid translocation and hydrogen production: Dynamic membrane biology during fermentation. *Biochimica et Biophysica Acta. Bioenergetics, 1864*(1), 148919. DOI: 10.1016/j.bbabio.2022.148919 PMID: 36152681

Plaga, W., Frank, R., & Knappe, J. (1988). Catalytic-site mapping of pyruvate formate lyase: Hypophosphite reaction on the acetyl-enzyme intermediate affords carbon-phosphorus bond synthesis (1-hydroxyethylphosphonate). *European Journal of Biochemistry*, *178*(2), 445–450. DOI: 10.1111/j.1432-1033.1988.tb14468.x PMID: 3061816

Rossmann, R., Sawers, G., & Böck, A. (1991). Mechanism of regulation of the formate-hydrogenlyase pathway by oxygen, nitrate and pH: Definition of the formate regulon. *Molecular Microbiology*, *5*(11), 2807–2814. DOI: 10.1111/j.1365-2958.1991.tb01989.x PMID: 1779767

Sambrook, J., Fritsch, E. F., & Maniatis, T. (1989). *Molecular Cloning: a Laboratory Manual*. Cold Spring Harbor Laboratory.

Sawers, G., & Böck, A. (1989). Novel transcriptional control of the pyruvate formate-lyase gene: Upstream regulatory sequences and multiple promoters regulate anaerobic expression. *Journal of Bacteriology*, *171*(5), 2485–2498. DOI: 10.1128/jb.171.5.2485-2498.1989 PMID: 2651404

Sawers, R. G. (2025). How FocA facilitates fermentation and respiration of formate by *Escherichia coli*. *Journal of Bacteriology*, *207*(2), e0050224. DOI: 10.1128/jb.00502-24 PMID: 39868885

Schlensog, V., Lutz, S., & Böck, A. (1994). Purification and DNA-binding properties of FHLA, the transcriptional activator of the formate hydrogenlyase system from *Escherichia coli*. *The Journal of Biological Chemistry*, *269*(30), 19590–19596. DOI: 10.1016/S0021-9258(17)32210-X PMID: 8034727

Steinhilper, R., Höff, G., Heider, J., & Murphy, B. J. (2022). Structure of the membrane-bound formate hydrogenlyase complex from *Escherichia coli*. *Nature Communications*, *13*(1), 5395. DOI: 10.1038/s41467-022-32831-x PMID: 36104349

Suppmann, B., & Sawers, G. (1994). Isolation and characterisation of hypophosphite-resistant mutants of *Escherichia coli*: Identification of the FocA protein, encoded by the *pfl* operon, as a putative formate transporter. *Molecular Microbiology*, *11*(5), 965–982. DOI: 10.1111/j.1365-2958.1994.tb00375.x PMID: 8022272

Trchounian, K., Poladyan, A., Vassilian, A., & Trchounian, A. (2012). Multiple and reversible hydrogenases for hydrogen production by *Escherichia coli*: Dependence on fermentation substrate, pH and the F_0F_1-ATPase. *Critical Reviews in Biochemistry and Molecular Biology*, *47*(3), 236–249. DOI: 10.3109/10409238.2012.655375 PMID: 22313414

Tyagi, A., Bhardwaj, S., Sinha, A. K., & Dhiman, M. (2025). Advances in hydrogen-driven bicarbonate conversion to formic acid and formate salts. *ChemSusChem*, *18*(14), e202500483. DOI: 10.1002/cssc.202500483 PMID: 40367116

Vanyan, L., Kammel, M., Sawers, R. G., & Trchounian, K. (2024). Evidence for bidirectional formic acid translocation *in vivo* via the *Escherichia coli* formate channel FocA. *Archives of Biochemistry and Biophysics*, *752*, 109877. DOI: 10.1016/j.abb.2023.109877 PMID: 38159898

Waight, A. B., Love, J., & Wang, D.-N. (2010). Structure and mechanism of a pentameric formate channel. *Nature Structural & Molecular Biology*, *17*(1), 31–37. DOI: 10.1038/nsmb.1740 PMID: 20010838

Wang, Y., Huang, Y., Wang, J., Cheng, C., Huang, W., Lu, P., Xu, Y.-N., Wang, P., Yan, N., & Shi, Y. (2009). Structure of the formate transporter FocA reveals a pentameric aquaporin-like channel. *Nature*, *462*(7272), 467–472. DOI: 10.1038/nature08610 PMID: 19940917

Chapter 2
Hydrogenase Activity and Polyhydroxyalkanoate Accumulation in Cupriavidus necator H16 Cultivated on Dairy and Biodiesel Industry By-Products

Anna Poladyan
http://orcid.org/0000-0002-8240-2382
Yerevan State University, Armenia

Ofelya Karapetyan
Yerevan State University, Armenia

Ela Minasyan
Yerevan State University, Armenia

Karen Trchounian
http://orcid.org/0000-0002-6624-8936
Yerevan State University, Armenia

ABSTRACT

Cupriavidus necator H16 expresses O_2-tolerant [NiFe]-hydrogenases (Hyds), which present promising anodic biocatalysts in enzymatic fuel cells (EFC). Bacteria also

DOI: 10.4018/979-8-3373-1524-9.ch002

produce polyhydroxyalkanates (PHA). Depolymerization of this polymer leads to the production of crotonic acid (CA), which was estimated by the production of CA. Cheese and curd whey (SW and AW) with a combination of glycerol (G) assessed for the growth of C. necator H16, PHA accumulation, and Hyds activity. Bacterial growth, PHA accumulation, and Hyd activity were observed under microaerophilic conditions for 7 days. Prolonged bacterial growth was observed in glycerol-supplemented samples. Maximum biomass formation was observed on the 5th day in AW (OD_{600}3.1) and on the 6th day in SW+G (OD_{600} 3.3). pH and ORP decrease and Hyd activity were detected under all conditions; however, SW+G sample showed enhanced Hyd activity (0.17±0.03 $U{\cdot}min^{-1}{\cdot}g^{-1}$CDW) compared with G-free samples. A clear correlation was shown between PHA accumulation and Hyd activity: with the Hyd activity decline, CA concentration increased, reaching up to 2.4 ± 0.05 mg/mL. The results will advance hydrogen-based EFC systems as well as the sustainable synthesis of biodegradable PHAs.

INTRODUCTION

Whey is generated as a byproduct during the manufacture of different types of cheeses and other dairy products. Important to notice that it is rich in proteins, lactose, various vitamins, and essential minerals. Global whey production is estimated at around 180– 190 million tons per year, yet less than half of this amount is currently utilized for producing valuable products (Zikmanis et al., 2020; Buchanan et al., 2023; Mirzakulova et al., 2025). There are two 2 groups of whey: sweet whey and acid whey. Sweet whey is produced during the manufacture of different types of cheese, with a pH of approximately 5.8. In contrast, acid sweet whey is a residue of the production of acidic dairy products, such as curd and yogurt, with a pH of approximately 5.1 (Usmani et al., 2022). The production of 1 kg of cheese or curd generates approximately 9 kg of whey as a dairy industry side-stream. In general, all types of whey are rich in diverse nutrients, various organic compounds, oils, mineral salts, as well as total suspended solids (Ganju & Gogate, 2017). Pescuma et al. (2015) demonstrated the main composition of whey: when sweet whey contains (in %) water (93– 94), dry matter (6.0 –6.5), lactose (4.5–5.0), lactic acid (traces), proteins (0.8 - 1.0), citric acid (0.1), and minerals (0.5–0.7) and has a pH value of 6.4 – 6.2. Meanwhile, acidic whey is composed of (in %) water (94 – 95), dry matter (5.0–6.0), lactose (3.8–4.3), and lactic acid (up to 0.8); pH =5.0– 4.6 (www .dairyforall.com). Dairy industry wastewater often exhibits significant acidity and salinity, which can be harmful to both the environment and public health (Zikmanis et al., 2020; Buchanan et al., 2023). These characteristics make whey to have a high biochemical oxygen demand (BOD, 30–50 g/L) and also chemical oxygen demand (COD, 60–80 g/L). These disadvantages make untreated whey as a huge environmental pollutant. Taken together, whey considers as a valuable substrate for a diverse group of functional microorganisms for fermentation. This biotechnological use of whey suggests an effective approach for the sustainable production of valuable metabolites and waste reduction (Poladyan et al., 2019; Iskandaryan et al., 2023; Malos et al., 2025). Moreover, it should be emphasized that such fermentation processes allowed functional microorganisms to efficiently transform whey or its components into significant metabolites that have great potential for large-scale application. Glycerol is a biodiesel production side-stream (Pirzadi and Meshkani, 2022). Biodiesel production amounts to approximately 47.1 billion liters in the world. During this process, a waste glycerol is produced as a co-product, accounting for over 12% of the total esters produced. Each gallon of biodiesel produced accumulates a 1.05 pounds of glycerol. Additionally, waste glycerol is produced during bioethanol manufacturing, typically corresponding to around 10% of the total sugar utilized (Yang et al., 2012; Kazimierowicz et al., 2024). Glycerol production has increased recently, and its effect on the market demand has turned it into a by-product. Sometimes, it is a waste material, posing both environmental and economic challenges. Furthermore, it should be emphasized that a sustainable solution, which suggests converting glycerol into high-value chemical compounds by biotechnological processes, enhances the value of glycerol as a waste and supports the biodiesel industry (Magalhães et al., 2024).

Glycerol's fermentative metabolism has already been reported in some species, like *Bacillus, Clostridium, Citrobacter, Enterobacter, Klebsiella, Lactobacillus, Propionibacterium,* etc. (Haft et al., 2014). However, the potential for using these bacteria in different fields is limited due to their ability to be pathogenic. *Cupriavidus necator* H16 is a facultative chemolithoautotrophic β- proteobacterium which well known as a promising biotechnological microorganism, as one of the best producers of valuable enzymes, as well as biodegradable polymers (bioplastics). Having metabolic flexibility, this bacterium is able to grow on a wide range of carbon and energy sources, including different organic substrates such as sugars. Even inorganic compounds like hydrogen and carbon dioxide can be utilized by this bacterium. This versatility makes *C. necator* H16 an ideal candidate for producing environmentally friendly materials, biofuels, and other high-value bio-chemicals. This bacterium is a good biotechnological object that, depending on different growth conditions able to synthesize biodegradable polyhydroxyalkanoates (PHAs), accumulating these biopolymers up to 80% of its cell dry weight (CDW).

This biopolymer's accumulating proficiency makes it valuable for large- scale industrial PHA production. Furthermore, *C. necator* H16 can grow on inexpensive carbon sources, including industrial waste materials, and this properties create an economically and environmentally sustainable method for producing competitive bioplastics in the modern market (Sohn et al., 2021; Ng et al., 2025). The PHAs, as a biopolymer, represent a promising raw material, due to the fact that these aliphatic polyesters are non- toxic and biodegradable, enabling a broad spectrum of applications (Koller, 2019; Puppi et al., 2019).

These biopolymers are produced through various living organisms, such as plants, microalgae, and especially bacteria, by biosynthetic metabolic pathways. Overall, it is important to note that the production of bioplastics represents a sustainable, low-cost, and eco- friendly approach, which allows us to make sufficient plastic materials while reducing carbon emissions and minimizing fossil fuel consumption (Mangal et al., 2023; Ng et al., 2025). In the context of the present findings, the poly (3-hydroxybutyrate) (PHB) is a homopolymer of 3-hydroxybutyrate (3HB). These are the most extensively studied PHAs. The depolymerization of these polymers leads to the production of crotonic acid (CA). Through chemical or thermal degradation methods, such as pyrolysis or acid hydrolysis, crotonic acid can be obtained from PHB. These depolymerization processes cut down the PHB polymer chain into smaller molecules like monomers. As a result, crotonic acid is produced either directly by thermal decomposition or after further chemical hydrolysis.

Additionally, *C. necator* H16 harbors four distinct O_2-tolerant [NiFe]-hydrogenases that participate in hydrogen metabolism (Schäfer et al., 2013; Lenz et al., 2015). Hydrogenase enzymes catalyze the reversible H_2 production and oxidation. *C. necator* H16 expresses O_2-tolerant [NiFe]-hydrogenases during lithoautotrophic growth by using H_2, CO_2, and O_2. There are different types of hydrogenases, including SH - a soluble cytoplasmic hydrogenase linked to NAD^+, MBH - a membrane-bound hydrogenase, AH-an actinobacterial - type hydrogenases, and RH - a regulatory hydrogenase. It is important to note that O_2-tolerant hydrogenases are synthesized in stress conditions, particularly when energy and nutrients are limited. Hydrogenase enzymes of *C. necator* mainly operate in H_2 oxidation direction. However, in some microorganisms, the role of hydrogenase enzymes can be different: hydrogenase enzymes operate in H_2 production mode. For example, *Escherichia coli* produces H_2 during mixed acid fermentation of different sugars and glycerol by membrane bound [NiFe] - hydrogenase enzymes (Iskandaryan et al., 2023). In *C. necator* bacterium, O_2-tolerant hydrogenases are expressed not only under lithoautotrophic conditions with a gas (H_2, CO_2, O_2) mixture, but also during energy- limited or other stressful environments (Iskandaryan et al., 2023). It is well known that [NiFe]- hydrogenases serve as anodic catalysts in bio-electrochemical systems. Enzymatic biofuel cells (EFCs) and microbial fuel cells (MFCs) facilitate the physiological coupling of H_2 oxidation with O_2 reduction. This ability makes fuel cells promising systems for further efficient electricity generation. Consequently, this bacterium has potential for future H_2- based biotechnological applications aimed at producing a variety of environmentally and medically valuable compounds. It was shown that organic wastes, such as glycerol, lignocellulose-containing brewery spent grain hydrolysate and dairy industry side streams are promising carbon and energy sources for the formation of biomass harboring significant amounts of the biotechnologically relevant hydrogenases. In this context, it is important to establish optimal conditions for using a mixture of dairy sidestreams and glycerol, which enable the production of biomass with active hydrogenases (Poladyan et al., 2023) and promote PHA accumulation.

In this study, acid whey and sweet whey, combined with glycerol, were assessed for their suitability in cultivating *C. necator* H16, and examining the whole cells total H_2-oxidizing hydrogenase activity and the accumulation of PHA. Microbial growth parameters under microaerophilic conditions (OD_{600}, oxidation– reduction potential and pH changes), along with H_2 - oxidizing hydrogenases activity and crotonic acid production, were investigated. The optimal cultivation conditions for the simultaneous enhancement of hydrogenase activity and PHA production were also evaluated. Such an approach may significantly contribute to the advancement of a bio-based economy through the conversion of industrial by-products into key enablers of economic growth and environmental sustainability.

MATERIALS AND METHODS

The Media of Bacterial Growth and Cultivation Conditions

C. necator H16, obtained from Dr. O. Lenz (Technical University of Berlin, TUB, Berlin, Germany), was initially grown heterotrophically in Fructose-Nitrogen (FN) minimal mineral medium (Iskandaryan et al., 2025). The basic FN medium was prepared by combining 100 mL of 10× H16 buffer with 850 mL of distilled water, along with the following sterile stock solutions: 10 mL NH_4Cl (20% w/v), 1 mL $MgSO_4 \cdot 7H_2O$ (20% w/v), 1 mL $CaCl_2 \cdot 2H_2O$ (1% w/v), 1 mL $FeCl_3 \cdot 6H_2O$ (0.5% w/v), and 10 mL fructose (40% w/v). The 10× H16 buffer consisted of 90 g $Na_2HPO_4 \cdot 12H_2O$ and 15 g KH_2PO_4 dissolved in 1 L of distilled water, adjusted to a final pH of 7.0.

In the beginning the pH values of acid whey and sweet whey were approximately 5.1 ± 0.30 and 6.3 ± 0.20, respectively. Bacterial growth parameters were evaluated using a mixture containing 7.5 g/L, sweet whey and 6.5 g/L acid whey, both obtained from the Marianna Dairy Products Factory (Yerevan, Armenia), supplemented with 0.4% (w/v) glycerol (Iskandaryan et al., 2025). The dairy waste materials were subjected to hydrolysis, followed by filtration, four-fold dilution with distilled water, and sterilization. The pH was subsequently adjusted using a concentrated $NaHCO_3$ solution.

A 3% (v/v) *C. necator* H16 pre-culture was inoculated into whey and incubated under microaerophilic conditions in 1000 mL baffled flasks containing 800 mL of medium during 7 days.

Bacterial Growth Characteristics Determination

By using a spectrophotometer (Spectro UV- VIS Auto, Labo Med, Los Angeles, CA, USA) bacterial growth was evaluate by measuring the optical density (OD) at 600 nm. Oxidation reduction potential of the culture medium was measured using platinum (Pt) electrodes (EPB- 1, Measuring Instruments Enterprise, Gomel, Belarus; or PT42BNC, Hanna Instruments, Portugal). While Pt electrodes respond to O_2 and H_2 present in the medium. Before analysis, the electrode were calibrated in a standard solution containing 0.049 M potassium ferricyanide ($K_3[Fe(CN)_6]$) and 0.05 M potassium ferrocyanide ($K_4[Fe(CN)_6]\cdot 3H_2O$) at pH 6.86. At 25 °C, electrode produced readings of +245 ± 10 mV.

Monitoring of H_2-Oxidizing Hyd Activity

During each day of cultivation, bacterial cells were isolated by centrifugation at 5000 rpm, 4°C for 15 min (Iskandaryan et al., 2023). Whole cell H_2-oxidizing activity was determined by measuring methylene blue (MB) reduction at 570 nm at 30 °C, employing a Cary 60 UV– Vis spectrophotometer (Agilent Technologies, USA) (Iskandaryan et al., 2025). A 20 –30 μL aliquot of cell extract was added to an H_2-saturated reaction mixture containing 1.9 mL of 50 mM K-phosphate buffer (pH 7.0) and 5 μL of MB, which served as the electron acceptor in this artificial electron transfer system. Enzyme activity was determined by relating the slope of the kinetic activity curve to the cuvette path length, the MB extinction coefficient (13.1 mM^{-1} cm^{-1}), and the protein concentration. One unit of enzyme activity was defined as the oxidation of 1 μmol H_2 (equivalently, the reduction of 1 μmol MB) per minute per mg of cell dry weight (CDW).

Preparation of Crude Cells and Qualification of Crotonic Acid

During 7 days of bacterial growth, the concentration of 3HB was studied using HPLC (High-Performance Liquid Chromatography). The 3HB concentrations of the samples were determined by crotonic acid and 2- pentenoic acid standard samples, respectively. Samples were taken at different stages of growth, and the cells were disrupted using a French press (Emulsiflex C3 high pressure homogenizer) to release the intracellular contents for further analysis. 10 ml samples were used, and the homogenizing pressure was 18- 20000 psi (≈ 1300 bar) for 5 min for each sample. After each sample, the French press complex was washed with 1 mol NaOH and double-distilled water to obtain clear results and avoid mistakes. To determine the hydroxyalkanoate concentration, 1 mL of culture broth was subjected to alkaline digestion, following the procedure described by Satoh et al. (2016). For accurate quantification of 3 HB using a reverse phase column, alkaline hydrolysis of the bacterial culture is necessary (Watanabe et al., 2012; Satoh et al., 2016). Unlike acidic hydrolysis, alkaline hydrolysis promotes the chemical conversion of 3 HB within the polymer due to the higher nucleophilicity of hydroxide ions. At first 1 mL sample was treated with 0.5 mL 0.5 N, 1 N and 2 N NaOH solution. After that the results show that the optimal NaOH concentration was 0.5 N. The sample and 0.5 N NaOH mixture were transferred to a glass tube and closed accurately. Then samples were heated at 105 ° C for 1 h. After that, the samples were cooled to room temperature and acidified with 0.5 mL of 2 N H_2SO_4. Solids were removed from the acidified samples by centrifugation and filtration (Duvigneau et al., 2021). Centrifugation was carried out for 5 min at 12,000 rpm (13,000 × g) using a model 3740 centrifuge (Kubota Corp., Tokyo, Japan). Filtration was performed using a DISMIC 25CS020AN 0.20 µm membrane filter.

High-Performance Liquid Chromatography (HPLC) Analysis of Crotonic Acid

HPLC analysis of crotonic acid was performed using a modified protocol based on Küchlin et al. (2021). The system consisted of a Waters Alliance 2695e Separation Module (Waters Corporation, USA) coupled with a photodiode array (PDA) detector set at 235 nm. Separation was achieved on a Hypersil Gold C18 column (250 × 4.6 mm, 5 µm particle size; Thermo Fisher Scientific, USA), maintained at 55 ° C (Küchlin et al., 2021). The mobile phase was composed of 90% water containing 0.01% sulfuric acid (v/v) and 10% acetonitrile (Carlo Erba, Italy), delivered isocratically at a flow rate of 0.9 mL/min. The injection volume was 10 µL. Quantification was execute using an external calibration curve for crotonic acid in the concentration range of 0.01 to 20 mg/mL. The calibration curve exhibited excellent linearity, with a correlation coefficient of $R^2 = 0.999$. The limit of detection (LOD) and limit of quantification (LOQ) for crotonic acid were determined to be 0.00005 mg/mL and 0.0002 mg/mL, respectively.

Data Analysis and Experimental Materials

Data analysis was performed using Student's t-test in Microsoft Excel 2016, with differences considered statistically significant at $P < 0.05$. Analytical-grade reagents from Carl Roth GmbH (Germany) and Sigma-Aldrich (USA) were employed. All HPLC standards used had a purity of ≥ 99%.

RESULTS

Bacterial Growth Parameters During the Utilization of Acid Whey and Sweet Whey Media, With and Without Glycerol Supplementation

Double filtration and dilution (ranging from 0 - to 5- fold) were applied to sweet whey (7.5 g/L) and acid whey (6.5 g/L), both with and without glycerol supplementation. Subsequently, the pH of the media was adjusted to 7.0 using 0.1 M $NaCO_3$ (Poladyan et al., 2023). To remove insoluble particles, the solutions were centrifuged at 6000 rpm for 10 minutes at 4 ° C using a Rotina 420 R centrifuge (Hettich, Germany). Overnight bacterial cultures in FN medium were used to inoculate autoclaved whey-containing growth media at 3% (v/v), with the medium pH adjusted to 7.0. Experiments with glycerol were carried out by introducing it into the samples before introducing the bacteria. Optimal conditions for the growth of *C. necator* H16, H_2- oxidizing Hyd activity and PHB accumulation were investigated. Bacterial growth (OD_{600}), Oxidation reduction potential, and pH changes, and H_2 oxidizing Hyd activity were monitored over 7 days. Results for *C. necator* H16 were evaluated in comparison with a glycerol–fructose–nitrogen (GFN) medium, which has been established as the optimal heterotrophic condition for Hyd enzyme synthesis (Poladyan et al., 2019). During bacterial growth, the pH of the medium decreases. For samples with glycerol, a pH decrease from 7 to 5, and for samples without glycerol from 7 to 6.3 is stated: Fig.1. illustrates the bacterial biomass formation (OD_{600}) on acid whey and sweet whey with and without glycerol supplementation, and GFN during 7 days.

Figure 1. Biomass formation (OD_{600}) of C. necator H16 was assessed in acid whey (AW) and sweet whey (SW) media, with and without glycerol (G) supplementation, as well as in GFN medium. Cultivations were performed under microaerobic conditions at 30 °C for 7 days in 250 mL flasks; $n = 5$, $p < 0.05$.

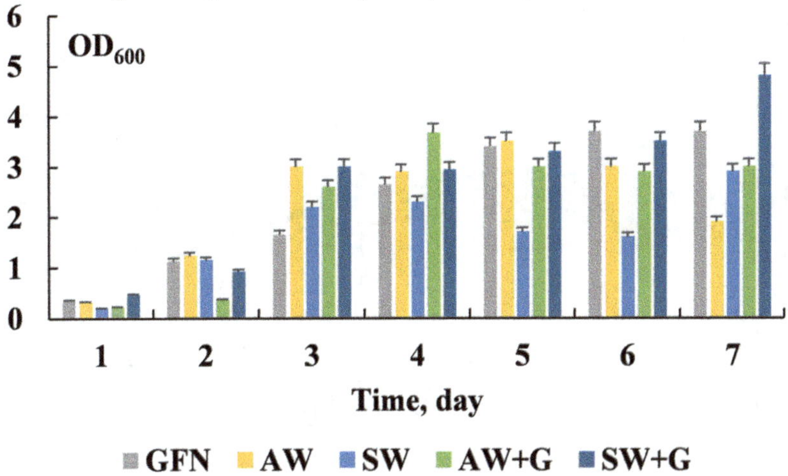

The results showed that glycerol addition stimulated bacterial growth: maximal growth was achieved on day 5th in acid whey (OD_{600} 3.13 ± 0.03), and day 7th in sweet whey + Glycerol (OD_{600} 4.8 ± 0.05) samples (Table 1). Compared to glycerol free samples, the addition of glycerol prolongs bacterial growth. Probably, glycerol serves as an additional source of carbohydrates and promotes the growth (Iskandaryan et al., 2025).

Oxidation- reduction reactions are fundamental for microorganisms. Its help them to adapt diverse environmental conditions and has important role for bacterial growth (Vassilian and Trchounian, 2009). Our results demonstrate that all samples exhibited a low oxidation reduction potential profile (Fig. 2). The maximum decrease, from a positive value of +250 ± 10 mV to a negative value of −113 ± 10 mV, was observed on the fifth day in the AW medium supplemented with glycerol.

Figure 2. Oxidation reduction potential kinetics of C. necator H16 on acid whey (AW) and sweet whey (SW) with and without glycerol (G) supplementations and GFN. Oxidation reduction potential was measured with a platinum redox electrode (Pt) and expressed in mV (Ag/ AgCl (saturated with KCl)); $n = 5$, $p < 0.05$.

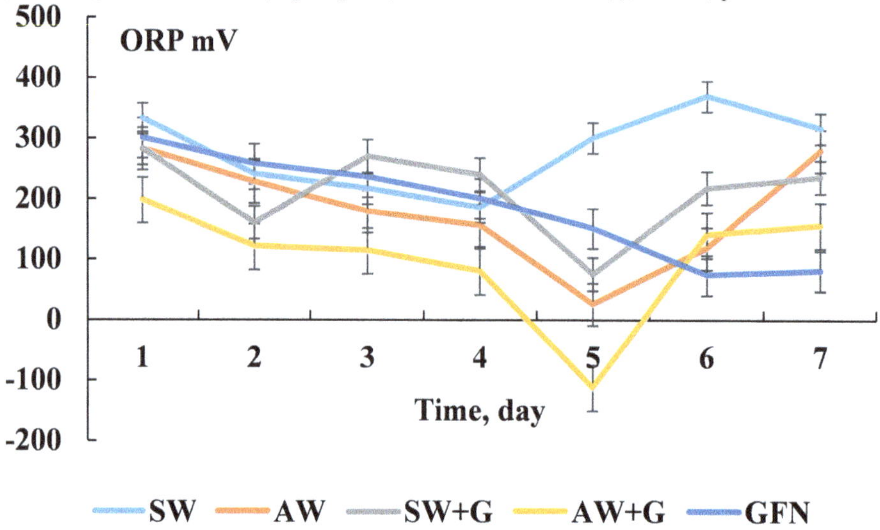

The decline of oxidation reduction potential might be due to a change in redox proteins on the cell surface of bacteria or the secretion of redox active metabolites into the medium (Gemünde et al., 2024). In comparison to glycerol free samples, the oxidation reduction potential values exhibited a greater reduction during bacterial growth in glycerol supplemented samples.

Determination of Crotonic Acid Concentration and Its Correlation With H_2-Oxidizing Hyd Activity

As mentioned, *C. necator* H16 expresses four hydrogenases with different roles (Lenz et al., 2015). The presence of these types of enzymes is really valuable and necessary to evaluate. Thus, for all samples, the total H_2 oxidizing hydrogenase activity of *C. necator* whole cells was investigated. Compared to acid whey and sweet whey, hydrogenase activity was observed earlier in samples with glycerol. H_2 oxidizing hydrogenase activity was significant: a maximal 0.15 U/min/g CDW value was shown for the sweet whey + glycerol sample.

Crotonic acid concentration was monitored during 7 days of growth under five different conditions: acid whey and sweet whey, acid whey + glycerol and sweet whey + glycerol, GFN. In almost all cases, after the 3rd day, crotonic acid steadily increases, reaching 2.4 ± 0.05 mg/mL. The maximum yield was recorded in acid whey + Glycerol on the 6th day (~ 2.44 ± 0.05 mg/mL) and in sweet whey on the 7th day of growth (~ 2.36 ± 0.05 mg/mL). Nevertheless, there is a correlation between crotonic acid (mg/mL) and H_2 oxidizing hydrogenase activity (Fig.3).

Figure 3. Correllation between crotonic acid (mg/mL) and H2 oxidizing Hyd activity of C. Necator H16 on the sweet whey (SW), acid whey (AW) with and without glycerol (G) and GFN. Bacteria were grown for 7 days. Respectively, SW(A), AW(B), AW+G(C), SW+G(D), GFN(E), n =5, p < 0.05

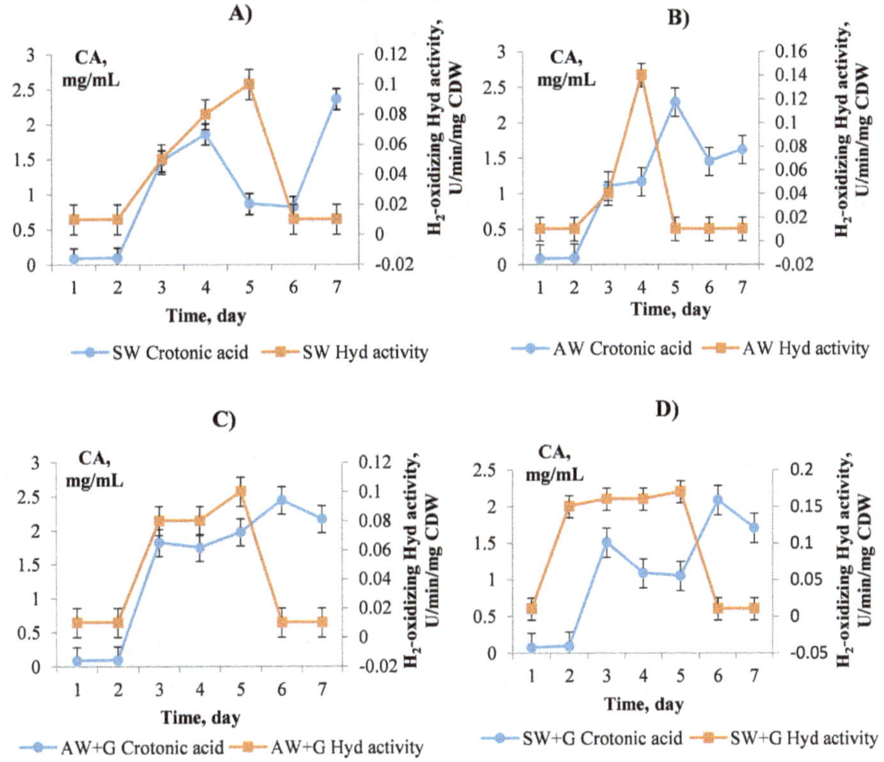

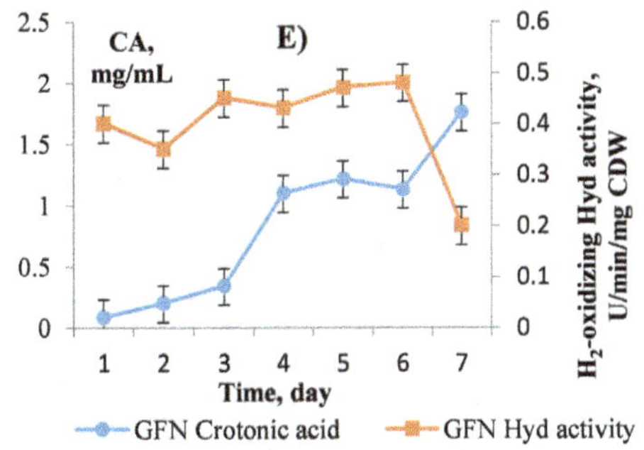

Fig. 3. Correlation between crotonic acid (mg/mL) and H_2 oxidizing Hyd activity of *C. necator* H16 on the sweet whey (SW), acid whey (AW) with and without glycerol (G) and GFN. Bacteria were grown for 7 days. Respectively, SW(A), AW(B), AW+ G(C), SW+ G(D), GFN (E); n=5, p<0.05.

Higher hydrogenase activity often coincided with reduced crotonic acid levels at corresponding time points. As the hydrogenase activity began to decline, the concentration of crotonic acid frequently exhibited an increase. Between days 5 and 7 of growth, during which hydrogenase activity was minimal or undetectable, elevated levels of crotonic acid were observed (Fig. 3). Notably, simultaneous high hydrogenase activity and crotonic acid concentrations were recorded in the sweet whey + glycerol (1.5 ± 0.05 mg/mL and 0.1 U/min/g CDW) and acid whey + glycerol (2 ± 0.05 mg/mL and 0.08 U/min/g CDW) samples on days 5 and 6 of growth (Table 1). Similarly, in the GFN samples, crotonic acid and hydrogenase activity reached 1.5 ± 0.05 mg/mL and 0.3 U/min/g CDW, respectively (Fig. 3). Table 1 summarizes the maximum values of hydrogenase activity and PHA synthesis in *C. necator* H16 under the growth conditions sweet whey, acid whey, acid whey + glycerol, sweet whey + glycerol, and GFN, as determined in this study.

Table 1. Maximum values of H_2-oxidizing hydrogenase (Hyd) activity, crotonic acid (CA) yield and OD_{600} under different C. necator H16 growth conditions.

Growth Conditions	Hyd activity (U/min/mg CDW)	CA yield (mg/mL)	OD_{600}
GFN	0.48 ± 0.05	1.75 ± 0.05	4.40 ± 0.05
AW	0.14 ± 0.05	2.28 ± 0.05	3.13 ± 0.05
AW+G	0.10 ± 0.02	2.44 ± 0.05	3.67 ± 0.05
SW	0.10 ± 0.03	2.36 ± 0.05	2.90 ± 0.05
SW+G	0.17 ± 0.05	2.08 ± 0.05	4.80 ± 0.05

Consequently, results suggest the inverse relationship between hydrogenase activity and PHB accumulation. The experimental data show that conditions promoting high hydrogenase activity tend to correspond with lower PHB accumulation during the same time frame.

DISCUSSION

It is well known that hydrogenases are not constitutively expressed; their expression is induced only under specific environmental conditions, such as the presence of a gas mixture containing hydrogen (H_2), oxygen (O_2), and carbon dioxide (CO_2) (Lenz et al., 2015). During heterotrophic growth, the activation of hydrogenases requires particular conditions, and their enzymatic activity is not sustained continuously. The supplementation of glycerol and its parallel utilization with whey has been suggested to be effective, as it may extend the duration of bacterial growth and enhance the hydrogenase activity and biosynthesis of polyhydroxyalkanoates (PHAs) (Fukui et al., 2014; Ng et al., 2025). Cellular redox balance increases hydrogenase activity during active bacterial growth and supports PHA accumulation. Under nutrition limitation conditions, when energy and reducing equivalents are redirected toward carbon storage, PHA accumulation is dominant. This relationship reflects a metabolic challenge between energy generation and storage during bacterial growth. From a biotechnological perspective, clarifying and designing this balance is necessary for optimizing both hydrogenase enzymes and PHA polymer production systems.

During the early stages of *C. necator* H16 growth, hydrogenase activity is increased to stabilize cellular redox balance; on the other hand, PHA accumulation is low due to the abundance of nutrients supporting cellular growth. During the exponential growth phase, hydrogenase activity may further increase to balance electron flow, especially under microaerobic conditions. At the same time, PHA accumulation started to increase gradually as cells prepared for the oncoming nutrient limitation. This clearly demonstrates the fact that under carbon excess conditions combined with nutrient limitation, the parental strain of *C. necator* H16 can accumulate up to 80% (w/w) of its cell dry weight (CDW) as polyhydroxybutyrate (PHB), which is one of the well-known polyhydroxyalkanoate (PHA) biopolymers (Steinbüchel, 1995; Riedel et al., 2015). There is a strong likelihood that conditions that promote high hydrogenase activity often, in parallel, reduce PHA accumulation, as electrons are probably directed to hydrogen metabolism rather than carbon storage. In an opposite manner, when hydrogenase activity is moderate, bacteria start to generate and accumulate huge quantities of PHB. In contrast to it, when hydrogenase activity is generated, bacteria start to accumulate huge quantities of PHB. On the other hand, samples with the addition of glycerol significantly promoted the growth and biomass formation of *C. necator* H16. The results demonstrate that there is higher biomass formation, lower oxidation reduction potential values, and stimulated hydrogenase activity. Moreover, it should be emphasized that the addition of glycerol prolongs the growth phase of bacteria by providing a supplemental carbon and energy source. Measurements demonstrate lower oxidation reduction potential values in samples with the addition of glycerol. A possible explanation for this observation is enhanced metabolic activity and redox adaptation. Maximum hydrogenase activity was observed in the sweet whey and glycerol samples; on the other hand, the higher crotonic acid accumulation was observed in acid whey and glycerol samples (Table 1). A clear inverse relationship was found between hydrogenase activity and crotonic acid concentration, as well as between hydrogenase activity and PHB accumulation. Overall, results suggest that high hydrogenase activity coincides with low PHB accumulation and the opposite. In the context of the present findings, demonstrated that substrate composition and redox conditions strongly affect bacterial growth, hydrogenase activity, and biopolymer accumulation.

From a global perspective, integrating hydrogenase enzymes and PHA production into a single bioprocess suggests significant ecological and biotechnological advantages. Utilizing organic waste materials to produce both valuable enzymes as a promising energy source and PHA as a biodegradable polymer enhances resource efficiency, promotes carbon circularity, and reduces environmental pollution. Developing gradually, low-energy reactor structures and promoting waste valorization together could stimulate the sustainability and commercial viability of these two production approaches.

CONCLUSION

To summarize the main points discussed above, a promising approach and new methods are offered for the enhancement of hydrogenase enzyme production by utilizing affordable industrial by-products, especially whey and glycerol. By using dairy and biodiesel industries' side-streams, the process sets up a sustainable and cost-effective platform for microbial cultivation, and in addition, simultaneously addresses the environmental issue of industrial waste management. The results clearly demonstrate the fact that supplementation of glycerol, simultaneously with a decrease in pH and oxidation reduction potential, significantly enhances H_2 - oxidizing hydrogenase activity and also promotes the accumulation of PHA in bacterial cells.

It is of significant importance to note that a correlation between H_2 - oxidizing hydrogenase activity and PHA production was observed. This fact indicates that effective optimization of bacterial growth conditions can strategically balance and enhance enzyme production with biopolymer accumulation. These beneficial advantages indicate the potential for developing this kind of bioprocesses capable of at the same time producing valuable enzymes and biomaterials at the same time. To sum up, the experiment demonstrates the effectiveness of the approach. In the final analysis, the results validate the theoretical framework that this study provides a sustainable, circular biotechnological approach with significant implications for waste utilization and biodegradable polymer manufacturing.

ACKNOWLEDGMENT

This work supported by the grant [22AA-1F016] to OK and grant [21AG-1F043] to AP from the State Committee of Science, Ministry of Education, Science, Culture and Sport of Armenia.

REFERENCES

Buchanan, D., Martindale, W., Romeih, E., & Hebishy, E. (2023). Recent advances in whey processing and valorisation: Technological and environmental perspectives. *International Journal of Dairy Technology, 76*(2), 291–312. DOI: 10.1111/1471-0307.12935

Duvigneau, S., Kettner, A., Carius, L., Griehl, C., Findeisen, R., & Kienle, A. (2021). Fast, inexpensive, and reliable HPLC method to determine monomer fractions in poly(3-hydroxybutyrate-co-3-hydroxyvalerate). *Applied Microbiology and Biotechnology, 105*(11), 4743–4749. DOI: 10.1007/s00253-021-11265-3 PMID: 34014345

Ganju, S., & Gogate, P. R. (2017). A review on approaches for efficient recovery of whey proteins from dairy industry effluents. *Journal of Food Engineering, 215*, 84–96. DOI: 10.1016/j.jfoodeng.2017.07.021

Gemünde, A., Gail, J., & Holtmann, D. (2024). Redox mediator interaction with Cupriavidus necator – spectroelectrochemical online analysis. *Electrochemistry Communications, 162*, 107705. DOI: 10.1016/j.elecom.2024.107705

Haft, R. J. F., Keating, D. H., Schwaegler, T., Schwalbach, M. S., Vinokur, J., Tremaine, M., Peters, J. M., Kotlajich, M. V., Pohlmann, E. L., Ong, I. M., Grass, J. A., Kiley, P. J., & Landick, R. (2014). Correcting direct effects of ethanol on translation and transcription machinery confers ethanol tolerance in bacteria. *Proceedings of the National Academy of Sciences of the United States of America, 111*(25). Advance online publication. DOI: 10.1073/pnas.1401853111 PMID: 24927582

Iskandaryan, M., Baghdasaryan, L., Minasyan, E., Trchounian, K., Antranikian, G., & Poladyan, A. (2025). A novel, cost-effective approach for the production of hydrogenase enzymes and molecular hydrogen from recycled whey-based by-products. *International Journal of Hydrogen Energy, 140*, 1191–1202. DOI: 10.1016/j.ijhydene.2024.10.256

Iskandaryan, M., Blbulyan, S., Sahakyan, M., Vassilian, A., Trchounian, K., & Poladyan, A. (2023). L-amino acids affect the hydrogenase activity and growth of Ralstonia eutropha H16. *AMB Express*, *13*(1), 33. DOI: 10.1186/s13568-023-01535-w PMID: 36932299

Kazimierowicz, J., Dębowski, M., Zieliński, M., Kasiński, S., & Cruz Sanchez, J. (2024). Biotechnological Valorization of Waste Glycerol into Gaseous Biofuels—A Review. *Energies*, *17*(2), 338. DOI: 10.3390/en17020338

Koller, M. (2019). Switching from petro-plastics to microbial polyhydroxyalkanoates (PHA): The biotechnological escape route of choice out of the plastic predicament? *The EuroBiotech Journal*, *3*(1), 32–44. DOI: 10.2478/ebtj-2019-0004

Küchlin, R., Huber, T., & Hausmann, R. (2021). A new class of medium-chain-length polyhydroxyalkanoates produced by Pseudomonas citronellolis. *Applied Microbiology and Biotechnology*, *105*, 7421–7435. https://doi.org/DOI%2520not%2520found

Lenz, O., Lauterbach, L., Frielingsdorf, S., & Friedrich, B. (2015). 4 Oxygen-tolerant hydrogenases and their biotechnological potential. In Rögner, M. (Ed.), *Biohydrogen* (pp. 61–96). GRUYTER. DOI: 10.1515/9783110336733.61

Magalhães, C. P., Alves, J. I., Duber, A., Oleskowicz-Popiel, P., Stams, A. J. M., & Cavaleiro, A. J. (2024). Metabolic versatility of anaerobic sludge towards platform chemical production from waste glycerol. *Applied Microbiology and Biotechnology*, *108*(1), 419. DOI: 10.1007/s00253-024-13248-6 PMID: 39012392

Mangal, M., Rao, C. V., & Banerjee, T. (2023). Bioplastic: An eco-friendly alternative to non-biodegradable plastic. *Polymer International*, *72*(11), 984–996. DOI: 10.1002/pi.6555

Mirzakulova, A., Sarsembaeva, T., Suleimenova, Z., Kowalski, Ł., Gajdzik, B., Wolniak, R., & Bembenek, M. (2025). Whey: Composition, Processing, Application, and Prospects in Functional and Nutritional Beverages—A Review. *Foods*, *14*(18), 3245. DOI: 10.3390/foods14183245 PMID: 41008217

Ng, H.-S., Chang, C.-Y., Chiu, C.-A., & Lan, J. C.-W. (2025). Metabolic engineering of Ralstonia eutropha for enhanced poly(3-hydroxybutyrate) production from crude glycerol. *Bioresource Technology*, *133531*. Advance online publication. DOI: 10.1016/j.biortech.2025.133531 PMID: 41139026

Pescuma, M., De Valdez, G. F., & Mozzi, F. (2015). Whey-derived valuable products obtained by microbial fermentation. *Applied Microbiology and Biotechnology*, *99*(15), 6183–6196. DOI: 10.1007/s00253-015-6766-z PMID: 26124070

Pirzadi, Z., & Meshkani, F. (2022). From glycerol production to its value-added uses: A critical review. *Fuel, 329*, 125044. DOI: 10.1016/j.fuel.2022.125044

Poladyan, A., Blbulyan, S., Sahakyan, M., Lenz, O., & Trchounian, A. (2019). Growth of the facultative chemolithoautotroph Ralstonia eutropha on organic waste materials: Growth characteristics, redox regulation and hydrogenase activity. *Microbial Cell Factories, 18*(1), 201. DOI: 10.1186/s12934-019-1251-5 PMID: 31739794

Poladyan, A., Trchounian, K., Paloyan, A., Minasyan, E., Aghekyan, H., Iskandaryan, M., Khoyetsyan, L., Aghayan, S., Tsaturyan, A., & Antranikian, G. (2023). Valorization of whey-based side streams for microbial biomass, molecular hydrogen, and hydrogenase production. *Applied Microbiology and Biotechnology, 107*(14), 4683–4696. DOI: 10.1007/s00253-023-12609-x PMID: 37289241

Puppi, D., Pecorini, G., & Chiellini, F. (2019). Biomedical Processing of Polyhydroxyalkanoates. *Bioengineering (Basel, Switzerland), 6*(4), 108. DOI: 10.3390/bioengineering6040108 PMID: 31795345

Satoh, H., Sakamoto, T., Kuroki, Y., Kudo, Y., & Mino, T. (2016). Application of the Alkaline-Digestion-HPLC Method to the Rapid Determination of Polyhydroxyalkanoate in Activated Sludge. *Journal of Water and Environment Technology, 14*(5), 411–421. DOI: 10.2965/jwet.16-027

Schäfer, C., Friedrich, B., & Lenz, O. (2013). Novel, Oxygen-Insensitive Group 5 [NiFe]-Hydrogenase in Ralstonia eutropha. *Applied and Environmental Microbiology, 79*(17), 5137–5145. DOI: 10.1128/AEM.01576-13 PMID: 23793632

Sohn, Y. J., Son, J., Jo, S. Y., Park, S. Y., Yoo, J. I., Baritugo, K.-A., Na, J. G., Choi, J., Kim, H. T., Joo, J. C., & Park, S. J. (2021). Chemoautotroph *Cupriavidus necator* as a potential game-changer for global warming and plastic waste problem: A review. *Bioresource Technology, 340*, 125693. DOI: 10.1016/j.biortech.2021.125693 PMID: 34365298

Usmani, Z., Sharma, M., Gaffey, J., Sharma, M., Dewhurst, R. J., Moreau, B., Newbold, J., Clark, W., Thakur, V. K., & Gupta, V. K. (2022). Valorization of dairy waste and by-products through microbial bioprocesses. *Bioresource Technology, 346*, 126444. DOI: 10.1016/j.biortech.2021.126444 PMID: 34848333

Vassilian, A., & Trchounian, A. (2009). Environment oxidation-reduction potential and redox sensing by bacteria. In *Bacterial Membranes* (pp. 163–195). Research Signpos.

Watanabe, Y., Ichinomiya, Y., Shimada, D., Saika, A., Abe, H., Taguchi, S., & Tsuge, T. (2012). Development and validation of an HPLC-based screening method to acquire polyhydroxyalkanoate synthase mutants with altered substrate specificity. *Journal of Bioscience and Bioengineering*, *113*(3), 286–292. DOI: 10.1016/j.jbiosc.2011.10.015 PMID: 22088761

Yang, F., Hanna, M. A., & Sun, R. (2012). Value-added uses for crude glycerol—A byproduct of biodiesel production. *Biotechnology for Biofuels*, *5*(1), 13. DOI: 10.1186/1754-6834-5-13 PMID: 22413907

Zikmanis, P., Kolesovs, S., & Semjonovs, P. (2020). Production of biodegradable microbial polymers from whey. *Bioresources and Bioprocessing*, *7*(1), 36. DOI: 10.1186/s40643-020-00326-6

Chapter 3

Exploring the Resilience of Traditional Barley (Hordeum vulgare L.) Genotype in Mitigating Salinity Stress for Sustainable Agriculture and Food Security

Anna Harutyunyan
http://orcid.org/0009-0002-4616-9170
Faculty of Biology, Yerevan State University, Armenia

Abhishek Singh
http://orcid.org/0000-0002-0049-6910
Faculty of Biology, Yerevan State University, Armenia

Hrant Khachatryan
Faculty of Biology, Yerevan State University, Armenia

Hasmik Movsesyan
Faculty of Biology, Yerevan State University, Armenia

Karen Ghazaryan
http://orcid.org/0000-0002-8973-2595
Faculty of Biology, Yerevan State University, Armenia

DOI: 10.4018/979-8-3373-1524-9.ch003

ABSTRACT

Understanding the salinity stress processes is crucial for crop development and sustainable agriculture. Salinity stress has the highest level of adverse impact on crop growth and development, agricultural yield, and food security compared to other abiotic stressors. Therefore, it is important to protect, investigate, and document traditional plants that can reduce salt damage and increase salt tolerance in newly grown crops for food security and sustainable agriculture. This study aimed to examine the response of the cereal crop barley (Hordeum vulgare L.) genotypes to salinity stress. The seeds of the local genotype of barley (Alashkert) were grown for 3 months in the greenhouse of Yerevan State University (Republic of Armenia) in perlite pots while applying various concentrations of NaCl (0, 100, 200, 300, 400, and 500 mM). In response to increased salt stress, barley stem length and diameter, root, stem, leaf biomass and water content, chlorophyll content, and gas exchange were all reduced. However, the data also demonstrate that the barley cultivar absorbs more Na^+ ions and increases the Na^+/K^+ ratio within its leaves, shoots, and roots under high-salinity conditions. The findings suggest that the "Alashkert" local barley genotype has potential as a crop for use in saline agriculture in Armenia and other countries. Therefore, our in-depth primary analysis of barley genotypes at the physiological, morphological, and biochemical levels provides comprehensive insights into the potential to improve agricultural practices for crop improvement and management programs.

1. INTRODUCTION

The global population is rapidly increasing and may exceed 10 billion by 2050, thereby increasing food demand (Mora et al., 2020; Zulfiqar et al., 2021). Simultaneously, constant population growth and climate change threaten agricultural productivity (Adhikary et al., 2022). Currently, about one-third of the world's agricultural lands are affected by salinization due to poor practices and contamination (Isayenkov et al., 2019). Therefore, it is essential to study the morphological and physiological responses of crops to salinity before applying genetic or environmental solutions. Salinity stress reduces plant growth by causing water and oxidative stress, ion toxicity, restricted mineral uptake, and nutritional imbalance. High soil salt concentrations lead to osmotic and ionic stress (Mirza Hasanuzzaman et al., 2020; Paul, 2013). Moreover, salt-induced nutritional imbalance decreases photosynthetic activity by affecting pigments (Fernando, 2020; Parida et al., 2005a). The primary mechanism is the controlgof the accumulationn of sodium (Na^+) and potassium (K^+) in certain organs. Because of the danger posed by excess Na^+, plants employ this defense mechanism to maintain a K^+/ Na^+ ratio high within their cells and regulate ion homeostasis (Rana Munns et al., 2008). This high K /Na ratio physiologically alters the leaf water potential. This causes plants to experience water stress, which in turn causes stomatal closure and a reduced gas exchange mechanism, increasing the likelihood of oxidative injury. When stomata close, gas exchange is reduced, which can have a negative impact on photosynthesis (Abideen et al., 2020).

Barley (*Hordeum vulgare L.*), is a crucial agricultural crop such as wheat and rice that is grown in a broad range of climates around the world and hence serves as a model crop for predicting how other crops will react to climate change (Abou-Zeid et al., 2014; Peltonen-Sainio et al., 2009). In Armenia, barley is the second most widely grown crop after wheat; therefore, understanding its salinity tolerance is essential for improving the lives of Armenian farmers and the population.

High salinity can inhibit barley development and yield (Pour-Aboughadareh et al., 2021a), and the growth response to salt stress may differ among cultivars. This study examined how NaCl affected the growth patterns of the undocumented barley *"Alashkert"* which is known for its varying salt resistance. Under salt stress, we measured morpho-physiological and biochemical changes in relation to growth, ion concentration, and photosynthesis. This study documents the *"Alashkert"* genotype at multiple levels and provides criteria for its improvement, offering valuable insights for future crop research and sustainable agriculture in Armenia and beyond.

2. MATERIAL AND METHODS

2.1 Experimental Sites

The pot experiment was conducted in 2022 in the greenhouse facilities available within the Department of Ecology and Nature Protection at the Faculty of Biology, Yerevan State University, Republic of Armenia.

2.2 Plant Material Collection

Armenian barley *"Alashkert"* seeds were collected from the Agriculture Scientific Center in Armenia. The barley used in this study is a rare and undocumented genotype grown mostly in saline soils in Armenia's Ararat Plain.

2.3 Growth Conditions

Barley seeds were surface-sterilised with 10% sodium hypochlorite for 15 min and washed with distilled water. Ten seeds were planted in clean PVC containers (height 13.5 cm, top diameter 16 cm, bottom diameter 15 cm, capacity 2 L). A completely randomised block (CRD) design with three replicates per treatment was used. The pot experiment used perlite instead of soil: 200g expanded perlite (density 120 kg/m^3) was placed in each pot, provided by "Aragats perlite" OJSC, Armenia. Expanded perlite is heat-treated, is a good insulator, and absorbs moisture. Plants were supplied with 20% Hoagland nutrient solution throughout the trial (The water-culture method for growing plants without soil by D. R. Hoagland | Open Library, n.d.) containing $MgSO_4$, KNO_3, KH_2PO_4, $MnCl_2{\cdot}4H_2O$, $CuSO_4{\cdot}5H_2O$, H_2MoO_4, $(NH_4)_2SO_4$, K_2SO_4, $Ca(NO_3)_2$, Fe citrate, $ZnSO_4{\cdot}7H_2O$, and H_3BO_3. After 7 weeks of germination, the plants were irrigated with 640 ml NaCl solution. To minimise osmotic shock, 80 ml was applied every other day over a 15-day period, with eight applications in total, at concentrations of 0, 100, 200, 300, 400, and 500 mM. The applied volume was sufficient to stabilise the NaCl concentration in the substrates. Throughout the experiment, the substrate EC and pH were maintained constant and regularly monitored. All plants were harvested 30 days after salt treatment, at the end of May 2022.

2.4 Morphological Indices

The landrace barley genotypes were rated based on two major morphological characteristics. When measuring shoot length, the distance from the top leaf tip to the base of the stem was measured. We also compared the diameters of the stems of untreated and treated plants.

2.5 Physiological Indices

2.5.1 Measurement of Biomass

Biomass accumulation was determined in both control and salt-stressed plants. Roots, stems, and leaves were collected at the time of harvest. The fresh weights of stems (SFW), roots (RFW), and leaves (LFW) were determined. For each plant, root dry weight (RDW), stem dry weight (SDW), and leaf dry weight (LDW) were determined by drying the samples in an oven at 70°C until a constant mass was reached.

2.5.2 Measured Roots, Stems, and Leaves Water Contents

RWC%, SWC%, and LWC% were calculated using the following formulas (Al Hassan et al., 2016a):

$$RWC\% = \left[(FW_{root} - DW_{root})/FW_{root}\right] \times 100 \tag{1}$$

$$SWC\% = \left[(FW_{stem} - DW_{stem})/FW_{stem}\right] \times 100 \quad (2)$$

$$LWC\% = \left[(FW_{leaf} - DW_{leaf})/FW_{leaf}\right] \times 100 \quad (3)$$

where FM_{root}, FM_{stem}, and FM_{leaf} are the fresh weights of the root, stem, and leaves, and DW_{root}, $DW_{stem,}$ and DW_{leaf} are the dry weights of the root, stem, and leaf, respectively.

2.5.3 Estimation of Chlorophyll Content Index (CCI)

The chlorophyll content in the fully developed leaves from the top third portion of plants under investigation was estimated from stressed plant growing in salt affected condition as per CCM-200 plus Chlorophyll Content Meter (Opti-sciences, USA). Ten readings were recorded for each plant, and the Chlorophyll Content Meter provided the average of all data instantly.

2.5.4 Gas exchange parameters

Gas exchange parameters, such as the transpiration rate (E) and photosynthetic net CO_2 assimilation rate (Pn), were measured using a portable photosynthesis system (CI-340). After salt treatment, the five youngest, completely developed leaves from each plant were measured for each replication. The water-use efficiency (WUE) was measured as the net carbon uptake divided by the water lost through transpiration in the leaf area, as described by Rabhi et al. (2012):

$$WUE = Pn/E \quad (4)$$

2.6 Biochemical Indices

Plant parts were dried, ground into powder, and then 1 g of plant sample (root, stem, and leaves) was digested in HNO_3 (0.5%) solution for 30 min at 100 °C to extract ions. After digestion, the solution was filtered through filter paper, and the ions were immediately analysed (Al Hassan et al., 2016b). The concentrations of Na^+ and K^+ were determined using a flame photometer (FP-I6431, Bioevopeak, PRC), and the concentration of Cl was measured using a laboratory ionometer (I-160 M, Anatech, Belarus). The tissue samples were autoclaved for 15 min at 121°C to remove residual ions. The material was cooled to room temperature, and the TDS was measured using a laboratory conductometer (MARK 603).

2.7 Statistical Analysis

The data were analysed using SPSS-19 and Microsoft Excel 2021. Fisher's least significant difference (LSD) test was used to determine statistically significant differences. The 95% confidence intervals are shown as error bars in the figures.

3. RESULTS AND DISCUSSION

3.1 Effect of Salinity Stress on Morphological Indices of Barley

Salinity can have positive or negative effects on plant growth parameters, such as length, biomass, stem diameter, viability, and threshold responses, depending on the salinity level and tolerance of the plant. To evaluate the effect of salinity on plant growth parameters, stem diameter, root, stem, and leaf FM and DM were measured after 30 d of NaCl treatment. In addition, shoot length was measured every 10 days after the end of NaCl treatment (three times). According to the survival and threshold responses to different NaCl stress regimes, the growth and biomass of the study crops were significantly affected by the different NaCl concentrations. The dynamics of barley shoot length growth under different salt treatments are shown in Figure 1.

Figure 1. Effect of NaCl stress on growth of barley during salt treatment (n=30)

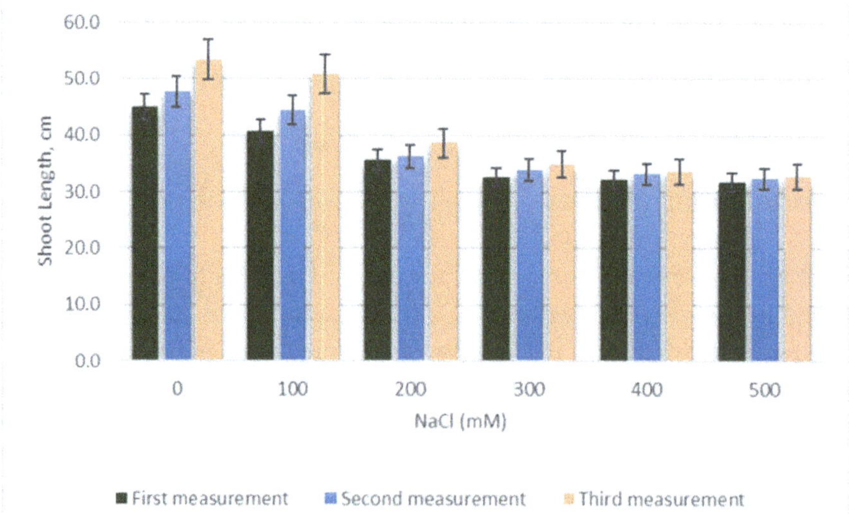

Barley height decreased significantly with increasing NaCl concentrations. Ten days after the end of salt addition, the shoot height of the plant grown in NaCl medium compared to the control was as follows: 100 mM - 90.5%, 200 mM - 79.3%, 300mM - 72.6%, 400mM - 71.6%, and 70.7% at 500mM concentration. After 30 d, when the third measurement was made, the dynamics of barley growth inhibition changed, and most decrease at a concentration of 500 mM NaCl was observed. The heights of barley compared to the control at concentrations of 100-500 mM were 95.3%, 72.4%, 65.3%, 63.1%, and 61.6%, respectively. NaCl treatment also affected the stem diameter of barley (Figure 2). Stem diameter increased normally up to 200mM concentration and then decreased more significantly at 400mM and 500mM NaCl concentrations compared to the control.

Figure 2. Effect of NaCl stress on stem diameter of barley (n=30)

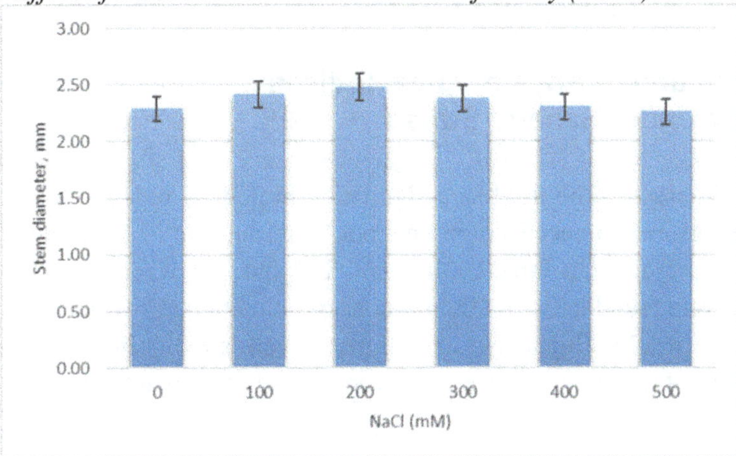

Plant cells subjected to salinity undergo significant as a result changes in morphology, physiology, and biochemical (Abou-Zeid et al., 2014; Flowers, 2004). The present study investigated thee morpho-physiological and biochemicalprofilese of previously undocumented barley plants after exposure to salt. Barley root, stem, and leaf length, diameter, weight (fresh and dry), water content, chlorophyll content, gas exchange all decreased as ionic content and total dissolved solids (TDS) increased due to the NaCl concentration increasing from 0 to 500 mM. This study confirmed previous research showing that barley plants experience considerable growth loss when subjected to salinity stress above their tolerance threshold. Previous research into barley plants found a similar decline in growth (Abrar et al., 2020; Ahmadi et al., 2020; Rajeswari et al., 2019). Both the length and diameter of the shoots were reduced as a result of stunted growth. Water and nutrient intake by xylem loading in the root is disrupted when roots are severely damaged which inhibits shoot growth. In addition, leaf area is immediately reduced by osmotic stress. Therefore, shoots are more susceptible to salt stress than roots. Our results showed that shoot growth was drastically reduced under saline conditions (Figure 2). Increasing salt concentrations also dramatically affected the root, shoot, and leaf weights of the plants. Root, shoot, and leaf weight were all observed to decrease with increasing salt concentrations, correlating with previous studies (Abdeen et al., 2015; Cope et al., 2022; Md Hasanuzzaman et al., 2017; Long et al., 2013). This is consistent with the findings of a previous study, which showed that total plant biomass decreased with increasing salt concentrations. Both the fresh and dry weights of the shoots, roots, and leaves were drastically reduced as the salt content increased. (Pour-Aboughadareh et al., 2021b) and (Ali et al., 2003) both discovered results that were compatible with the present studies. (Gill et al., 2001) and (Özdemir et al., 2004) found the same result for sorghum and rice, respectively.

3.2 Evaluation of Physiological Indices of Barley Under Salinity Stress

Plant biomass can be used as a significant biological index to evaluate plant tolerance to salinity-induced stress. The root, stem, and leaf FM and DM results for the crop are presented in Table 1. Maximum FM and DM of root, shoot and leaf of the barley were found in 0 and 100mM concentration of NaCl while a decrease for FM followed by DM at higher NaCl treatment was observed. It should be mentioned that the descent was, in most cases, not proportional and that interruptions and new increases in fresh and dry mass also occurred.

The reduction of the fresh and dry masses of the roots of plants growing at 500 mM compared to 0 was 1.96 and 1.57, respectively. times, in the stem 3.91 and 4.34 times, and in the leaf 1.63 and 1.52 times. The ratios of FM_r/FM_s and DM_r/DM_s first decreased and then increased significantly, that is, as an adaptive reaction of the plant, there was an increase in the value of root mass/stem mass. Along with the increase in salinity, there was also a decrease in RWC% and LWC% value (compared to 0 mM in plants growing at 500 mM decreased by 4.00% and 2.02% respectively). The SWC% value increased (500 mM compared to 0 mM, it increased by 2.64%).

Table 1. Impact of NaCl stress on growth parameters and tissue water content in barley (n=30).

NaCl (mM)	RFW	RDW	SFW	SDW	LFW	LDW	RFW/ SFW	RDW/ SDW	RWC%	SWC%	LWC%
0	1.842	0.255	1.670	0.349	1.315	0.277	0.617	0.407	86.18	79.09	78.92
100	1.340	0.205	1.972	0.403	1.144	0.242	0.430	0.318	84.69	79.57	78.89
200	1.199	0.200	1.418	0.287	0.919	0.196	0.513	0.415	83.29	79.76	78.64
300	1.140	0.195	0.861	0.156	0.875	0.194	0.656	0.559	82.85	81.90	77.87
400	1.141	0.195	0.524	0.100	0.824	0.187	0.846	0.681	82.87	81.01	77.27
500	0.938	0.162	0.427	0.080	0.806	0.183	0.760	0.616	82.73	81.18	77.33

The osmotic effect of salinity stress reduces the amount of water that can be absorbed by plants. A decrease in water potential results from salt accumulation in the root zone, which reduces the osmotic potential. Consequently, the root system of the plant absorbs less water. It has been suggested that plants achieve water balance due to losing water, resulting in a reduction in osmotic potential (Acosta-Motos et al., 2017; Gupta et al., 2014). As a result, plants may develop a tolerance to salt stress by decreasing their water intake levels. Both shoot and root water content were shown to decrease in this investigatio. Several more investigations came to a similar conclusion (Al-Maskri et al., 2010; Rajput et al., 2021; Rasel et al., 2021). Increasing salt concentrations additionally has a deleterious effect on the water content of barley root, shoot and leaves (Table 2). The effects of salinity stress on the relative water content of undocumented barley plants have not been investigated. However, studies on rice *(Oryza sativa* L.), sorghum (*Sorghum bicolor* L.) and buffaloberry (*Shepherdia argentea* L.) have investigated into the possibility that salinity stress treatment may reduce intercellular water status in plants that support our finding (Gill et al., 2001; Pattanagul et al., 2008; Qin et al., 2010).

CCI values changed depending on both the change in NaCl content of the root medium and the growth stage of the crops (Figure 3). In particular, along with the growth of barley, depending on the change in the vegetation stages (sprouting, flowering, earing, etc.), an increase in the CCI value was observed for barley at first, and then a decrease. Nevertheless, as salinity increased, the CCI value decreased significantly.

From the measurements taken 10 days after the completion of salt treatment, it was found that the decrease in CCI value of plants grown in 500 mM NaCl medium compared to those grown in 0 mM NaCl medium was 73.93%.

Figure 3. Effect of NaCl stress on CCI of barley (n=30)

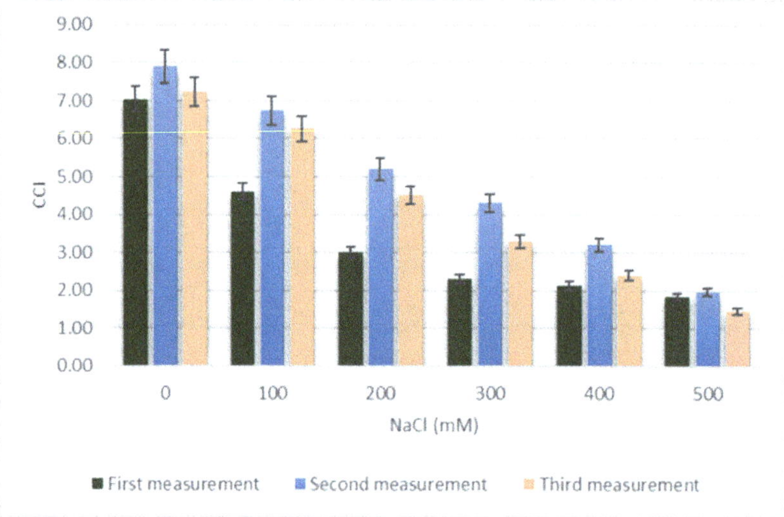

The highest CCI values were recorded during the measurements made after 20 days, which is due to both plant growth stages and adaptation. In particular, the increase in the value of CCI at 0mM was 12.32%, and in the range of 100-400mM, where the increase of CCI was 46.38-86.96%. Even at 500mM, the adaptation potential of the plant was not satisfied, and the increase in CCI was only 7.27%. During the third measurement, the CCI value decreased by 79.72% (500 mM compared to 0 mM).

The results of the study on the photosynthetic rate, transpiration rate, and WUE under different salt treatment conditions are shown in Table 2.

Table 2. Effect of NaCl stress on gas exchange of barley

Salinity degree NaCl (mM)	Pn	E	WUE
0	12.39	2.89	4.29
100	6.34	1.53	4.14
200	4.94	1.05	4.69
300	5.27	1.12	4.70
400	4.74	1.16	4.08
500	4.21	0.97	4.36

As can be seen in the table, the intensity of photosynthesis at the concentration of 0mM of NaCl was quite high, then it decreased sharply to the concentration of 200mM (60.13%), and no significant differences were observed in the range of

200-500mM (57.50%-65.99%). In the crop grown in 500 mM NaCl medium, the transpiration rate decreased compared to that grown in 0 mM NaCl medium, it was 66.55%. With increasing NaCl concentration, no significant changes in WUE were observed in barley.

Essential physiological activities involving plant health and development under adverse growing conditions include chlorophyll and its related physicochemical processes, such as the chlorophyll content index (CCI), photosynthetic rate (Pn), transpiration rate (E), and water-use efficiencyy (WUE). Salinity stress was shown to reduce the proportion of chlorophyll content and other photosynthesis-related characteristics in the current study (Table 2). The chlorophyll content in undocumented genotype barley was dramatically affected by salt stress as saline concentrations increased (Figure 4). Barley chlorophyll content was minimally affected at 100 mM NaCl, but it declined as salinity stress increased from 200 mM to 500 mM. (Higbie et al., 2010), (Zhang et al., 2014), (Pour-Aboughadareh et al., 2021a) and (Shin et al., 2020) all experimental results confirm this finding. An increase in salt solutions in the leaf, which produces a corresponding reduction in leaf area has become a key factor in the development of chlorosis (Sohan et al., 1999). Another case of increased chlorophyll content under salt conditions may be the correlation between increased leaf thickness and decreased leaf area, as shown by Papp et al. (PAPP et al., 1983). As shown in Table 2, salt treatment considerably reduced the photosynthetic rate (Pn), transpiration rate (E), and water usage efficiency (WUE) of treated plants. Two symmetric guard cells regulate gas exchange between the interior and exterior of leaves. The density and function of its pores, as well as the capacity of guard cells to move water over the leaf surface, all play a role in regulating stomatal conductance (X. Zhu et al., 2018). Pn is used to estimate the CO_2 rate, while the resistance to gas flow within the leaf essentially determines the transpiration rate (E). While salinity stress conditions did not significantly impact water usage efficiency (WUE), the rate of decrease for Pn and E was considerable in this study. Current results confirm those from (Pour-Aboughadareh et al., 2021a), finding that show found that salt treatment led to a drop in Gs and TE.

3.3 Biochemical Indices and Phytodesalination Capacity of Barley under Salt Stress Barley

The analysis of some ions and TDS of roots, stems and leaves of the barley is shown in Table 3. Ion accumulation in roots, stems, and leaves of the crop varied with increasing NaCl concentration. Basically, along with the increase of NaCl concentration in the root medium, the content of K^+ in the roots, stems and leaves of the crop first increased and then significantly decreased. Along with the increase of NaCl concentration, the content of Na^+ and Cl^- in the roots, stems and leaves of the barley increased. Only in some cases, at high concentrations of NaCl, a certain decrease in the content of Na^+ and Cl^- was observed, which is probably due to the activation of the protective mechanisms of the crop, in particular, the highest value in the root was observed at 300 mM (increased 4.74 times compared to 0 mM), after which it started to decrease. A stem increase in Na^+ content was observed in the stem up to 300 mM (increased 7.38 times compared to 0 mM), and no significant change was observed in the range of 300-500. and a leaf increase in Na^+ content in the leaf was observed up to 400mM (increased 11.42 times compared to 0mM), and a slight increase was recorded at 500mM. In the case of Cl^-, the result is slightly different, the highest value in the root was observed at 400 mM (increased 223 times compared to 0 mM), after which a drastic decrease was recorded. And in the stem and leaf up to 400mM concentration, it increased drastic again (increased 872 and 452 times compared to 0mM, respectively), then a slight change was observed.

The change of the Na^+ /K^+ ratio in the roots, stems and leaves of the barley showed a dynamic increase along with the increase in NaCl concentration (in plants grown at 500mM compared to 0mM, the increase in root, stem and leaf was 9.30, 13.60, 9.96 times, respectively). The analysis of the roots, stems and leaves of the crop showed that an increase in the TDS value was mainly observed along with the increase in NaCl concentration (in plants grown at 500mM compared to 0mM, the increase in root, stem and leaf was 2.35, 2.81, 2.05 times, respectively). There was an increase in TDS value in roots and leaves up to 400mM concentration, then it decreased slightly at 500mM concentration. The greatest increase in TDS was observed near the stem at a concentration of 300 mM.

In roots, stems and leaves of barley harvested after 30 days of NaCl treatment, a significantly higher accumulation of Na^+ and Cl^- was observed than in control plants (Figure 4). From the point of view of phytosalination, the accumulated amount of Na^+ and Cl^- in the above-ground mass of plants is the most important. As a result of the studies, it was found that the largest accumulation of Na^+ was observed in the leaf, and in the case of Cl^-, in the stem.

Table 3. Effect of NaCl stress on ionic content of barley root, stem, and leaf (mg/g, n = 5)

Salinity degree (mM)	K^+			Na^+			Cl^-			TDS		
	root	stem	leaf	root	stem	leaf	root	stem	leaf	root	stem	leaf
0	3.37	19.25	12.10	4.66	4.00	3.26	0.06	0.05	0.07	19.07	36.19	41.39
100	3.67	23.77	19.30	13.60	20.83	15.30	1.55	11.50	4.00	51.53	84.61	66.29
200	2.33	23.06	18.63	18.35	23.20	18.48	5.16	22.00	4.31	47.59	81.82	70.04
300	2.16	20.05	18.84	22.10	29.50	31.00	11.21	31.20	19.00	49.06	104.80	91.45

continued on following page

Table 3. Continued

Salinity degree (mM)	K^+			Na^+			Cl^-			TDS		
	root	stem	leaf	root	stem	leaf	root	stem	leaf	root	stem	leaf
400	1.72	11.02	17.91	21.91	29.10	37.23	14.30	41.00	29.50	52.55	97.31	99.85
500	1.45	10.70	14.03	18.65	30.23	37.65	11.52	42.00	28.00	44.81	101.60	84.82

In general, with the increase of NaCl content, the above-ground mass of the barley decreased, but the content of Na^+ and Cl^- increased there. In particular, depending on the degree of salinity of the soil, 9.31-12.09 mg of sodium and 5.60-9.61 mg of chlorine can be removed from the soil through the above-ground mass of per plant. And with all this, it can be stated that the barley has the potential of phytodesalination.

Salinity stress causes a decrease in plant nutrients concentrations (Parida et al., 2005b; Taffouo et al., 2009). Plant concentrations of Na^+, K^+, Cl^-, and TDS were all found to be increased or decreased in the treated plants compared to the control plants after exposure to NaCl at concentrations of 0 mM, 100 mM, and 500 mM in the current investigation. Ion homeostasis becomes disturbed when there is an excess of sodium or other salts in the environment (Arif et al., 2020; Ashraf et al., 2009; J. K. Zhu, 2016). Our results showed that under salt stress, barley cultivar absorbed more Na^+ and Cl^- ions by nonspecific ion uptake as well as membrane leakage in their leaves, shoot and roots (Table 3). Na^+ and Cl^- levels both increased with increasing NaCl concentration, which suggests that both cultivars may not differ in Na^+ uptake and its transportation to leaves and that the observed rise cannot be explained by an ionic exclusion mechanism (Figure 4).

Figure 4. The mass of Na^+ and Cl^- (mg) accumulated in the above-ground parts of barley (per plant)

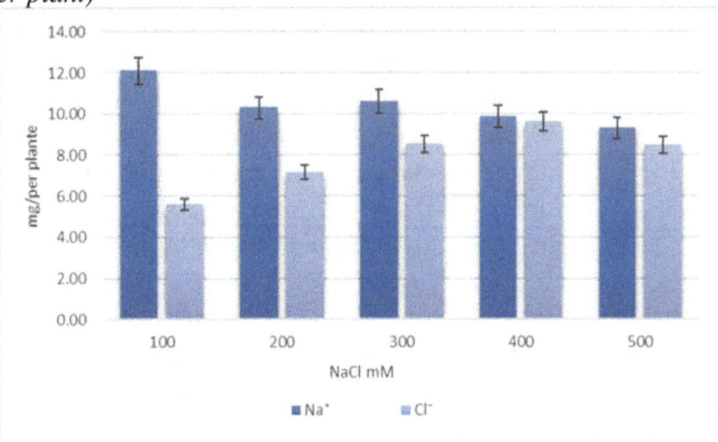

Furthermore, at all salinity levels, the concentration of Na^+ ions in the leaves of barley were greater compared to that in the roots and stem, which suggests an inability of such cultivar to inhibit Na^+ ion transit from the roots to the leaves via shoot. Despite the greater salinity, the distribution of chlorine ions was very equivalent to the distribution of sodium ions. Plants rapidly absorb Na^+ and Cl^- and transfer them into their leaves, where they probably function as *osmotica*. Our finding was similar to experimental result of (Zhang et al., 2014) in cotton (*Gossypium hirsutum* L.) plant. However, Na^+ and Cl^- ions can only be tolerated at moderate amounts before they stunt plant development and inhibit photosynthesis (Fig 4 and Table 3). Due to Cl/NO_3 antagonism, higher uptake, and buildup of Cl^- under saline conditions might reduce total shoot nitrogen intake (Muchate et al., 2016; R. Munns et al., 1986; Wang et al., 2003). The intake of phosphorus, which is essential for photosynthesis, storage, and energy transfer, is also impacted by salt stress. High ionic strength of the media and limited solubility of the Ca P minerals may be to responsible for decreased phosphorous uptake when soil contains significant Cl and SO ² (Bose et al., 2014; Chaudhuri et al., 1997; Chen et al., 2005; J. K. Zhu, 2003). Protein synthesis and water balance both rely on the presence of potassium, an essential inorganic solute. A high intake of Na^+ and Cl^- may lead to a deficit of other nutritional ions like K^+ and Ca^{2+} and an imbalance of cations (Parida et al., 2004). Barley K^+ levels dropped dramatically over the same period of time and this trend continued with increasing NaCl concentration, while the loss was most pronounced in the leaves compared to the shoot and root (Fig. 4 and Table 3). The uptake of K^+ is inhibited by a high external Na^+ concentration because of the chemical similarity between these two ions (Asgaria et al., 2012; Borsani et al., 2003; Kaya et al., 2007). Both K^+ and Na^+ are in high demand in an environment with salt. Plants cannot thrive in saline soil unless the Na^+/K^+ ratio in their cells is perfectly balanced. Although Na^+ cannot perform the role of K^+ in cellular activities because their chemical similarities with K^+ to be replaced by Na^+. Since Na^+ and K^+ compete for root absorption sites which increase the Na^+ concentration that causes a reduction in both K^+ concentrations. Na^+ inhibits K^+ uptake in plants by a competitive mechanism, which occurs whether Na^+, Cl^- or SO_4^{2-} salts predominate in the solution. Numerous studies on plants like *Manihot esculenta, Zea mays, G. hirsutum and* wheat show that salinity lowers crop yields and leads to nutritional imbalances (Cruz et al., 2018; Grattan et al., 1998; Hasana et al., 2017; Higbie et al., 2010; Hu et al., 2005; Niste et al., 2014).

Salt tolerance requires not only adaptation to Na^+ influx but also acquisition of K^+. In fact, many plant species' salt tolerance is thought to be due in large part to their selective uptake of K^+ as opposed to Na^+ (Asgaria et al., 2012). Losses of K^+ have been recorded under salt stress and K^+ efflux has been implemented as an indication of cellular toxicity for various kinds of hazardous substances (Asgaria et al., 2012; Borsani et al., 2003). It is generally accepted that limiting membrane injury results in a massive and persistent outflow of K^+ (Deinlein et al., 2014; Rasel et al., 2021; Van Zelm et al., 2020).

4. CONCLUSION

In conclusion, our investigation into the previously undocumented barley genotype, *"Alashkert"*, and its response to saline conditions (0–500 mM) has yielded invaluable insights into its adaptability and resilience under salinity stress. This research contributes to understanding the potential applications of *"Alashkert"* in saline-prone regions, while the examination of its physiological, morphological, and biochemical responses provides knowledge to enhance agricultural practices and crop management. The documented data will aid plant breeders, researchers, and stakeholders in developing salt-tolerant barley varieties and optimizing cultivation techniques. By rigorous research, we enhance capacity to address agricultural challenges and ensure food security. Studying barley genotypes like *"Alashkert"* is vital for both agriculture and understanding salinity resilience, making this experiment a key step toward sustainable agriculture benefiting Armenia and the European Union.

ACKNOWLEDGMENT

KG is supported by Higher education and science committee MESCS RA, grant number 21AG-4C075.

REFERENCES

Abdeen, S. A., Rizk, A. H., & El-Hamamsy, S. M. (2015). Effect of salinity stress on growth and macronutrients uptake of some Egyptian barley landraces ((Hordeum vulgare L.). *J.Soil Sci. and Agric. Eng. Mansoura Univ*, *6*(11), 1355–1364.

Abideen, Z., Koyro, H. W., Huchzermeyer, B., Ahmed, M. Z., Zulfiqar, F., Egan, T., & Khan, M. A. (2020). Phragmites karka plants adopt different strategies to regulate photosynthesis and ion flux in saline and water deficit conditions. DOI: 10.1080/11263504.2020.1762783

Abou-Zeid, H. M., & Moustafa, Y. (2014). *Physiological and Cytogenetic Responses of Wheat and Barley to Silver Nanopriming Treatment.*

Abrar, M. M., Saqib, M., Abbas, G., Atiq-Urrahman, M., Mustafa, A., Shah, S. A. A., Mehmood, K., Maitlo, A. A., ul-Hassan, M., Sun, N., & Xu, M. (2020). Evaluating the Contribution of Growth, Physiological, and Ionic Components Towards Salinity and Drought Stress Tolerance in Jatropha curcas. *Plants*, *9*(11), 1–18. DOI: 10.3390/plants9111574 PMID: 33203052

Acosta-Motos, J. R., Ortuño, M. F., Bernal-Vicente, A., Diaz-Vivancos, P., Sanchez-Blanco, M. J., & Hernandez, J. A. (2017). Plant Responses to Salt Stress: Adaptive Mechanisms. *Agronomy 2017, Vol. 7, Page 18, 7*(1), 18. DOI: 10.3390/agronomy7010018

Adhikary, S., Biswas, B., Chakraborty, D., Timsina, J., Pal, S., Chandra Tarafdar, J., Banerjee, S., Hossain, A., & Roy, S. (2022). Seed priming with selenium and zinc nanoparticles modifies germination, growth, and yield of direct-seeded rice (Oryza sativa L.). *Scientific Reports, 12*(1), 1–14. DOI: 10.1038/s41598-022-11307-4

Ahmadi, J., Pour-Aboughadareh, A., Fabriki Ourang, S., Khalili, P., & Poczai, P. (2020). Unraveling salinity stress responses in ancestral and neglected wheat species at early growth stage: A baseline for utilization in future wheat improvement programs. *Physiology and Molecular Biology of Plants : an International Journal of Functional Plant Biology*, *26*(3), 537–549. DOI: 10.1007/s12298-020-00768-4 PMID: 32205929

Al-Maskri, A., Al-Kharusi, L., Al-Miqbali, H., & Mumtaz Khan, M. (2010). Effects of Salinity Stress on Growth of Lettuce (Lactuca sativa) under Closed-recycle Nutrient Film Technique. *Int. J. Agric. Biol*, *12*, 377–380.

Ali, R. M., & Abbas, H. M. (2003). Response of salt stressed barley seedlings to phenylurea. *Https://Pse.Agriculturejournals.* DOI: 10.17221/4107-PSE

Arif, Y., Singh, P., Siddiqui, H., Bajguz, A., & Hayat, S. (2020). Salinity induced physiological and biochemical changes in plants: An omic approach towards salt stress tolerance. *Plant Physiology and Biochemistry*, *156*, 64–77. DOI: 10.1016/j.plaphy.2020.08.042 PMID: 32906023

Asgaria, H. R., Cornelisb, W., & Dammec, P. V. (2012). *Salt stress effect on wheat (Triticum aestivum L.) growth and leaf ion concentrations.*

Ashraf, M., & Akram, N. A. (2009). Improving salinity tolerance of plants through conventional breeding and genetic engineering: An analytical comparison. *Biotechnology Advances*, *27*(6), 744–752. DOI: 10.1016/j.biotechadv.2009.05.026 PMID: 19500659

Borsani, O., Valpuesta, V., & Botella, M. A. (2003). Developing salt tolerant plants in a new century: A molecular biology approach. *Plant Cell, Tissue and Organ Culture*, *73*(2), 101–115. DOI: 10.1023/A:1022849200433

Bose, J., Rodrigo-Moreno, A., & Shabala, S. (2014). ROS homeostasis in halophytes in the context of salinity stress tolerance. *Journal of Experimental Botany*, *65*(5), 1241–1257. DOI: 10.1093/jxb/ert430 PMID: 24368505

Chaudhuri, K., & Choudhuri, M. A. (1997). Effects of short-term NaCl stress on water relations and gas exchange of two jute species. DOI: 10.1023/A:1001013913773

Chen, Z., Newman, I., Zhou, M., Mendham, N., Zhang, G., & Shabala, S. (2005). Screening plants for salt tolerance by measuring K+ flux: A case study for barley. *Plant, Cell & Environment*, *28*(10), 1230–1246. DOI: 10.1111/j.1365-3040.2005.01364.x

Cope, J. E., Norton, G. J., George, T. S., & Newton, A. C. (2022). Evaluating Variation in Germination and Growth of Landraces of Barley (Hordeum vulgare L.) Under Salinity Stress. *Frontiers in Plant Science*, *13*, 863069. DOI: 10.3389/fpls.2022.863069 PMID: 35783948

Cruz, J. L., Coelho, E. F., Filho, M. A. C., & dos Santos, A. A. (2018). Salinity reduces nutrients absorption and efficiency of their utilization in cassava plants. *Ciência Rural*, *48*(11), e20180351. DOI: 10.1590/0103-8478cr20180351

Deinlein, U., Stephan, A. B., Horie, T., Luo, W., Xu, G., & Schroeder, J. I. (2014). Plant salt-tolerance mechanisms. *Trends in Plant Science, 19*(6), 371–379. DOI: 10.1016/j.tplants.2014.02.001 PMID: 24630845

Fernando, V. C. D. (2020). Major transcription factor families involved in salinity stress tolerance in plants. *Transcription Factors for Abiotic Stress Tolerance in Plants*, 99–109. DOI: 10.1016/B978-0-12-819334-1.00007-1

Flowers, T. J. (2004). Improving crop salt tolerance. *Journal of Experimental Botany, 55*(396), 307–319. DOI: 10.1093/jxb/erh003 PMID: 14718494

Gill, P. K., Dev Sharma, A., Singh, P., & Singh Bhullar, S. (2001). Effect of various abiotic stresses on the growth. *Soluble Sugars and Water Relations of Sorghum Seedlings Grown in Light and Darkness, 27*(2), 72–84.

Grattan, S. R., & Grieve, C. M. (1998). Salinity–mineral nutrient relations in horticultural crops. *Scientia Horticulturae, 78*(1–4), 127–157. DOI: 10.1016/S0304-4238(98)00192-7

Gupta, B., & Huang, B. (2014). Mechanism of salinity tolerance in plants: Physiological, biochemical, and molecular characterization. *International Journal of Genomics, 2014*, 1–18. Advance online publication. DOI: 10.1155/2014/701596 PMID: 24804192

Hasana, R., & Miyake, H. (2017). Salinity Stress Alters Nutrient Uptake and Causes the Damage of Root and Leaf Anatomy in Maize. *KnE Life Sciences, 3*(4), 219. DOI: 10.18502/kls.v3i4.708

Hasanuzzaman, M., Bhuyan, M. H. M. B., Zulfiqar, F., Raza, A., Mohsin, S. M., Al Mahmud, J., Fujita, M., & Fotopoulos, V. (2020). Reactive Oxygen Species and Antioxidant Defense in Plants under Abiotic Stress: Revisiting the Crucial Role of a Universal Defense Regulator. *Antioxidants, 9*(8), 681. DOI: 10.3390/antiox9080681

Hasanuzzaman, M., Davies, N. W., Shabala, L., Zhou, M., Brodribb, T. J., & Shabala, S. (2017). Residual transpiration as a component of salinity stress tolerance mechanism: A case study for barley. *BMC Plant Biology, 17*(1), 1–12. DOI: 10.1186/s12870-017-1054-y PMID: 28629324

Higbie, S. M., Wang, F., Stewart, J. McD., Sterling, T. M., Lindemann, W. C., Hughs, E., & Zhang, J. (2010). Physiological Response to Salt (NaCl) Stress in Selected Cultivated Tetraploid Cottons. *International Journal of Agronomy, 2010*, 1–12. DOI: 10.1155/2010/643475

Hu, Y., & Schmidhalter, U. (2005). Drought and salinity: A comparison of their effects on mineral nutrition of plants. *Journal of Plant Nutrition and Soil Science, 168*(4), 541–549. DOI: 10.1002/jpln.200420516

Isayenkov, S. V., & Maathuis, F. J. M. (2019). Plant salinity stress: Many unanswered questions remain. *Frontiers in Plant Science, 10*, 80. DOI: 10.3389/fpls.2019.00080 PMID: 30828339

Kaya, C., Tuna, A. L., Ashraf, M., & Altunlu, H. (2007). Improved salt tolerance of melon (Cucumis melo L.) by the addition of proline and potassium nitrate. *Environmental and Experimental Botany, 60*(3), 397–403. DOI: 10.1016/j.envexpbot.2006.12.008

Long, N. V., Dolstra, O., Malosetti, M., Kilian, B., Graner, A., Visser, R. G. F., & van der Linden, C. G. (2013). Association mapping of salt tolerance in barley (Hordeum vulgare L.). *Theoretical and Applied Genetics, 126*(9), 2335–2351. DOI: 10.1007/s00122-013-2139-0 PMID: 23771136

Mora, O., Le Mouël, C., De Lattre-Gasquet, M., Donnars, C., Dumas, P., Réchauchère, O., Brunelle, T., Manceron, S., Marajo-Petitzon, E., Moreau, C., Barzman, M., Forslund, A., & Marty, P. (2020). Exploring the future of land use and food security: A new set of global scenarios. *PLoS One, 15*(7), e0235597. DOI: 10.1371/journal.pone.0235597 PMID: 32639991

Muchate, N. S., Nikalje, G. C., Rajurkar, N. S., Suprasanna, P., & Nikam, T. D. (2016). Plant Salt Stress: Adaptive Responses, Tolerance Mechanism and Bioengineering for Salt Tolerance. *The Botanical Review, 82*(4), 371–406. DOI: 10.1007/s12229-016-9173-y

Munns, R., & Termaat, A. (1986). Whole-Plant Responses to Salinity. *Australian Journal of Plant Physiology, 13*(1), 143–160. DOI: 10.1071/PP9860143

Munns, R., & Tester, M. (2008). Mechanisms of salinity tolerance. *Annual Review of Plant Biology, 59*(1), 651–681. DOI: 10.1146/annurev.arplant.59.032607.092911 PMID: 18444910

Niste, M. G., Stoian, V., Monica, N., Vidican, R., Rotar, I., Pop, R., & Miclea, R. (2014). *Plant Nutrition Affected by Soil Salinity and Response of Rhizobium Regarding the Nutrients Accumulation Ioan Rotar Plant Nutrition Affected by Soil Salinity and Response of Rhizobium Regarding the Nutrients Accumulation.*

Özdemir, F., Bor, M., Demiral, T., & Türkan, I. (2004). Effects of 24-epibrassinolide on seed germination, seedling growth, lipid peroxidation, proline content and antioxidative system of rice (Oryza sativa L.) under salinity stress. *Plant Growth Regulation*, *42*(3), 203–211. DOI: 10.1023/B:GROW.0000026509.25995.13

Papp, J. C., Ball, M. C., & Terry, N.PAPP. (1983). A comparative study of the effects of NaCl salinity on respiration, photosynthesis, and leaf extension growth in Beta vulgaris L. (sugar beet). *Plant, Cell & Environment*, *6*(8), 675–677. DOI: 10.1111/1365-3040.ep11589273

Parida, A. K., & Das, A. B. (2004). Effects of NaCl stress on nitrogen and phosphorous metabolism in a true mangrove Bruguiera parviflora grown under hydroponic culture. *Journal of Plant Physiology*, *161*(8), 921–928. DOI: 10.1016/j.jplph.2003.11.006 PMID: 15384403

Parida, A. K., & Das, A. B. (2005a). Salt tolerance and salinity effects on plants: A review. *Ecotoxicology and Environmental Safety*, *60*(3), 324–349. DOI: 10.1016/j.ecoenv.2004.06.010 PMID: 15590011

Parida, A. K., & Das, A. B. (2005b). Salt tolerance and salinity effects on plants: A review. *Ecotoxicology and Environmental Safety*, *60*(3), 324–349. DOI: 10.1016/j.ecoenv.2004.06.010 PMID: 15590011

Pattanagul, W., & Thitisaksakul, M. (2008). Effect of salinity stress on growth and carbohydrate metabolism in three rice (*Oryza sativa* L.) cultivars differing in salinity tolerance. *IJEB, 46*, 736–742.

Paul, D. (2013). Osmotic stress adaptations in rhizobacteria. *Journal of Basic Microbiology*, *53*(2), 101–110. DOI: 10.1002/jobm.201100288 PMID: 22581676

Peltonen-Sainio, P., Jauhiainen, L., & Laurila, I. P. (2009). Cereal yield trends in northern European conditions: Changes in yield potential and its realisation. *Field Crops Research*, *110*(1), 85–90. DOI: 10.1016/j.fcr.2008.07.007

Pour-Aboughadareh, A., Sanjani, S., Nikkhah-Chamanabad, H., Mehrvar, M. R., Asadi, A., & Amini, A. (2021). Identification of salt-tolerant barley genotypes using multiple-traits index and yield performance at the early growth and maturity stages. *Bulletin of the National Research Centre, 45*(1), 1–16. DOI: 10.1186/s42269-021-00576-0

Qin, J., Dong, W. Y., He, K. N., Yu, Y., Tan, G. D., Han, L., Dong, M., Zhang, Y. Y., Zhang, D., Li, A. Z., & Wang, Z. L. (2010). NaCl salinity-induced changes in water status, ion contents and photosynthetic properties of Shepherdia argentea (Pursh) Nutt. seedlings. *Plant, Soil and Environment, 56*(7), 325–332. DOI: 10.17221/209/2009-PSE

Rajeswari, S., Sood, N., Swarup, T. G., Subramanian, R., Salcedo-Reyes, J. C., Rajeswari, S., Sood, N., Swarup, T. G., Subramanian, R., & Salcedo-Reyes, J. C. (2019). Assessing salt-stress tolerance in barley. *Universitas Scientiarum, 24*(1), 91–109. DOI: 10.11144/Javeriana.SC24-1.asst

Rajput, V. D., Minkina, T., Fedorenko, A., Chernikova, N., Hassan, T., Mandzhieva, S., Sushkova, S., Lysenko, V., Soldatov, M. A., & Burachevskaya, M. (2021). Effects of Zinc Oxide Nanoparticles on Physiological and Anatomical Indices in Spring Barley Tissues. *Nanomaterials, 11*(7), 1722. DOI: 10.3390/nano11071722

Rasel, M., Tahjib-Ul-Arif, M., Hossain, M. A., Hassan, L., Farzana, S., & Brestic, M. (2021). Screening of Salt-Tolerant Rice Landraces by Seedling Stage Phenotyping and Dissecting Biochemical Determinants of Tolerance Mechanism. *Journal of Plant Growth Regulation, 40*(5), 1853–1868. DOI: 10.1007/s00344-020-10235-9

Shin, Y. K., Bhandari, S. R., Jo, J. S., Song, J. W., Cho, M. C., Yang, E. Y., & Lee, J. G. (2020). Response to Salt Stress in Lettuce: Changes in Chlorophyll Fluorescence Parameters, Phytochemical Contents, and Antioxidant Activities. *Agronomy, 10*(11), 1627. DOI: 10.3390/agronomy10111627

Sohan, D., Jasoni, R., & Zajicek, J. (1999). Plant–water relations of NaCl and calcium-treated sunflower plants. *Environmental and Experimental Botany, 42*(2), 105–111. DOI: 10.1016/S0098-8472(99)00027-1

Taffouo, V. D., Kouamou, J. K., Ngalangue, L. M. T., Ndjeudji, B. A. N., & Akoa, A. (2009). Effects of salinity stress on growth, ions partitioning and yield of some cowpea (Vigna unguiculata L. Walp.) Cultivars. *International Journal of Botany, 5*(2), 135–143. DOI: 10.3923/ijb.2009.135.143

The water-culture method for growing plants without soil by D. R. Hoagland | Open Library. (n.d.). Retrieved from https://openlibrary.org/books/OL25240089M/The_water-culture_method_for_growing_plants_without_soil

Van Zelm, E., Zhang, Y., & Testerink, C. (2020). Salt Tolerance Mechanisms of Plants. DOI: 10.1146/annurev-arplant-050718-100005

Wang, W., Vinocur, B., & Altman, A. (2003). Plant responses to drought, salinity and extreme temperatures: Towards genetic engineering for stress tolerance. *Planta*, *218*(1), 1–14. DOI: 10.1007/s00425-003-1105-5 PMID: 14513379

Zhang, L., Ma, H., Chen, T., Pen, J., Yu, S., & Zhaoe, X. (2014). Morphological and Physiological Responses of Cotton (Gossypium hirsutum L.) Plants to Salinity. *PLoS One*, *9*(11), e112807. DOI: 10.1371/journal.pone.0112807 PMID: 25391141

Zhu, J. K. (2003). *Salt and drought stress signal transduction in plants*. DOI: 10.1146/annurev.arplant.53.091401.143329

Zhu, J. K. (2016). Abiotic Stress Signaling and Responses in Plants. *Cell*, *167*(2), 313–324. DOI: 10.1016/j.cell.2016.08.029 PMID: 27716505

Zhu, X., Cao, Q., Sun, L., Yang, X., Yang, W., & Zhang, H. (2018). Stomatal conductance and morphology of arbuscular mycorrhizal wheat plants response to elevated co2 and NaCl Stress. *Frontiers in Plant Science*, *9*, 410525. DOI: 10.3389/fpls.2018.01363 PMID: 30283478

Zulfiqar, F., & Ashraf, M. (2021). Nanoparticles potentially mediate salt stress tolerance in plants. *Plant Physiology and Biochemistry*, *160*, 257–268. DOI: 10.1016/j.plaphy.2021.01.028 PMID: 33529801

Chapter 4
Evaluation of Wheat Straw for Biofuel Production Using Various Microbial Strains

Anna Galstyan
http://orcid.org/0009-0009-0173-6136
Research Institute of Biology, Armenia

Nare Patvakanyan
Department of Biochemistry, Microbiology, and Biotechnology, Armenia

Anahit Shirvanyan
http://orcid.org/0000-0002-8753-3218
Research Institute of Biology, Armenia

Liana Vanyan
Research Institute of Biology, Armenia

Ayaulym Daniyarova
Research School of Food Engineering, Kazakhstan

Kamila Baichiyeva
http://orcid.org/0009-0001-8691-4625
Research School of Food Engineering, Kazakhstan

Kairat Bekbayev
Research School of Food Engineering, Kazakhstan

Karen Trchounian
http://orcid.org/0000-0002-6624-8936
Research Institute of Biology, Armenia

ABSTRACT

Wheat straw (WS) is one of the most widely generated agro-industrial residues, which has high-carbohydrate content; however, its complex structure requires optimized processing and efficient microbial strains to achieve high conversion yields. Thermal-treated (TT) and acid-treated (AT) WS were used as substrates for S. cerevisiae to assess bioethanol production, while untreated WS as substrate for hydrogen, ethanol, and butanol production by Clostridium species. The highest ethanol yield (80 ±

DOI: 10.4018/979-8-3373-1524-9.ch004

2 g L^{-1}) for S. cerevisiae ATCC 13007 strain was produced using non-diluted AT hydrolysates during oxygen-limited, while the highest ethanol yield (92 ± 3 g L^{-1}) for S. cerevisiae ATCC 9804 observed in non-diluted AT hydrolysates under aerobic conditions. Six percent WS with glucose supplementation maximizes and sustains hydrogen yields in C. beijerinckii producing up to 30 mmol L^{-1} butanol and 72 mmol L^{-1} ethanol. The findings contribute to the valorization of agro-industrial residues and the advancement of renewable biofuel production strategies.

1. INTRODUCTION

The rapid reduction of global energy supplies, combined with global warming, requires the development of alternative energy sources. While electrification represents a cleaner alternative to fossil fuels, liquid biofuels offer a sustainable solution in sectors where electrification faces challenges, e.g. aviation, long-haul transport or marine shipping (Ge et al., 2025). Biofuels are renewable and sustainable sources that represent a promising alternative to traditional energy sources. (Jhariya et al., 2021), moreover, biofuels emit significantly fewer green gas emissions than traditional fossil fuels, reducing emissions by up to 88% (Ge et al., 2025). Recently published studies (Ceaser et al., 2024; Fernández-Blanco et al., 2023; Kasaeian et al., 2025; Tse et al., 2021) have focused on the manufacture of second-generation biofuel; however, several challenges remain, including the efficient pretreatment, cost-effective enzymatic hydrolysis, strain optimisation for robust fermentation, scalability and economic feasibility. Therefore, further study of liquid biofuel development, optimisation, and large-scale adoption is essential for a holistic strategy to global carbon reduction and energy sustainability.

Wheat straw (WS) is a lignocellulosic biomass (LB) typically containing 33–40% cellulose, 26–32% hemicellulose, and 17–19% lignin(Ziaei-Rad et al., 2023). It is one of the most widely generated LB and by-products of cereal crop cultivation. From 2015 to 2024, global wheat production averaged 768.16 million metric tons annually. Armenia ranks 55th in wheat production (~170,000 metric tons annually), while Kazakhstan ranks 11th (~18.58 million metric tons annually)—accounting for 2.3% of the global output [USDA, 2024, https://www.fas.usda.gov/data/production/commodity/0410000]. Wheat straw was selected not only for its abundance but also for its balanced composition and moderate recalcitrance, which make it a suitable model for lignocellulosic bioconversion. Its cellulose, hemicellulose, and lignin fractions can be efficiently valorized through integrated biorefinery approaches that enable near-zero-waste utilization(Kaur & Goyal, 2026).Compared with other residues such as rice straw or corn stover, wheat straw offers favorable processing characteristics and compatibility with circular bioeconomy models. Recent reviews emphasize that such lignocellulosic valorization strategies significantly improve the sustainability and techno-economics of biofuel production (Abida Nisar et al., 2023; Nazar et al., 2025)

Saccharomyces cerevisiae is widely used and the most efficient bioethanol-producing organism, extensively applied in various industrial processes due to its unicellular structure, safe status, rapid growth in relatively inexpensive media, high biomass production yield, indifference to bacteriophage infection, easily genetically manipulable nature, broad pH growth, as well as resistance to high ethanol and inhibitory compounds concentrations. (A. Shirvanyan & Trchounian, 2024). Moreover, *S. cerevisiae* cells contain high levels of proteins and micronutrients, reaching up to 65% of the cell mass (Oliveira et al., 2025)

Clostridia strains have attracted significant attention due to their high hydrogen yields and ability to metabolize a broad spectrum of carbon sources. During dark fermentation, these microorganisms convert organic matter into hydrogen, butanol, ethanol and other metabolites. (Wang & Yin, 2019). *Clostridium beijerinckii* is a well-studied solventogenic bacterium performing acetone, butanol, and ethanol (ABE) fermentation from a wide range of carbon sources, including LB hydrolysates. As a versatile anaerobe, *C. beijerinckii* possesses an array of hydrolytic and fermentative enzymes that enable the conversion of complex polysaccharides into fermentable sugars and subsequently into biofuels (An et al., 2014).

This study investigates the WS bioconversion potential using various microbial strains under different processing conditions. Thermal-treated (TT) and acid-treated (AT) WS were used as substrates for *S. cerevisiae* to assess optimal bioethanol production, while Clostridium strains fermented raw (untreated) WS to determine hydrogen, ethanol, and butanol yields. The findings contribute to the valorization of agro-industrial residues and the advancement of renewable biofuel production strategies.

2. MATERIALS AND METHODS

2.1. Strains and Growth Conditions

The strains *Clostridium beijerinckii* DSM 791 was obtained from the Leibniz Institute DSMZ – German Collection of Microorganisms and Cell Cultures (Leibniz Institute DSMZ), and *S. cerevisiae* ATCC 13007 (prototroph, *pof+ sta*, diploid, European nucleotide archive (ENA browser) genome sequence #ERR5358805) and ATCC 9804 (GenBank: Z95939.1; ENA browser genome sequence #ERR5285363) wild-type strains were purchased from the Microbial Depository Centre of ''Armbiotechnology" Scientific and Production Centre of the National Academy of Sciences of RA. Strain selection was based on their biotechnological usage: the *S. cerevisiae ATCC 13007 strain was isolated from an* ale super-attenuated Irish beer. (Pedersen, 1986)(Pedersen, 1986) and was found to exhibit extracellular glucoamylase activity (Favaro et al., 2012)(Favaro et al., 2012), while the *S. cerevisiae* ATCC 9804 strain was isolated from palm wine (Hurley Jr. & Kenneth Eldon, 2021). Both strains are promising for high-yield production of bioethanol from LB.

Monoclonal cells from single colonies were transferred to YPD (grams per liter: Yeast extract-10; Peptone-20; D-glucose-20) liquid media (pH 6.5) and grown overnight in aerobic conditions with the medium filling one-third of the total flask volume, incubated at 30 °C; 180 rpm on a rotary shaker (Infors Multitron Standard (Infors HT, Switzerland). During the experiments, culture medias inoculated with overnight culture cells (initial 0.2 OD concentration) was grown in aerobic or under oxygen-limited conditions (medium filling five-sixths of the total flask volume, incubated without shaking in a thermostat (WiseCube WIS-50, South Korea)). The pH was adjusted using potassium monohydrophosphate or 0.1N HCl using HI3220 pH-meter combined with HI1131 pH electrode (Hanna Instruments, USA) (A. Shirvanyan & Trchounian, 2024; Zhu et al., 2020).

The composition of the growth medium for *Clostridium* strains was as follows (per liter): K_2HPO_4 – 1.50 g, KH_2PO_4 – 1.50 g, $MnSO_4 \cdot H_2O$ – 75 mg, $FeSO_4 \cdot 7H_2O$ – 40 mg, para-aminobenzoic acid – 10 mg, yeast extract – 0.50 g, peptone – 0.50 g, resazurin sodium salt – 0.4 mg, L-cysteine HCl – 50 mg, biotin – 0.2 mg, thiamine HCl – 5 mg. (Trchounian, Müller, et al., 2017).

For anaerobic cultivation, the prepared media were flushed with nitrogen gas (N_2) to remove oxygen (O_2). After nitrogen flushing, resazurin sodium salt was added as a redox indicator (blue in oxidized form, turning pink upon reduction). The medium was dispensed into special bottles sealed with butyl rubber stoppers and aluminum caps to minimise gas permeability. Oxygen removal was achieved by repeated vacuum–nitrogen cycles (10–15 min vacuum with a Rocker 300 pump, followed by 2–3 min N_2 flushing, repeated 2–3 times). Media were autoclaved at 121 °C for 20 min, which also removed residual O_2.

Batch fermentations of bacteria were carried out in 120 mL sealed serum bottles containing 40 ml medium incubated at 37°C.

2.2. Waste Collection and Pretreatment

WS was collected from the fields of the crop grown by LLP "A.I. Barayev research and production centre for grain farming" located in the Shortandy district of Akmola region, Republic of Kazakhstan. Collected waste materials were stored at -20°C before the experiments.

During the experiment, WS was manually shredded and dried at room temperature for 2 hours in a clean tray lined with paper towels to remove surface moisture. To ensure consistent moisture content and reproducibility of the medium, all waste materials were dried to a constant weight at room temperature.

2.2.1 Waste Media Preparation for *S. cerevisiae* Growth and Bioethanol Production

WS was used to prepare 10% (w/v in distilled water) solutions with or without 0.5% sulfuric acid. Waste solutions were hydrolysed via autoclaving (WiseClave WACS 1100, South Korea) at 121°C for 45 min. Waste solutions treatment without sulfuric acid are considered as TT samples, while the solutions with 0.5% sulfuric acid undergo physicochemical pretreatment and are considered hereafter as AT samples. Consequently, the solutions were filtered through cotton and two layers of cheesecloth, and the resulting hydrolysate was used in subsequent experiments. The pH of the hydrolysate was adjusted in the same manner as the growth media.

To discard all sediments, the hydrolysates were boiled at 100°C for 15 minutes, then centrifuged at 7500xg for 15 minutes using a Sorvall LYNX 6000 Centrifuge (Thermo Scientific, USA). To test concentration-dependent growth and bioethanol production changes, a non-diluted and 2-fold diluted supernatant was used as a growth medium, which was sterilized at 121°C for 20 minutes (Awogbemi & Kallon, 2022; Dunuweera et al., 2021).

2.2.2 Waste Media Preparation for *Clostridia ssp.* Growth and Biohydrogen Production

Waste at a final concentration of 20–100 g L^{-1} was placed into bottles, which were sealed with butyl rubber stoppers and crimped with aluminium caps. Air was removed using a vacuum pump for 10–15 minutes, after which the bottles were flushed with nitrogen gas for 5 minutes.

This procedure was repeated 3 times. The bottles containing the waste were then sterilized at 121 °C for 20 minutes (WiseClave WACS 1100, South Korea). After sterilization, the pre-prepared medium in which the microorganisms had been grown was added to the WS-containing bottles using sterile syringe.

2.3. Oxidation-Reduction Potential and Extracellular pH Monitoring

The hydrolysates were inoculated with mid-exponentially grown cells from an overnight culture and cell growth was monitored under aerobic or oxygen-limited conditions as described above. (Olivares-Marin et al., 2018). Yeast growth in hydrolysates was assessed by optical density changes over time using a UV-Vis spectrophotometer (Cary 60, Agilent Technologies, Germany) at 600 nm. (Trchounian, Mirzoyan, et al., 2017)(Trchounian, Mirzoyan, et al., 2017).

Oxidation-reduction potential (ORP) was determined at 0-120 hours of yeast growth using an ORP platinum electrode (HI3131B, Hanna Instruments, USA) at 25°C. Extracellular (hereafter referred to as external pH) pH was determined at the same growth hour using an HI3220 pH meter combined with an HI1131 pH electrode (Hanna Instruments, USA) (Shirvanyan et al., 2023).

2.4. Determination of Butanol and Ethanol

Two mL of each culture was transferred to a sterile Eppendorf tube at 0, 6, 24, 48, 72 and 96th h of yeast growth in hydrolysates and stored at -20°C. During the experiment, the samples were thawed entirely at room temperature and centrifuged (Daihan Scientific, South Korea) at 13500 rpm for 10 minutes. Ethanol concentration in supernatants was determined using the potassium dichromate oxidation method. (Seo et al., 2009; Zoecklein B. W. & Fugelsang K. C., 1990). Results were calculated using an ethanol standard curve and expressed in g L^{-1}.

GC-MS analysis for the quantitative determination of butanol and ethanol was performed as follows: samples of approximately 200 μL were collected under sterile conditions and centrifuged to remove solids. The supernatant was diluted 10-fold with methanol and stored at 4 °C until analysis. The GC-MS was operated in split injection mode with a split ratio of 100:1. The inlet temperature was set to 220 °C and the carrier gas flow rate to 0.5 mL/min. The oven temperature program started at 40 °C with a hold time of 0.5 minutes, followed by a temperature ramp of 1 °C per minute up to 45 °C with no hold. A second ramp increased the temperature at 50 °C per minute up to 200 °C with a hold time of 3 minutes. Auxiliary heaters were maintained at 150 °C. Quantification of butanol and ethanol was achieved by comparing peak areas to a standard calibration curve prepared from known concentrations.

2.5 Hydrogen Determination

The amount of H_2 produced during fermentation was determined by GC system (7820A, Agilent Technologies, Germany) equipped with a thermal conductivity detector (TCD). High-purity nitrogen gas (Nitrogen generator, F-DGSi, France) was used as the carrier gas. A 200 μL gas sample was injected manually using a hermetic syringe (5190-1531, Agilent, Australia) into an Agilent Technologies Chromosorb 750 packed column (1/8" × 2.0 mm × 2 m). The operational parameters were as follows: carrier gas flow rate of 10 mL min^{-1}, inlet temperature 150 °C, detector temperature 250 °C, column flow rate 30 mL min^{-1}, and carrier gas (N_2) flow rate 3 mL min^{-1}. H_2 quantification was performed using a calibration curve prepared with known concentrations of pure hydrogen (Hydrogen generator, F-DGSi, France), yielding a correlation coefficient (R) of 0.995. The accumulated H_2 concentration was calculated from the calibration curve and expressed in mmol L^{-1} (Vanyan et al., 2025).

2.6 Chemicals, Data Analysis, and Statistics

All chemicals and reagents of analytical grade (>98%) were used in the study. Each result presented is the mean value gained from three individual replicates; the standard deviation did not exceed 5%. Results calculation was done using Microsoft Excel and Graphpad Prism 8.0.2 (GraphPad Software, Inc., USA) programs. The statistical analysis was conducted using 2-way ANOVA Tukey's test (A. Shirvanyan & Trchounian, 2024). Results visualization, graph construction was done using Graphpad Prism 8.0.2.

3. RESULTS AND DISCUSSION

3.1. Growth Dynamics of *S. cerevisiae* Strains in WS Hydrolysates

Previously, several studies (Deng et al., 2025; Favaro et al., 2012; Ilanidis et al., 2021; Ji et al., 2015; Rahmani et al., 2023; Rusănescu et al., 2024 were focused on the hydrolysis of complex matrixes of WS using various hydrothermal, acid or mixed methods. The chemical composition and hydrolysis efficiency were determined using high-performance analytical methods, e.g. high-performance anion-exchange chromatography (HPAEC), HPLC, ultra-high performance liquid chromatography-electrospray ionization-triple quadrupole-mass spectrometry (UHPLC-ESI-QqQ-MS), and pyrolysis-gas chromatography/mass spectrometry (Py-GC/MS). The wheat straw contained 39.8% glucan, 24.3% hemicellulose, 22.8 ± 0.1% lignin and 5.1 ± 0.3% ash and it was shown that temperature increase facilitates the hydrolysis process (Ilanidis et al., 2021).The maximum reducing sugar yield ratio after hydrothermal hydrolysis was about 60% (Ji et al., 2015) and it was shown that the hydrothermal pretreatment degraded predominantly the hemicellulose by about 55% (Rahmani et al., 2023 al.,)and the maximum 4.3 g/L of xylooligosaccharide and 0.6 g/L of arabinooligosaccharide was achieved during wheat straw slurry pretreatment at 180 °C for 20 min (Deng et al., 2025). However, the hydrolysis of lignocellulosic biomass at high temperatures can increase the production of inhibitory compounds such as furfural and 5-HMF, which are highly toxic for microorganisms (Jönsson & Martín, 2016; Luo et al., 2019).

To evaluate yeast performance changes depending on treatment method employed or cultivation conditions, yeast growth was studied in TT and AT WS solutions under aerobic or oxygen-limited conditions for 72 hours. Results indicate (Figure 1a) that under aerobic conditions, the 2-fold diluted samples of *S. cerevisiae* ATCC 13007 cells exhibit higher OD in TT WS at the 8th hour of growth compared to non-diluted samples. In TT WS hydrolysates, the growth of *S. cerevisiae* ATCC 13007 cells continues up to 8 hours, followed by a stationary phase, indicating that the limitation of nutrients may be due to limited hydrolysis. No growth of *S. cerevisiae* ATCC 13007 was observed in undiluted or 2-fold diluted TT WS hydrolysates under oxygen-limited conditions (Figure 1c), which indicates the role of oxygen in adaptation to nutrient-limited conditions. Interestingly, growth patterns in AT are similar to those in TT under aerobic conditions (Figure 1b); however, in AT waste hydrolysates, growth continues up to 48 hours, indicating a high concentration of nutrients due to acid hydrolysis. At the 8th hour of growth in 2-fold diluted AT hydrolysates, the OD of *S. cerevisiae* ATCC 13007 culture was 2-fold higher compared to non-diluted samples. No growth of *S. cerevisiae* ATCC 13007 was observed under oxygen-limited conditions (Figure 1d) in 2-fold diluted AT hydrolysates, as in TT hydrolysates (Figure 1c); however, the growth in non-diluted hydrolysates continues up to 48 hours. This indicates that high carbohydrate content, formed due to acid hydrolysis, allows overcoming growth inhibition due to oxygen limitation in this strain.

S. cerevisiae ATCC 9804 exhibits similar growth dynamics to the *S. cerevisiae* ATCC 13007 strain; higher growth performance was found. Thus, aerobic growth of *S. cerevisiae* ATCC 9804 cells in TT non-diluted hydrolysates stimulates 2-fold higher OD at 8th hours of growth compared to *S. cerevisiae* ATCC 13007. In contrast, in 2-fold diluted hydrolysates the growth dynamics are similar (Figure 1a). Higher growth rates of *S. cerevisiae* ATCC 9804 can be due to higher adaptation capabilities to the nutrient limitation compared to the *S. cerevisiae* ATCC 13007 strain.

There was no observed significant growth of *S. cerevisiae* ATCC 9804 in 2-fold diluted TT hydrolysates under oxygen-limited conditions; however, a higher biomass yield was recorded in non-diluted hydrolysates under the same cultivation conditions (Figure 1C). Under aerobic conditions, the growth of *S. cerevisiae* ATCC 9804 in AT hydrolysates was like that of *S. cerevisiae* ATCC 13007. However, a significant increase was not observed in oxygen-limited conditions.

Summarizing, the highest biomass yield of *S. cerevisiae* ATCC 13007 and ATCC 9804 strains were observed in 2-fold diluted AT hydrolysates under aerobic conditions, while the growth of both strains was mainly inhibited under oxygen-limited conditions, especially in 2-fold diluted samples. ORP changes dynamics (Figure 2) show that the ORP in all samples at initial growth phases was (-150) - (-200) mV, which corresponds to a strongly reducing environment. At initial growth phases of yeast (up to 8th hours or up to mid exponential growth phase), the ORP values stay stable or ORP decrease were observed, especially in AT hydrolysates, which correlates with pH drop and indicates activation of fermentation and accumulation of acids, such as acetic acid, in the extracellular medium.

Figure 1. Growth curves of S. cerevisiae ATCC 9804 and ATCC 13007 cells in thermal-treated (TT; a, c) or acid-treated (AT; b, d) WS under aerobic (a, b) or oxygen-limited (c, d) conditions at pH 6.5. Abbrv.: n. d.-non-diluted, 2x d.-2-fold diluted.

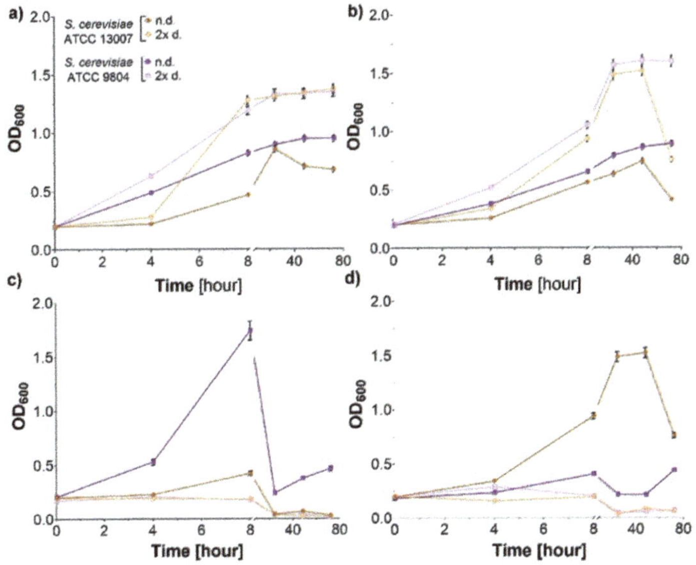

Consequently, the ORP values increase, which indicates activation of respiration and oxidation of compounds via oxygen. Both ORP and pH changes were in reasonable conformity with growth dynamics. Thus, the maximal ORP change (~150 ±5 mV) was observed in non- and 2-fold diluted TT hydrolysates for both strains (Figure 2a, c, e, g). The maximum external pH drop (0.54 ± 0.02) was observed at 4-8 hours of aerobic growth of *S. cerevisiae* ATCC 9804 in non- and 2-fold diluted TT hydrolysates, while for S. cerevisiae ATCC 13007 strain, the pH drop was 2-fold lower. Different pH change patterns highlight fermentation efficiency and acid production differences of the strains used. Previously, we have shown (A. H. Shirvanyan, 2023) that during growth in YPD medium both - ATCC 13007 and 9804, strains exhibit similar redox changes in YPD medium but the changes dynamics were highly depending on oxygen availability. This can be also due to differences in metabolite productions. For example, it was shown that both strains had similar substrate assimilation rate, however the ethanol production yield was 1.4-fold higher in ATCC 13007. On the other hand, ATCC 9804 was found to highly produce glycerol. The maximum ethanol yield (mM) per glucose (mM) in YPD medium was 0.28 and 0.74 for ATCC 9804 and, 0.37 and 1.06 for ATCC 13007 during aerobic and microaerophilic growth respectively.

Figure 2. Monitoring ORP and external pH changes dynamics during the growth of S. cerevisiae ATCC 13007 (a-d) and ATCC 9804 (e-h) strains cultivated thermal-treated (TT; a, c, e, g) or acid-treated (AT; b, d, f, h) WS at pH 6.5. Abbrv.: n. d.-non-dilute.

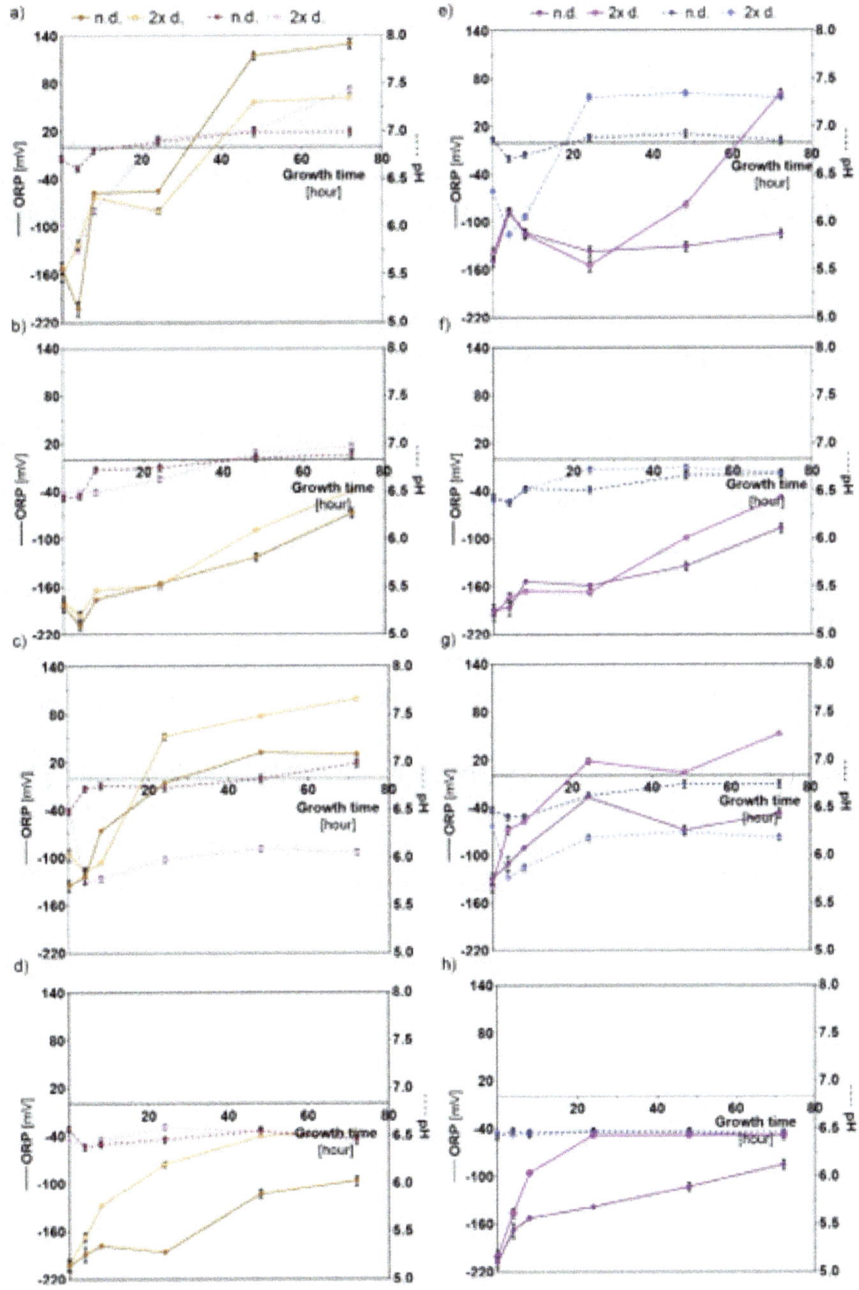

Glycerol production rate was 4 μM/h and 1.3 μM/h in *S. cerevisiae* ATCC 9804 and ATCC 13007 respectfully, whereas, during microaerophilic growth, it reached up to 98 μM/h and 175 μM/h for same strains (Anahit Shirvanyan et al., 2023). Acetate maximal production was observed during 24 h of growth ATCC 13007 at pH 5.0 with maximal yield of 14 mM, which correlated with specific growth rate inhibitions and highly correlates with extracellular pH and ORP changes dynamics (A. H. Shirvanyan, 2023;). These metabolite production yields and rate differences can have they print on differences in ORP, and pH profiles observed, however, the complex structure and initial reductive conditions of waste hydrolysate make it interesting for further studies of metabolite profiling for precise fermentation and strain construction strategies.

3.2. Bioethanol Production From WS Hydrolysates Depending on Oxygen Availability

S. cerevisiae ATCC 13007 shows no significant ethanol production in non-diluted TT hydrolysates, neither in aerobic, nor in oxygen-limited conditions (Figure 3), which correlates with growth changes. However, aerobic, as well as oxygen-limited growth in non-diluted AT hydrolysates, yield up to peak of 80 ± 2 g L^{-1} ethanol. Moreover, ethanol yield in oxygen-limited conditions was 35% higher compared to aerobic conditions.

Interestingly, the growth of *S. cerevisiae* ATCC 13007 in 2-fold diluted TT or AT hydrolysates under oxygen-limited conditions inhibits ethanol production, which correlates with poor growth dynamics under these conditions. Despite that, aerobic growth in these conditions supports high ethanol yield with 2-fold higher in AT hydrolysates compared to TT hydrolysates. There were no significant observed differences in ethanol production by *S. cerevisiae* ATCC 13007 between non- and 2-fold diluted AT hydrolysates under aerobic conditions.

Figure 3. Maximum ethanol production yields of S. cerevisiae ATCC 13007 and ATCC 9804 cultivated in thermal-treated (TT) or acid-treated (AT) WS under aerobic or oxygen-limited conditions at pH 6.5. Abbrv.: n. d.-non-diluted, 2x d.-2-fold diluted.

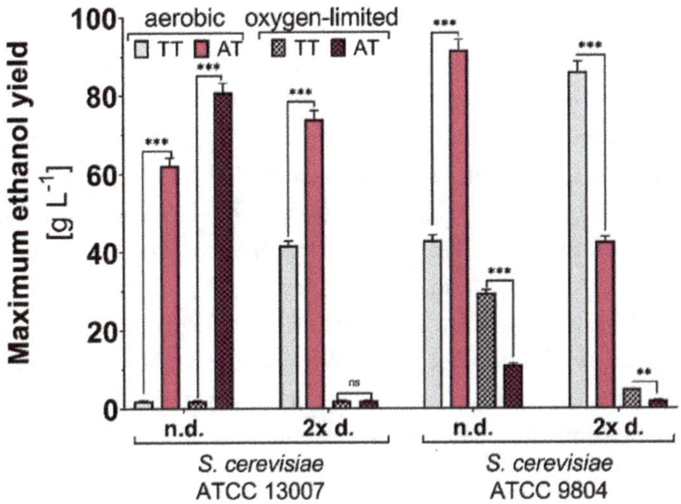

Ethanol production studies show that *S. cerevisiae* ATCC 9804 has significantly higher ethanol production potential while growing in TT or AT hydrolysates compared to the *S. cerevisiae* ATCC 13007 strain (Figure 3). This phenotype is in reasonable conformity with growth changes. It indicates higher adaptation potential of the strain for growth and ethanol metabolism in *S. cerevisiae* ATCC 9804 compared to the *S. cerevisiae* ATCC 13007 strain. *S. cerevisiae* ATCC 9804 exhibits 2-fold higher ethanol production yield from AT hydrolysates compared to TT hydrolysates under aerobic conditions, while oxygen-limited growth supports the opposite phenotype. The ethanol yield in TT hydrolysates is only 30% higher under aerobic conditions compared to oxygen-limited conditions. In contrast, the same parameter in AT hydrolysates under aerobic conditions is about 9-fold higher than that under oxygen-limited conditions.

Surprisingly, the ethanol production results for *S. cerevisiae* ATCC 9804 from 2-fold diluted hydrolysates significantly differ from those in non-diluted samples but are similar to those for *S. cerevisiae* ATCC 13007 strain. While oxygen-limited growth again inhibits ethanol production, which correlates with growth total inhibition, aerobic conditions stimulate ethanol production by this strain. Moreover, ethanol yield in 2-fold diluted TT hydrolysate is 2-fold higher than that in AT hydrolysate and 10-fold higher than that under oxygen-limited conditions.

Summarizing, the highest maximal ethanol yield (80 ± 2 g L^{-1}) for *S. cerevisiae* ATCC 13007 strain can be obtained during growth in non-diluted AT hydrolysates under oxygen-limited conditions. In contrast, the highest ethanol yield (80 ± 2 g L^{-1}) for *S. cerevisiae* ATCC 9804 can be obtained during growth in non-diluted AT hydrolysates under aerobic conditions.

The increased ethanol yield can be due to higher hydrolysis yield of lignocellulosic biomass which yields increased concentration of reducing sugars serving as fermentation substrates (Deng et al., 2025; Ji et al., 2015; Rusănescu et al., 2024). Earlier reported data (Choi et al., 2015; Kasaeian et al., 2025; Prasertwasu et al., 2014; Turhan et al., 2010) Shows that high yields (29.5-119.0 g L^{-1}) of ethanol can be produced from LB during the growth of different *S. cerevisiae* strains. However, the growth conditions, duration, and strain properties vary among the data reported, which makes it difficult to standardise for the large-scale production. Compared to the data, our ethanol production data from WS is at the higher end of the reported values and represents a promising alternative to first-generation ethanol. Further optimization of the procedure can enhance production efficiency and overall performance.

3.3 Bioethanol, Biobutanol, and Biohydrogen Production by *Clostridia*

The production of certain biofuels by *C. beijerinckii* was investigated during the utilization of WS at concentrations ranging from 2-10%, in the presence or absence of glucose. As a control, a clostridial growth medium with glucose was selected. After the inoculation, the first measurement started from the 24th hour of growth until the 120th hour, and in each datapoint, biohydrogen and both bioethanol and biobutanol were measured. It is worth mentioning that in the case of butanol and ethanol, better yields were observed when both WS and glucose were supplemented, while in more cases with sole WS, total absence of alcohol production was monitored (Fig. 4a, b).

The heat map analysis revealed that *C. beijerinckii* exhibited distinct solvent production patterns under different fermentation conditions. Butanol production reached its highest level, constituting approximately 30 mmol L^{-1} (Fig. 4 b) when 2% of waste together with glucose was used, indicating a shift toward solventogenesis.

Figure 4. Ethanol and butanol production by different Clostridium strains under various fermentation conditions. The color intensity corresponds to the concentration of the produced bioethanol (mmol L^{-1}), with darker shades indicating higher production levels.

a.

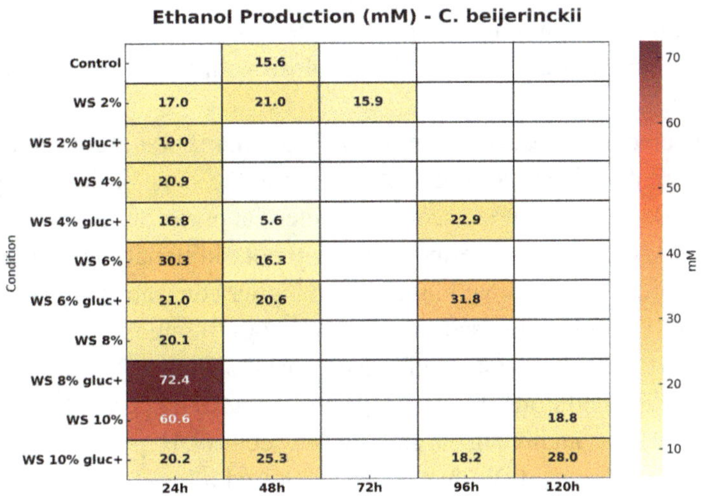

b.

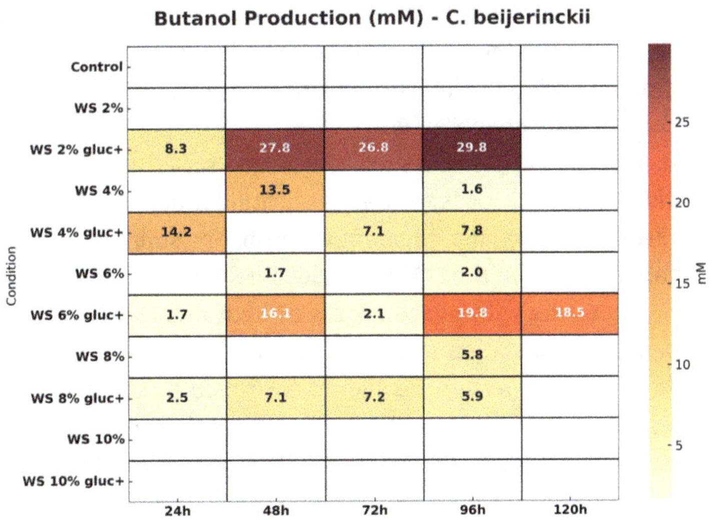

To enable direct comparison, literature-reported butanol concentrations originally expressed in g L^{-1} were converted to mmol L^{-1} using the molecular weight of n-butanol (74.12 g mol^{-1}). In a recent study using *Clostridium beijerinckii* cultivated on mixed glucose–xylose media, butanol concentrations ranged from 0.192–0.806 g L^{-1} (≈ 1.9–10.9 mmol L^{-1}) in a BioLector® system and 0.513–1.480 g L^{-1} (≈ 5.7–20.0 mmol L^{-1}) in serum flask experiments, showing that higher sugar availability enhanced solvent yields(Birgen et al., 2021). Another report demonstrated that a synthetic consortium of *C. beijerinckii* with *C. cellulovorans* produced 23.3 g L^{-1} of total solvents, including 14.2 g L^{-1} (≈ 192 mmol L^{-1}) butanol, from biologically pretreated wheat straw enriched in cellulose within five days(Valdez-Vazquez et al., 2015)

Compared with these benchmarks, the maximum butanol concentration observed in the present study (≈ 30 mmol L^{-1}) is within the same order of magnitude as the serum flask yields and underscores the potential of raw wheat straw as an effective substrate for solventogenic fermentation without additional pretreatment. In contrast, ethanol production was maximized when higher concentrations of sole WS or the mixture of glucose were used, making up to 60-70 mmol L-1 (Fig. 4a), reflecting conditions that favour a more acidogenic metabolic state.

Across all conditions (Fig. 5a, b), *C. beijerinckii* displayed time- and substrate-dependent H_2 production. In the glucose-only control, H_2 peaked at ~48 h (~38 mmol L^{-1}) and then declined. With WS, peaks shifted a bit at 2–4% WS maximum H_2 yield was observed at 72 hour of growth constituting ~15 mmol L^{-1}, at the presence of 6–8% WS, maximal yield were observed at ~72–96 h (~25 mmol L^{-1}), and at 10% WS, production started with significantly lower rates. However, it increased steadily to ~72 h (~28 mmol L^{-1}), consistent with a more extended adaptation/solubilization phase on lignocellulosic substrates. When glucose was supplemented, totally different data were observed.

Overall, glucose supplementation (Fig. 5 b) raised and sustained H_2 output most clearly at mid WS loads (4–6%), indicating a beneficial co-substrate effect where a readily fermentable sugar most probably supports redox balance and cell growth. At the same time, WS provides a slower, prolonged carbon release. Similar improvements with mixed sugars or hydrolysate co-utilization have been reported in dark fermentation systems (Zagrodnik & Sobociński, 2024). Thus, in our experiments, *C. beijerinckii* achieved hydrogen production peaks of ~15–28 mmol/L depending on WS concentration and supplementation with glucose. By comparison, the bioaugmentation of native wheat straw microbial communities with *Clostridium cellulovorans* reported a maximum H_2 yield of 115 mL/L, equivalent to ~4.7 mmol/L. This indicates that *C. beijerinckii* alone can produce roughly **3–6** times more hydrogen per liter than the augmented native communities under similar lignocellulosic conditions (Valdez-Vazquez et al., 2019)

Figure 5. Hydrogen production (mmol L^{-1} H_2) by Clostridium beijerinckii cultivated on WS hydrolysate at different concentrations (2–10% w/v) over 120 h: (a) without glucose supplementation and (b) with glucose supplementation. Bars represent the mean.

a

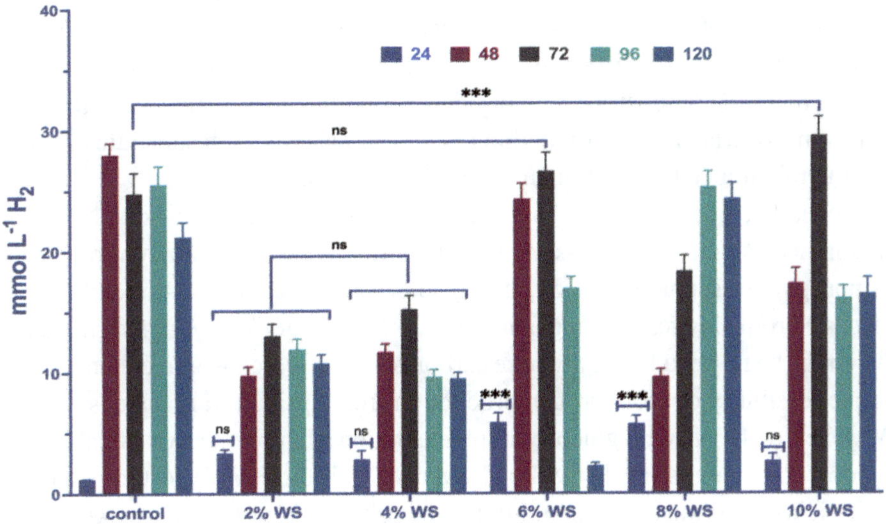

b

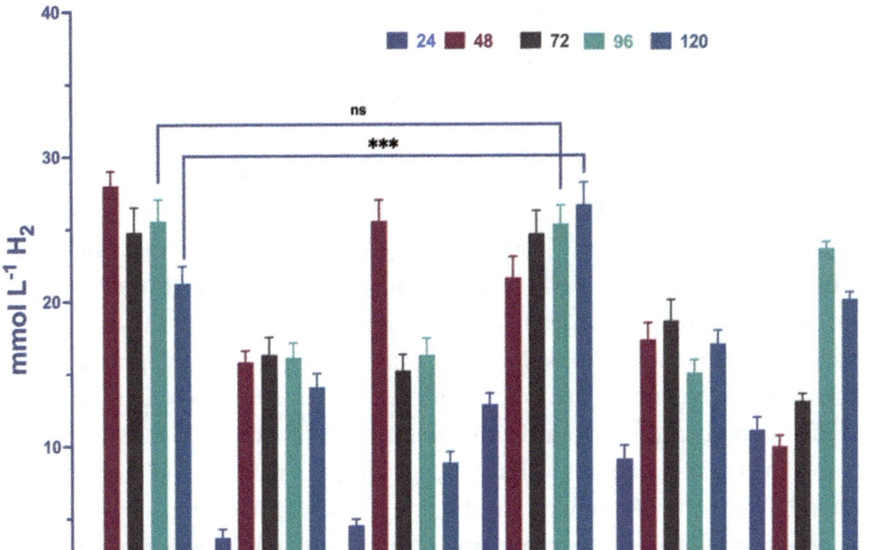

Significantly, increasing WS concentration did not monotonically increase H_2, especially in the presence of glucose (Fig. 5b). At the highest WS loads (8–10%), the delayed peaks and lower initial rates represent a result of combination of factors: the physical complexity of raw lignocellulosic material, which limits microbial access to fermentable sugars; the gradual enzymatic breakdown of cellulose and hemicellulose; and the accumulation of naturally occurring phenolic compounds that can exert inhibitory effects. Increased viscosity at higher solids content may further slow mass transfer, collectively delaying peak hydrogen output. These patterns align with reports that the composition and physical properties of raw lignocellulosic waste can strongly influence H_2 production kinetics by shaping both substrate availability and microbial metabolism.(Quéméneur et al., 2012).

Generally, the most optimal hydrogen production is at the 72nd hour. Notably, the data indicates that it was possible to obtain almost the same amount of hydrogen without glucose supplementation, relying solely on untreated WS as the carbon source. This is a promising result, suggesting that in the case of *Clostridium* fermentations, there may be no need for costly pure carbon sources, which would enhance the economic feasibility of biohydrogen production from agricultural residues. Moderate WS levels (≈4–6%) with glucose supplementation still provide a practical operating window for maximizing H_2 without severe inhibition. However, the potential for high yields from waste alone highlights the robustness of *C. beijerinckii*.

Previous studies have highlighted the potential for co-culture strategies integrating *Clostridial* species with yeast. For instance, Zhang et al. (2013) demonstrated that artificial co-cultures of different *Clostridium* species enhanced substrate consumption and hydrogen production compared to single-strain cultures (Zuroff et al., 2013). More recently, Wang et al. (2025) reported that co-cultivation of *Clostridium tyrobutyricum* and *Saccharomyces cerevisiae* allowed simultaneous production of butyric acid and ethanol, showing synergistic benefits in mixed fermentation systems (Oehlenschläger et al., 2025).

Thus, the present study is not primarily aimed at optimizing single-product outcomes such as ethanol, biohydrogen, or solvents, nor at wheat straw valorization in isolation. Rather, it provides insight into mixed strategies that leverage complementary metabolic capabilities, supporting the development of integrated sequential fermentation or co-culture approaches. Moreover, results demonstrate the potential for scalable biofuel production from untreated wheat straw, generating hydrogen, ethanol, and butanol, thereby valorizing agricultural residues and contributing to a circular bioeconomy. Achieving substantial yields without extensive pre-treatment or supplementation suggests promising techno-economic feasibility, with reduced operational costs and simplified processing. Future strategies to enhance production could include generating microbial mutations to improve tolerance to inhibitory compounds, optimizing hydrolytic enzymes to accelerate lignocellulose breakdown, and bioreactor scale-up to support high solids loading and efficient mass transfer. Integration of these approaches could enable industrial-scale fermentation processes that simultaneously produce multiple biofuels while converting lignocellulosic waste into value-added energy carriers.

4. CONCLUSION

This study demonstrates the feasibility of WS as a substrate for the production of biofuels using various *S. cerevisiae* and Clostridia spp. strains depending on growth conditions. Main findings include:

- Growth conditions and yeast strain selection can significantly affect biomass and ethanol yield from TT and AT hydrolysates. Aerobic culture of *S. cerevisiae* ATCC 9804 shows higher biomass and ethanol production capability compared to the ATCC 13007 strain; moreover, AT is more effective than TT in maintaining extended growth and long-term ethanol production. Furthermore, oxygen availability is the critical factor affecting yeast growth and ethanol production yield from WS hydrolysates.
- The highest peak ethanol yield (80 ± 2 g L^{-1}) for the *S. cerevisiae* ATCC 13007 strain can be obtained during growth in non-diluted AT hydrolysates under oxygen-limited conditions. In contrast, the highest ethanol yield (92 ± 3 g L^{-1}) for *S. cerevisiae* ATCC 9804 can be obtained during growth in non-diluted AT hydrolysates under aerobic conditions.
- *C. beijerinckii* can utilize untreated wheat straw to produce hydrogen yields comparable to glucose-supplemented setups, indicating that this strain can efficiently process raw lignocellulosic waste without relying on costly pure carbon sources.
- Moderate wheat straw level (6%) with glucose supplementation maximises and sustains hydrogen yields, benefiting from a co-substrate effect where glucose supports rapid metabolism and WS provides prolonged carbon release.
- *C. beijerinckii* produced notable amounts of ethanol and butanol from untreated wheat straw, demonstrating its dual capability for biohydrogen and bioethanol generation, which enhances the overall valorization potential of raw lignocellulosic waste.

AVAILABILITY OF DATA AND MATERIALS

The datasets generated during and analyzed during the current study are available from the corresponding author on reasonable request.

COMPETING INTERESTS

The authors declare that they have no competing interests

FUNDING

This research has been funded by the Science Committee of the Ministry of Science and Higher Education of the Republic of Kazakhstan (Grant No. BR24992914) and by Basic support and a research grant from the Higher Education and Science Committee of the of the Ministry of Education, Science, Culture and Sports of Armenia (24FP-3D020).

AUTHOR'S CONTRIBUTIONS

AS, LV and KT designed the experiments; NP, AG, KB, AD performed the experiments, AS, LV did a literature search, wrote the first draft, and made the figures; AG, and AS were responsible for data acquisition and analysis; KB-Funding acquisition, Project administration; Writing-review and editing; KT-Funding acquisition, Writing-review and editing, Validation, Supervision, Conceptualization.

REFERENCES

An, D., Li, Q., Wang, X., Yang, H., & Guo, L. (2014). Characterization on hydrogen production performance of a newly isolated Clostridium beijerinckii YA001 using xylose. *International Journal of Hydrogen Energy*, *39*(35), 19928–19936. DOI: 10.1016/j.ijhydene.2014.10.014

Awogbemi, O., & Von Kallon, D. V. (2022). Pretreatment techniques for agricultural waste. *Case Studies in Chemical and Environmental Engineering*, *6*, 100229. DOI: 10.1016/j.cscee.2022.100229

Ceaser, R., Montané, D., Constantí, M., & Medina, F. (2024). Current progress on lignocellulosic bioethanol including a technological and economical perspective. *Environment, Development and Sustainability, 27*(11), 25993–26038. Advance online publication. DOI: 10.1007/s10668-024-04792-2

Choi, I. S., Lee, Y. G., Khanal, S. K., Park, B. J., & Bae, H.-J. (2015). A low-energy, cost-effective approach to fruit and citrus peel waste processing for bioethanol production. *Applied Energy, 140*, 65–74. DOI: 10.1016/j.apenergy.2014.11.070

Deng, Z., Xia, A., Zhang, Z., Zhu, X., Huang, Y., Zhu, X., & Liao, Q. (2025). A continuous hydrothermal pretreatment system for efficient sugar production from wheat straw. *Chemical Engineering Journal, 508*, 160867. DOI: 10.1016/j.cej.2025.160867

Dunuweera, A. N., Nikagolla, D. N., & Ranganathan, K. (2021). Fruit Waste Substrates to Produce Single-Cell Proteins as Alternative Human Food Supplements and Animal Feeds Using Baker's Yeast (Saccharomyces cerevisiae). *Journal of Food Quality, 2021*, 1–6. DOI: 10.1155/2021/9932762

Favaro, L., Basaglia, M., & Casella, S. (2012). Processing wheat bran into ethanol using mild treatments and highly fermentative yeasts. *Biomass and Bioenergy, 46*, 605–617. DOI: 10.1016/j.biombioe.2012.07.001

Fernández-Blanco, C., Robles-Iglesias, R., Naveira-Pazos, C., Veiga, M. C., & Kennes, C. (2023). Production of biofuels from <scp> C_1 </scp> -gases with *Clostridium* and related bacteria—Recent advances. *Microbial Biotechnology, 16*(4), 726–741. DOI: 10.1111/1751-7915.14220 PMID: 36661185

Ge, L., Ali, M. M., Osman, A. I., Elgarahy, A. M., Samer, M., Xu, Y., & Liu, Z. (2025). A critical review on conversion technology for liquid biofuel production from lignocellulosic biomass. *Renewable & Sustainable Energy Reviews, 217*, 115726. DOI: 10.1016/j.rser.2025.115726

Ilanidis, D., Stagge, S., Jönsson, L. J., & Martín, C. (2021). Hydrothermal Pretreatment of Wheat Straw: Effects of Temperature and Acidity on Byproduct Formation and Inhibition of Enzymatic Hydrolysis and Ethanolic Fermentation. *Agronomy (Basel), 11*(3), 487. DOI: 10.3390/agronomy11030487

Jhariya, U., Dafale, N. A., Srivastava, S., Bhende, R. S., Kapley, A., & Purohit, H. J. (2021). Understanding Ethanol Tolerance Mechanism in Saccharomyces cerevisiae to Enhance the Bioethanol Production: Current and Future Prospects. *BioEnergy Research, 14*(2), 670–688. DOI: 10.1007/s12155-020-10228-2

Ji, W., Shen, Z., & Wen, Y. (2015). Hydrolysis of wheat straw by dilute sulfuric acid in a continuous mode. *Chemical Engineering Journal, 260*, 20–27. DOI: 10.1016/j.cej.2014.08.089

Jönsson, L. J., & Martín, C. (2016). Pretreatment of lignocellulose: Formation of inhibitory by-products and strategies for minimizing their effects. *Bioresource Technology, 199*, 103–112. DOI: 10.1016/j.biortech.2015.10.009 PMID: 26482946

Kasaeian, A., Fereidooni, L., Ahmadbeigi, A., Kahedi, S., Koohshoori, M. S., Ghafarian, S., Tajmousavilangerudi, A., & Salaripoor, H. (2025). Review on Bioethanol Production from Fruit Peels. *Waste and Biomass Valorization, 16*(6), 2669–2691. Advance online publication. DOI: 10.1007/s12649-025-02938-5

Luo, Y., Li, Z., Li, X., Liu, X., Fan, J., Clark, J. H., & Hu, C. (2019). The production of furfural directly from hemicellulose in lignocellulosic biomass: A review. *Catalysis Today, 319*, 14–24. DOI: 10.1016/j.cattod.2018.06.042

Oehlenschläger, K., Lorenz, M., Schepp, E., Di Nonno, S., Holtmann, D., & Ulber, R. (2025). Integrated co-cultivation and subsequent esterification: Harnessing Saccharomyces cerevisiae and Clostridium tyrobutyricum for streamlined ester production. *Biotechnology for Biofuels and Bioproducts, 18*(1), 98. DOI: 10.1186/s13068-025-02698-3 PMID: 40890840

Olivares-Marin, I. K., González-Hernández, J. C., Regalado-Gonzalez, C., & Madrigal-Perez, L. A. (2018). <em>Saccharomyces cerevisiae</em> Exponential Growth Kinetics in Batch Culture to Analyze Respiratory and Fermentative Metabolism. *Journal of Visualized Experiments, 139*(139). Advance online publication. DOI: 10.3791/58192 PMID: 30320748

Oliveira, A. S., Ferreira, C., Pereira, J. O., Pintado, M. E., & Carvalho, A. P. (2025). Valorisation of protein-rich extracts from spent brewer's yeast (Saccharomyces cerevisiae): An overview. *Biomass Conversion and Biorefinery, 15*(2), 1771–1793. DOI: 10.1007/s13399-022-02636-5

Pedersen, M. B. (1986). DNA sequence polymorphisms in the genus saccharomyces III. Restriction endonuclease fragment patterns of chromosomal regions in brewing and other yeast strains. *Carlsberg Research Communications, 51*(3), 163–183. DOI: 10.1007/BF02907322

Prasad, S., Singh, A., & Joshi, H. C. (2007). Ethanol as an alternative fuel from agricultural, industrial and urban residues. *Resources, Conservation and Recycling, 50*(1), 1–39. DOI: 10.1016/j.resconrec.2006.05.007

Prasertwasu, S., Khumsupan, D., Komolwanich, T., Chaisuwan, T., Luengnaruemitchai, A., & Wongkasemjit, S. (2014). Efficient process for ethanol production from Thai Mission grass (Pennisetum polystachion). *Bioresource Technology*, *163*, 152–159. DOI: 10.1016/j.biortech.2014.04.043 PMID: 24811442

Quéméneur, M., Hamelin, J., Barakat, A., Steyer, J.-P., Carrère, H., & Trably, E. (2012). Inhibition of fermentative hydrogen production by lignocellulose-derived compounds in mixed cultures. *International Journal of Hydrogen Energy*, *37*(4), 3150–3159. DOI: 10.1016/j.ijhydene.2011.11.033

Rahmani, A. M., Tyagi, V. K., Kazmi, A. A., & Ojha, C. S. P. (2023). Hydrothermal and thermal-acid pretreatments of wheat straw: Methane yield, recalcitrant formation, process inhibition, kinetic modeling. *Energy*, *283*, 129083. DOI: 10.1016/j.energy.2023.129083

Rusănescu, C. O., Ciobanu, M., Rusănescu, M., & Dinculoiu, R. L. (2024). Pretreatments Applied to Wheat Straw to Obtain Bioethanol. *Applied Sciences (Basel, Switzerland)*, *14*(4), 1612. DOI: 10.3390/app14041612

Seo, H.-B., Kim, H.-J., Lee, O.-K., Ha, J.-H., Lee, H.-Y., & Jung, K.-H. (2009). Measurement of ethanol concentration using solvent extraction and dichromate oxidation and its application to bioethanol production process. *Journal of Industrial Microbiology & Biotechnology*, *36*(2), 285–292. DOI: 10.1007/s10295-008-0497-4 PMID: 18989713

Shirvanyan, A., Mirzoyan, S., & Trchounian, K. (2023). Relationship between proton/potassium fluxes and central carbon catabolic pathways in different Saccharomyces cerevisiae strains under osmotic stress conditions. *Process Biochemistry (Barking, London, England)*, *133*, 309–318. DOI: 10.1016/j.procbio.2023.09.015

Shirvanyan, A., & Trchounian, K. (2024). Sodium transport and redox regulation in Saccharomyces cerevisiae under osmotic stress depending on oxygen availability. *Scientific Reports*, *14*(1), 23982. DOI: 10.1038/s41598-024-75108-7 PMID: 39402154

Shirvanyan, A. H. (2023). Evaluation of ethanol and biomass production rate by different $saccharomyces~cerevisiae strains depending on external ph and temperature. *Proceedings of the YSU B: Chemical and Biological Sciences*, *57*(2 (261)), 141–153. DOI: 10.46991/PYSU:B/2023.57.2.141

Trchounian, K., Mirzoyan, S., Poladyan, A., & Trchounian, A. (2017). Hydrogen production by Escherichia coli growing in different nutrient media with glycerol: Effects of formate, pH, production kinetics and hydrogenases involved. *International Journal of Hydrogen Energy*, *42*(38), 24026–24034. DOI: 10.1016/j.ijhydene.2017.08.032

Trchounian, K., Müller, N., Schink, B., & Trchounian, A. (2017). Glycerol and mixture of carbon sources conversion to hydrogen by Clostridium beijerinckii DSM791 and effects of various heavy metals on hydrogenase activity. *International Journal of Hydrogen Energy*, *42*(12), 7875–7882. DOI: 10.1016/j.ijhydene.2017.01.011

Tse, T. J., Wiens, D. J., & Reaney, M. J. T. (2021). Production of Bioethanol—A Review of Factors Affecting Ethanol Yield. *Fermentation (Basel, Switzerland)*, *7*(4), 268. DOI: 10.3390/fermentation7040268

Turhan, I., Bialka, K. L., Demirci, A., & Karhan, M. (2010). Ethanol production from carob extract by using Saccharomyces cerevisiae. *Bioresource Technology*, *101*(14), 5290–5296. DOI: 10.1016/j.biortech.2010.01.146 PMID: 20189805

Vanyan, L., Aghekyan, H., Vassilian, A., Poladyan, A., & Trchounian, K. (2025). Biotechnological potential of spent coffee grounds for biohydrogen production by Escherichia coli. *International Journal of Hydrogen Energy*, *142*, 1121–1131. DOI: 10.1016/j.ijhydene.2024.12.380

Wang, J., & Yin, Y. (2019). Progress in microbiology for fermentative hydrogen production from organic wastes. *Critical Reviews in Environmental Science and Technology*, *49*(10), 825–865. DOI: 10.1080/10643389.2018.1487226

Zagrodnik, R., & Sobociński, D. (2024). Continuous dark-fermentative H2 production using carbon components of lignocellulose hydrolysates: Insight into the difference between mixed and single substrates. *International Journal of Hydrogen Energy*, *52*, 378–388. DOI: 10.1016/j.ijhydene.2023.06.119

Zhu, Z., Hu, Y., Teixeira, P. G., Pereira, R., Chen, Y., Siewers, V., & Nielsen, J. (2020). Multidimensional engineering of Saccharomyces cerevisiae for efficient synthesis of medium-chain fatty acids. *Nature Catalysis*, *3*(1), 64–74. DOI: 10.1038/s41929-019-0409-1

Zuroff, T. R., Xiques, S. B., & Curtis, W. R. (2013). Consortia-mediated bioprocessing of cellulose to ethanol with a symbiotic Clostridium phytofermentans/yeast co-culture. *Biotechnology for Biofuels*, *6*(1), 59. DOI: 10.1186/1754-6834-6-59 PMID: 23628342

Chapter 5
Effect of Nanoparticles on Germination and Seedling Tolerance Traits of Armenian Genotypes of Wheat (Triticum aestivum L.) Under Salinity Stress

Armine Chakhmakhchyan
http://orcid.org/0009-0004-0377-6477
Faculty of Biology, Yerevan State University, Armenia

Abhishek Singh
http://orcid.org/0000-0002-0049-6910
Faculty of Biology, Yerevan State University, Armenia

Nare Darbinyan
Faculty of Biology, Yerevan State University, Armenia

Sakshi Singh
Faculty of Biology, Yerevan State University, Armenia

Karen Ghazaryan
http://orcid.org/0000-0002-8973-2595
Faculty of Biology, Yerevan State University, Armenia

DOI: 10.4018/979-8-3373-1524-9.ch005

ABSTRACT

Soil salinity is considered to be one of the major abiotic stresses, worldwide which disrupts water uptake, nutrient balance, and physiological processes and limits wheat growth and productivity leading to poor germination, stunting, and loss in yields. Nanoparticles (NPs) have great potential in improving plant salt stress tolerance. The effects of zinc oxide and silicon dioxide nanoparticles (ZnO-NPs and SiO_2 NPs) and (50 mg/L) on seed germination in two Armenian wheat genotypes, Vars and Rima, at various levels of salt stress, as well as their early growth and physiological responses were studied.

1. INTRODUCTION

The world's population is growing rapidly and is projected to reach nearly 9.6 billion by 2050, posing a potential threat to food security (Kumar and Sharma 2020). To ensure food security, agricultural production must increase by 70–100% (FAO, 2009). However, crop quality and yield have been declining in recent decades due to increasing abiotic stresses caused by climate change and global warming. One of the abiotic factors that inhibits crop yield is soil salinity. (Johnson et al. 2022). High soluble salt content impairs water absorption, causes ionic toxicity, and disrupts key physiological and biochemical processes in plants, which is especially relevant for arid and semi-arid regions such as the Ararat Valley in Armenia. Salinity and drought are among the major factors limiting agricultural productivity and causing significant economic losses worldwide (Shabala et al. 2014);(Salam et al. 2022);(Hayat et al. 2022). Wheat (*Triticum aestivum* L.) is one of the most important cereal crops in the world and a staple food for more than 50% of the world's population, providing a significant portion of the daily energy and protein intake (FAO 2009; Shewry 2009). Increasing wheat yields is a key challenge in the face of increasing demand. However, this crop is relatively tolerant to the effects of salinity (Grote et al. 2021) and is particularly sensitive to salt stress during early stages of seedling ontogenesis. Therefore, the development of efficient and environmentally friendly technologies to improve the salt tolerance of wheat is of paramount importance. One of the promising areas is the use of nanotechnology, which opens up new opportunities for sustainable agriculture, improving product quality and food security (Singh et al. 2023). Therefore, seed priming with ZnO-NPs stimulates active growth of plants as a zinc has an essential function in synthesis of auxin and cell division. Zinc takes part in chlorophyll synthesis, membrane integrity and production of generative organs promoting the growth and yield of wheat (Al-Salama 2022). The intensity of the response to zinc oxide nanoparticles is dictated by the dose and crop genotype. Besides, the root system and shoots are also stimulated by silicon oxide nanoparticles (SiO_2-NPs) enhanced with chlorophyll content and high photosynthetic activity. They improve antioxidant enzyme activity and decrease oxidative stress, thereby improving the adaptability of plants to an adverse growth environment (Karimi and Mohsenzadeh 2016). Overall, seed treatment or foliar application with SiO_2-NPs enhanced vigorous growth and advanced stress tolerance at the early growth stage of wheat plants. In this study, we have aimed to evaluate silicon oxide and zinc oxide nanoparticles in enhancing seed germination, early seedling growth and overcoming salt stress by improving physiological parameters among two Armenian wheat genotypes, Vars and Rima.

2. MATERIALS AND METHODS

2.1 Characterization of NPs

The ZnO-NPs and SiO 2 -NPs were obtained from Sisco Research Laboratories (SRL). The average sizes of ZnO-NPs and SiO_2-NPs were 30 nm and 40 nm, respectively. Purity of ZnO-NPs and SiO_2-NPs was 99.9% and 99.0%, respectively.

2.2 Preparation of NPs Solution

To obtain modified Hogland solution with 50 mg/L ZnO-NPs and SiO_2-NPs were added using magnectric mixer according to the procedure previously used (Rajput et al. 2021; Bhandari et al. 2023; Bishwakarma et al. 2022). Details of the treatments are presented in Table 1.

Table 1. Combination of nanoparticle and salinity treatments

Treatment Code	*Nanoparticles*	*NaCl Concentration (mM/L)*
T_1	—	0
T_2	ZnO-NPs (50 mg/L)	0
T_3	SiO_2-NPs (50 mg/L)	0
T_4	—	100
T_5	—	200
T_6	—	300
T_7	ZnO-NPs (50 mg/L)	100
T_8	ZnO-NPs (50 mg/L)	200

continued on following page

Table 1. Continued

Treatment Code	*Nanoparticles*	*NaCl Concentration (mM/L)*
T_9	ZnO-NPs (50 mg/L)	300
T_{10}	SiO_2-NPs (50 mg/L)	100
T_{11}	SiO_2-NPs (50 mg/L)	200
T_{12}	SiO_2-NPs (50 mg/L)	300

2.3 Seed Treatment Protocol

Wheat genotypes seeds were obtained from Gyumri breeding station and Armenian Scientific Center of Agriculture. To clean the seeds of Armenian wheat genotypes (V 1: Vars; V 2: Rima) from contaminants, they were washed three times with deionized water and then sterilized with 4% sodium hypochlorite solution for 30 minutes. The seeds were placed in sterile Petri dishes (10 seeds in each) on filter paper. To each dish, 4 ml of pre-prepared Hogland solution with the appropriate treatment with NaCl (100, 200 and 300 mM) and NPS (ZnO and SiO_2 at a concentration of 50 mg/L) were added in accordance with the above-described order (Table 1); a total of 12 treatments were carried out. For each treatment, 3 replicates (3 dishes) were carried out. After adding the solution, the Petri dishes were wrapped in parafilm and placed in an incubator at 25 °C under a 12-h light and 12-h dark photoperiod for 8 days.

2.4 Evaluation of Seed Germination Indices and Resistance to Germination Stress

2.4.1 Germination Indices

The initial count of germinated seeds was performed when a seed showed a healthy white rootlet of approximately 2 mm in length (Hasanaklou et al. 2023). Germinated seeds were counted every 24 hours for 8 days to assess the germination parameters GP and GE (Ghaleb et al. 2022). GR, MDG, and MGT were assessed after eight days. The corresponding formulas were used to calculate these germination indices (Khajeh-Hosseini et al. 2009; Pathak et al. 2023; Singh et al. 2023; Domingues Neto et al. 2024).

2.4.2 Germination Stress Tolerance Indices

Environmental stress tolerance indicators are also used for the evaluation of plants with respect to abiotic factors. PI values are a measure of seed germination percentage, GSTI indicates the general germination efficiency at stress. And also, PHSI is the ability of plant to keep height under stress and RLSI indicates rool length stability in undesirable condition. Biomass indicators such as SFSI or RFSI represent the ability of the plant to maintain fresh biomass under stress, whereas SDSI and RDSI are related to dry weight accumulation. GR, SLR and RLR are adopted to measure the extent of growth reduction under stress. In conjunction, these indices form a systemic way to fully evaluate plant tolerance against adverse environmental conditions. All of these indices are applied to evaluate the plant adaptability to unfavorable environments, and were computed using following formulas as per the previous research (Partheeban et al. 2017; Tarchoun et al. 2022).

2.5 Analysis of Morpho-Physiological Traits of Wheat Seedlings

Wheat plants physiological parameter measurements, root length and shoot length, fresh weight of root and shoot as well as the dry weight of them were accurately taken.

2.6 Statistical Analysis

The obtained data were processed using Microsoft Excel 2021 and SPSS 19. Statistical significance of differences between groups was evaluated by applying Fisher's Least Significant Difference (LSD) test. All comparisons were considered significant at $p \leq 0.05$.

3. RESULTS AND DISCUSSION

3.1 Effect of ZnO-NPs and SiO_2-NPs on Germination Indices Under Salinity Stress

As previously noted, soil salinization is one of the main factors contributing to the decline in crop yields, as it restricts plant growth and productivity, reduces seed germination, disrupts osmotic balance, and causes various other physiological disturbances. Previous studies have confirmed that when ZnO-NPs is applied at a concentration of 50 mg/L in pea (*Pisum Sativum*), an increase in root length and other growth parameters is observed (Mustafa et al. 2024). Also, some studies have shown that seed priming is effective for seed storage, changes in physiological conditions improve seed germination, germination rate and growth, and also increases resistance to abiotic and biotic stresses.(Bourioug et al. 2020); (Salam et al. 2022). Overall, our results show that ZnO and SiO_2 nanoparticles treatment (50 mg/L) enhanced salt tolerance (100, 200 and 300 mM NaCl) in both wheat genotypes, as reflected by improvements in GR, GP, MDG, MGT and GE (Fig. 1). During our studies on the effects of nanoparticles, the following data were obtained.

(A) Germination rate (GR)

Comparative analysis of the germination rate (GR) showed that the treatment with ZnO and SiO_2 nanoparticles had a positive effect on both genotypes compared to the control. The greatest effect was observed with the ZnO nanoparticles treatment, especially for genotype V1. The maximum increase in GR relative to the corresponding control for genotype V1 reached 48.69% at 200 mM NaCl, whereas for genotype V2 at the same salinity level the increase was only about 4%. At a salinity of 100 mM NaCl, genotype V2 showed a higher increase in GR (ZnO-NPs: 28.6%; SiO_2-NPs: 25%) compared to genotype V1 (ZnO-NPs about 8%, SiO_2-NPs - 7%). Under high salinity conditions (300 mM NaCl), when treated with ZnO nanoparticles, the V1 genotype showed an increase in GR relative to the corresponding control by 12.5%, and the V2 genotype showed an increase by 11.5%. When treated with SiO_2 nanoparticles (300 mM NaCl), the effect was insignificant: the V1 genotype showed an increase in GR by approximately 1.5%, and in V2 by 2.5%.

(B) Germination percentage (GP%)

Analysis of the effect of nanoparticles on the GP% under salt stress conditions showed a consistently positive effect of ZnO nanoparticles on GP% at all salinity levels in both genotypes, while SiO_2 nanoparticles had a significant positive effect, only under higher salinity conditions (200 and 300 mM NaCl). Thus, genotype V1 showed a pronounced increase in the GP% compared to the untreated control, especially at 200 mM NaCl, where ZnO-NPs treatment resulted in a 23.3% increase in GP% compared to the control, while SiO_2-NPs treatment resulted in a 12% increase. This suggests that the combination of 200 mM NaCl and ZnO-NPs provides the highest GP% for this genotype. Similarly, in genotype V2, ZnO-NPs treatment in 200 mM NaCl increased GP% by about 21.67%, while SiO_2-NPs treatment caused a smaller increase of 8.33%. At high salinity (300 mM NaCl), both genotypes showed a decrease in the effectiveness of the nanoparticles compared to that at 200 mM NaCl.

(C) Mean Daily Germination index (MDG)

Thus, the effect on MDG of ZnO-NPs in all levels of salinity was stable positive for both genotypes and for SiO_2-NPs significant positive effect was observed only under conditions of high 200 and 300 mM NaCl salt. The most notable increases in MDG relative to the control for genotype V1 were observed at 200 mM NaCl: treatment with ZnO-NPs by 46.7% with SiO2-NPs by 29.7%. For genotype V2, MDG was increased at the indicated level of NaCl 200 for ZnO-NPs by 18.5% for SiO_2-NPs by 7.8%. At 300 mM NaCl, the effect with the help of ZnO-NPs for genotype V1 was 26.7%, and for genotype V2 was 21.89%. Overall, nanoparticles increased MDG in moderate salinity and high salinity, ZnO-NPSs had a more positive effect than SiO_2-NPs.

(D) Mean Germination Time (MGT)

Under saline-free conditions (0 mM NaCl), in genotype V1, ZnO-NPs reduced MGT by 21.9% and SiO_2-NPs by 4%, whereas in genotype V2, ZnO-NPs had no effect and SiO_2-NPs resulted in only a minor reduction of 2%. For genotype V1, the most pronounced reduction in MGT by ZnO-NPs was observed under high salinity conditions (300 mM NaCl), where MGT decreased by 23.4%. For genotype V2, the most pronounced positive effect of ZnO-NPs was found at 200 mM NaCl, where MGT decreased by 31.3%, while SiO_2-NPs decreased MGT by 10.5%. Overall, these results indicate that ZnO-NPs effectively reduced the mean germination time in both genotypes. SiO_2-NPs were more effective in genotype V2 at all salinity levels (the maximum reduction in MGT was observed at 100 mM NaCl - 12.6%).

(E) Germination Energy (GE)

According to the obtained results, ZnO-NPs had a consistent positive effect on germination vigor (GE) in both genotypes at all salinity levels, with the most pronounced increase observed at high salinity (200–300 mM NaCl). For genotype V1, the maximum increase in GE was approximately 41% at 200 mM NaCl, while for genotype V2, the maximum increase was approximately 39% at 300 mM NaCl. The effect of SiO_2-NPs on GE was weaker and less consistent compared to ZnO-NPs. At higher salinity (200–300 mM NaCl), their effect on GE was generally positive or neutral. SiO_2-NPs were most effective at 300 mM NaCl: genotype V1 showed a 12% increase in GE, whereas genotype V2 showed a 15.5% increase.

Figure 1. Effect of 50 mg/L ZnO and SiO_2 nanoparticles on the parameters of wheat seed germination under both control and salt stress condition for varieties V1: Vars, V2: Rima. Values represent three replicates means. Error bars represent least significant difference (LSD) at $p \leq 0.05$ among treatments.

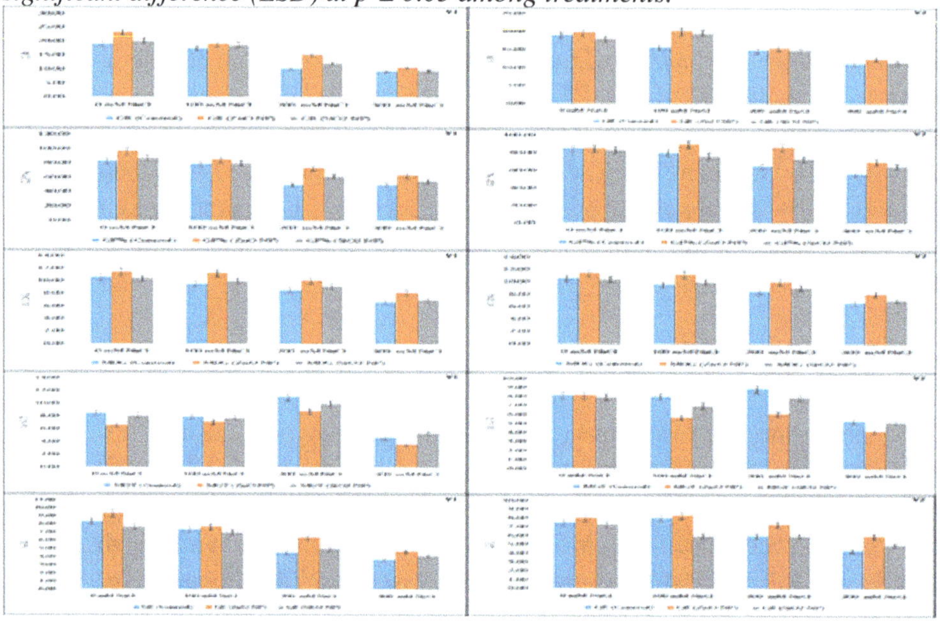

3.2 Effects of ZnO-NPs and SiO_2-NPs on Germination Stress Tolerance Indices

To investigate the effects of ZnO and SiO_2 nanoparticles on wheat germination under salt stress, various stress tolerance indices including PI, GSTI, PHSI, RLSI, SFSI, RFSI, SDSI, RDSI, GR%, SLR and RLR were calculated (Fig. 2). One of the early indicators of salt stress in plants is a decrease in shoot growth, shoot and root fresh weight, or total plant biomass.(Jha et al. 2022). Previous studies showed that priming of Stevia rebaudiana Bertoni seeds with SiO_2 nanoparticles at a concentration of 10 mg/L was effective in increasing the germination percentage and rate, root/shoot/dry weight, biochemical characteristics and enzyme activity.(Hasanaklou et al. 2023). The effect of Zn in the form of ZnO nanoparticles on maize (*Zea mays L.*) was studied, which resulted in improvement in plant growth parameters such as root and shoot length and plant biomass.(Adhikari et al. 2015).

(A) PI

PI shows the efficiency of seed germination, reflecting the speed and uniformity of germination.

According to the results obtained, ZnO-NPs effectively stimulated GI in both genotypes, with the most pronounced effect observed in genotype V1 at 200 mM NaCl. ZnO-NPs increased GI by 48.5%, while SiO_2-NPs led to an increase of 20.4% compared to the corresponding control. For genotype V2, the maximum effect was recorded at 100 mM NaCl: an increase of 38.3% for ZnO-NPs treatment and 29.7% for SiO_2-NPs treatment. Under the same conditions (100 mM NaCl), the V1 genotype showed an increase of only 8%, indicating a greater sensitivity of the V2 genotype to ZnO-NPs at this salinity level (38% versus 8%). The effectiveness of SiO_2-NPs was weaker than that of ZnO. -NPs. In the V2 genotype, SiO_2 NPs had a positive effect on PI growth at all salinity levels, while in the V1 genotype only at 200 and 300 mM NaCl. In both genotypes, the smallest effect of SiO_2 NPs was observed under high salinity conditions (300 mM NaCl), where PI increased by only 1%.

(B) GSTI

GSTI shows the ability of seeds to maintain their germination capacity under stressful conditions.

ZnO-NPs at all salinity levels, and especially at 200 and 300 mM NaCl, contributed to the increase in GSTI compared to the control. For example, at 200 mM NaCl, GSTI in genotype V1 increased by 48.6%, and in genotype V2 by 14.7%. In contrast, the effect of SiO_2-NPs on GSTI was less consistent, and in some cases even negative. The greatest positive effect of SiO_2-NPs was observed at 200 mM NaCl in genotype V1 with a maximum increase in GSTI by 19.9% compared to the corresponding control.

(C) PHSI

PHSI evaluates the ability of plants to maintain growth under stressful conditions.

In genotype V1 at 0 mM NaCl, ZnO-NPs treatment increased PHSI by 22.3% compared to the control, and in genotype V2 by 14.68%. The maximum efficiency of ZnO-NPs in genotype V1 is observed at 200 mM NaCl: PHSI increases by 145% compared to the control. In genotype V2 at 300 mM NaCl, PHSI increases by 124% compared to the control (in contrast to genotype V1), indicating higher innate salt tolerance. In contrast, the effect of SiO_2 nanoparticles was less constant and generally weaker. SiO_2 NPs show maximum efficiency only in genotype V1 at 200 mM NaCl, increasing PHSI by 81.6%.These results are consistent with previous reports suggesting that ZnO-NPs can enhance salt stress tolerance by improving physiological parameters such as chlorophyll content, osmotic adaptation, and antioxidant activity (Mahawar et al. 2024).

(D) RLSI

RLSI is an index representing the ability of plants to resist against stress directly related through root length. Treatment with ZnO-NPs increased RLSI in both wheat genotypes to a higher extent at all saline treatments. The highest increase was 156% as compared to the control (genotype V1 at 200 mM of NaCl), suggesting a strong potential of ZnO-NPs to enhance wheat salinity tolerance. SiO2-NPs demonstrated the greatest response in genotype V1 at 200 mM NaCl with a maximum RLSI increase of 58.87%. In genotype V2, the maximum increase in RLSI compared with control upto 300 mM NaCl: 93.96% by ZnO-NPs and 35.23% by SiO_2-NPS were observed. This reflects the intrinsic tolerance of V2 genotype toward salinity; which is also consistent with promising application of ZnO-NPs for enhancing wheat root's growth in saline environment.

(E) SFSI

SFSI based on fresh shoot weight evaluates plant tolerance to adverse environmental conditions. When treated with ZnO-NPs, both genotypes at all salinity levels showed an increase in SFSI compared to the respective control. For genotype V1, the maximum efficiency of nanoparticle treatment was observed at 200 mM NaCl: an increase in SFSI compared to the control by 111.4% with ZnO-NPs and by 91% with SiO_2-NPs. However, at 300 mM NaCl, the positive effect of ZnO and SiO_2 nanoparticles on genotype V1 was less pronounced: SFSI increased by 65% with ZnO-NPs and by 14.34% with SiO_2-NPs compared to the control. In contrast, genotype V2 showed the highest efficiency of nanoparticle treatment at 300 mM NaCl: ZnO-NPs increased SFSI by 301.56% and SiO_2 - NPS by 150% compared to the control variant, which is significantly higher than that of genotype V1 under the same conditions. Under mild salt stress conditions (100 mM NaCl), ZnO and SiO_2 nanoparticle treatments were also more effective for genotype V2: SFSI increased by 67.9% with ZnO-NPs treatment and by 52.38% with SiO_2-NPs. At 100 mM NaCl, genotype V1 showed a slight increase in SFSI (9.91%) only with ZnO-NPs treatment. According to the obtained data, ZnO-NPs have a more effective effect on shoot fresh weight under salt stress conditions, especially in the more tolerant V2 genotype and under severe stress conditions.

(F) RFSI

RFSI evaluates the impact of adverse environmental conditions based on the fresh weight of plant roots. According to the obtained data, ZnO-NPs treatment had a positive effect on increasing RFSI compared to the control sample at all levels of salt stress. The maximum effect of nanoparticles for genotype V1 was observed at 200 mM NaCl: RFSI increased by 241.4% with ZnO-NPs treatment and by 131.8% with SiO_2-NPs treatment compared to the untreated control. For genotype V2, the highest effectiveness of nanoparticles was observed at 300 mM NaCl: RFSI increased by 70.38% with ZnO-NPs treatment and by 45.1% with SiO_2-NPs treatment. The obtained results confirm the positive effect of nanoparticle treatment, in particular ZnO-NPs, on increasing RFSI of wheat under salinity conditions.

(G) SDSI

SDSI is used to quantify the effect of adverse conditions on the dry weight of plant shoots.

According to the data obtained, ZnO-NPs have a positive effect on SDSI at all levels of salt stress. The effect is noticeable even under optimal conditions (0 mM NaCl), where the V1 genotype showed a significant increase in SDSI by 26.57%, and the V2 genotype by 8.24% compared to the control sample. Under mild salt stress conditions (100 mM NaCl), SDSI increased by 12% in the V1 genotype and by 55% in the V2 genotype relative to the control. At 200 mM NaCl, the increase reached 33.8% in the V1 genotype and 14.5% in the V2 genotype. Under high salinity (300 mM NaCl) conditions, the SDSI index increased by 64% in genotype V1 and by 130% in genotype V2. The effect of SiO_2-NPs on SDSI differed between genotypes. In genotype V1, under non-stressed conditions (0 mM NaCl), SiO_2-NPs slightly decreased SDSI by 5% compared to the control. However, at higher salinity levels, the positive effect of SiO_2-NPs on SDSI gradually increased. In genotype V2, the maximum positive effect of SiO_2- NPs was observed under high salinity conditions, resulting in an increase in SDSI by 59.2% compared to the control. In conclusion, ZnO-NPs effectively increased SDSI of wheat under both non-salinity and salinity stress conditions, while SiO_2-NPs showed a more ambiguous and variable effect.

(H) RDSI

RDSI is a dry weight measurable trait reflecting root tolerance to unfavorable environments. The findings revealed that ZnO-NPs had a uniform positive impact on RDSI by increasing the salinity tolerance of roots at all salt stress levels, while the greatest effect was observed at moderate to high salinity (200–300 mM NaCl).

The highest augmentation of RDSI is observed in both genotypes at 300 mM NaCl, increasing by up to 80.8% (genotype V1) and by 27% (genotype V2) in comparison with control. The application of SiO_2-NPs in genotype V1 did not show significant reduction at low salinity (0–100 mM NaCl); however, the highest favorable effect of SiO_2-NPs was under high salinity (300 mM NaCl) where RDSI increased up to 28.9% as compared with control. The results showed that the ZnO nanoparticles likely have more potential to improve roots tolerance to salt stress.

(I) GR

GR shows the percentage reduction in seed germination under the stress factor. The results show that SiO_2-NPs alleviate the negative impact of salinity, but their effectiveness is lower compared to ZnO-NPs. Increasing the NaCl concentration leads to a decrease in the germination percentage. The control variants without nanoparticles treatment showed the greatest reduction in germination. ZnO-NPs treatment significantly mitigated the reduction in germination at higher salinity levels (200–300 mM NaCl), reducing this reduction by about 23% in the V1 genotype and by about 7% in the V2 genotype, indicating a significant protective effect. SiO_2-NPs treatment also mitigated the reduction in germination, but to a lesser extent than ZnO-NPs treatment. In addition, in the control variants, the V2 genotype showed a lower germination rate than V1 at all NaCl concentrations, which may indicate its inherently higher salt tolerance. Overall, these results indicate that the application of ZnO and SiO_2 nanoparticles can improve seed germination and enhance plant tolerance to salinity.

(J) SLR

SLR shows how much the shoot length of plants decreases under adverse conditions.

Effective treatment with ZnO-NPs reduced shoot length reduction (SLR) by approximately 12–52% (the maximum reduction was observed in genotype V2 at 100 mM NaCl and reached 52%) compared to the corresponding controls. The greatest effect of ZnO-NPs was observed at salinity levels of 100–200 mM NaCl.

Treatment with SiO_2-NPs had a more positive effect on genotype V2 and resulted in a decrease in shoot length reduction at all salinity levels (the maximum reduction in SLR was observed at 100 mM NaCl and reached approximately 35%). However, the effect on genotype V1 was less significant (~0–5%) or even negative (21% at 100 mM NaCl). The results show that ZnO-NPs treatment is more effective, especially under low and moderate salinity levels.

(K) RLR

RLR shows how much the root length is shortened under the influence of a stress factor. The obtained data show that ZnO-NPs are more effective in reducing the negative impact of salt stress on root growth compared to SiO_2-NPs. Thus, at 100 mM NaCl, the RLR decrease in the V2 genotype is about 25% compared to the control, and in the V1 genotype it is about 16%. Under moderate and strong salinity, an almost identical decrease in RLR is observed in both genotypes: at 200 mM - about 10%, and at 300 mM - about 7%. SiO_2-NPs treatment demonstrated a smaller effect: at 100 mM NaCl, the decrease was only about 8%, at 200 mM - about 4%, and at 300 mM a slight increase in RLR was observed compared to the control.

Figure 2.

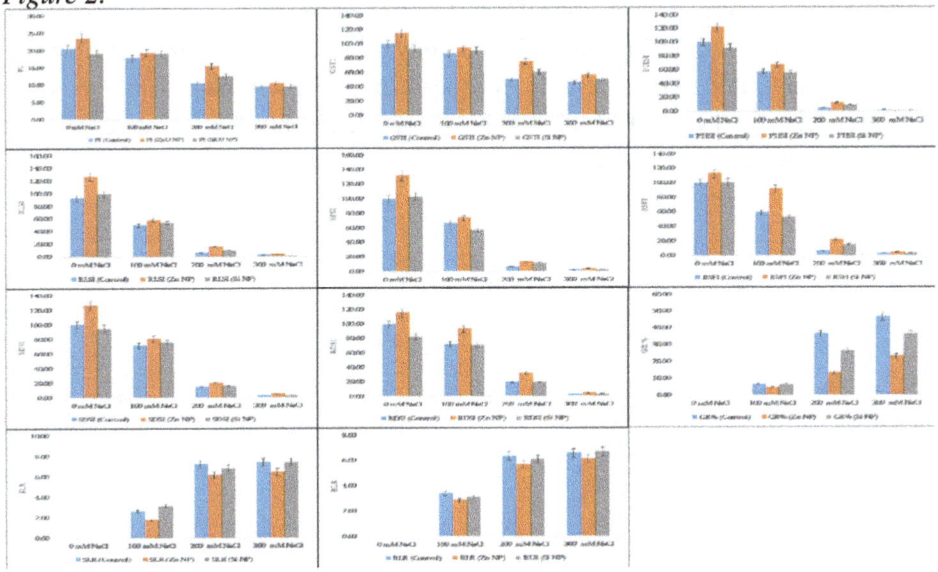

4. CONCLUSION

This study focused on the effects of ZnO-NPs and SiO_2-NPs on the performance of Vars (V1) and Rima (V2) wheat genotypes under different salinity conditions. Four types of treatments were conducted: control, salt treatment (100, 200, 300 mM NaCl), nanoparticles treatment (ZnO-NPs and SiO_2-NPs at 50 mg/L) and combined treatment (NPS-NaCl). Overall, morphological and physiological analysis of nanoparticles-treated plants showed enhanced stress tolerance. ZnO-NPs application had significant stress-mitigating effects on both genotypes, especially on the more sensitive Vars genotype under high salinity conditions, as evidenced by the decrease in GR, SLR, RLR and MGT. After analyzing the growth and biomass improvement parameters (GR, MDG, GE, PI, GSTI, PHSI, RLSI, SFSI, RFSI, SDSI, RDSI), it can be said that ZnO-NPs had a stable positive effect on both genotypes, but under high salinity (300 mM NaCl) conditions, PHSI, RLSI, SFSI, RFSI, SDSI showed significantly higher growth of the Rima genotype, which is explained by its greater innate salt tolerance. When the Rima genotype was treated with SiO_2-NPS under high salinity conditions, the PHSI, RLSI, SFSI, RFSI, SDSI parameters also showed higher growth than the Vars genotype. However, in the control, when treated with ZnO-NPs under low and medium salinity conditions, the Vars genotype showed better results than Rima, since it is probably more susceptible to the effect of zinc oxide.Overall, the effectiveness of SiO_2-NPs varied depending on the genotype and salinity level and requires further study. To study the effect of nanoparticles in more detail, it is advisable to conduct a field experiment, as it will be possible to study the long-term effect and, if necessary, change the concentration of nanoparticles and the method of treatment with them.

ACKNOWLEDGMENT

Higher education and science committee MESCS Republic of Armenia provided the 24FP-4C013 grant to support KG.

REFERENCES

Adhikari, T., Kundu, S., Biswas, A. K., Tarafdar, J. C., & Subba Rao, A. (2015). Characterization of Zinc Oxide Nano Particles and Their Effect on Growth of Maize (*Zea Mays L.*) Plant. *Journal of Plant Nutrition, 38*(10), 1505–1515. DOI: 10.1080/01904167.2014.992536

Al-Salama, Y. (2022). Effect of Seed Priming with ZnO Nanoparticles and Saline Irrigation Water in Yield and Nutrients Uptake by Wheat Plants. *The 2nd International Laayoune Forum on Biosaline Agriculture*, 37. DOI: 10.3390/environsciproc2022016037

Bhandari, K. P., Sapkota, D. R., Jamarkattel, M. K., Stillion, Q., & Collins, R. W. (2023). Zinc Oxide Nanoparticles—Solution-Based Synthesis and Characterizations. *Nanomaterials (Basel, Switzerland), 13*(11), 1795. DOI: 10.3390/nano13111795 PMID: 37299698

Bishwakarma, H., Kumar, N., Anand, M., Das, A. K., Singh, P. K., & Mishra, S. K. (2022). Magnetic Assisted Synthesis of ZnO Nanoparticle by Electrochemical Discharge Method. *Materials Today: Proceedings, 56*, 857–861. DOI: 10.1016/j.matpr.2022.02.511

Bourioug, M., Ezzaza, K., Bouabid, R., Alaoui-Mhamdi, M., Bungau, S., Bourgeade, P., Alaoui-Sossé, L., Alaoui-Sossé, B., & Aleya, L. (2020). Influence of Hydro- and Osmo-Priming on Sunflower Seeds to Break Dormancy and Improve Crop Performance under Water Stress. *Environmental Science and Pollution Research International, 27*(12), 13215–13226. DOI: 10.1007/s11356-020-07893-3 PMID: 32016860

Ghaleb, W., Ahmed, L. Q., Wagner, M.-H., Eprinchard-Ciesla, A., Olivares-Rodríguez, W. E., Perrot, C., Chenu, K., Norton, M., & Escobar-Gutiérrez, A. J. (2022). The Concepts of Seed Germination Rate and Germinability: A Re-Evaluation for Cool-Season Grasses. *Agronomy (Basel), 12*(6), 1291. DOI: 10.3390/agronomy12061291

Grote, U., Fasse, A., Nguyen, T. T., & Erenstein, O. (2021). Food Security and the Dynamics of Wheat and Maize Value Chains in Africa and Asia. *Frontiers in Sustainable Food Systems, 4*(February), 617009. DOI: 10.3389/fsufs.2020.617009

Hasanaklou, N. T., Mohagheghi, V., Hasanaklou, H. T., Ma'mani, L., Malekmohammadi, M., Moradi, F., & Dalvand, Y. (2023). Seed Nano-Priming Using Silica Nanoparticles: Effects in Seed Germination and Physiological Properties of Stevia Rebaudiana Bertoni. *Chemical and Biological Technologies in Agriculture*, *10*(1), 96. Advance online publication. DOI: 10.1186/s40538-023-00445-0

Hayat, K., Menhas, S., & Hayat, S. (2022). Stress-Tolerant Species of Medicinal Plants and Phytoremediation Potential. In *Environmental Challenges and Medicinal Plants*. Springer International Publishing. DOI: 10.1007/978-3-030-92050-0_18

Jha, S., Singh, J., Chouhan, C., Singh, O., & Srivastava, R. K. (2022). Evaluation of Multiple Salinity Tolerance Indices for Screening and Comparative Biochemical and Molecular Analysis of Pearl Millet [Pennisetum Glaucum (L.) R. Br.] Genotypes. *Journal of Plant Growth Regulation*, *41*(4), 1820–1834. DOI: 10.1007/s00344-021-10424-0

Johnson, R., Vishwakarma, K., Hossen, M. S., Kumar, V., Shackira, A. M., Puthur, J. T., Abdi, G., Sarraf, M., & Hasanuzzaman, M. (2022). Potassium in Plants: Growth Regulation, Signaling, and Environmental Stress Tolerance. *Plant Physiology and Biochemistry*, *172*(February), 56–69. DOI: 10.1016/j.plaphy.2022.01.001 PMID: 35032888

Kandil, A. A., Sharief, A. E., & Fatma, M. (2019). Evaluation of Some Rice Cultivars to Salt Tolerance under Antioxidant Using Physiological Indices. *International Journal of Environment, Agriculture and Biotechnology*, *4*(4), 1069–1079. DOI: 10.22161/ijeab.4427

Karimi, J., & Mohsenzadeh, S. (2016). Effects of Silicon Oxide Nanoparticles on Growth and Physiology of Wheat Seedlings. *Физиология растений*, *63*(1), 126–130. DOI: 10.7868/S0015330316010103

Khajeh-Hosseini, M., Lomholt, A., & Matthews, S. (2009). Mean Germination Time in the Laboratory Estimates the Relative Vigour and Field Performance of Commercial Seed Lots of Maize (Zea Mays L.). *Seed Science and Technology*, *37*(2), 446–456. DOI: 10.15258/sst.2009.37.2.17

Kumar, P., & Sharma, P. K. (2020). Soil Salinity and Food Security in India. *Frontiers in Sustainable Food Systems*, *4*(October), 533781. Advance online publication. DOI: 10.3389/fsufs.2020.533781

Kumar, S., Li, G., Yang, J., Huang, X., Ji, Q., Liu, Z., Ke, W., & Hou, H. (2021). Effect of Salt Stress on Growth, Physiological Parameters, and Ionic Concentration of Water Dropwort (Oenanthe Javanica) Cultivars. *Frontiers in Plant Science, 12*(June), 660409. Advance online publication. DOI: 10.3389/fpls.2021.660409 PMID: 34234795

Lungoci, C., Motrescu, I., Filipov, F., Jitareanu, C. D., Teliban, G.-C., Ghitau, C. S., Puiu, I., & Robu, T. (2022). The Impact of Salinity Stress on Antioxidant Response and Bioactive Compounds of Nepeta Cataria L. *Agronomy (Basel), 12*(3), 562. DOI: 10.3390/agronomy12030562

Mahawar, L., Živčák, M., Barboricova, M., Kovár, M., Filaček, A., Ferencova, J., Vysoká, D. M., & Brestič, M. (2024). Effect of Copper Oxide and Zinc Oxide Nanoparticles on Photosynthesis and Physiology of Raphanus Sativus L. under Salinity Stress. *Plant Physiology and Biochemistry, 206*(January), 108281. DOI: 10.1016/j.plaphy.2023.108281 PMID: 38157834

Mustafa, G., Chaudhari, S. K., Manzoor, M., Batool, S., Hatami, M., & Hasan, M. (2024). Zinc Oxide Nanoparticles Mediated Salinity Stress Mitigation in Pisum Sativum: A Physio-Biochemical Perspective. *BMC Plant Biology, 24*(1), 835. Advance online publication. DOI: 10.1186/s12870-024-05554-y PMID: 39243061

Neto, D., José, F., Carneiro, D. C. D. S., & Putti, F. F. (2024). Physiological Indexes in Seed Germination and Seedling Growth of Rangpur Lime (Citrus Limonia L. Osbeck) under Plant Growth Regulators. *Agronomy (Basel), 14*(9), 2066. DOI: 10.3390/agronomy14092066

Partheeban, C., Chandrasekhar, C. N., Jeyakumar, P., Ravikesavan, R., & Gnanam, R. (2017). Effect of PEG Induced Drought Stress on Seed Germination and Seedling Characters of Maize (Zea Mays L.) Genotypes. *International Journal of Current Microbiology and Applied Sciences, 6*(5), 1095–1104. DOI: 10.20546/ijcmas.2017.605.119

Pathak, Mehta, & Thakur. (2023). In Vitro Seed Germination Studies in Barley Exposed to Zinc Oxide Nanoparticles. *Agricultural Reviews*. .DOI: 10.18805/ag.R-2637

Rajput, V. D., Minkina, T., Fedorenko, A., Chernikova, N., Hassan, T., Mandzhieva, S., Sushkova, S., Lysenko, V., Soldatov, M. A., & Burachevskaya, M. (2021). Effects of Zinc Oxide Nanoparticles on Physiological and Anatomical Indices in Spring Barley Tissues. *Nanomaterials (Basel, Switzerland), 11*(7), 1722. DOI: 10.3390/nano11071722 PMID: 34208886

Ro ca, M., Mihalache, G., & Stoleru, V. (2023). Tomato Responses to Salinity Stress: From Morphological Traits to Genetic Changes. *Frontiers in Plant Science*, *14*(February), 1118383. Advance online publication. DOI: 10.3389/fpls.2023.1118383 PMID: 36909434

Salam, A., Ali, A., Afridi, M. S., Ali, S., & Ullah, Z. (2022). Agrobiodiversity: Effect of Drought Stress on the Eco-Physiology and Morphology of Wheat. In Öztürk, M., Khan, S. M., Altay, V., Efe, R., Egamberdieva, D., & Khassanov, F. O. (Eds.), *Biodiversity, Conservation and Sustainability in Asia*. Springer International Publishing. DOI: 10.1007/978-3-030-73943-0_33

Shabala, S., Bose, J., & Hedrich, R. (2014). Salt Bladders: Do They Matter? *Trends in Plant Science*, *19*(11), 687–691. DOI: 10.1016/j.tplants.2014.09.001 PMID: 25361704

Shewry, P. R. (2009). Wheat. *Journal of Experimental Botany*, *60*(6), 1537–1553. DOI: 10.1093/jxb/erp058 PMID: 19386614

Singh, A., Sengar, R. S., Rajput, V. D., Agrawal, S., Ghazaryan, K., Minkina, T., Tawaha, A. R. A., Mahmoud Al Zoubi, O., & Habeeb, T. (2023). Impact of Zinc Oxide Nanoparticles on Seed Germination Characteristics in Rice (*Oryza Sativa* L.) Under Salinity Stress. *Journal of Ecological Engineering*, *24*(10), 142–156. DOI: 10.12911/22998993/169142

Tarchoun, N., Saadaoui, W., Mezghani, N., Pavli, O. I., Falleh, H., & Petropoulos, S. A. (2022). The Effects of Salt Stress on Germination, Seedling Growth and Biochemical Responses of Tunisian Squash (Cucurbita Maxima Duchesne) Germplasm. *Plants*, *11*(6), 800. DOI: 10.3390/plants11060800 PMID: 35336682

Thangjam, U., & Sahoo, U. K. (2016). Effect of Seed Mass on Germination and Seedling Vigour of Parkia Timoriana (DC.) Merr. *Current Agriculture Research Journal*, *4*(2), 171–178. DOI: 10.12944/CARJ.4.2.06

Włodarczyk, K., & Smolińska, B. (2022). The Effect of Nano-ZnO on Seeds Germination Parameters of Different Tomatoes (Solanum Lycopersicum L.) Cultivars. *Molecules (Basel, Switzerland)*, *27*(15), 4963. DOI: 10.3390/molecules27154963 PMID: 35956913

Chapter 6
Physiological and Biochemical Responses of Salt-Tolerant Cultivated Amaranthus "Ultra" to Salinity Stress

Gohar Margaryan
http://orcid.org/0009-0005-5072-0107
Faculty of Biology, Yerevan State University, Armenia

Abhishek Singh
Faculty of Biology, Yerevan State University, Armenia

Hrant Khachatryan
Faculty of Biology, Yerevan State University, Armenia

Karen Ghazaryan
http://orcid.org/0000-0002-8973-2595
Faculty of Biology, Yerevan State University, Armenia

ABSTRACT

In arid and semiarid areas of the world, drought and soil salinity are the main reasons for low production. The Ararat Plain, an important agricultural zone in Armenia, suffers strongly from salinity stress and the potential for agriculture in this region is largely reduced. Hence, it is important to select salt-tolerant crops and plant them in these areas where they can grow while helping in land reclamation to cope with food security challenges. Due to the increasingly serious soil salinity worldwide, in

DOI: 10.4018/979-8-3373-1524-9.ch006

this part of the study the authors tested salinity tolerance ability of specific cultivar of Amaranthus "Ultra". This study therefore attempted to characterize the salt stress resistance of A. "Ultra" and examine to what extent these responses are associated with salt tolerance mechanisms.

INTRODUCTION

Soil salinity is one of the most severe abiotic stress conditions due to the reduction in crop yield and geographic range. It is particularly true in countries where agriculture depends on irrigation and accounts for a significant portion of the national economy. Salinization of soil through accumulation of soluble salts is one of the most crucial constraints adversely affecting plant productivity in arid ecologies (Hammer et al., 2011). High salinity disturbs plant's crucial physiological and biochemical processes, including but not limited to development of the plants, metabolic process as well as photosynthesis (Fig 1) (Evelin et al., 2009).

About 30.000 hectares of land in Armenia, have been affected by salinity, affecting the sustainability of agriculture and food provision of the country. The Ararat Plain, is the most vulnerable region due to salinization and has significantly lost its potential for agriculture. This increasing problem once more points out to the necessity of sustainable alternative solutions, such as halophytoremediation for enhancing soil health in vulnerable areas (Ghazaryan et al., 2020).

Soil salinity is one of the most important limiting factors for horticultural productivity (Wahid et al., 2006). In order to devise measures to properly minimize soil salinization there is a need for joint research on physiological and molecular effects of salinity in plants, as well as for the identification of salt-tolerant cultivars (Rengasamy et al., 2006). The level of salt tolerance breaks down plant species and genotypes into two completely distinct groups (Dasgan et al., 2002). Only a small proportion of vascular terrestrial plants are halophytes the overwhelming majority including almost all major crops are glycophytes, i.e. intrinsically salt sensitive. The exact definition of halophyte is somewhat unclear. Halophytes are generally characterized, on the basis of a wide range of terms (e.g., an endothat performed life cycles in environments with a salinity concentration equal to or greater than 200 mM NaCl; bosaic and structural features and physiological attributes) that provide tolerance mechanisms (Flowers et al., 2015). Apart from their ecological adaptation, halophytes have garnered attention due to their potential use in a range of applications, such as demineralization of soil, removal of contaminants at toxic sites and as novel bioenergy feedstock (Tıpırdamaz et al., 2021). So, their alternative uses as food and feed sources provide additional agronomic value (Nikalje et al., 2017; Liu et al., 2021).

Figure 1. Effect of salinity stress on plant growth and developments

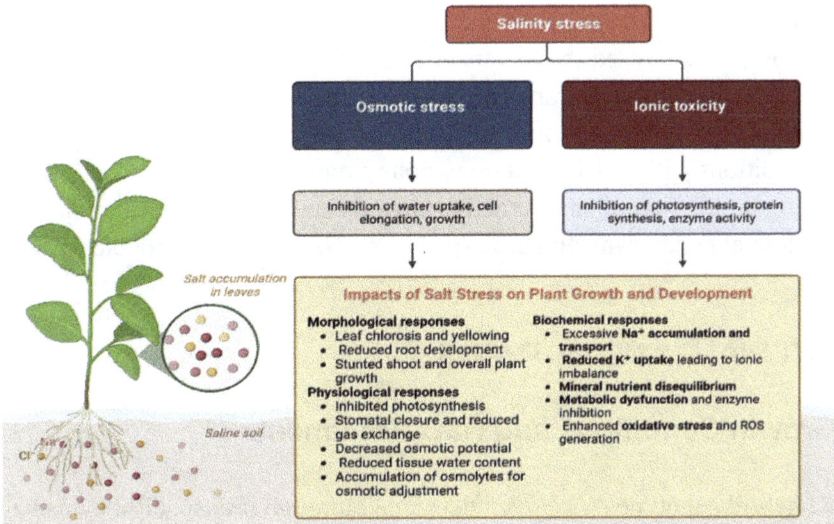

The *Amaranthaceae* (which includes the former *Chenopodiaceae*) is the dominant family in terms of representation of salt-tolerant species, representing c. 44% of all halophytes in plant families Adapted from Adolf et al. (2013). Many species of the genus *Amaranthus*, which belongs to this family, have shown significant salt tolerance at the seed germination and early seeding stage (Wouyou et al., 2017). Some *Amaranthus* species reported to tolerate a saline environment up to 300 mM NaCl at the vegetative stage (Estrada et al., 2021). As a result of their tolerance to harsh environmental conditions such as high salinity, temperature variation and drought, species of this genus are attracting greater attention worldwide for their ecological and agronomic value (Riggins et al., 2021). *A. "Ultra"* is an interspecific hybrid cultivar derived from *Amaranthus paniculatus* and *Amaranthus nutans*. Using this type of *Amaranth* may result in greater biomass production and stress resistance from *A. nutans* and a better harvest quality, higher leaf yield, and more phenotypic uniformity from *A. paniculatus*. The resultant hybrid is a strong upright grower that grows 1.5 to 2.5 m in height with broad lanceolate leaves and terminal panicles bearing dense reddish to orange inflorescences. The *"Ultra"* type is predominantly bred for high green biomass quality, and is useful as a forage, silage and soil improvement in the warm-temperate to subtropical areas. Recently it has been shown that this species is remarkably tolerant to drought, poor soil and salinity as well as has excellent regrowth ability after cutting (Temel et al., 2022). Indeed, the cultivar is intermediate to its parents in terms of both floral and vegetative morphology.

Given the rising global challenge of soil salinity, we conducted research to evaluate the salinity resistance of a particular cultivar of *A. "Ultra"*. The aim of this investigation was to assess the resistance of *A. "Ultra"* to salt stress and evaluate the extent to which these reactions are linked to the mechanisms involved in salt tolerance. For this purpose, we analyzed the effects of soil salinization ranging from non-saline to high saline, on various parameters, including morphological, physiological and biochemical. The results demonstrate that this plant grow well in slight to moderate saline conditions with low impact on various parameters which can contribute to a deeper comprehension of an alternative phytotechnology for remediation of saline soils by tolerant and promising crop species *A. "Ultra"* in sustainable agriculture.

2. MATERIALS AND METHODS

2.1 Growth Conditions and NaCl Treatments

The hybrid genotype *A. "Ultra"* developed through interspecific hybridization between *A. paniculatus* and *A. nutans*, was used in this study. The *"Ultra"* cultivar was selected for its superior biomass productionn and adaptability to saline and drought-prone environments. Seeds were obtained from Armenian organization of "Arm Eraz Amaranth Association". To sterilize and disinfect the seed surface, sodium hypochlorite (2%) was used for 10 min, followed by three rinses with distilled water. Afterwards, they were sown in the pots filled with clay loam soil (20 L, one seed per pot), which is classified as non-saline soil, since EC_e value was not reached ˂2.0 dS m^{-1}) (Allison et al., 1954). Five replicates (pots) of each treatment were used in a complete random block design. Salt treatments were initiated once the plants developed their first fully expanded true leaf. The experiment included four salinity treatments: (non-saline, slightly saline, moderately saline and high saline). NaCl was incrementally applied to the soil to prevent osmotic shock and achieve the target salinity levels. After NaCl treatment, plants were irrigated regularly with tap water. Some parameters including plant height, chlorophyll content index (CCI), photosynthetic (P_n) and transpiration rate (E) were monitored at 10-day intervals, after a stable adaptation period. The duration of the experiment was 120 days.

2.2 Morphological Traits Analysis

Key morphological parameters at fixed intervals throughout the vegetative growth period was assessed. The height of each plant was recorded by measuring the distance from the soil surface to the tip of the main shoot using a calibrated ruler. Stem diameter was measured after harvest, at the base using a vernier caliper.

2.3 Biomass and Harvest

Upon completion of the experiment, plants were harvested to assess total biomass yield. The aboveground parts stems and leaves were collected, while roots were gently washed to eliminate soil and then separated from the shoots.

Immediately after harvest, the fresh weight (FW) of each plant part was measured using a precision digital scale. Plants were subsequently dried in an oven at 70 °C for 72 hours to obtain dry weight (DW). Total biomass was calculated by summing the dry weights of aboveground and belowground parts. Biomass accumulation in plants grown under saline conditions was then compared to that of control (non-saline) plants to evaluate the effects of salinity on growth performance.

2.4 Leaf Succulence Index (LS)

To quantify the water storage capacity within leaf tissues under saline conditions, the leaf succulence index (LS) was calculated. At the end of the study, leaves that had reached full maturity were harvested from each plant. Their fresh weight was promptly determined using a high-precision analytical balance. The LS values were then computed following the methodology described by Agarie et al. (2007):

$$LS = LFW/LA \quad (1)$$

where LFW is the leaf fresh weight (g) and leaf area (LA) is expressed in cm^2.

2.5 Measurement of Photosynthetic Pigments

The chlorophyll content index (CCI) of *A. "Ultra"* leaves was assessed using a CCM-200 plus device (Opti-Sciences, USA) that measures leaf chlorophyll non-invasively. Ten fully mature leaves from both salt-treated and control groups were selected for evaluation. The chlorophyll content was measured thrice during the vegetative period and the measurements averaged to provide an overall estimate of chlorophyll content.

2.6 Gas Exchange Parameters

Gas exchange variables, including net photosynthetic (P_n), transpiration (E), and water use efficiency (WUE), were assessed on fully expanded leaves of *A. "Ultra"* plants grown under both saline and control conditions. A portable CI-340 analyzer was used to collect data between 09:00 and 11:00 a.m., under controlled environmental conditions: air temperature (25–30 °C), air pressure values (88.45-90.76 kPa), and air CO_2 concentration (450-480 μmol mol^{-1}). Measurements were performed thrice. Water use efficiency (WUE) was derived by dividing the net photosynthetic rate by the transpiration rate, following the methodology outlined by Chaves et al. (2009):

$$WUE = P_n / E \quad (2)$$

where P_n is expressed in μmol CO_2 m^{-2} s^{-1} and E in mmol H_2O m^{-2} s^{-1}.

2.7 Analysis of K^+/Na^+ Ratio

To determine the influence of salinity on the ion uptake, especially in correlation with ionic balance and salt tolerance, concentration of some ions have been estimated in shoot and root of *A. "Ultra"*. Ions such as sodium (Na^+), potassium (K^+), and chloride (Cl^-) received special attention, as they have a primary function in osmotic adjustment and ionic balance under salt-stress conditions.

The dried material was ground to a fine powder, and a 0.5% HNO_3 solution employed for digestion at 100°C for 30 min until complete dissolution (Hassan et al., 2016). Na^+ and K^+ were assayed by Flame Photometry (FP-I6431) and the amount of Cl^- was measured using an I-160 M lab ionometer. The K^+/Na^+ ratio was determined according to the following formula to assess ionic selectivity and salt tolerance capacity:

$$K^+/Na^+ \text{ ratio} = [K^+] / [Na^+] \quad (3)$$

where concentrations are expressed in mg g^{-1} DW.

2.8 Relative Electrolytic Leakage (REL)

Relative electrolyte leakage (REL) was measured to evaluate membrane stability and cellular injury in *A. "Ultra"* under salt stress conditions. The method is based on the quantification of ions leaked from leaf tissues into deionized water, which reflects the degree of membrane disruption.

Fully expanded healthy leaves from control and salt-treated plants were rinsed with deionized water and incubated in 50 mL deionized water at 25 ± 2 °C for 24 hours. Initial electrical conductivity (C_1) was measured, then samples were autoclaved at 121 °C for 20 minutes to release all electrolytes. After cooling, final conductivity (C_2) was recorded. Elevated REL values correspond to greater membrane injury and lower stress resilience. REL was calculated following Bajji et al. (2001):

$$REL (\%) = (C_1 / C_2) \times 100 \quad (4)$$

2.9 Evaluating Plants Growth and Phyto-Biochemical Parameters

2.9.1 Plants Growth Parameter

Inhibition Percentage of Shoot Height

With several adjustments, the inhibitory effect of the *A. "Ultra"* genotypes' shoot height was evaluated (Ali et al., 2014).

Shoot Height Inhibition Percentage = 100 – (Response of treatments × 100 / Response of control) (5)

2.9.2 Phyto-Biochemical Parameters

Chlorophyll Stability Index

Chlorophyll Stability Index = $C_s/C_c \times 100$ (6)

where: C_s is the Chlorophyll content of plants under stress and C_c is the Chlorophyll content of the control plant.

Chlorophyll Inhibition Percentage

With certain modifications, the inhibitory impact on the plants was computed using Ali and Ashraf's (2011) methodology.

Chlorophyll Inhibition Percentage = 100 – (Response of treatments × 100 / Response of control) (7)

3. RESULTS AND DISCUSSION

3.1 Morphological Traits Analysis

Salinity had a decreasing impact on *A." Ultra"* growth, and plant height was reduced as salt concentration increased. The best growth occurred under non-saline conditions (plants reached maximum height). With the increase in salinity, a continuous and significant decrease in shoot elongation was observed, leading to a 63.3% reduction in plant height at high salt levels compared with the control. These results demonstrate the salt sensitivity of *A. "Ultra,"* particularly with respect to morphological growth. Changes in plant height are shown in Fig. 2.

Figure 2. Impact of salt stress on the plant height of A. "Ultra", n = 5, P < 0.05.

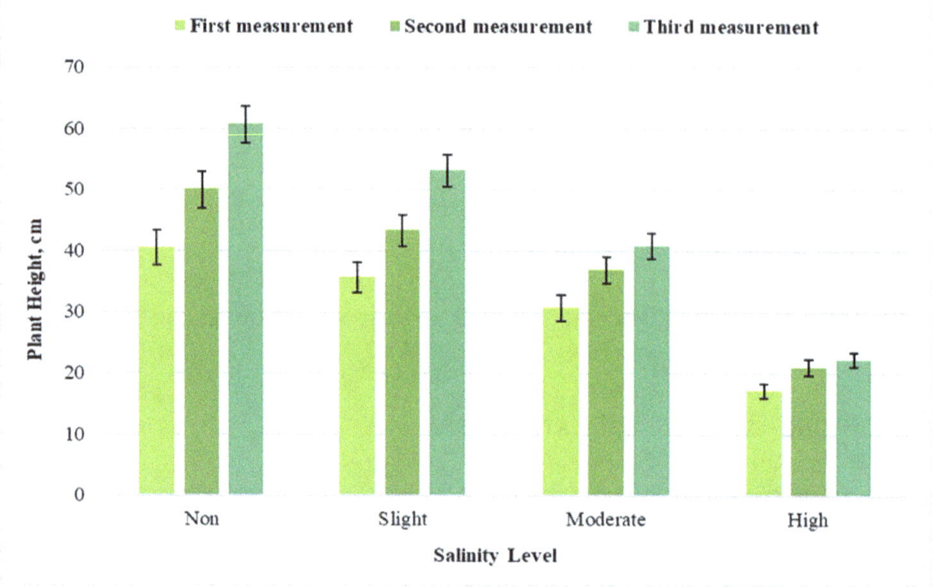

The growth and productivity of plants are negatively affected by a wide range of abiotic stressors and most species are either sensitive or only modestly tolerant. Salinity is one of the most serious constraints that largely hampers the growth and yield of agricultural crops. In response to salinity, stress is primarily characterized by osmotic stress and ion toxicity, which cause increased oxidative stress. Salinity stress also inhibits nutrient uptake, resulting in a deficiency that impairs chlorophyll biosynthesis and reduces fresh and dry biomass accumulation in plants (Loudari et al., 2020; A.H. Alalaf et al., 2020). All *Amaranth* stages are highly tolerant to salinity (Estrada et al., 2021). This adaptation is mostly associated with their intrinsically low stomatal function due to decreased stomatal density and reduced aperture, which efficiently restricts water loss via transpiration under salt stress. Together with their C_4 photosynthesis and favorable anatomical traits (e.g., a highly efficient root system), *Amaranths* are more resilient to different abiotic stresses, especially during salinity-induced drought (Assad et al., 2017). These characteristics allow *Amaranth*, next to quinoa, to be regarded as a promising pseudocereal crop for cultivation in marginal areas (Ismail et al., 2019). *Amaranth* species are well known for their salt tolerance (Gandonou et al., 2018; Pulvento et al., 2022), and the *A. "Ultra"* variety tested in this study survived under high salinity, suggesting that it is a crop of choice in saline-endemic environments. Our research indicated that the height and stem diameter of *A. "Ultra"* plants at low saline levels were not significantly different from those of the control. Nevertheless, these morphological characteristics declined with an increase in salinity from moderately saline to high. High doses of NaCl have been reported to inhibit plant growth and development, where the height and diameter are reduced in different plants due to lethal concentrations of NaCl (confirmed by several studies) (Borsai et al., 2020; Pulvento, Sellami & Lavini, 2022).

The *"Ultra"* stem diameter in the NaCl treatment decreased markedly. The average stem diameter under control and slight salinity levels was 7.7 and 7.8 mm, respectively, and it decreased continuously with increasing salinity, resulting in an average reduction of the stem diameter of 53.8% compared to thee control (Fig. 3). The decrease in stem diameter followed the same pattern as plant height, implying that salinity limits vegetative growth.

Figure 3. Impact of salt stress on the stem diameter of A. "Ultra", n = 5, P < 0.05.

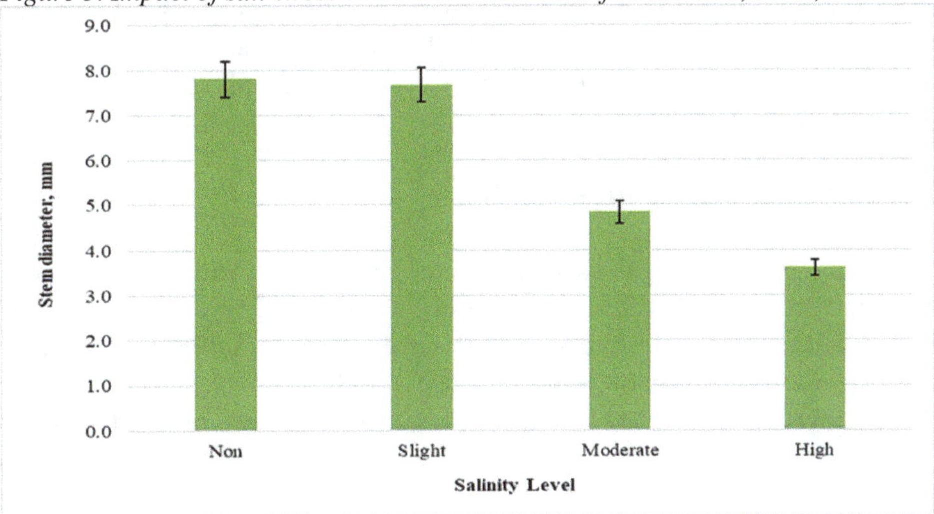

3.2 Biomass and Harvest

Biomass accumulation was significantly reduced by salinity stress in *A. "Ultra"* and the fresh and dry weights of shoots and roots decreased with increasing NaCl levels. The highest biomass occurred in plants under non-saline and slightly saline conditions, whereas high salinity significantly decreased biomass. At high salinity levels, the fresh weights of the roots, stems, and leaves decreased. 80.2%, 80.8%, and 68.5%, respectively, while dry weights decreased by 77.0%, 76.3%, and 65.5% compared to the controls, respectively. This decrease indicates metabolic and growth impairment generated by osmotic and ionic stress, as well as water deficit, which declines photosynthesis. Reductions in the biomass of plant organs are depicted in Fig. 4 A-C.

Stress caused by salinity in plants has been reported to produce morphological and physiological changes, such as stunted growth, leaf fall, and decreased photosynthetic efficiency and membrane permeability (Munns 2005), making biomass a good indicator for assessing the salinity tolerance of crops (Huang et al.2012). While some species buffer the effects of stress by reducing growth and through strategic reallocation of biomass (Grotkopp et al., 2002), our results indicate that increased NaCl progressively suppressed the growth of *A. "Ultra"* on higher salinity levels, and *A. "Ultra"* showed significant reductions in biomass under high saline conditions, while no obvious damage was observed under slightly saline conditions. Salim et al. (A. Loudari et al., 2020; A.H. Alalaf et al., 2020) also reported that high NaCl contents hindered plant growth, which is consistent with reduced FW values under saline conditions compared across canola (M. Tunçtürk et al., 2011), *Aeluropus littoralis* (B. Barzegargolchini et al., 2017), basil (M. Heidari, 2012), and lettuce (J.L. Andriolo et al., 2005).

3.3 Leaf Succulence Index (LS)

Salinity stress had a significant effect on the LS of *A. A. "Ultra"*, which showed an accumulation of LS with increased levels of NaCl (Fig. 5). Plants under high salinity exhibited significantly higher LS than the control, suggesting that the water content per leaf area increased. This response is likely a physiological mechanism to maintain water balance and cope with osmotic and ionic stress imposed by NaCl.

Figure 4. Effect of salt stress on the LS of A. "Ultra" (n = 10, P < 0.05).

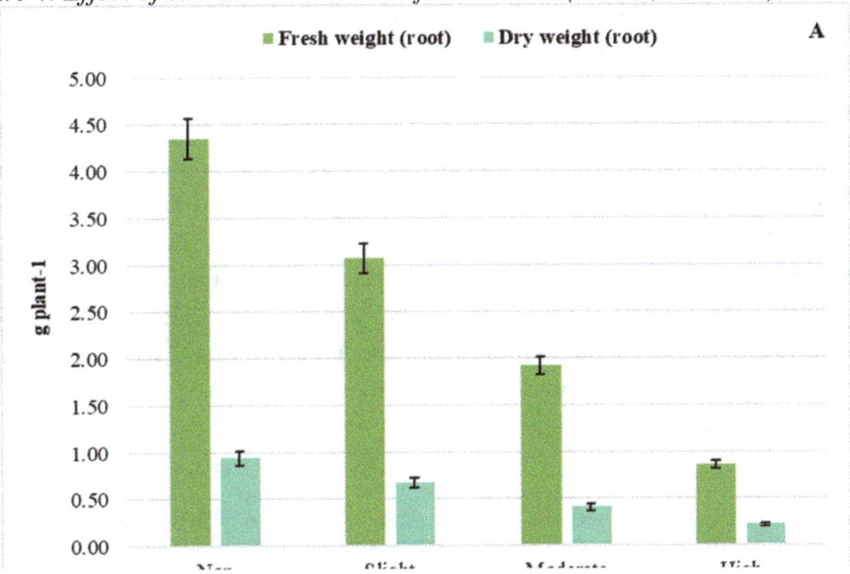

The Leaf Succulence Index (LSI) is a useful trait for evaluating plant water-holding ability under saline conditions. Higher contents of LSI were observed at rising salinity levels in *A."Ultra"* in this study. This indicates that the cultivar responds to osmotic stress by increasing the water content in leaf tissues, a normal feature present in salt-tolerant species. Increased leaf succulence could also be an important strategy to decrease the concentration of internal ions, and hence reduce the toxic effects of Na^+ and Cl^- accumulation. Similar results were reported by other authors, who found high succulence in *Brassica oleracea* and *Chenopodium album* under salinity stress and attributed these findings to a common adaptive strategy (Lema et al., 2019; Ghazaryan et al., 2024).

3.4 Measurement of Photosynthetic Pigments

Salinity stress had an apparent impact on photosynthetic pigment content in *A. "Ultra"* and CCI measurements exhibited a steady decrease in all three measurements (vegetative stage) under salinity treatment (Fig. 6). The CCI was highest in the control plants and lowest in the salt-stressed plants. In particular, CCI under high salinity dropped by 73.2% relative to that of plants grown in non-saline soil. The decrease in pigment content could be ascribed, to the degradation of chloroplasts, impaired synthesis of pigments, and oxidative stress caused by ion toxicity.

Photosynthesis is a central process that supports plant growth and development by producing assimilatory carbon compounds and metabolic energy. Under salt stress, photosynthetic repression becomes a major physiological restraint that directly reduces biomass productivity. Chlorophyll, which plays an indispensable role in light capture and energy saving, is an acceptable indicator of salt tolerance (Wang et al., 2022). In this study, a significant decrease in CCI was observed with increasing salinity, especially under high saline conditions. This reduction in photosynthetic pigments may be due to structural impairment of the chloroplast and changes in the thylakoid membrane (P.R. Sudhir et al., 2005), higher activity of chlorophyllase and diminished lipid levels by stress signals which in turn cause the down-regulation of important biosynthesis enzyme (A.A. Murkute et al., 2006). In agreement with this, notable chlorophyll decrease was found in high salt (M Heidari 2012; MdN. Hossain et al., 2022). A comparable reduction in total chlorophyll has been observed in radish and eggplant grown under high salinity, which is consistent with our results (B.B. Salim et al., 2019; K.Mahdavian, 2023).

Figure 5. Effect of salt stress on the CCI of A. "Ultra", n = 50, P < 0.05

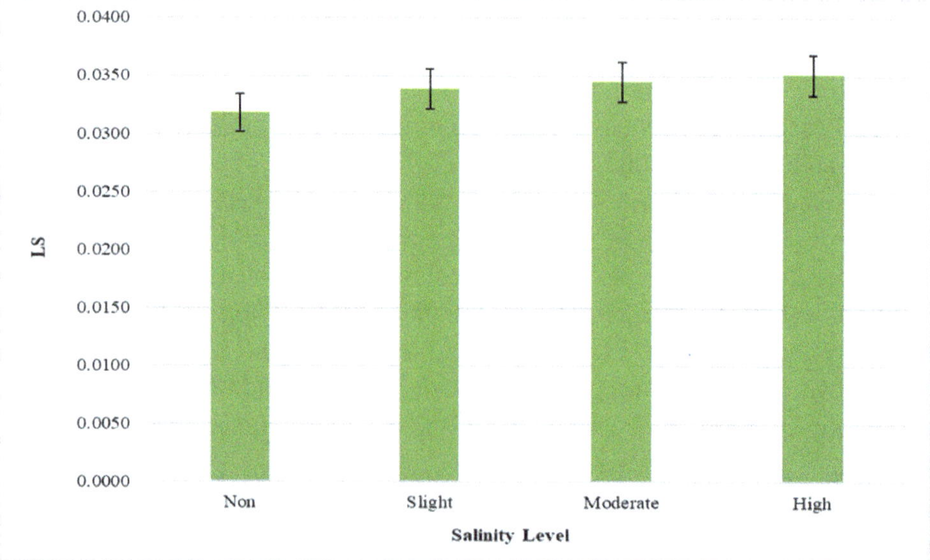

Figure 6. Effect of salt stress on the fresh and dry weights of shoots, stems, and leaves of A. "Ultra", n = 5, P < 0.05.

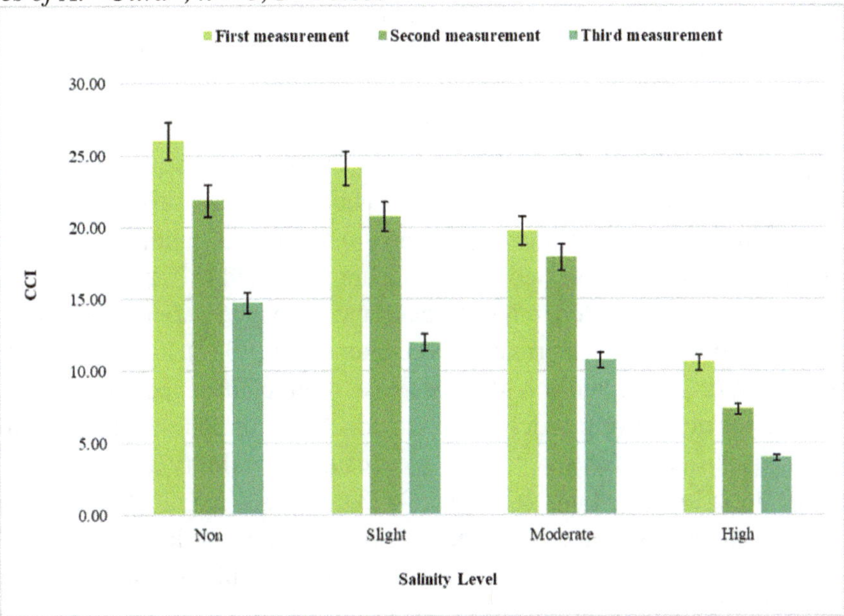

3.5 Gas Exchange Parameters

Salinity exerted a pronounced inhibitory effect on the gas exchange parameters of *A. "Ultra"*. Both P_n and E diminished progressively with increasing NaCl levels, with the steepest decrease recorded at high salinity levels. Compared to plants in non-saline conditions, P_n and E under high salt stress decreased by factors of 2.3 and 2.8, respectively. This reduction is primarily linked to stomatal closure, which restricts CO_2 entry and potentially impairs the photosynthetic apparatus due to osmotic and ionic disturbances.

WUE reacted in a complex manner to salinity but increased by 1.2-fold in the absence of high salt stress control. The lower reduction in transpiration compared to photosynthesis suggests a water-saving behavior. This increase in WUE indicates that *A. "Ultra"* can use water more efficiently under neutral salinity and salt stress conditions. The gas exchange and WUE data are presented in Table 1.

Salinity has been reported to reduce plant gas exchange parameters, leading to decreased photosynthetic performance (Hnilickova et al., 2021). In the current study, P_n and E in *"Ultra"* suffered a marked decrease when plants experienced salinization, which was most severe under a high salt concentration. These effects are likely more related to osmotic and water deficit stress caused by salinity (Pulvento, Sellami & Lavini, 2022; Luyckx, Lutts & Quinet, 2023a; Liu, An & Lai, 2024). Increased salinity also induces stomatal closure, leading to a decrease in transpiration and intercellular CO_2 concentration, which aggravates the inhibition of P_n via Vcmax (Tarin et al., 2020). Omamt et al. (Omami et al., 2006) also showed that salinity tolerance was associated with a ~ 50% higher intrinsic water use efficiency (WUE) in *A. cruentus* plants after exposure to 100 mM NaCl, which could be attributed to salt-induced stomatal closure and lower stomatal density. Similar results were reported by Gandonou et al. (Gandonou et al., 2018) in *A. cruentus* cv. Locale. Enhanced WUE is generally accepted as an important mechanism of salt tolerance in plants that allows them to reduce water loss and maintain photosynthesis under saline conditions (Tarin et al., 2020). Liao et al. (Liao et al., 2022) also reported improved WUE in maize when it was subjected to both salt and drought stress. *"A.Ultra"* in our research also showed an increase in WUE under salinity exposure, indicating a robust physiological acclimation for carbon acquisition with less water loss. As all *Amaranthus* species share C_4 photosynthesis (Sage et al., 2007), the high WUE determined here might also be representative of the inherent efficiency of this carbon-fixing mechanism, which is widely attributed to halophytic grass behavior (Bromham et al., 2014).

Table 1. Effect of salt stress on gas exchange parameters and WUE of A. "Ultra" plants, n = 15, P < 0.05.

Salinity Level	P_n (*µmol*/m²/s)			E (*mmol*/m²/s)			WUE		
	First	Second	Third	First	Second	Third	First	Second	Third
Non	26.50	9.37	8.52	2.91	1.67	1.59	9.09	5.61	5.34
Slight	16.59	8.90	7.44	1.01	1.58	1.22	16.43	5.63	6.08
Moderate	11.12	7.06	5.94	0.99	1.14	0.96	11.23	6.19	6.22
High	8.83	5.94	3.72	0.87	0.95	0.58	10.15	6.25	6.45

3.6 Uptake Ion Content, K^+/Na^+ Ratio

Salt stress significantly modified ion acquisition and reallocation in *A."Ultra"* with increasing NaCl concentrations, resulting in very high accumulation of sodium (Na^+) and chloride ions (Cl^-) in roots, stems, and leaves. Na^+ concentrations increased by 3.2, 31.8, and 31.3 times and Cl^- by 6.6, 5.0, and 13.7 times in the root, stem, and leaves respectively, with respect to the control. Such ion accumulation, also observed under high salinity, may suggest excessive Na^+ and Cl^- influx, leading to ionic toxicity and osmotic disturbance. However, K^+ also decreased by 3.8, 1.1, and 1.5 fold in roots, stems, and leaves, respectively, probably because of the Na^+-induced competitive inhibition of transport sites on membrane surfaces. This simultaneous reduction in the K^+/Na^+ ratio among treatments indicates disruption of ionic homeostasis and possible effects on vital physiological processes. These results emphasize the limited capability of the *A. "Ultra"* plant to support selective ion transport under saline conditions (Table 2).

Salt tolerance is closely linked to ion homeostasis, particularly the regulation of Na^+, K^+, and Cl^- levels in plants. Salt-tolerant species mitigate ionic stress by minimizing cellular Na^+ accumulation while maintaining elevated K^+ concentrations, thereby sustaining a low K^+/Na^+ ratio, which is an essential indicator of ionic balance (Munns & Tester, 2008). Excessive Na^+ influx disrupts membrane integrity, causes K^+ and Ca^{2+} efflux, and impairs metabolic function, ultimately suppressing growth (Carter et al., 2006). Thus, cytosolic K^+ retention is a critical determinant of salinity tolerance. Salinity stress increases Na^+ influx, leading to its accumulation in the cytosol and displacement of K^+, which disrupts metabolic functions, enzyme activity and cell growth (C. Kaya et al., 2007). As a key osmotic regulator, K^+ supports protein synthesis, photosynthesis, and turgor maintenance and acts as a Na^+ antagonist under saline conditions (S.A.A.E. Sayed et al., 2019). However, studies have shown that increasing salinity commonly reduces tissue K^+ levels (B.R.S. Silva et al., 2020). In the current study, the potassium content decreased in all plant organs, whereas the sodium content increased overall (it was higher in the roots).

Another parameter often considered a reliable physiological index for tolerance to salt stress is the K^+/Na^+ ratio, which significantly decreased in all organs in our study (Tao et al., 2021). In the present study, the K^+/Na^+ ratio in A. "Ultra" declined significantly with increasing salinity, indicating impaired ion homeostasis under NaCl stress. This reduction suggests either increased Na^+ accumulation, decreased K^+ uptake, or both, which are commonly associated with disrupted cellular metabolism, enzyme inhibition, and reduced growth performance. Maintaining a high K^+/Na^+ ratio is a well-documented salt tolerance mechanism that supports enzymatic activity, osmotic balance, and cellular turgor (Munns & Tester, 2008). Consistent with our findings, numerous previous studies have reported that increasing salinity levels lead to a significant decline in the K^+/Na^+ ratio across various crops, including Amaranthus species (Zeeshan et al., 2020b; Chourasia et al., 2021b).

Table 2. Effect of salt stress on ion content in the roots, stems, and leaves of A."Ultra", n = 5, P < 0.05.

Salinity Level	K^+			Na^+			Cl^-			K^+/Na^+ ratio		
	root	stem	leaf	root	stem	leaf	root	stem	leaf	root	stem	leaf
Non	19.00	31.55	31.45	8.75	0.22	0.23	0.14	0.42	0.68	2.17	143.41	139.78
Slight	15.03	34.20	39.67	14.70	0.35	0.47	0.36	0.66	0.82	1.02	97.71	83.80
Moderate	6.92	30.55	21.00	24.85	2.95	1.41	0.61	1.53	1.73	0.28	10.37	14.95
High	5.07	28.63	20.37	28.40	6.99	7.03	0.90	2.11	9.37	0.18	4.10	2.90

3.7 Relative Electrolytic Leakage (REL)

REL, a marker of membrane integrity, increased markedly in *A. "Ultra"* under salinity stress (Fig. 7). Low leakage under non-saline conditions indicated stable membranes, whereas high salinity led to an elevated REL, reflecting membrane damage likely caused by oxidative stress and ion toxicity. These results confirm REL as a sensitive indicator of salt-induced cellular injury.

Figure 7. Effect of salt stress on the REL of A. "Ultra", n = 10, P < 0.05.

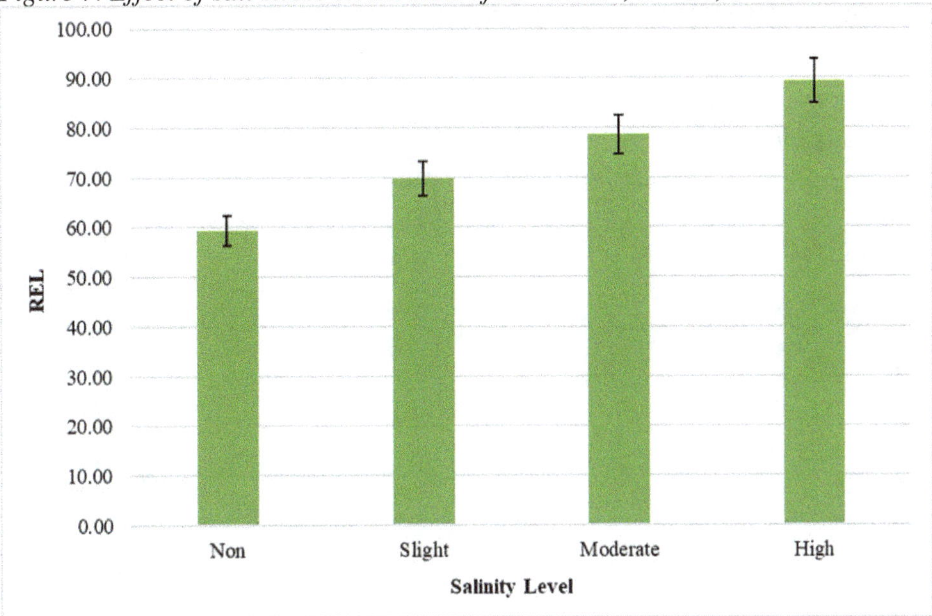

REL is a well-established indicator of membrane damage caused by salinity, reflecting altered membrane permeability (Ghorbani et al., 2019). In *A. "Ultra"*, the REL rose steadily with increasing soil salinity. Similar patterns occur in halophytes such as *Salicornia europaea*, *Suaeda salsa*, *Limonium bicolor*, and *Tamarix chinensis*, although species differ in tolerance levels (Liu et al., 2013). Salt-tolerant plants often counter osmotic stress by accumulating compatible solutes that maintain cellular stability despite elevated REL (Al Murad et al., 2020).

3.8 Evaluating Plants Growth and Phyto-Biochemical Parameters

3.8.1 Plants Growth Parameter

Inhibition Percentage of Shoot Height

Increasing salinity resulted in a progressive increase in the percentage inhibition of shoot height in *A. "Ultra"*. The highest inhibition was recorded at the high salinity level, where shoot height was suppressed by up to 63.3%, indicating a severe restriction of vegetative growth (Fig. 8). This increasing trend in the inhibition percentage reflects the enhanced impact of osmotic stress and ionic toxicity on cell expansion and division. These data emphasize the sensitivity of shoot development to salt-induced physiological disruption.

Figure 8. Effect of salt stress on the Shoot Height Inhibition Percentage of A. "Ultra", n = 10, P < 0.05.

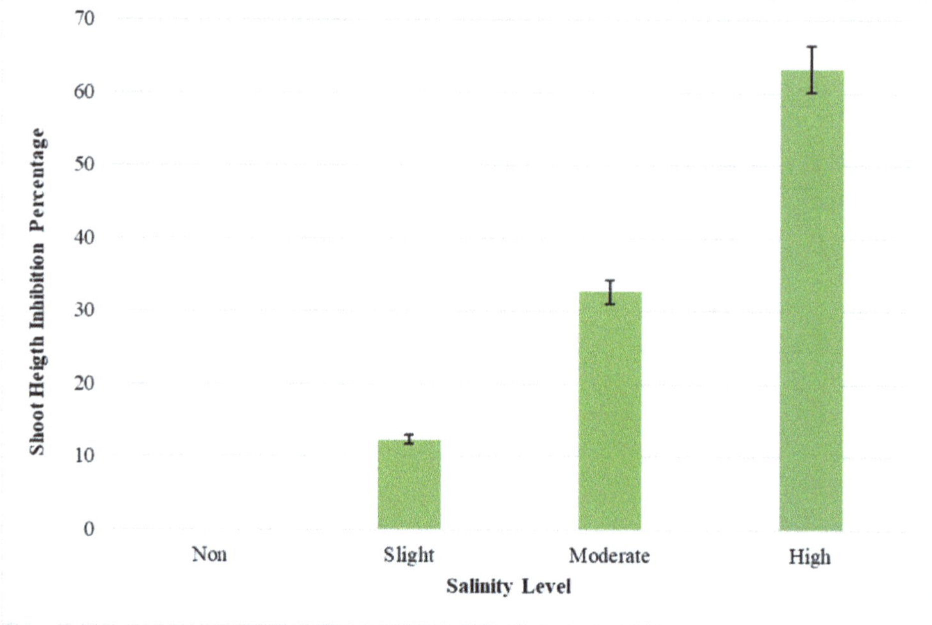

3.8.2 Phyto-Biochemical Parameters

Chlorophyll Stability Index

In *A. "Ultra"*, the Chlorophyll Stability Index, a key indicator of stress tolerance, declined significantly with increasing salinity, decreasing by 26.8% under high salinity conditions compared to non-saline conditions (Fig. 9). This decline suggests that salt stress compromises the structural integrity and functional efficiency of chlorophyll-containing complexes, likely because of oxidative damage. The observed reduction in the Chlorophyll Stability Index was consistent with the overall decline in chlorophyll content and photosynthetic efficiency under saline conditions.

Figure 9. Effect of salt stress on the Chlorophyll Stability Index of A. "Ultra", n = 10, P < 0.05.

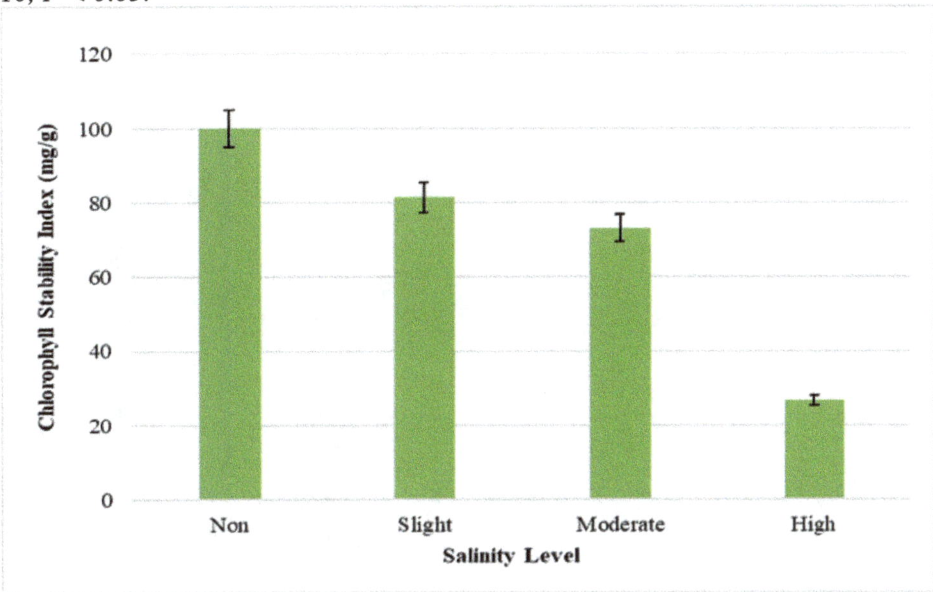

Chlorophyll Inhibition Percentage

Salinity stress significantly inhibited chlorophyll biosynthesis in *A. "Ultra"*, as reflected by the progressive increase in chlorophyll inhibition percentage with increasing NaCl concentrations (Fig. 10). The highest inhibition was observed under high-salinity conditions, where chlorophyll inhibition reached 73.2%, indicating severe disruption of pigment synthesis and degradation of chloroplast structures. This pattern aligns with the increased membrane instability and oxidative stress observed under salt stress.

Figure 10. Effect of salt stress on the Chlorophyll Inhibition Percentage of A. "Ultra", n = 10, P < 0.05.

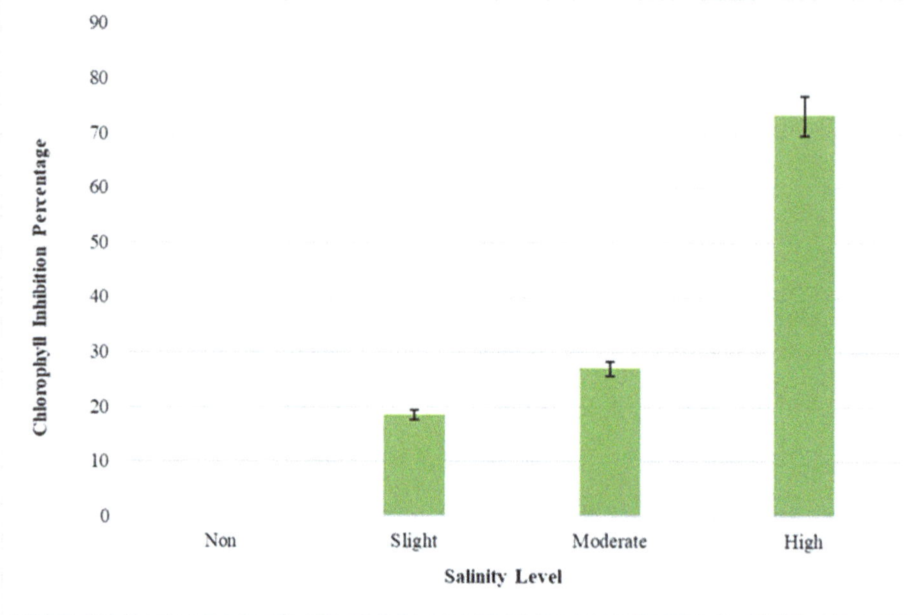

4. CONCLUSION

The results showed that *A. "Ultra"* could tolerate mild and moderate soil salinity effectively, maintaining plant growth and physiological activities under salt stress. Although biomass and photosynthetic activity were continually inhibited as higher NaCl concentrations were applied, the hybrid responded adaptively such that water use efficiency increased and ionic balance was regulated by altering K^+ and Na^+ levels in particular. Notably, the K^+/Na^+ ratio decreased dramatically under severe salinity, suggesting difficulties in ion homeostasis and constituting a typical salt-induced response. The high relative electrolyte leakage is associated with membrane damage, however the resistance of the plant indicates a response of osmotic adjustment mechanisms. Conclusively, *A. "Ultra"* is the potential candidate for sustainable agriculture in salt-affected environments; it can be used for phytoremediation and as a source of forage under stress conditions. The mechanistic and genetic aspects of its salt tolerance warrant further investigation, which may impact the breeding of salt-resistant crops.

ACKNOWLEDGMENT

KG is supported by Higher Education and Science Committee MESCS RA, grant number 21AG-4C075.

REFERENCES

Adolf, V. I., Jacobsen, S., & Shabala, S. (2012). Salt tolerance mechanisms in quinoa (Chenopodium quinoa Willd.). *Environmental and Experimental Botany*, *92*, 43–54. DOI: 10.1016/j.envexpbot.2012.07.004

Agarie, S., Shimoda, T., Shimizu, Y., Baumann, K., Sunagawa, H., Kondo, A., Ueno, O., Nakahara, T., Nose, A., & Cushman, J. C. (2007). Salt tolerance, salt accumulation, and ionic homeostasis in an epidermal bladder-cell-less mutant of the common ice plant Mesembryanthemum crystallinum. *Journal of Experimental Botany*, *58*(8), 1957–1967. DOI: 10.1093/jxb/erm057 PMID: 17452753

Alalaf, A., & Fekry, W. (2020). Effect of saline stress on growth of fruit plants (review article). *Mağalla Zirā a Al-rāfidayn*, *48*(4), 51–59. DOI: 10.33899/magrj.2020.128155.1070

Ali, M. N., Ghosh, B., Gantait, S., & Chakraborty, S. (2014). Selection of rice genotypes for salinity tolerance through Morpho-Biochemical Assessment. *Rice Science*, *21*(5), 288–298. DOI: 10.1016/S1672-6308(13)60189-4

Ali, Q., & Ashraf, M. (2011). Induction of Drought Tolerance in Maize (Zea mays L.) due to Exogenous Application of Trehalose: Growth, Photosynthesis, Water Relations and Oxidative Defence Mechanism. *Journal Agronomy & Crop Science*, *197*(4), 258–271. DOI: 10.1111/j.1439-037X.2010.00463.x

Andriolo, J. L., Da Luz, G. L., Witter, M. H., Godoi, R. D. S., Barros, G. T., & Bortolotto, O. C. (2005). Growth and yield of lettuce plants under salinity. *Horticultura Brasileira*, *23*(4), 931–934. DOI: 10.1590/S0102-05362005000400014

Assad, R., Reshi, Z. A., Jan, S., & Rashid, I. (2017). Biology of amaranths. *Botanical Review*, *83*(4), 382–436. DOI: 10.1007/s12229-017-9194-1

Bajji, M., Lutts, S., & Kinet, J. (2001). Water deficit effects on solute contribution to osmotic adjustment as a function of leaf ageing in three durum wheat (Triticum durum Desf.) cultivars performing differently in arid conditions. *Plant Science*, *160*(4), 669–681. DOI: 10.1016/S0168-9452(00)00443-X PMID: 11448742

Bartels, D., & Sunkar, R. (2005). Drought and salt tolerance in plants. *Critical Reviews in Plant Sciences*, *24*(1), 23–58. DOI: 10.1080/07352680590910410

Barzegargolchini, B., Movafeghi, A., Dehestani, A., & Mehrabanjoubani, P. (2017). Morphological and anatomical changes in stems of Aeluropus littoralis under salt stress. *Journal of Plant Molecular Breeding*, *5*(1), 40–48. DOI: 10.22058/jpmb.2017.63945.1133

Blair, M. W., Riggins, C., & De La Rosa, A. P. B. (2023). Editorial: Amaranthus: naturally stress-resistant resources for improved agriculture and human health, volume II. *Frontiers in Plant Science*, *14*, 1329377. Advance online publication. DOI: 10.3389/fpls.2023.1329377 PMID: 38152149

Borsai, O., Hassan, M. A., Negru ier, C., Raigón, M. D., Boscaiu, M., Sestra , R. E., & Vicente, O. (2020). Responses to Salt Stress in Portulaca: Insight into Its Tolerance Mechanisms. *Plants*, *9*(12), 1660. DOI: 10.3390/plants9121660 PMID: 33260911

Bromham, L., & Bennett, T. H. (2014). Salt tolerance evolves more frequently in C4 grass lineages. *Journal of Evolutionary Biology*, *27*(3), 653–659. DOI: 10.1111/jeb.12320 PMID: 24494637

Carter, J. L., Colmer, T. D., & Veneklaas, E. J. (2005). Variable tolerance of wetland tree species to combined salinity and waterlogging is related to regulation of ion uptake and production of organic solutes. *The New Phytologist*, *169*(1), 123–134. DOI: 10.1111/j.1469-8137.2005.01552.x PMID: 16390424

Chaves, M. M., Flexas, J., & Pinheiro, C. (2008). Photosynthesis under drought and salt stress: Regulation mechanisms from whole plant to cell. *Annals of Botany*, *103*(4), 551–560. DOI: 10.1093/aob/mcn125 PMID: 18662937

Chourasia, K. N., Lal, M. K., Tiwari, R. K., Dev, D., Kardile, H. B., Patil, V. U., Kumar, A., Vanishree, G., Kumar, D., Bhardwaj, V., Meena, J. K., Mangal, V., Shelake, R. M., Kim, J., & Pramanik, D. (2021). Salinity stress in potato: Understanding physiological, biochemical and molecular responses. *Life (Chicago, Ill.)*, *11*(6), 545. DOI: 10.3390/life11060545 PMID: 34200706

Christophe, B. G., Hermann, P., Séraphin, A. Z., Agapit, D. W., Stanley, L., David, H. M., Françoise, A. K., & Armel, C. G. M. (2018). Effects of salinity stress on growth in relation to gas exchanges parameters and water status in amaranth (Amaranthus cruentus). *International Journal of Plant Physiology and Biochemistry*, *10*(3), 19–27. DOI: 10.5897/IJPPB2018.0280

Dasgan, H., Aktas, H., Abak, K., & Cakmak, I. (2002). Determination of screening techniques to salinity tolerance in tomatoes and investigation of genotype responses. *Plant Science*, *163*(4), 695–703. DOI: 10.1016/S0168-9452(02)00091-2

Estrada, Y., Fernández-Ojeda, A., Morales, B., Egea-Fernández, J. M., Flores, F. B., Bolarín, M. C., & Egea, I. (2021). Unraveling the strategies used by the under-exploited amaranth species to confront salt stress: Similarities and differences with quinoa species. *Frontiers in Plant Science*, *12*, 604481. Advance online publication. DOI: 10.3389/fpls.2021.604481 PMID: 33643343

Evelin, H., Kapoor, R., & Giri, B. (2009). Arbuscular mycorrhizal fungi in alleviation of salt stress: A review. *Annals of Botany*, *104*(7), 1263–1280. DOI: 10.1093/aob/mcp251 PMID: 19815570

Flowers, T. J., & Colmer, T. D. (2015). Plant salt tolerance: Adaptations in halophytes. *Annals of Botany*, *115*(3), 327–331. DOI: 10.1093/aob/mcu267 PMID: 25844430

Ghazaryan, K., Movsesyan, H., Gevorgyan, A., Minkina, T., Sushkova, S., Rajput, V., & Mandzhieva, S. (2020). Comparative hydrochemical assessment of groundwater quality from different aquifers for irrigation purposes using IWQI: A case-study from Masis province in Armenia. *Groundwater for Sustainable Development*, *11*, 100459. DOI: 10.1016/j.gsd.2020.100459

Ghazaryan, K. A., Margaryan, G. H., Minkina, T. M., Rajput, V. D., & Movsesyan, H. S. (2024). Chenopodium album L.: A promising NaCl accumulating and tolerant plant for mitigation of salinity stress. *Revista Brasileira de Botânica. Brazilian Journal of Botany*, *47*(3), 837–856. DOI: 10.1007/s40415-023-00974-w

Ghorbani, A., Omran, V. O. G., Razavi, S. M., Pirdashti, H., & Ranjbar, M. (2019). Piriformospora indica confers salinity tolerance on tomato (Lycopersicon esculentum Mill.) through amelioration of nutrient accumulation, K+/Na+ homeostasis and water status. *Plant Cell Reports*, *38*(9), 1151–1163. DOI: 10.1007/s00299-019-02434-w PMID: 31152194

Grotkopp, E., Rejmánek, M., & Rost, T. L. (2002). Toward a causal explanation of plant invasiveness: Seedling Growth and Life-History Strategies of 29 pine (Pinus) species. *American Naturalist*, *159*(4), 396–419. DOI: 10.1086/338995 PMID: 18707424

Hammer, E. C., Nasr, H., Pallon, J., Olsson, P. A., & Wallander, H. (2010). Elemental composition of arbuscular mycorrhizal fungi at high salinity. *Mycorrhiza*, *21*(2), 117–129. DOI: 10.1007/s00572-010-0316-4 PMID: 20499112

Hassan, M. A., Pacurar, A., López-Gresa, M. P., Donat-Torres, M. P., Llinares, J. V., Boscaiu, M., & Vicente, O. (2016). Effects of salt stress on three ecologically distinct plantago species. *PLoS One*, *11*(8), e0160236. DOI: 10.1371/journal.pone.0160236 PMID: 27490924

Heidari, N. M. (2011). Effects of salinity stress on growth, chlorophyll content and osmotic components of two basil (Ocimum basilicum L.) genotypes. *African Journal of Biotechnology*, *11*(2). Advance online publication. DOI: 10.5897/AJB11.2572

Hnilickova, H., Kraus, K., Vachova, P., & Hnilicka, F. (2021). Salinity Stress Affects Photosynthesis, Malondialdehyde Formation, and Proline Content in Portulaca oleracea L. *Plants*, *10*(5), 845. DOI: 10.3390/plants10050845 PMID: 33922210

Hossain, M. N., Sarker, U., Raihan, M. S., Al-Huqail, A. A., Siddiqui, M. H., & Oba, S. (2022). Influence of Salinity Stress on Color Parameters, Leaf Pigmentation, Polyphenol and Flavonoid Contents, and Antioxidant Activity of Amaranthus lividus Leafy Vegetables. *Molecules (Basel, Switzerland)*, *27*(6), 1821. DOI: 10.3390/molecules27061821 PMID: 35335185

Huang, Z., Long, X., Wang, L., Kang, J., Zhang, Z., Zed, R., & Liu, Z. (2012). Growth, photosynthesis and H+-ATPase activity in two Jerusalem artichoke varieties under NaCl-induced stress. *Process Biochemistry (Barking, London, England)*, *47*(4), 591–596. DOI: 10.1016/j.procbio.2011.12.016

Ismail, S., Rao, N. K., & Dagar, J. C. (2019). Identification, evaluation, and domestication of alternative crops for saline environments. In *Springer eBooks* (pp. 505–536). DOI: 10.1007/978-981-13-5832-6_17

Kaya, C., Tuna, A. L., Ashraf, M., & Altunlu, H. (2007). Improved salt tolerance of melon (Cucumis melo L.) by the addition of proline and potassium nitrate. *Environmental and Experimental Botany*, *60*(3), 397–403. DOI: 10.1016/j.envexpbot.2006.12.008

Lema, M., Ali, M. Y., & Retuerto, R. (2019). Domestication influences morphological and physiological responses to salinity in Brassica oleracea seedlings. *AoB Plants*, *11*(5), plz046. Advance online publication. DOI: 10.1093/aobpla/plz046 PMID: 31579110

Liao, Q., Gu, S., Kang, S., Du, T., Tong, L., Wood, J. D., & Ding, R. (2021). Mild water and salt stress improve water use efficiency by decreasing stomatal conductance via osmotic adjustment in field maize. *The Science of the Total Environment*, *805*, 150364. DOI: 10.1016/j.scitotenv.2021.150364 PMID: 34818800

Liu, L., & Wang, B. (2021). Protection of halophytes and their uses for cultivation of Saline-Alkali soil in China. *Biology (Basel), 10*(5), 353. DOI: 10.3390/biology10050353 PMID: 33922035

Liu, S., An, Z., & Lai, Z. (2024). Amaranth's growth and physiological responses to salt stress and the functional analysis of ATRTCP1 gene. *International Journal of Molecular Sciences, 25*(10), 5437. DOI: 10.3390/ijms25105437 PMID: 38791475

Loudari, A., Benadis, C., Naciri, R., Soulaimani, A., Zeroual, Y., Gharous, M. E., Kalaji, H. M., & Oukarroum, A. (2020). Salt stress affects mineral nutrition in shoots and roots and chlorophyll a fluorescence of tomato plants grown in hydroponic culture. *Journal of Plant Interactions, 15*(1), 398–405. DOI: 10.1080/17429145.2020.1841842

Luyckx, A., Lutts, S., & Quinet, M. (2023). Comparison of Salt Stress Tolerance among Two Leaf and Six Grain Cultivars of Amaranthus cruentus L. *Plants, 12*(18), 3310. DOI: 10.3390/plants12183310 PMID: 37765474

Mahdavian, K. (2023). Application of salicylic acid on chlorophyll, carotenoids, and proline in radish under salinity stress. *Proceedings of the National Academy of Sciences. India. Section B, Biological Sciences, 93*(4), 809–818. DOI: 10.1007/s40011-023-01484-1

Munns, R. (2005). Genes and salt tolerance: Bringing them together. *The New Phytologist, 167*(3), 645–663. DOI: 10.1111/j.1469-8137.2005.01487.x PMID: 16101905

Munns, R., & Tester, M. (2008). Mechanisms of salinity tolerance. *Annual Review of Plant Biology, 59*(1), 651–681. DOI: 10.1146/annurev.arplant.59.032607.092911 PMID: 18444910

Murad, M. A., Khan, A. L., & Muneer, S. (2020). Silicon in Horticultural Crops: Cross-talk, Signaling, and Tolerance Mechanism under Salinity Stress. *Plants, 9*(4), 460. DOI: 10.3390/plants9040460 PMID: 32268477

Murkute, A. A., Sharma, S., & Singh, S. K. (2006). Studies on salt stress tolerance of citrus rootstock genotypes with arbuscular mycorrhizal fungi. *Horticultural Science (Prague), 33*(2), 70–76. DOI: 10.17221/3742-HORTSCI

Nikalje, G. C., Srivastava, A. K., Pandey, G. K., & Suprasanna, P. (2017). Halophytes in biosaline agriculture: Mechanism, utilization, and value addition. *Land Degradation & Development, 29*(4), 1081–1095. DOI: 10.1002/ldr.2819

Omamt, E. N., Hammes, P. S., & Robbertse, P. J. (2006). Differences in salinity tolerance for growth and water-use efficiency in some amaranth(Amaranthusspp.) genotypes. *New Zealand Journal of Crop and Horticultural Science, 34*(1), 11–22. DOI: 10.1080/01140671.2006.9514382

Pulvento, C., Sellami, M. H., & Lavini, A. (2021). Yield and quality of Amaranthus hypochondriacus grain amaranth under drought and salinity at various phenological stages in southern Italy. *Journal of the Science of Food and Agriculture*, *102*(12), 5022–5033. DOI: 10.1002/jsfa.11088 PMID: 33448400

Rengasamy, P. (2006). World salinization with emphasis on Australia. *Journal of Experimental Botany*, *57*(5), 1017–1023. DOI: 10.1093/jxb/erj108 PMID: 16510516

Richards, L. A. (1954). Diagnosis and improvement of saline and alkali soils. *Soil Science*, *78*(2), 154. DOI: 10.1097/00010694-195408000-00012

Sage, R. F., Sage, T. L., Pearcy, R. W., & Borsch, T. (2007). The taxonomic distribution of C4 photosynthesis in Amaranthaceae sensu stricto. *American Journal of Botany*, *94*(12), 1992–2003. DOI: 10.3732/ajb.94.12.1992 PMID: 21636394

Salim, B. B., Hikal, M. S., & Osman, H. S. (2019). Ameliorating the deleterious effects of saline water on the antioxidants defense system and yield of eggplant using foliar application of zinc sulphate. *Annals of Agricultural Science*, *64*(2), 244–251. DOI: 10.1016/j.aoas.2019.12.008

Sayed, S. a. a. E., Hellal, F. A., El–Rab, N. G., & Zewainy, R. M. (2019). Ameliorative effects of potassium on the salinity stress in plants: a review. *Asian Journal of Soil Science and Plant Nutrition*, 1–15. DOI: 10.9734/ajsspn/2019/v4i230040

Silva, B. R. S., Batista, B. L., & Lobato, A. K. S. (2020). Anatomical changes in stem and root of soybean plants submitted to salt stress. *Plant Biology*, *23*(1), 57–65. DOI: 10.1111/plb.13176 PMID: 32841475

Sudhir, P., Pogoryelov, D., Kovacs, L., Garab, G., & Murthy, S. D. (2005). The Effects of Salt Stress on Photosynthetic Electron Transport and Thylakoid Membrane Proteins in the Cyanobacterium Spirulina platensis. *BMB Reports*, *38*(4), 481–485. DOI: 10.5483/BMBRep.2005.38.4.481 PMID: 16053716

Tao, R., Ding, J., Li, C., Zhu, X., Guo, W., & Zhu, M. (2021). Evaluating and screening of Agro-Physiological indices for salinity stress tolerance in wheat at the seedling stage. *Frontiers in Plant Science*, *12*, 646175. Advance online publication. DOI: 10.3389/fpls.2021.646175 PMID: 33868346

Tarin, T., Nolan, R. H., Medlyn, B. E., Cleverly, J., & Eamus, D. (2019). Water-use efficiency in a semi-arid woodland with high rainfall variability. *Global Change Biology*, *26*(2), 496–508. DOI: 10.1111/gcb.14866 PMID: 31597216

Temel, S., & Keskin, B. (2022). Determination of forage quality properties of plant parts in different amaranth varieties cultivated under irrigated and rainfed conditions. *Atatürk University Journal of Agricultural Faculty*, *53*(2), 122–132. DOI: 10.54614/AUAF.2022.1034402

Tıpırdamaz, R., Karakas, S., & Dikilitas, M. (2021). Halophytes and the future of agriculture. In *Springer eBooks* (pp. 2225–2239). DOI: 10.1007/978-3-030-57635-6_91

Tunccedil, M., Tuuml, Rk, Ruuml, Tunccedil, V., Buuml, Yildirim, N., Ccedil, V., & Iftccedil. (2011). Effect of salinity stress on plant fresh weight and nutrient composition of some Canola (Brassica napus L.) cultivars. *African Journal of Biotechnology*, *10*(10), 1827–1832. DOI: 10.5897/ajb10.1618

Wahid, A., & Ghazanfar, A. (2005). Possible involvement of some secondary metabolites in salt tolerance of sugarcane. *Journal of Plant Physiology*, *163*(7), 723–730. DOI: 10.1016/j.jplph.2005.07.007 PMID: 16616583

Wang, Y., Chen, X., Li, X., Song, Y., Wang, J., Huang, J., & Xue, R. (2021). Exogenous application of 5-aminolevulinic acid alleviated damage to wheat chloroplast ultrastructure under drought stress by transcriptionally regulating genes correlated with photosynthesis and chlorophyll biosynthesis. *Acta Physiologiae Plantarum*, *44*(1), 12. Advance online publication. DOI: 10.1007/s11738-021-03347-6

Wouyou, A., Gandonou, C., Komlan, F., Montcho, D., Zanklan, A., Lutts, S., & Gnancadja, S. (2017). Salinity Resistance of Five Amaranth (Amaranthus cruentus) Cultivars at Young Plants Stage. *International Journal of Plant and Soil Science*, *14*(3), 1–11. DOI: 10.9734/IJPSS/2017/31611

Xianzhao, L., Chunzhi, W., & Qing, S. (2013). Screening for Salt Tolerance in Eight Halophyte Species from Yellow River Delta at the Two Initial Growth Stages. *ISRN Agronomy*, *2013*, 1–8. DOI: 10.1155/2013/592820

Zeeshan, M., Lu, M., Sehar, S., Holford, P., & Wu, F. (2020). Comparison of biochemical, anatomical, morphological, and physiological responses to salinity stress in wheat and barley genotypes deferring in salinity tolerance. *Agronomy (Basel)*, *10*(1), 127. DOI: 10.3390/agronomy10010127

Chapter 7
The Influence of Dicranum scoparium Moss Extract on the Growth of Yeasts Candida guilliermondii NP-4

Gayane Henrik Semerjyan
http://orcid.org/0009-0006-0820-5088
Yerevan State Unisversity, Armenia

Inessa Henrik Semerjyan
http://orcid.org/0009-0008-3840-2642
Yerevan State University, Armenia

Syuzan Ashot Marutyan
Yerevan State University, Armenia

Hripsime Masis Ter-Sahakyan
http://orcid.org/0009-0003-0898-7129
Yerevan State University, Armenia

Hasmik Maenlest Karapetyan
Yerevan State University, Armenia

Seda Victor Marutyan
http://orcid.org/0000-0002-8019-9093
Yerevan State University, Armenia

ABSTRACT

This study aimed to evaluate the effect of D. scoparium moss extract on biomass accumulation, colony formation ability, and growth kinetics of yeasts Candida guilliermondii NP-4, as well as its effect on the activity of ATPases, antioxidant enzymes catalase, and superoxid dismutase. The antiradical activity of the moss extract was determined using the free radical method. The total content of phenolic substances and total content of flavonoids of moss was determined. D. scoparium moss extracts contained 17.53 µg GAE g-1 of total phenolic compounds and 7.88 µg QE mL-1 of total flavonoids, with an IC_{50} value of 51.487 µg mL-1. Treatment with the moss extract inhibited yeast colony formation and mitotic activity. The moss extract led

DOI: 10.4018/979-8-3373-1524-9.ch007

to a 64% reduction in malondialdehyde levels and enhanced superoxide dismutase activity by 43%. There was a significant increase in total and H+-ATPase activity in the cell homogenates and mitochondria. The data obtained prove the potential application of D. scoparium extract as a natural antifungal agent with additional antioxidative properties.

INTRODUCTION

Figure 1.

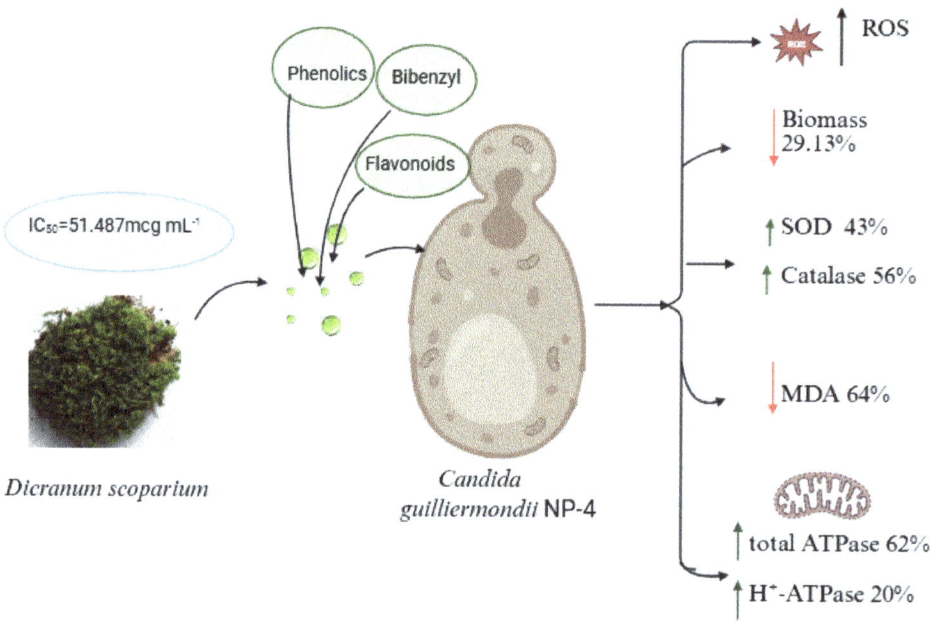

In recent years, there has been a concerning increase in fungal infections caused by *Candida* species, particularly in newborns in medical settings. This trend underscores the challenges associated with resistance to conventional antifungal medications, such as fluconazole and voriconazole, which can also cause significant side effects and contribute to drug resistance. To address these issues, researchers have turned their attention to natural products, particularly those derived from bryophytes such as mosses and liverworts. Bryophytes have been discovered to produce a diverse array of compounds with potent antifungal properties, including bibenzyls, sesquiterpenoids, steroids, acetophenones, stilbenes, and essential oils. These natural compounds offer several advantages, including lower environmental impact owing to biodegradability and potentially reduced toxicity compared to synthetic drugs

(Alyousef., 2021). The advance in technology have played a crucial role in identifying and characterizing phytochemicals across different bryophyte species worldwide. Furthermore, the induced production of these phytoconstituents under specific stress conditions highlights their potential as effective antimicrobial agents. Among the identified compounds, bibenzyls and bisbibenzyls exhibited notable inhibitory effects against various fungal strains. Particularly noteworthy are the marcantine macrocyclic bis-bibenzyls, which exhibit robust biological activity. These compounds are not only relevant in medicine but also haved applications in agriculture and horticulture. Overall, the exploration and utilization of natural products from bryophytes represent a promising avenue in the ongoing quest to effectively combat fungal infections. This approach is becoming increasingly significant as the challenges of drug resistance grow and the limitations of current antifungal treatments become more apparent (Commisso et al., 2021).

Bryophytes can be found in all types of habitats, except oceans. The necessary conditions for colonization and diversification on land were provided by the variety of biochemical adaptations. For example, the potency to form close interdependance with bacteria and fungi allows, access to limited and dispersed nutrients. It has been established that arbuscular mycorrhizal fungi regularly flowering plants, liverworts, and hornworts (Rimington et al., 2018). Lacking the mechanical defenses typical of leaf-bearing plants, which are protected from pathogenic microbes by a cuticle or shell, bryophytes are characterized by a poikilogenetic life cycle, for example, the secretion of phenolic compounds from moistened thalli effectively inhibits the germination of fungal spores that sometimes land on their surface, so it has allowed them to survive for over 350 million years. Thus, bryophytes, which are necessarily tend to synthesize secondary metabolites capable of limiting biotic stressors, are comparatively unchanged by microbial diseases, albeit they typically grow in moist areas and are capable of producing constitutive or inducible small molecules with antimicrobial properties. Therefore, they represent a valuable source of bioactive compounds: a number of terpenoids (the largest group of secondary metabolites of bryophytes, numbering more than 1600 described structures) (Ludwiczuk and Asakawa, 2020) and isolated aromatic compounds (flavonoids, phenylpropanoids, benzoids and bibenzyl derivatives). Some of these showed important antimicrobial potential: several liverwort species contained metabolites, such as Marchantin A and lunularin, which show expressed antifungal activity. Some species contain prenylphenyl ethers, which have a mild fungicidal effect. Bryophytes have an extensive chemical arsenal, with over 3000 different metabolites. The most representative class, with numbering 2200 different molecules, is terpenoids, followed by phenolic compounds (several hundred) and other molecules such as saccharides, lipids, and nitrogen- and sulfur-containing compounds (Martinez et al., 2022; Martínez-Abaigar et al., 2021). Over the decades, the number of antifungal compounds found in bryophytes has steadily increased with focused research efforts. Biological activity has been reported from both crude plant extracts in various solvents and purified compounds. In the first case, the effectiveness of a bryophyte extract in exerting a fungicidal/fungistatic effect is highly dependent on its balsamic period, as well as on seasonal, territorial (latitude, altitude and substrate composition) and climatic changes, which specifically affect the phytochemical composition of plant tissues and, in turn, determine their pharmacological potential (Ludwiczuk and Asakawa, 2020).

In recent years, the importance of searching for plants with antimicrobial, and antifungal activity has increased owing to growing concerns about the increase in infections caused by antibiotic-resistant microorganisms. Some authors have proven the fungicidal and insecticidal effect of bryophyte extracts, which is due to the presence of flavonoids (Uyar et al, 2025). Bryophytes have adapted their biochemistry to synthesize various compounds to combat pathogenic bacteria and fungi. Numerous studies have focused on isolating and identifying antibacterial compounds from bryophytes responsible for their antimicrobial effects. These include: macrocyclic (bis)bibenzyls such as marchantin A, perrottetin, and bibenzylunulari; glycosylated flavonoids like luteolin-7-O-neohesperoside, apigenin-7-O-triglycoside, and saponarin; flavonoids such as apigenin and lucenin-2; sesquiterpenoids (Cianciullo et al., 2022) which are notable in medicine and pharmacology. These bioactive secondary metabolites of plant origin serve as alternative compounds for the treatment of diseases such as cancer and type 2 diabetes. Many flavonoids also interact with ATP-dependent drug transporters involved in multidrug resistance in cancer cells, making them potential candidates for overcoming resistance to cytostatic drugs. The antibacterial activity of flavonoids is structure-dependent, particularly influenced by substitutions in the aromatic rings (Semerjyan et al., 2020).Yeast *Candida guilliermondii*, as single-celled eukaryotes, whose pathogenicity, virulence, and taxonomy have been under recent investigation for details, is considered uncommon, with low incidence even among immunocompromised hosts. While these yeasts exhibit lower innate virulence than *C. albicans*, their role in causing serious pathologies, particularly fungemia and deep-seated infections in cancer patients, has been highlighted in the literature. Moreover, these species are notably more prone to multidrug resistance than other *Candida* species. Therefore, investigations into *Candida guilliermondii* are crucial for suppressing its survival and preventing their development during complex and combined pathologies. The composition and structure of yeasts closely resemble those of higher eukaryotes; they share common morphological and functional properties and exhibit similarities in genetic structure. Additionally, yeasts can multiply rapidly and can be cultivated under precisely reproducible and controlled conditions, making them highly valuable in applied microbiology and biotechnology. Therefor, yeasts are increasingly employed as a convenient model for studying eukaryotic cells in various research applications (Marutyan et a.l, 2021).

This study imed to evaluate the effect of the extract obtained from *D. scoparium*, a common moss species in Armenia, on biomass accumulation, colony formation ability and growth kinetics of *C. guilliermondii* NP-4, as well as on the antioxidant enzymes and ATPase activity. The data obtained can be used as antioxidants for the treatment of chronic metabolic disorders, in the fight against infectious diseases, and as antifungal agents against pathogens that affect fruits and vegetables during storage.

MATERIALS AND METHODS

Plant Materials, Extract Preparation

The bryophyte *D. scoparium* (Hedw) was collected from Armenia (Gyulagarak Arboretum, at a height of 1450 m). The plants were identified by Dr. A.Poghosyan (Depertment of Botany and Mycology, Yerevan State University, Armenia) and deposited in the Takhtadjan Herbarium of the Department of Botany and Mycology, YSU (Vouchers no.13452). Collected bryophytes (100g. were collected, they were used in replicates). The mosses were dried and extracted under cold conditions (20 g in 200 ml of 80% water-ethanol solution). Filtered extracts were used for the experiments.

Determination of Antiradical Activity of Moss Extract

The antiradical activity of the alcoholic extract of moss was determined by the free radical method (Cianciullo et al, 2022), using the following concentrations: 1000, 500, 100, 50, 10 mg mL^{-1}. Catechin was used as a positive control. The sample solution contained 125 μL (1 mM) DPPH, 375 μL ethanol and 500 μL of test-solution (extract or catechin with different concentrations (1000 - 10 μg mL^{-1}). The test-solution was replaced by ethanol in the negative control sample. The absorbance was measured at the wavelength of 517 nm using a spectrophotometer Genesys 10S UV-Vis (Thermo Scientific, USA). Antiradical activity was calculated using the following formula:

Antiradical activity (%)=(Ac-As)/(Ac x 100),

where:

Ac – control (optical density of DPPH, excluding the test solution),
As - the optical density of the test samples.

As a result, IC_{50} was determined, which is the concentration of the solution under the influence of which 50% of the DPPH contained in the solution is reduced.

Quantitative Determination of Total Phenolic Compounds and Flavonoids

The total content of phenolic compounds in the D. scoparium moss extract was measured using the F–C method. (Genwali et al., 2013). The total flavonoid content of the moss was determined using the colorimetric aluminum chloride ($AlCl_3$) assay (Tadevosyan et al, 2025).

Investigated Strain and Growth Conditions, Growth Dinamics, Colony Formation, and Biomass Accumulation of *Candida guilliermondii* NP-4

Yeast cells were grown in a liquid medium of the following composition: 3.8 mM $(NH_4)_2HPO_4$, 17.5 mM $NH_4H_2PO_4$, 1 mM K_2SO_4, 1.6 mM $MgSO_4$ x $7H_2O$, 100 mM D-glucose, (pH=5.5) (Marutyan et al., 2021) in a shaker (BioSan Shaker-Incubator ES-20/80, Latvia, 200-250 rpm). Separation of biomass from the culture medium was carried out by centrifugation at a controlled temperature.speed of 3500 x g, for 10 minutes. The amount of biomass was determined spectrophotometrically (GENESYS 10S UV-VIS, λ= 590 nm). Yeast growth kinetic parameters were measured over the first 24 hours by recording the optical density of the culture medium hourly. The specific growth rate of yeast, μ, was calculated during the logarithmic phase where the optical density changes linearly with time, using the equation $dN/dt = \mu N(1 - N/Nmax)$, where N is the biomass amount of the yeasts at time t, N_{max} is the maximum amount of biomass (Marutyan et al., 2021). The cell doubling time, t_d, was determined using $\mu=\ln 2/t_d$. The duration of the latent (lag) phase was assessed graphically by identifying the point of intersection of the tangent with the abscissa axis (growth time), occurring within the segment corresponding to the logarithmic growth phase (Olivares-Marin et al., 2018).

To count the formed colonies, various dilutions of the original yeast sample were prepared until achieving a dilution of 10^7. Subsequently, 0.1 mL of yeast suspension from test tubes with this dilution was inoculated onto Petri dishes containing 2% wort agar and placed in a thermostat at 30°C for 48 hours, allowing macrocolonies visible to the naked eye to develop.

C. guilliermondii NP-4 strain was used. The yeast was stored at a temperature of 4 ^{0}C, on the wort agar solid medium. Before each experiment, yeast cells were inoculated into test tubes containing wort agar at 30^{0}C for 48 h (Marutyan et al, 2021).

Investigation of the Yeast Growth Inhibition Zones Under the Influence of Moss Extract

Moss extract concentrations of 10, 15 and 20 μg mL^{-1} were tested by placing them into wells on agar plates inoculated with microorganisms. Control plates without extract were also prepared. After allowing for diffusion at a low temperature for 60-90 min., the plates were incubated at 30°C for 24 h. After incubation the inhibition zones were observed on the agar plates.

Determination of the Activities of Antioxidant Enzymes

In yeast homogenates, catalase activity (EC 1.11.1.6) was assessed using the method described by Marutyan et al. (Marutyan et al., 2024). The amount of hydrogen peroxide (H_2O_2) and hydrogen peroxide degraded - milligram of protein (μmol H_2O_2 mg^{-1} protein) were determined. Superoxide dismutase (SOD) activity was determined by its ability to inhibit the rate of adrenaline autoxidation. SOD activity was determined based on the inhibition of adrenaline autooxidation, recorded by the increase in absorption at 480 nm and a temperature of 30 °C. The reaction was carried out in 50 mM/L sodium carbonate buffer, pH 10.2, and initiated by the addition of 0.1 mM/L adrenaline (Marutyan et al., 2024).

Quantitative Determination of MDA in Yeasts

To prepare the yeast homogenate, 0.025 M Tris–HCl buffer (pH 7.4) was used in a ratio of 3 ml of buffer per 1 g of yeast biomass. For protein precipitation, 1 ml of 17% YMC was added to 2.5 ml of the extract, centrifuged for 10 min at 4000 rpm. 1 ml of 0.8% aqueous thiobarbituric acid (TBA) solution was added to 2 ml of the supernatant was heating for 10 min. At high temperature, in an acidic environment, malonic dialdehyde forms a trimethine complex with thiobarbituric acid, the OD of which is measured spectrophotometrically at a wavelength of 532 nm. The amount of malonic dialdehyde was determined by the formula $D = \varepsilon cl$, where: D – optical density, c – concentration of MDA, l – thickness of the cuvette, ε – extinction coefficient of the dyed trimethine complex ($\varepsilon = 1.56 \times 10^5$ cm^{-1} M^{-1}) (Wu et al., 2024).

Mitochondrial Isolation From Yeast and Assessment of ATPase Activity

Mitochondria were isolated from yeast, frozen (-10°C) biomass of *C. guilliermondii* NP-4 was pressed using Avestin EmulsiFlex-C3. The yeast homogenate was subjected to differential centrifugation in a 0.25 M sucrose solution as follows: first at 3,000

× g for 20 minutes, and the resulting supernatant was further centrifuged at 15,000 × g for 15 minutes. The resulting supernatant was centrifuged at 30,000 × g for 20 minutes, and the pellet was considered the mitochondrial fraction. ATPase activity in yeast homogenates and isolated mitochondria was determined by measuring the amount of inorganic phosphate (Pi) released during the hydrolysis of 5 mM ATP at pH 7.5, as described by Marutyan et al. (2024). The reaction mixture contained 50 mM Tris–HCl buffer (pH 7.5) and 1 mM $MgSO_4$. Enzyme activity was expressed as µg Pi per mg protein. Quantification of Pi was performed spectrophotometrically according to the cited method. N,N′-Dicyclohexylcarbodiimide (DCCD, 10 µl/ml) was used as a specific inhibitor of yeast F_0F_1-ATPase. (Marutyan et al., 2024)

Statistical Analysis of Data

Statistical analyzes were conducted using GraphPad Prism 8 software (GraphPad Software, Inc., USA). A p-value of less than 0.05 was considered statistically significant. Results are expressed as means ± standard deviation (SD). All reagents used in the study were of analytical grade.

RESULTS AND DISCUSSION

Phytotherapy occupies a significant place in traditional medicine worldwide, with widespread use of herbal remedies for treating various ailments. There is a growing trend to associate the phytochemical constituents of plants with their pharmacological activities (Joshi et al., 2022). In recent years, natural antioxidants derived from medicinal and edible plants have gained significant attention for their potential to reduce the risk of neurological diseases associated with oxidative stress. These antioxidants hold promise for improving overall health and well-being through their antioxidant and therapeutic properties. Natural phenolic compounds of plant origin are particularly valued for their redox properties, which enable them to act as reducing agents, hydrogen donors, and singlet oxygen quenchers (Muscolo et al., 2024).

Bryophytes have a large amount of bioactive compounds, including sesquiterpenoids, bibenzyls, and phenols, thereby they are well-adapted to oxidative stress. Among these, phenols - non-enzymatic antioxidants - play a key role in mitigating oxidative damage. Of all the metabolites, phenolic compounds are the most abundant in mosses. They function as electron donors in the case of vacuolar peroxidase enzymes or as activators in the case of other antioxidant enzymes, thereby contributing to cellular protection against oxidative stress. *D. scoparium*, a species of the *Dicranaceae* family, can serve as a natural source of secondary metabolites with redox properties. The predominant phenol in bryophytes is p-coumaric acid, a key precursor in the biosynthesis of phenolic acids and flavonoids (Commisso et al., 2021; Marija et al., 2023). In general, bryophytes produce a variety of secondary metabolic compounds that serve as a defense mechanism against pathogenic microbes and insects.

The Total Phenolic and Flavonoid Composition of *D. scoparium* Moss Extracts

The *D. scoparium* moss extract examined in this study contains secondary antioxidant metabolites, which exhibit significant antioxidant activity. The overall phenolic content of moss extract of *D. scoparium* was determined by a gallic acid calibration curve. It has shown to be 17.53 GAE μg mL^{-1}. Using a quercetin calibration curve, it has been shown that the overall flavonoid content in the moss extract was determined to be QE 7.88 μg mL^{-1} as indicate at Table 1. The extract of *D.scoparium* exhibits an IC_{50} of 51.487 μg mL^{-1} (Table 1) (Shivaprakash et al., 2019).

Table 1. The amount of phenolic compounds and flavonoids in the aqueous extract of D.scopatium moss. (n=5, $p < 0.05$)

Total fenols (μg GAE mL^{-1})	Total flavonoids (μg QE mL^{-1})
17.53±0.96	7.88 ±0.24

The Antifungal Activity of *D.scoparium* Moss Extracts

The antifungal activity of moss extracts was assessed using the agar well diffusion method (Marutyan et al., 2024). Moss extracts were tested at concentrations of 10 - 20 μg mL^{-1} by placing them into wells on agar plates that had been previously inoculated with microorganisms (refer to Table 2). No yeast growth was observed where the moss solution had dripped onto the plate surfaces. Zones of inhibition of yeast cell growth of varying sizes appeared, depending on the concentration of the moss extract (Fig.1). As indicated in Table 2, there is a direct correlation between the concentration of moss extract (c) and the size of the inhibition zones and the extent of fungal growth inhibition on the agar plates.

Figure 2. The diameters (r) of growth inhibition zones of C. guilliermondii NP-4 yeast by various concentrations of aqueous extract of D.scoparium (a - 20 μg mL^{-1} - r=15mm; b – c=15 μg mL^{-1} – r=13.5mm; c- c=10 μg mL^{-1} – r=10mm, d-control).

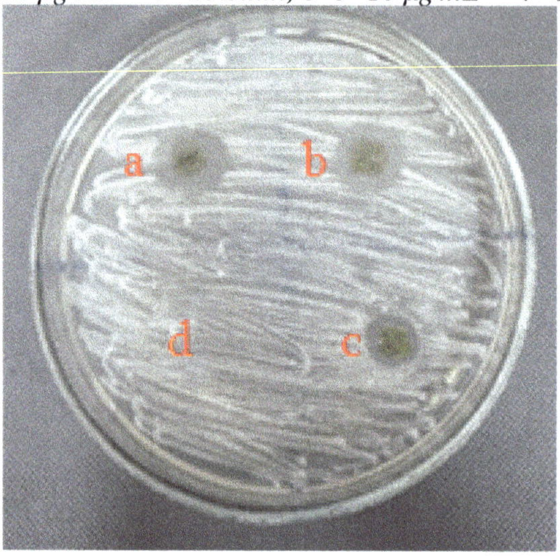

Table 2. Influence of varying concentrations of D. scoparium aqueous extract on the growth of C. guilliermondii NP-4 (n = 5, p < 0.05).

Moss extract concentration, μg mL^{-1}	Inhibition zones	
	Diameter, mm	Surface, mm^2
10	10.0 ± 0.5	78.5
15	13.5 ± 0.6	143.0
20	15.0 ± 0.9	176.6

Growth Dynamics of *C.guilliermondii* NP-4 Yeast Influneced by *D. scoparium* Extract

The aqueous extract of *D. scoparium* demonstrated antifungal activity, particularly in its effects on the growth dynamics of yeast *C. guilliermondii* NP-4. The growth kinetic parameters of yeasts in the presence of moss extract (10 – 500 μg mL^{-1}) were determined over the first 24 hours, encompassing the lag, logarithmic, and stationary phases of the life cycle (Figure 3). All concentrations of the aqueous moss extract inhibited yeast growth. Specifically, the presence of the moss extract initially extended the lag phase by 2 hours.

Figure 3. Growth dynamics of yeasts C.guilliermondii NP-4 influenced by D.scoparium extract: native yeasts, • added 85 µg mL-1 moss extract

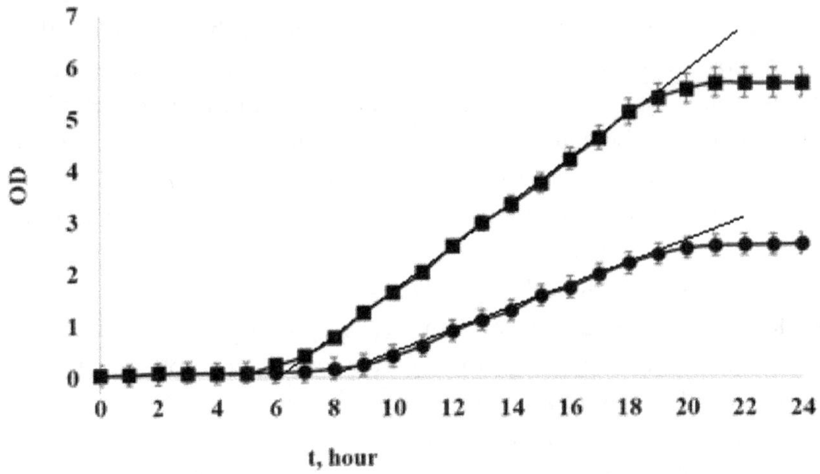

The duration of the lag phase was determined graphically by drawing a tangent to the linear growth part of the curve until it intersected the X-axis. In the presence of *D. scoparium* moss extract, the lag phase of yeast cell cycle lasted 8 hours, compared to the typical 6 hours for native yeast *C. guilliermondii* NP-4. This extension of the lag phase is known in the literature as delayed mitosis or G2 block (Navasardyan et al., 2017). Consequently, it was observed that under the influence of moss extract, the normal progression of the yeast cell cycle was disrupted, with cells accumulating in the G2 phase. Such an effect of lag phase delay is also observed in the yeast life cycle under the influence of other stressors such as X-radiation, millimeter waves, nitrogen starvation, etc. (Marutyan et al., 2024). Thus, this process represents an adaptive response of yeast cells to unfavorable environmental conditions in order to survive. A two-hour delay in the yeast's lag phase may allow cells to adapt to unfavorable conditions, mitigate the negative effects of the extract's secondary metabolites, and transition into the logarithmic growth phase. During this phase, the specific growth rate of both treated and untreated yeast was nearly identical. However, total biomass accumulation in the stationary phase, as well as colony-forming ability, decreased with increasing extract concentration.

The moss extract hinders the accumulation of yeast biomass (Table 3): yeast biomass decreased with increasing the aqueous moss extract concentration. Correspondingly, as the concentration of moss extract increases, the colony-forming ability of yeasts decreases, too.

It is noteworthy that a concentration of 85 μg mL^{-1} of moss extract exhibits a fungistatic effect: the accumulated biomass and colony forming ability of yeasts decrease by approximately 50%. In contrast, a concentration of 500 μg mL^{-1} of moss extract reduces yeast biomass and colony forming ability by more than 70%.

Furthermore, there were observable changes in the kinetic parameters of yeast growth. Although the specific growth rate during the linear subphase of the logarithmic phase is slightly reduced in the presence of moss extract compared to native yeast, the difference is not significant, amounting to no more than 7%.

This decline in growth rate corresponds to an increase in cell doubling time, which is minimal for native yeasts. Such a minor difference in the specific growth rate can be attributed to the delay in the lag phase.

The logarithmic phase of growth in yeasts exposed to moss extract begins 2 hours later than in native yeasts, allowing time for the repair of damage caused by the moss extract before normal multiplication ensues.

The influence of an aqueous extract of *D. scoparium* moss on processes of lipid peroxidation and the activities of antioxidant enzymes of yeast *C.guilliermondii* NP-4.

To gain a deeper understanding of the antifungal effect of the moss extract, the amount of malondialdehyde was determined in the yeast as an indicator of lipid peroxidation intensity. It has been shown that the moss extract suppressed lipid peroxidation in yeast cells.

Table 3. Growth properties of C.guilliermondii NP-4 yeast exposed to different concentration of D.scoparium moss extract (n=3, P<0,05)

Concentration of the extract, μg mL^{-1}	**Number of colonies, x 10^8 colony mL^{-1}**	**Yeast biomass, mg mL^{-1}**	**Kinetic properties of growth in the log phase of growth**		**Effect on growth inhibition, %**
			Yeast specific growth rate (μ), h^{-1}	**Cell doubling time (t_d), h**	
0	15.31 ± 0.09	1.27 ± 0.08	0.075 ± 0.02	9.24	-
10	13.3 ± 0.08	1.11± 0.06	0.070 ± 0.04	9.9	12.5
25	12.94 ± 0.09	1.08± 0.08	0.072 ± 0.05	9.96	15.35
50	10.84 ± 0.07	0.90 ± 0.02	0.071 ± 0.10	9.76	29.13

continued on following page

Table 3. Continued

Concentration of the extract, µg mL^{-1}	Number of colonies, x 10^8 colony mL^{-1}	Yeast biomass, mg mL^{-1}	Kinetic properties of growth in the log phase of growth		Effect on growth inhibition, %
			Yeast specific growth rate (µ), h^{-1}	Cell doubling time (t_d), h	
85	6.92 ± 0.01	0.58 ± 0.03	0.070 ± 0.04	9.9	54.72
170	4.96 ± 0.02	0.41± 0.02	0.070 ± 0.03	9.9	67.55
500	4.22 ± 0.01	0.35 ± 0.02	0.071 ± 0.09	9.76	72.44

The data obtained indicate that the addition of 25 µg mL^{-1} of aqueous moss extract to the nutrient medium for yeast growth resultsed in an approximately 50% reduction in the MDA level in the yeast homogenate (Figure 4a).

Figure 4. The effect of different concentrations of the extract of D. scoparium moss on yeast C.guilliermondii NP-4: a - the lipid peroxidation proceses, b - catalase activity, c - SOD activity (n=3, p<0.05)

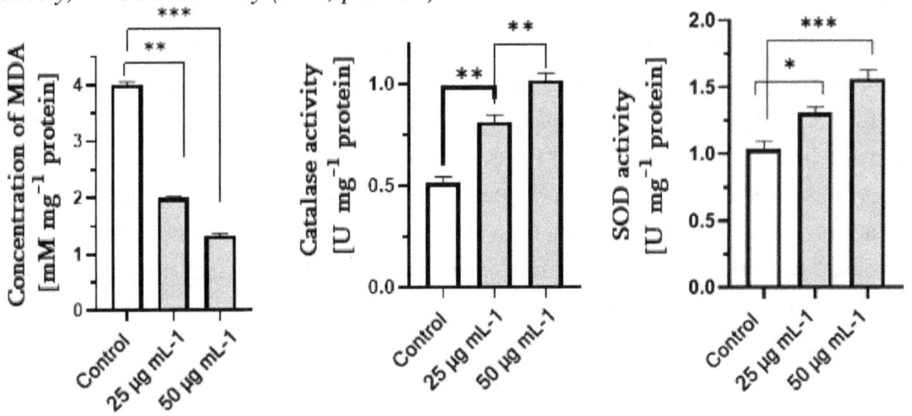

When the concentration of the moss extract was increased to 50 µg mL-1, the MDA level decreased by 64%. These findings suggest that the moss extract inhibits lipid peroxidation, likely due to the presence of biologically active secondary metabolites.

Oxidative stress in living cells arises when there is an imbalance between the production of ROS and reactive nitrogen species (RNS) in living cells. This imbalance can lead to damage to cellular components such as proteins, lipids, and nucleic acids, ultimately resulting in cell death. The body's enzymatic and non-enzymatic antioxidant systems - such as SOD and catalase - play a crucial role in maintaining the balance between ROS and antioxidants. It was found out that the moss extract significantly suppressed lipid peroxidation processes and tgereby the oxidative stress in yeast cells. The activities of enzymes SOD and catalase were stimulated by the moss extract, but the most notable changes observed in yeast strains grown in medium containing moss extract at a concentration of 50 µg mL^{-1}. Under these conditions, catalase activity increased by 56% and SOD activity by 43% compared to the control. The stimulation of antioxidant enzyme activity leads to a reduction in the levels of ROS, which may also contribute to the alleviation of oxidative stress in the yeast cells (Kozlov et al., 2024).

Influence of an aqueous extract of *D. scoparium* moss on total and H^+-ATPase activity in *C.guilliermondii* NP-4 yeast homogenate and mitochondria.

This led to a reduction in ROS levels and a corresponding decrease in MDA content. The presence of bioactive compounds such as terpenes, sesquiterpenes, and bibenzyls in the moss extract can induce oxidative stress in fungal cells by increasing the levels of reactive oxygen species (Nandy, Dey., 2020; Kakali et al., 2023). Overall, these effects contribute to enhancing the antioxidant defenses of yeast and stimulating their adaptive cellular response to oxidative stress. These protective antioxidant reactions, which occur in stressed cells, are highly energy-intensive and require additional energy expense. ATP is the primary energy source for all living cells, and in the production of ATP the critical role play ATPases. Thereby we next recorded the change in ATPase activity in yeast cells exposed to moss extract. The results showed that in yeast homogenate treated with 25 µg mL^{-1} aqueous moss extract, ATPase activity increased by ˜3 times, and H^+-ATPase activity – by ˜3.5 times. In the mitochondria of yeasts, the ATPase activity increased by approximately 48%, and H^+-ATPase activity – by 2 times (Figure 5). In the presence of 50 µg mL^{-1} of moss extract in yeasts growth medium, the ATPase activities in yeast homogenates were increased by ˜3times; in mitochondria, the total and H^+-ATPase activity increased by 62% and 20%, respectively.

Mitochondrial ATPsynthase is a membrane-bound enzyme complex and a member of the F-type ATPsynthase family, which is found in mitochondria (eukaryotes), chloroplasts (plants), and the plasma membrane of bacteria. Yeast mitochondrial ATPase is often used as a model system for studying mitochondrial function, bioenergetics, and mitochondrial-related diseases (Marutyan et al., 2024).

Figure 5. Changes in C.guilliermondii NP-4 yeast ATP-ase activity under the influence of moss extract (n=3, p<0.05): a-homogenat, b-mitochondria

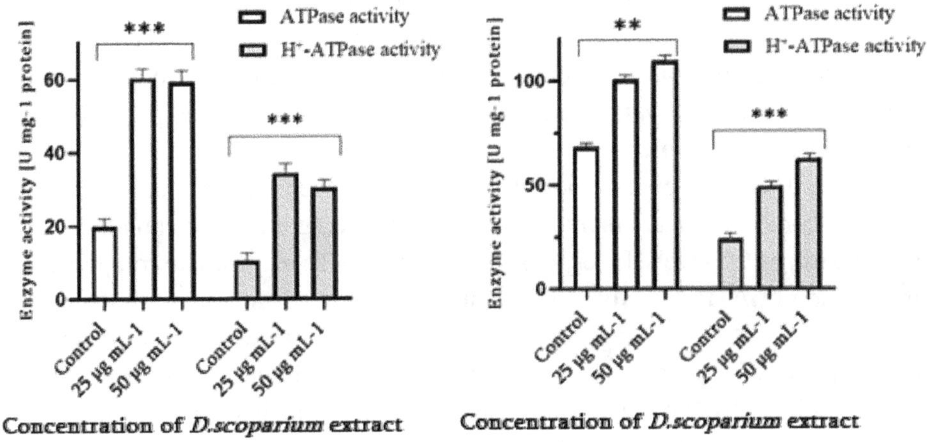

Summarizing the data we obtained, we can draw the following conclusions. At a moss extract concentration of 50 mg/mL, which caused approximately a 30% reduction in biomass, the antioxidant system of yeast (SOD, and catalase) was activated under oxidative stress conditions, enabling the yeast to survive. Activation of the antioxidant system requires additional energy, which is fulfilled through ATP hydrolysis. To meet the energy demands of cells, ATP synthase acts as an ATPase, hydrolyzing ATP into ADP and inorganic phosphate.

Research on yeast ATPase is crucial for understanding cellular energy metabolism and drug interactions, especially given that mitochondrial dysfunction is implicated in various diseases, including neurodegenerative and metabolic disorders. Fungal infections - particularly candidiasis and aspergillosis - remain major causes of mortality in humans. Fungal growth and survival depend heavily on ATP production, primarily through oxidative phosphorylation. A key enzyme in this process is FoF_1-ATP synthase. Inhibition of H^+-ATPase has been shown to play a significant role in antifungal activity (Li et al., 2020).

It can therefore be concluded that the increase in ATPase activity observed in this study is a result of the antifungal action of bioactive compounds present in the moss extract. These compounds may compromise the integrity of yeast cell membranes by increasing membrane permeability, leading to leakage of ions and intracellular contents, osmotic imbalance, and ultimately cell death. Certain compounds - such as the flavonoid luteolin (Dziwak et al., 2022), terpenoids, and bibenzyls - can also interfere with mitochondrial function by disrupting ATP synthesis, the respiratory chain, and oxidative phosphorylation, thereby depleting the cell's energy reserves.

The results of this study demonstrate that aqueous bryophyte extracts inhibit the growth of C. guilliermondii NP-4 in a dose-dependent manner. The diverse bioactive compounds found in bryophytes (e.g., terpenoids and phenolics) make them a promising bioresource in addressing drug resistance. These findings could be applied in the development of antifungal therapies, the treatment of chronic metabolic disorders, and the preservation of agricultural products by targeting pathogens that affect fruits and vegetables during storage.

Moss extracts also contain polysaccharides and phenolic compounds that can alter lipid composition and membrane fluidity, inducing mild mitochondrial stress and increasing ROS levels. In response to this stress, cells may initiate compensatory upregulation of ATPase activity to maintain pH balance and energy homeostasis (Quejada et al., 2024). Changes in membrane fluidity are known to influence H^+-ATPase activity by enhancing the proton gradient and improving substrate availability. Additionally, bioactive compounds in moss may activate plasma membrane H^+-ATPase by modulating its conformation, thereby enhancing proton transport and ATPase activity (Zhao et al., 2024).

CONCLUSION

Extracts from *D. scoparium* moss exhibit antioxidant activity. The aqueous extracts of the moss showed antifungal activity against yeasts *C.guilliermondii* NP-4, inhibiting mitotic processes and impairing colony formation.

Moss extracts can be utilized as antifungal agents on a large scale through advanced techniques, such as tissue culture. This approach allows for the environmentally safe treatment of candidiasis and the elimination of plant pathogens.

REFERENCES

Adele, Mariateresa, Giulio, & Mariateresa. (2024). Oxidative Stress: The Role of Antioxidant Phytochemicals in the Prevention and Treatment of Diseases. *Int J Mol Sci., 25*(6), 3264. DOI: 10.3390/ijms25063264

Agnieszka & Asakawa. (2020). *Terpenoids and Aromatic Compounds from Bryophytes and their Central Nervous System Activity. Current Organic Chemistry, 24*(1), 113 – 128. DOI: 10.2174/1385272824666200120143558

Alyousef, A.A. (2021) Antifungal Activity and Mechanism of Action of Different Parts of Myrtus communis Growing in Saudi Arabia against Candida Spp. *Hindawi Journal of Nanomaterials*. DOI: 10.1155/2021/3484125

Cianciullo, P., Maresca, V., Sorbo, S., & Basile, A. (2022). Antioxidant and Antibacterial Properties of Exstracts and Bioactive Compounds in Bryophytes. *Applied Sciences (Basel, Switzerland), 12*(1), 160. DOI: 10.3390/app12010160

Commisso, M., Guarino, F., Marchi, L., Muto, A., Piro, A., & Degola, F. (2021). Bryo-Activities: A Review on How Bryophytes Are Contributing to the Arsenal of Natural Bioactive Compounds against Fungi. *Plants, 10*(2), 203. DOI: 10.3390/plants10020203 PMID: 33494524

Dziwak, M., Wróblewska, K., Szumny, A., & Galek, R. (2022). Modern Use of Bryophytes as a Source of Secondary Metabolites. *Agronomy (Basel), 12*(6), 1456. DOI: 10.3390/agronomy12061456

Genwali, G. R., Acharya, P. P., & Rajbhandari, M. (2013). Isolation of gallic acid and estimation of total phenolic content in some medicinal plants and their antioxidant activity. *Nepal Journal of Science and Technology, 14*(1), 95–102. DOI: 10.3126/njst.v14i1.8928

Güray, U., Sibilay, K., Nazife, A., Gülden, V., & Mihrican, B. (2025). Investigation of the Antimicrobial and Antifungal Activities of Some Moss Species (*Scleropodium touretii* (Brid.) L.F.Koch, *Hypnum cupressiforme* Hedw. var. *lacunosum* Brid., *Brachythecium glareosum* (Bruch ex Spruce) Schimp.). *Anatolian Bryology., 11*(1), 72–82. DOI: 10.26672/anatolianbryology.1718549

Inesa, Gayane, Henrikh, & Armen. (2020). Antibacterial Properties and Flavonoids Content of Some Mosses Common in Armenia. *Iranian Journal of Pharmaceutical Sciences, 16*(4), 31-42. www.ijps.ir

Kozlov, A. V., Javadov, S., & Sommer, N. (2024). Cellular ROS and Antioxidants: Physiological and Pathological Role. *Antioxidants, 13*(5), 602. DOI: 10.3390/antiox13050602 PMID: 38790707

Li, J.-L., Wie, L.-L., Chen, C., Liu, D., Gu, Y.-Q., Duan-Mu, J.-X., Chen, G.-T., & Song, Y. (2020). Bioactive Constituents from the Bryophyta Hypnum Plumaeforme. *Chemistry & Biodiversity, 17*(12), e2000552. DOI: 10.1002/cbdv.202000552 PMID: 33098214

Marija, V. (2023). Analysis of the Qualitative and Quantitative Content of the Phenolic Compounds of Selected Moss Species under NaCl Stress. *Molecules (Basel, Switzerland), 28*(4), 1794. DOI: 10.3390/molecules28041794 PMID: 36838781

Martínez-Abaigar, J., & Núñez-Olivera, E. (2021). Novel biotechnological substances from bryophytes. In Sinha, R. P., & Häder, D.-P. (Eds.), *Natural Bioactive Compounds* (pp. 233–248). Academic Press., DOI: 10.1016/B978-0-12-820655-3.00011-2

Marutyan, S., Marutyan, S., Navasardyan, L., Hovnanyan, K., & Trchounian, A. (2021). Changes in growth kinetic parameters, morphology and mitotic activity of yeast Candida guilliermondii exposed to the low-intensity waves of 51.8-GHz frecuency. *Archives of Microbiology, 203*(6), 3707–3714. DOI: 10.1007/s00203-021-02336-0 PMID: 33938972

Navasardyan, L., Marutyan, S., Hovnanyan, K., & Trchounyan, A. (2017). Survival and changes in morphology, mitotic and metabolic activity of yeast *Candida guilliermondii* exposed to X-irradiation. *Indian Journal of Biochemistry & Biophysics, 54*, 273–280. http://nopr.niscair.res.in/handle/123456789/43366

Olivares-Marin, I. K., González-Hernández, J. C., Regalado-Gonzalez, C., & Madrigal-Perez, L. A. (2018, September 30). Saccharomyces cerevisiae Exponential Growth Kinetics in Batch Culture to Analyze Respiratory and Fermentative Metabolism. *Journal of Visualized Experiments, 58192*(139). DOI: 10.3791/58192 PMID: 30320748

Priyanka, S., Gautam, N., Arora, A., & Ghanghoriya, H. (2024). A Review of Phytochemical Constituents and their Pharmacological Activities. *Frontiers in Health Informatics, 14*(2), 2645-2658. DOI: 10.63682/fhi2726

Quejada, L. F., Hernandez, A. X., & Chitiva, L. C. (2024) Unmasking Antifungal Might:Anacardium occidentale Leaf ExtractUnleashes Mitochondrial Disruption,Thwarting Candida albicans withHeightened Reactive Oxygen Species. *Preprints*.org. DOI: 10.20944/preprints202405.0579.v1

Rimington, W. R., Pressel, S., Duckett, J. G., Field, K. J., Read, D. J., & Bidartondo, M. I. (2018). Ancient Plants with Ancient Fungi: Liverworts Associate with Early-Diverging Arbuscular Mycorrhizal Fungi. *Proceedings. Biological Sciences, 285*(1888), 20181600. DOI: 10.1098/rspb.2018.1600 PMID: 30305437

Rudramurthy, Paul, Chakrabarti, Mouton., & Meis. (2019). Invasive Aspergillosis by *Aspergillus flavus*: Epidemiology, Diagnosis, Antifungal Resistance, and Management. DOI: 10.3390/jof5030055

Samapika & Dey. (2020). Bibenzyls and bisbybenzyls of bryophytic origin as promising source of novel therapeutics: pharmacology, synthesis and structure-activity. *Daru Journal of Pharmaceutical Sciences, 28*(2), 701–734. DOI: 10.1007/s40199-020-00341-0

Seda, Karapetyan, Khachatryan, Muradyan, Marutyan, Poladyan, & Trchounian. (2024). The antimicrobial effects of silver nanoparticles obtained through the royal jelly on the yeasts *Candida guilliermondii* NP-4. *Scientific Reports, 14*, 19163. DOI: 10.1038/s41598-024-70197-w

Sen, K. (2023). Recent Advances in the Phytochemistry of Bryophytes: Distribution. *Structures and Biological Activity of Bibenzyl and Bisbibenzyl Compounds Plants, 12*, 4173. DOI: 10.3390/plants12244173 PMID: 38140499

Sidonia, Fuentes, & Carballo. (2022). Antioxidant Activity, Total Phenolic Content and Total Flavonoid Content in Sweet Chestnut (Castanea sativa Mill.) Cultivars Grown in Northwest Spain under Different Environmental Conditions. *Foods, 11*, 3519. DOI: 10.3390/foods11213519

Silvard, T., Babayan, A., & Sahakyan, N. (2025). Antioxidant and Antibiotic-modulatory Potential of *Teucrium polium* Ethanolic Extract Against Genetically-modified Antibiotic-resistant *E. coli* Strains. *Majallah-i Mikrub/Shinasi-i Pizishki-i Iran, 19*(2), 107–117. DOI: 10.30699/ijmm.19.2.107

Supriya, J. (2022). Bryophytes as a safeguard of fruits from postharvest fungal diseases: A Review. *Natr Resour Human Health, 2*(3), 327–334. DOI: 10.53365/nrfhh/145476

Wu, Yuwen, Wenjie, & Feng. (2024). Measurement of cellular MDA content through MTBE-extraction based TBA assay by eliminating cellular interferences. *Journal of Pharmaceutical and Biomedical Analysis, 248*, 116332 DOI: 10.1016/j.jpba.2024.116332

Zhao, C.-R., You, Z.-L., & Bai, L. (2024). Fungal Plasma Membrane H^+-ATPase: Structure, Mechanism, and Drug Discovery. *Journal of Fungi (Basel, Switzerland)*, *10*(4), 273. DOI: 10.3390/jof10040273 PMID: 38667944

Chapter 8
Recellularized Rat Liver Scaffold:
An In Vitro Model for Biosafety Assessments

Olga Kalinovskaia
http://orcid.org/0000-0001-8013-4874
L.A. Orbeli Institute of Physiology, Armenia

Mikhail Poliakov
http://orcid.org/0009-0009-4863-6028
L.A. Orbeli Institute of Physiology, Armenia

Zaruhi Karabekian
https://orcid.org/0000-0002-8693-7288
L.A. Orbeli Institute of Physiology, Armenia

ABSTRACT

At present, fully decellularized animal organ scaffolds are widely used in research for cell biology, tissue engineering, and regenerative medicine. These scaffolds serve as enhanced 3D models, providing natural microenvironments that support and guide cell interactions. In this study, the authors aimed to generate a reproducible 3D in vitro liver model for drug metabolism and toxicity testing that is based on decellularized rat liver scaffolds and to validate the model's suitability for toxicity tests. To restore liver-specific functions, various rat and human cell lines were used, including hepatocytes and macrophages. To date, the authors have achieved high reproducibility in the preparation of the scaffolds and demonstrated low cytotoxicity of liver scaffold material. Immunogenicity assays and experiments on recellularization of rat liver scaffolds are ongoing. The model's metabolic functionality is to

DOI: 10.4018/979-8-3373-1524-9.ch008

be confirmed by assessing liver-specific markers. The development of such a model should enable efficient assessment of drug candidates while reducing the need for animal testing.

INTRODUCTION

To date, organ transplantation remains a crucial field in regenerative medicine because of the high demand. In the case of end-stage liver diseases, such as cirrhosis, chronic viral hepatitis, hepatocellular carcinoma, injuries related to alcohol addiction, or inborn metabolic disorders, liver transplantations remain necessary and, in some cases, appear to be the only solution for patient survival. Increasing morbidity and a shortage of donor organs worldwide, as well as the need for lifelong immunosuppressive therapy, make it necessary to search for alternatives (Sabetkish et al., 2015). One possible alternative is tissue engineering (TE). In recent TE studies, attempts have been made to generate artificial, de novo functional liver tissue (Dai et al., 2022; Hussein et al., 2024). Nowadays, liver-like organoids are created that recapitulate the complex functions of the whole liver by combining cells with artificial scaffolds using mechanical techniques and/or chemical agents. These organoids are used as important in vitro models in research on liver biology and drug metabolism (Robertson et al., 2018). These models have been shown to increase the efficiency of drug screening, thereby accelerating preclinical studies in the drug development process. Therefore, such models may significantly reduce the production cost of certain drugs in the near future (Godoy et al., 2013).

Under suitable conditions, artificial liver-like organoids can exhibit high stability and display some liver-specific functions, including regeneration (Chhabra et al., 2022). Despite these advances, there is still a need for a tissue-engineered liver model that would reproduce the in vivo microenvironment of the liver (Robertson et al., 2018). Modern TE techniques have enabled the production of an extracellular matrix (ECM) that allows for cell adhesion, increases rates of their proliferation and maturation, and is suitable for cell differentiation inducement and the development of intercellular contacts. Research has demonstrated that interactions between cells and the surrounding ECM can promote cell migration and tissue-specific gene expression (Sabetkish et al., 2015).

The advances we have mentioned can be achieved through various methods, including decellularization of human or animal liver with subsequent repopulation of the resulting ECM scaffolds with human liver cells. During the decellularization, cell antigens are removed, and immunogenicity is significantly reduced, thus improving the biocompatibility of the derived scaffolds (Kasravi et al., 2023). Repopulation of a 3D decellularized liver ECM with liver cells in coculture can provide an in vitro model that is suitable for biomedical assays and drug screening (Thanapirom et al., 2021).

Improving such ECM-based models is crucial if we are to achieve long-term maintenance of biocompatible organ-like systems. Although there is a wide array of research works dedicated to improving 2D and 3D in vitro models for bioassessments, such as hepatotoxicity testing (see, e.g., the review article by Kim et al., 2025), the potential of the models based on decellularized natural scaffolds remains underexplored. Hybrid 3D printed models and organ-on-a-chip systems would be perfect for research of this type, but their use in practice remains difficult because such approaches are expensive and technically demanding.

In our study, we suggest simpler, cheaper, and well-established experimental models that are still well-suited for answering the main research questions. These models make it possible to obtain reliable data while keeping the methods cost-effective.

The aim of the study was to reproducibly generate a 3D recellularized liver in vitro model for drug metabolism and toxicity testing. Because of regulatory restrictions on the procurement and handling of human tissue in nonclinical experimental biology laboratories, human biomaterials could not be used in this study. The final model is based on decellularized rat liver scaffolds (DRLSs) repopulated with animal or human liver cell cocultures. The reproducibility and stability of the model is to be evaluated on the basis of several parameters, such as cell viability in coculture upon the 3D recellularization and the ability of the cells to exhibit liver-specific functions. The model's functionality would also be evaluated in hepatotoxicity tests, using acetaminophen as a test agent with well-known hepatotoxic effects, such as glutathione depletion, protein adduct formation, and mitochondrial dysfunction leading to cell death (Ramachandran & Jaeschke, 2019), and metamizole as a test agent causing oxidative stress-mediated hepatocyte necrosis, alteration of intracellular enzyme levels (Ciftel et al., 2024), and immune-mediated responses, including various allergic reactions and, in one clinical case, a severe lobular inflammation (Krisai et al., 2019).

MATERIALS AND METHODS

Animals

Adult outbred albino rats (Rattus norvegicus) derived from the Wistar strain were provided by L.A. Orbeli Institute of Physiology animal facility. We randomized the animal selection by first assigning numbers to animals suitable for the research and then by choosing the number via a random number generator.

A total of 18 animals were used to establish the reproducibility of the liver vein perfusion, liver extraction, and decellularization process at the desired level (≥90%). For each type of biochemical assessment, a minimum of three scaffolds were used to demonstrate the test's required reproducibility. We empirically determined that the amount of ECM derived from one rat liver was sufficient to conduct one to two experiments involving scaffold homogenate or recellularization, so three animals were used to assess the scaffold cytotoxicity, six for immunogenicity assessments, and six more for the recellularization experiments on thin sections.

Liver Preparation

An adult rat, male or female, 150–300 g in weight, was taken and euthanized using an overdose of pentobarbital solution (100 mg/kg body weight, intraperitoneally) administered by injection or using chloroform by inhalation following the regulations of the European Convention for the Protection of Vertebrate Animals Used for Experimental and Other Scientific Purposes (Council of Europe, 1986; Louhimies, 2002). The rat's body was then fixed on a surgical table in dorsal recumbency. Its abdominal cavity was opened, and then an intravenous injection of 0.5 ml (2,500 IU) of heparin solution was made into the vena cava inferior to prevent blood clotting (Figure 1).

Figure 1. Rat Autopsy (Left Panel) and Heparin Injection into Vena Cava Inferior (Right Panel)

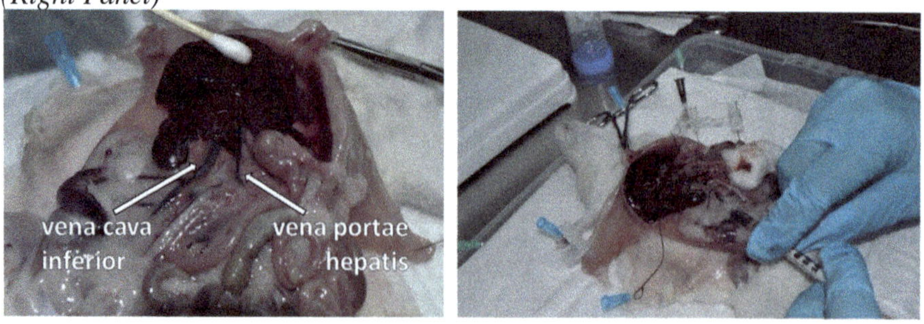

Depending on the diameter of the veins, the vena portae hepatis (VPH) was then perfused using a 22- or 24-gauge catheter, and 5 ml of the solution containing 4.5 ml of normal saline and 2,500 IU of heparin was infused into the VPH (Figure 2), and then the remaining blood was washed out of the liver with 20–50 ml of normal saline. The catheter was then fixed using two ligatures, each tied on both the vein and the tissue surrounding the VPH, and the liver was extracted by dissecting the spleen and vessels connecting it to other organs of the abdominal cavity and to the diaphragm (Dubko et al., 2021).

Figure 2. Ligature Fixation on the Catheter (Left Panel) and Heparin Injection into the Vena Portae Hepatis (Right Panel)

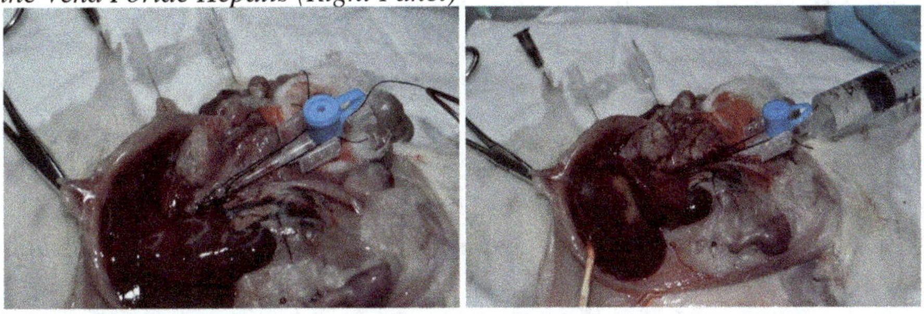

Note. Infusion of the heparin solution causes lightening of the liver lobes.

Washing Process

Among existing washing agents, including ionic and non-ionic detergents, we chose sodium dodecyl sulphate (SDS) for its ability to effectively remove cells and their fragments from tissues without washing collagen out, despite some important ECM components being affected by SDS (Gilbert et al., 2006; Mendibil et al., 2020). Preliminary experiments showed effective liver tissue decellularization by SDS at concentrations of 0.1%–1% and good conservation of the ECM main structure.

The completely isolated rat liver was placed into a container and suspended via the catheter to the connector embedded into the container lid. Then the catheter opening was connected with a peristaltic pump. SDS solution (0.1% in distilled water) and distilled water were infused at a speed of 2–5 L/hr, switched from one to another after every 0.5 L of the fluid. The total amount of SDS solution used for a liver scaffold was 3–5 L, depending on the decellularization process, which was controlled visually. The washing was conducted until all lobes became uniformly semi-transparent without any spots of remaining tissue discernible, with the blood vessel architecture clearly visible. At the final step, 2 L of distilled water were infused to wash the remaining SDS out of the scaffold. To confirm the absence of cell material in the scaffold, tissue samples were taken before and after the decellularization and fixed in 10% formalin, dehydrated in alcohols of ascending concentrations, and embedded in paraffin. The prepared sections with a thickness of 5 μm were stained with hematoxylin and eosin to reveal cell and nuclei remnants.

To ensure the scaffold quality after the washing, the residual DNA was measured therein. The DNA was extracted from lyophilized DRLS fragments using the QIAamp DNA Mini Kit™ (QIAGEN, 2024) according to the manufacturer's instructions. The total DNA content was quantified using a Quant-iT™ dsDNA BR Assay Kit (Thermo Fisher Scientific, 2023) according to the supplied protocols to determine the residual DNA in the decellularized scaffolds.

SDS Cytotoxicity

Residual SDS content in the DRLS washes was quantified spectrophotometrically by chloroform extraction of the methylene blue-SDS complex with a Synergy H1 plate reader, as previously described (Ghorbani et al., 2022). A standard calibration curve was generated using SDS solutions prepared in distilled water at 11 defined concentrations (0, 0.0005, 0.001, 0.002, 0.004, 0.008, 0.016, 0.032, 0.064, 0.128, 0.256, and 0.512 mg/ml). Preliminary cytotoxicity assays in HeLa and HepG2 cells confirmed a dose-dependent response to SDS and were used to determine 70% cell viability and 90% cell viability thresholds. Scaffolds with residual SDS levels below 0.216 mg/ml, corresponding to 90% cell viability, were considered acceptable and selected for subsequent experiments.

We evaluated the cytotoxicity of the residual detergent in the decellularizing solution according to ISO 10993-5:2009 (International Organization for Standardization, 2009). An MTT (thiazolyl blue tetrazolium bromide; Sigma-Aldrich, M5655) cytotoxicity test that quantifies cell viability on the basis of metabolic activity was used. In this assay, the yellow, water-soluble MTT dye is reduced by metabolically active cells to insoluble blue-violet formazan, the amount of which is directly proportional to the number of viable cells and is quantified photometrically after dissolution in alcohol.

Eight SDS concentrations (1.024, 0.512, 0.256, 0.128, 0.064, 0.032, 0.016, and 0.008 mg/ml) prepared in Dulbecco's Modified Eagle Medium (DMEM), with 10% fetal bovine serum and 1% penicillin/streptomycin mixture, were tested. According to ISO 10993-5 criteria, SDS concentrations that maintained cell viability ≥70% were classified as noncytotoxic, and the concentration corresponding to 90% viability was additionally calculated as indicative of negligible cytotoxicity. Cells cultured without SDS served as negative controls, and cells treated with 0.09% Triton X-100 solution in phosphate-buffered saline (PBS) served as positive controls.

After 24 hr of cell culture, SDS dilutions in the culture medium were added to the wells and incubated for an additional 24 hr. The reaction was stopped by adding the solvent. Absorbance was measured at 570 nm using a spectrophotometric plate reader, and metabolic activity was calculated using Equation 1:

$$Metabolic\ activity\ (\%) = \frac{OD_{SDS} \times CD_{SDS}}{OD_{ctrl-} \times CD_{ctrl-}} \times 100\%\ (1),$$

where OD denotes the average optical density, and CD is the average cell density for the corresponding conditions.

Cell Culture and Scaffold Homogenate Preparation

To assess the potential immunogenicity of the obtained scaffold material, we performed initial cytotoxicity testing to define nontoxic concentration ranges (cell viability ≥90%). Three human cell lines—HeLa (cervical carcinoma), HEK (human embryonic kidney), and HepG2 (hepatocellular carcinoma)—were used for evaluation. Cells were seeded into 96-well tissue culture–treated plates at a density of 20,000–30,000 cells/well in triplicate. After overnight incubation at 37 °C in a humidified incubator with 5% CO_2 to allow for attachment, cells were treated with scaffold homogenate. The scaffold was mechanically homogenized using a BioMasher® II system (Nippi, Inc., 2024) and diluted in complete Roswell Park Memorial Institute (RPMI) 1640 medium (Gibco™, Cat. No. 11875093) supplemented with 10% fetal bovine serum, 1% penicillin-streptomycin, and 2 mM L-glutamine to achieve final concentrations of 6.25, 12.5, 25, and 50 mg/ml. The homogenate was added to the wells, and cells were incubated for 24 hr at 37 °C under standard culture conditions.

Scaffold Cytotoxicity

Wet scaffold pieces (0.1–0.2 g) were washed three times in sterile saline containing double-strength penicillin–streptomycin by centrifugation (5 min, 8,600 × g), then reweighed. Complete DMEM was added at a 19:1 volume-to-weight ratio, and samples were incubated overnight at 37 °C. The supernatant, collected after centrifugation, was used for indirect cytotoxicity testing. For direct testing, scaffold pieces were homogenized in fresh DMEM using a BioMasher II to obtain a concentration of 50 mg/ml. Both the overnight supernatant and homogenate were applied to HeLa, HEK, and HepG2 cells (~50,000 cells/well) in duplicate at 100 µl/well and tested in the concentrations 0, 6.25, 12.5, 25, and 50 mg/ml in DMEM. After 24 hr incubation, cytotoxicity was assessed by MTT assay, and the results were expressed as mean optical density (triplicates) normalized to the control.

Data for the cytotoxicity part and for other paragraphs of the "Results and Discussion" section are presented as mean ± standard deviation (*M*±*SD*). We performed group comparisons using a one-way analysis of variance. When only two groups were compared, we used an unpaired two-tailed Student's t test. We considered p values $<.05$ statistically significant. All analyses and graphing were conducted using Microsoft Excel software.

Obtaining Primary Cell Cultures

Rat blood mixed with heparin solution was sampled from the vena cava inferior at the time of liver extraction. The whole human blood was purchased from the Blood Bank of Yeolyan Hematology and Oncology Center (Yerevan, Armenia). Both rat and human peripheral blood mononuclear cells (PBMCs) were processed as described in Bøyum's (1976) article.

Interleukin-2 Measurement in Human Lymphocytes

We conducted a set of five experiments to evaluate levels of cytokines in the culture medium of lymphocytes after exposure to the DRLS homogenate at concentrations of 0.5–25 mg/ml of wet scaffold. An enzyme-linked immunosorbent assay was performed using the ELISA MAX™ Standard Set (BioLegend, 2024).

Proliferation Analysis by Carboxyfluorescein Succinimidyl Ester

Cells (Jurkat and human PBMCs) were harvested and resuspended at a density of 1×10^6 cells/ml in prewarmed PBS, then stained with carboxyfluorescein succinimidyl ester (CFSE) according to the kit protocol (BioLegend, 2024). A 5-μM working solution was freshly prepared by diluting the stock solution in PBS (Quah et al., 2007). Cells were incubated with CFSE at a final concentration of 5 μM for 10 min at 37 °C in the dark to allow covalent labelling of intracellular amine groups. The staining reaction was quenched by adding an equal volume of culture medium RPMI supplemented with 10% fetal bovine serum, followed by incubation for 5 min at room temperature. Labelled cells were washed twice with the complete culture medium to remove excess unbound dye. Next, cells were seeded in culture wells with various concentrations of rat liver scaffold material (0, 0.5, 0.9, 1.65, 3, 5.5, and 10 mg/ml) and incubated under standard culture conditions (37 °C, 5% CO_2) for 6 days.

After incubation, cells were harvested, labelled with T-cell marker CD3, and analysed by flow cytometry (CyFlow Cube 6, Partec). CFSE fluorescence intensity was measured in the FITC channel (530/30 nm bandpass filter) for CD3-positive cells. Nondivided cells displayed the highest fluorescence intensity, whereas each generation showed approximately half the fluorescence of the previous one.

We analysed the data using CyFlow software (Sysmex Partec, 2014) to determine the percentage of nondivided and divided cells, enabling assessment of scaffold-induced effects on cell proliferation (Quah et al., 2007).

Scaffold Slice Recellularization

The three cell lines used in scaffold cytotoxicity assessments - HEK 293, HeLa, and HepG2 - were also used for recellularization of the DRLS thin slices. These lines were chosen because one is a noncancer cell line (HEK293), and the other (HeLa) is a popular cancer cell line often used in cytotoxicity assessment. Once the HepG2 cell line was obtained, the results were confirmed using a relevant type of cell: liver cells.

All the three lines showed good attachment and viability during the first 48 hr of incubation (data not shown). The HepG2 line, which is of hepatocyte origin and retains some liver-specific functions, showed the best adherence and viability and was selected for further experiments.

Although the recellularization of the DRLS with human hepatocytes in coculture with other types of human liver cells was the ultimate goal, we selected the HepG2 culture for mastering the models and for the proof of principle. Working with HepG2 was also considered to be the best option because HepG2 cells demonstrate good attachment to the scaffold and quick proliferation, and because they are of less ethical concern compared with human primary hepatocytes. The next step to optimizing the model relies on animal primary hepatocytes being used for allogeneic and xenogeneic recellularization experiments. The experiments are currently being conducted using rodent primary hepatocytes, PBMCs, and mesenchymal stem cells.

To obtain DRLS slices suitable for recellularization, a scaffold piece with a size $<1\ cm^3$ was cut and transferred to a plate filled with optimal cutting temperature compound (Tissue-Tek®). The scaffold was then frozen at -25 °C and cut into slices 30- to 50-μm thick. The slices were transferred into 24-well plates and defrosted. Then, the remaining compound was washed out with sterile PBS, and the slices were incubated in complete DMEM culture medium for 2 hr. The HepG2 cells were seeded onto scaffolds afterwards at a concentration of approximately 10^5/well. (The plates were incubated at 37 °C, and the pictures were made after 24 and 72 hr of exposition.) A set of recellularized slices was selected for imaging.

Recellularized Scaffold Staining

After culturing on liver scaffold constructs for 2 and 6 days, the cells were subjected to a triple-staining procedure to visualize cell viability, nuclear morphology, and collagen content. Viable cells were stained with CellTracker Orange CMRA

(Thermo Fisher Scientific, C34551; excitation 548 nm, emission 576 nm), nuclei were labelled with 4',6-diamidino-2-phenylindole (DAPI; Thermo Fisher Scientific), and Sirius Red (DirectRed 80, Sigma Aldrich) was used to stain collagen of the ECM.

Cells were incubated with a 5-μM working solution of CellTracker Orange CMRA in complete DMEM at 37 °C for 30 min in the dark. After incubation, samples were washed twice with prewarmed medium to remove excess dye.

For nuclear visualization, samples were fixed with 4% paraformaldehyde for 10 min, followed by two washes in PBS. Samples were then incubated with 1 μg/ml DAPI in PBS for 10 min at room temperature, protected from light, and washed twice with PBS.

To detect ECM collagen, samples were processed for Sirius Red staining. Fixed samples were incubated with 0.1% Sirius Red in saturated aqueous picric acid for 1 hr at room temperature. After the staining, samples were rinsed in 0.5% acetic acid to remove unbound dye and thoroughly washed in distilled water. Stained samples were mounted in Mowiol mounting medium, coverslipped, and imaged using fluorescent and brightfield microscopy (Axiovert, Zeiss).

3D Scaffold Recellularization

We conducted experiments on 3D recellularization using the DRLSs and the perfusion system described earlier. Before cell seeding, DRLSs were infused with complete DMEM and then incubated overnight in a Petri dish filled with the same medium to ensure thorough wetting of the matrix and to remove air pockets from the vascular network. HepG2 cells (~5 million) suspended in complete DMEM culture medium were infused into the DRLS via the catheter, allowing homogeneous cell distribution through the preserved vasculature, and/or injected using a 5-ml syringe to increase local cell density.

Ethical Compliance

All the procedures involving animals were approved by the Institutional Animal Care and Use Committee (IACUC) of the L. A. Orbeli Institute of Physiology (Approved Protocol No. 29.08.2025/1) and were performed in accordance with international guidelines for animal care and use. Human blood samples were purchased from a public source (Blood Bank of Yeolyan Hematology and Oncology Center, Yerevan, Armenia).

RESULTS AND DISCUSSION

Decellularized Scaffolds Obtained

The decellularization method described shows high reproducibility and allows for the DRLS yield of >95% in case the perfusion is performed successfully. In our experience, perfusion with 3 L of 0.1% SDS alternated with distilled water, followed by washing with 2 L of distilled water to remove residual detergent, is sufficient to produce a fully decellularized scaffold (Figure 3) with minimal residual SDS content that is suitable for subsequent recellularization. Superior results, including reduced residual DNA content, were obtained by subjecting the scaffolds that had been decellularized according to the protocol described earlier to agitation for 1 hr three times, each in fresh distilled water, effectively clearing cell remnants from poorly perfused liver regions, such as liver lobe edges.

Figure 3. A Completely Decellularized Rat Liver with the Perfusion System Remaining

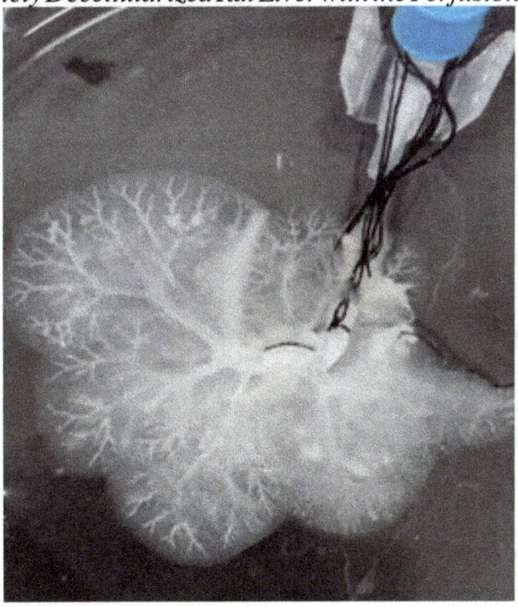

Note. The blood vessel architecture is clearly visible.

An examination of the sections made from the resulting scaffolds shows the completely decellularized liver tissue with no cell material visible (Figure 4).

Figure 4. Microphotographs of Native (Left Panel) and Completely Decellularized (Right Panel) Liver Sections

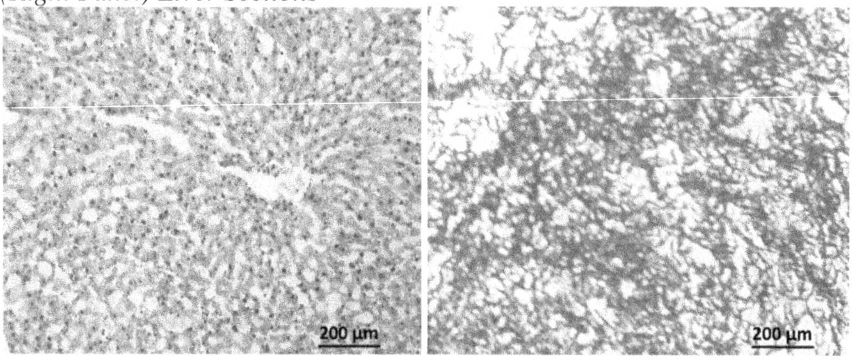

Note. Multiple black dots present in the left panel and absent in the right panel are cell nuclei (magnification 200×)

Residual SDS Cytotoxicity Assessment

The MTT assay demonstrated a clear, concentration-dependent decrease in cell viability in response to increasing SDS levels. SDS dilutions that maintained cell viability ≥70% were classified as noncytotoxic according to ISO 10993-5, and concentrations corresponding to 90% viability were interpreted as having negligible cytotoxicity and were used to define the acceptable residual SDS threshold for scaffold use. Threshold cytotoxic SDS concentrations were determined to be 0.328 mg/ml for 70% viability and 0.216 mg/ml for 90% viability. SDS concentration in all scaffolds tested was below the 90% viability threshold (0.216 mg/ml), ranging from 0.002 to 0.040 mg/ml. Thus, the cytotoxicity of residual SDS concentration in scaffold washing water was found to be <10%, indicating that the washing protocol used was appropriate and met the research objectives.

Scaffold Cytotoxicity

Indirect testing showed that the supernatant at the highest concentration did not significantly affect cell viability (data not shown). In contrast, the MTT assay performed with scaffold homogenates on HeLa, HEK293, and HepG2 cells demonstrated a clear, concentration-dependent reduction in metabolic activity. This indicates that the homogenate from the used scaffold demonstrated a noticeable cytotoxic effect. There were no substantial differences in sensitivity among the three cell lines. The cytotoxicity threshold, defined as 70% cell survival, was reached on average at approximately 40 mg/ml of wet scaffold for all of them, and the ED_{90} (effective dose causing a 10% reduction in viability) ranged from 5 to 20 mg/ml (Figure 5).

Figure 5. Dependence of the Viability of the Tested Cells on the Concentration of Rat Liver Scaffold Homogenate

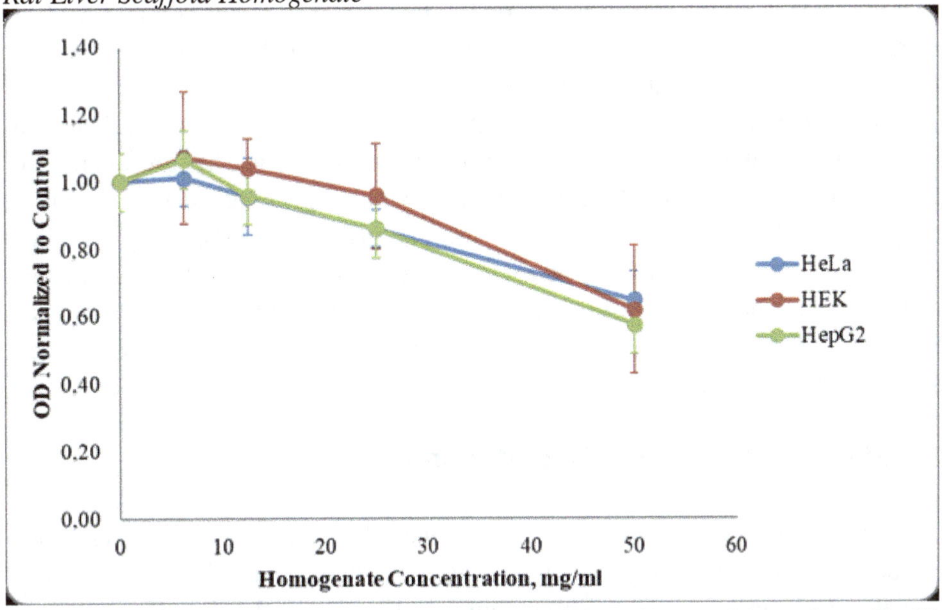

Note. Cervical carcinoma (HeLa), human embryonic kidney (HEK), and hepatocellular carcinoma (HepG2) cells were exposed to rat liver scaffold homogenate at the indicated concentrations and incubated for 24 hr. The viability of the cells was assessed with the MTT assay (thiazolyl blue tetrazolium bromide; Sigma-Aldrich, M5655). OD = optical density.

On the basis of these observations, we selected scaffold homogenate concentrations of ≤10 mg/ml for subsequent immunogenicity experiments to minimize confounding cytotoxic effects. The cytotoxic effect of the decellularized scaffold may be attributed to several groups of factors. Those include degradation products of the ECM, such as low-molecular–weight fragments of collagen, elastin, or glycosaminoglycans, which in turn can act as damage-associated molecular patterns and trigger an inflammatory response. In addition, residual DNA and RNA fragments, as well as protein antigens, may activate Toll-like receptors on cells, thereby inducing an immune reaction.

Interleukin-2 Level Evaluation

As a test of T-cell activation, we assessed Interleukin-2 (IL-2) levels in the culture supernatants. Although IL-2 levels showed a dose-dependent trend, the results were not reproducible. This suggests that the applied approach may not be suitable under the current experimental conditions. IL-2 alone without T-cell stimulation is insufficiently sensitive to scaffold-induced innate immune activation because it potentially misses innate-driven scaffold effects and residual DNA effects. IL-2 primarily drives adaptive T-cell proliferation and T-regulatory cell maintenance, peaking later in antigen-specific responses than the innate phase crucial for biomaterial evaluation. Thus, IL-2 is less sensitive for detecting material-triggered inflammation than pro-inflammatory cytokines.

To obtain more reliable information, IL-2 production is now being evaluated by intracellular cytokine staining with specific antibodies followed by flow cytometric analysis, which allows direct identification of cytokine-positive T cells. In parallel, enzyme-linked immunosorbent assay quantification of pro-inflammatory cytokines IL-1β and IL-6 is being implemented. These cytokines are secreted by macrophages and monocytes and help measure acute responses from innate immune cells directly because, among PBMCs, macrophages are activated before T cells and thus indicate stimulation or lack thereof.

Proliferation Level by CFSE Staining

The CFSE data indicate that both the presence of a scaffold and its concentration affect the proliferation of lymphocytes and Jurkat cells. Control samples without a scaffold (0 mg/ml) had a high percentage of nondivided cells, approximately 90%–99%, which means most of the untreated cells had not divided (Table 1).

Table 1. Proliferation Level of Carboxyfluorescein Succinimidyl Ester (CFSE)-Stained Human Lymphocytes and Jurkat Cells in Response to Exposure to Scaffold Homogenate

Scaffold Conc (mg/ml)	Nondivided (%, $M \pm SD$)		Divided (%, $M \pm SD$)	
	Jurkat	Human Lymphocytes	Jurkat	Human Lymphocytes
0.00	96.34±2.36	93.36±3.98	3.67±2.72	6.65±3.98
0.50	96.64±1.71	84.15±13.75	3.37±2.43	15.86±13.75
0.90	91.81±3.87	85.95±9.06	8.19±6.74	14.06±9.06
1.65	85.33±2.80	77.64±6.21	14.67±5.60	22.36±6.21
3.00	82.87±3.93	80.10±2.45	17.13±7.86	19.91±2.45
5.50	71.40±6.61	77.78±6.59	28.60±13.22	22.22±6.59
10.00	80.93±6.14	76.26±3.39	19.07±12.29	23.74±3.39

Note. CFSE-loaded Jurkat or human peripheral blood mononuclear cells (PBMC) were cultured with decellularized rat liver scaffold (DRLS) homogenates at the indicated concentrations for 6 days.

The percentage of divided cells is calculated as 100 minus the percentage of non-divided cells. Conc = concentration. M±SD = mean ± standard deviation.

In Jurkat cells, the percentage of divided cells increased gradually with scaffold concentration, from approximately 3%–4% at 0 mg/ml to roughly 15%–20% at 1.65-3 mg/ml and peaking at around 29% at 5.5 mg/ml. This pattern suggests a weak-to-moderate, concentration-dependent proliferative response at intermediate doses. The trend is not perfectly linear; at the highest scaffold concentration (10 mg/ml), the proliferation levels decrease, possibly because of scaffold toxicity at this concentration.

Primary human lymphocytes showed higher baseline division (approximately 7% at 0 mg/ml) and a more variable response, with proliferation increasing to approximately 16%–22% across 0.5–5.5 mg/ml, but with large standard deviations and no clear monotonic dose–response relationship. At 10 mg/ml, the fraction of divided lymphocytes decreased again.

A CFSE proliferation analysis showed only modest T-cell stimulation by the scaffold homogenate in both Jurkat cells and primary human lymphocytes, indicating low direct immunostimulatory potential of the scaffold material.

The slight reduction in proliferation at the highest scaffold dose suggests potential toxicity. These findings show that the scaffold environment can influence lymphocyte proliferation and may have important implications for immune system studies and TE.

Recellularized DRLS Slices

Our preliminary experiments showed that all the three cell cultures chosen—HeLa, HEK, and HepG2—showed good viability and adhesion to the ECM during the first week of culture maintenance. Among the cultures, HepG2 showed the best proliferation in the first 48 hr, covering all the visible ECM surface and reaching 100% confluence under the optical microscope at the end of the observation period (6 days). These results suggest that specific liver proteins, such as growth factors, may remain within the scaffold in substantial quantities, thereby promoting hepatic cell proliferation.

The next set of DRLS slices was made for recellularization with HepG2 cells, followed by the triple staining. Images made with the slices confirm the intensive attachment of the HepG2 cells to both the scaffold and to the culture plate well surface (Figure 6). DAPI staining shows the cell nuclei (Figure 7, Panel B). CellTracker Orange shows the cytoplasm of the living cells (Figure 7, Panel C). The overlapped image shows that most of the visible cells are live (Figure 7, Panel D). The scaffold is visible in all the modes because of its autofluorescence.

Figure 6. Hepatocellular Carcinoma (HepG2) Cells Attached to the Decellularized Rat Liver Scaffold (DRLS) and to the Culture Plate Well Surface, After 48 Hr of Incubation in Complete Dulbecco's Modified Eagle Medium (DMEM)

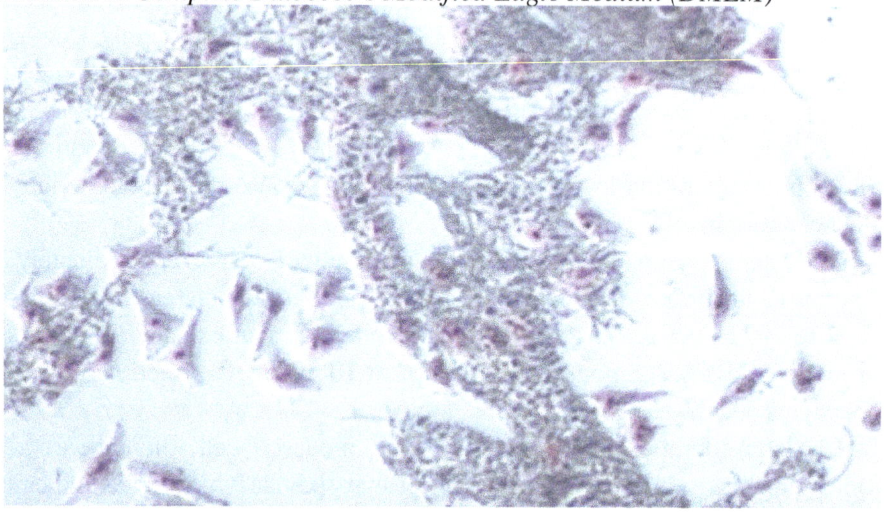

Note. Cells were stained with Sirius Red (DirectRed 80, Sigma Aldrich), magnification 200×.

We then estimated average cell coverage by HepG2 cells after 48 hr of incubation on the scaffold thin sections using ImageJ (v1.54; NIH, https://imagej.nih.gov/ij) software (Schneider et al., 2012). On the basis of five sets of fluorescent microscope images, we noted that the cell coverage was approximately 13.18%±4.25% of the scaffold surface area.

Figure 7. Hepatocellular Carcinoma (HepG2) Cell Viability After 48 Hr of Incubation on the Decellularized Rat Liver Scaffold

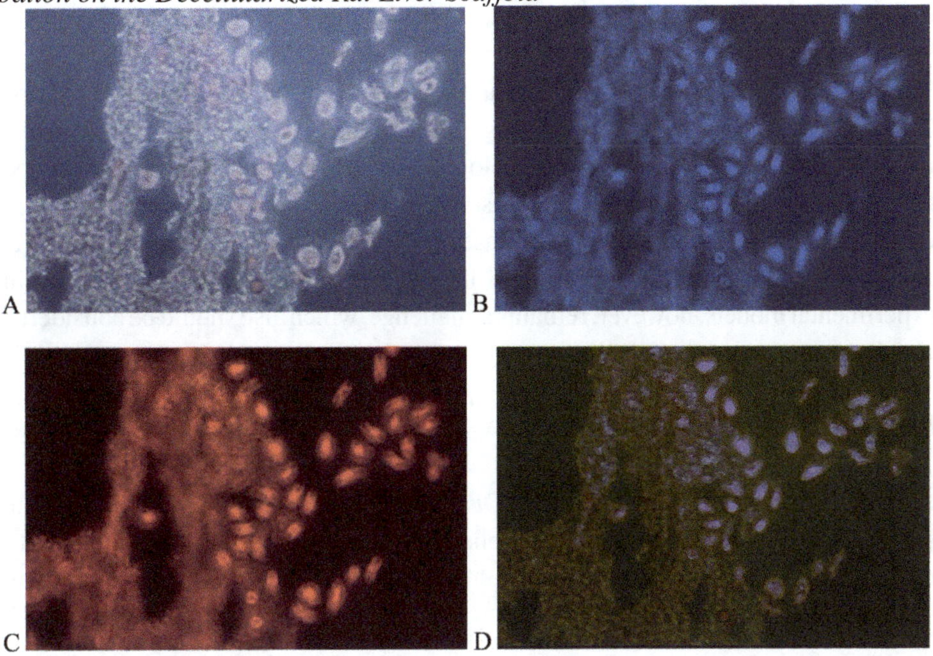

Note. Panel A: Brightfield image. Panel B: 4',6-diamidino-2-phenylindole (DAPI) staining highlights the cell nuclei. Panel C: CellTracker Orange CMRA staining highlights the cytoplasm of living cells. Panel D: Overlapped image. Magnification 100×.

Limitations of xenogeneic cells in scaffold recellularization mainly involve immunogenicity, incomplete cell integration, and challenges in maintaining cell viability and functionality within the scaffold. Effective recellularization depends on good cell–scaffold adhesion, survival, and proliferation. Xenogeneic cells often face problems like poor adhesion that is due to species-specific ECM differences, inadequate support from the scaffold microenvironment, and limited capability for appropriate differentiation. However, the human cell lines we used demonstrated good viability and cell adhesion on the rat liver scaffolds, allowing for the reproduction of the experiments on human cell cocultures.

Cell lines considered for developing the 3D recellularized model include commercially available human hepatocyte and endotheliocyte lines because endothelial cell involvement is known to improve hepatocyte functional maintenance (Salerno et al., 2011). Other cell lines, such as human liver stem-like cells and human primary monocytes, are also being considered. Although such a scheme lacks some types of liver-specific cells, such as Kupffer cells or stellate cells, it would still allow for evaluation of hepatocyte metabolic activity and of the whole model's viability. The translatability of the tests conducted using the model may be estimated by making comparisons with data from previous experimental and clinical research on primary liver cultures. Such an estimation would ideally be made before further attempts to improve the model by including primary liver cell cocultures therein.

The functional stability of the recellularized scaffolds depends on many factors, but usually remains at a high level during only a limited period of time. For example, in Butter et al.'s (2017) study, the physiological activity of primary hepatocytes used for recellularization was severely decreased after 1 week. In the case of immortalized cell cultures, the model longevity increases up to several weeks (see, e.g., Mazza et al., 2015). Prolonging the exhibition of liver-specific functions by hepatocytes in experimental models, however, remains a challenge, which also should be considered in subsequent experiments.

Recellularized 3D Scaffolds

The first set of experiments on 3D DRLS recellularization is still ongoing. The scaffolds recellularized with HepG2 cells are frozen after 3 days of incubation to make the slices suitable for evaluation of cell viability, adhesion, and proliferation.

CONCLUSION

The results of this study show that rat liver decellularization is a highly reproducible technique that is applicable in various TE studies. The DRLSs appear to be a suitable model for experiments on repopulation with xenogeneic cells that provides appropriate conditions for cell survival and adhesion. However, further assessments of the model's biosafety and additional recellularization experiments are needed for the use of the model in toxicological studies. Such coculture recellularization experiments will be conducted after good reproducibility of DRLS recellularization with liver cells in a mono culture is obtained.

As a future outcome, the resulting 3D recellularized model should be able to serve as a substitute for existing human liver ex vivo and in vitro models in hepatotoxicity tests that yield results that are translatable to native human liver physiology. Such a model would also reduce animal usage in drug or toxicity tests because a smaller sample size will be necessary to demonstrate the significance of the test results compared with classic in vivo tests. Furthermore, the objectives of this study required the use of animals only for tissue and organ sampling, which were performed after humane euthanasia, so the model may also prove to be a refinement improvement in terms of the 3Rs principle (Replacement, Reduction, Refinement; Russell & Burch, 1959) compared with any in vivo model for toxicity assessment on mammals. Other possible applications of the model include a useful platform for drug testing, disease modelling, toxicity screening, and the study of human hepatocyte functionality with perspectives for use in personalised medicine.

FUNDING STATEMENT

This research was supported by L. A. Orbeli Institute of Physiology.

COMPETING INTERESTS

The authors of this publication declare there are no competing interests.

REFERENCES

BioLegend. (2024). *CFSE Cell Division Tracker Kit: Protocol* (Cat. No. 9396). https://www.biolegend.com/en-us/products/cfse-cell-division-tracker-kit-9396

BioLegend. (2024). *ELISA MAX™ Standard Set Human IL-2: Data sheet and protocol* (Cat. No. 431801). https://www.biolegend.com/Files/Images/media_assets/pro_detail/datasheets/431801_V02_JYH.pdf

Bøyum, A. (1976). Isolation of lymphocytes, granulocytes and macrophages. *Scandinavian Journal of Immunology*, *5*(s5, Suppl. 5), 9–15. DOI: 10.1111/j.1365-3083.1976.tb03851.x PMID: 1052391

Butter, A., Aliyev, K., Hillebrandt, K., Raschzok, N., Kluge, M., Seiffert, N., Tang, P., Napierala, H., Muhamma, A. I., Reutzel-Selke, A., Andreou, A., Pratschke, J., Sauer, I. M., & Struecker, B. (2017). Evolution of graft morphology and function after recellularization of decellularized rat livers. *Journal of Tissue Engineering and Regenerative Medicine*, *12*(2), e807–e816. DOI: 10.1002/term.2383 PMID: 27957815

Chhabra, A., Song, H. G., Grzelak, K. A., Polacheck, W. J., Fleming, H. E., Chen, C. S., & Bhatia, S. N. (2022). A vascularized model of the human liver mimics regenerative responses. *Proceedings of the National Academy of Sciences of the United States of America*, *119*(28), e2115867119. DOI: 10.1073/pnas.2115867119 PMID: 35763565

Ciftel, S., Suleyman, B., Mammadov, R., Coskun, R., Coban, T. A., Mokhtare, B., Suleyman, H., Cerrah, S., Cicek, B., & Suleyman, Z. (2024). Adverse effects of metamizole on heart, lung, liver, kidney, and stomach in rats. *BMC Pharmacology & Toxicology*, *25*(1), 55. Advance online publication. DOI: 10.1186/s40360-024-00780-4 PMID: 39175070

Council of Europe. (1986). *European convention for the protection of vertebrate animals used for experimental and other scientific purposes* (ETS No. 123). https://rm.coe.int/168007a67b

Dai, Q., Jiang, W., Huang, F., Song, F., Zhang, J., & Zhao, H. (2022). Recent Advances in Liver Engineering With Decellularized Scaffold. *Frontiers in Bioengineering and Biotechnology*, *10*, 831477. DOI: 10.3389/fbioe.2022.831477 PMID: 35223793

Dubko, A. D., Yurkevich, M. Y., Labai, M. V., Svirskaya, A. V., Yukhanov, N. G., Savitskaya, T. V., Nizheharodava, D. B., & Zafranskaya, M. M. (2021). Engineering a whole rat liver scaffold. *Doklady Natsional'noi Akademii Nauk Belarusi*, *65*(4), 448–455. DOI: 10.29235/1561-8323-2021-65-4-448-455

European Convention for the Protection of Vertebrate Animals used for Experimental and Other Scientific Purposes. (1987). In *Medical entomology and zoology*.

Ghorbani, F., Ekhtiari, M., Chaghervand, B. M., Moradi, L., Mohammadi, B., & Kajbafzadeh, A. (2022). Detection of the residual concentration of sodium dodecyl sulfate in the decellularized whole rabbit kidney extracellular matrix. *Cell and Tissue Banking*, *23*(1), 119–128. DOI: 10.1007/s10561-021-09921-z PMID: 33909237

Gilbert, T., Sellaro, T., & Badylak, S. (2006). Decellularization of tissues and organs. *Biomaterials*, *27*(19), 3675–3683. DOI: 10.1016/j.biomaterials.2006.02.014 PMID: 16519932

Godoy, P., Hewitt, N. J., Albrecht, U., Andersen, M. E., Ansari, N., Bhattacharya, S., Bode, J. G., Bolleyn, J., Borner, C., Böttger, J., Braeuning, A., Budinsky, R. A., Burkhardt, B., Cameron, N. R., Camussi, G., Cho, C., Choi, Y., Rowlands, J. C., Dahmen, U., & Hengstler, J. G. (2013). Recent advances in 2D and 3D in vitro systems using primary hepatocytes, alternative hepatocyte sources and non-parenchymal liver cells and their use in investigating mechanisms of hepatotoxicity, cell signaling and ADME. *Archives of Toxicology*, *87*(8), 1315–1530. DOI: 10.1007/s00204-013-1078-5 PMID: 23974980

Hussein, K. H., Ahmadzada, B., Correa, J. C., Sultan, A., Wilken, S., Amiot, B., & Nyberg, S. L. (2024). Liver tissue engineering using decellularized scaffolds: Current progress, challenges, and opportunities. *Bioactive Materials*, *40*, 280–305. DOI: 10.1016/j.bioactmat.2024.06.001 PMID: 38973992

International Organization for Standardization. (2009). *Biological evaluation of medical devices - Part 5: Tests for in vitro cytotoxicity* (ISO Standard No. 10993-5:2009/(R)2014). https://www.iso.org/standard/36406.html

Kasravi, M., Ahmadi, A., Babajani, A., Mazloomnejad, R., Hatamnejad, M. R., Shariatzadeh, S., Bahrami, S., & Niknejad, H. (2023). Immunogenicity of decellularized extracellular matrix scaffolds: A bottleneck in tissue engineering and regenerative medicine. *Biomaterials Research*, *27*(1), 10. Advance online publication. DOI: 10.1186/s40824-023-00348-z PMID: 36759929

Kim, Y., Kim, H., & Kim, Y. (2025). Advancing hepatotoxicity assessment: Current advances and future directions. *Toxicological Research*, *41*(4), 303–323. DOI: 10.1007/s43188-025-00289-w PMID: 40612519

Krisai, P., Rudin, D., Grünig, D., Scherer, K., Pichler, W., Terracciano, L., & Krähenbühl, S. (2019). Acute liver failure in a patient treated with metamizole. *Frontiers in Pharmacology*, *10*, 996. Advance online publication. DOI: 10.3389/fphar.2019.00996 PMID: 31572180

Louhimies, S. (2002). Directive 86/609/EEC on the protection of animals used for experimental and other scientific purposes. *Alternatives to Laboratory Animals*, *30*(2, Suppl.), 217–219. DOI: 10.1177/026119290203002S36 PMID: 12513679

Mazza, G., Rombouts, K., Rennie Hall, A., Urbani, L., Vinh Luong, T., Al-Akkad, W., Longato, L., Brown, D., Maghsoudlou, P., Dhillon, A. P., Fuller, B., Davidson, B., Moore, K., Dhar, D., De Coppi, P., Malago, M., & Pinzani, M. (2015). Decellularized human liver as a natural 3D-scaffold for liver bioengineering and transplantation. *Scientific Reports*, *5*(1), 13079. Advance online publication. DOI: 10.1038/srep13079 PMID: 26248878

Mendibil, U., Ruiz-Hernandez, R., Retegi-Carrion, S., Garcia-Urquia, N., Olalde-Graells, B., & Abarrategi, A. (2020). Tissue-specific decellularization methods: Rationale and strategies to achieve regenerative compounds. *International Journal of Molecular Sciences*, *21*(15), 5447. Advance online publication. DOI: 10.3390/ijms21155447 PMID: 32751654

Nippi, Inc. (2024). *BioMasher® II: Protocol manual* (Cat. No. BM-02N). https://www.nippi-inc.com/en/products/biomashers/biomasher-ii/

QIAGEN. (2024). *QIAamp DNA Mini and Blood Mini handbook* (Cat. No. 51304). https://www.qiagen.com/us/resources/resourcedetail?id=bf602afb-cc1e-4032-b762-6e364adf39ff&lang=en

Quah, B. J. C., Warren, H. S., & Parish, C. R. (2007). Monitoring lymphocyte proliferation in vitro and in vivo with the intracellular fluorescent dye carboxyfluorescein diacetate succinimidyl ester. *Nature Protocols*, *2*(9), 2049–2056. DOI: 10.1038/nprot.2007.296 PMID: 17853860

Ramachandran, A., & Jaeschke, H. (2019). Acetaminophen hepatotoxicity. *Seminars in Liver Disease*, *39*(2), 221–234. DOI: 10.1055/s-0039-1679919 PMID: 30849782

Robertson, M. J., Soibam, B., O'Leary, J. G., Sampaio, L. C., & Taylor, D. A. (2018). Recellularization of rat liver: An in vitro model for assessing human drug metabolism and liver biology. *PLoS One*, *13*(1), e0191892. DOI: 10.1371/journal.pone.0191892 PMID: 29377912

Russell, W. M. S., & Burch, R. L. (1959). *The principles of humane experimental technique*. Methuen.

Sabetkish, S., Kajbafzadeh, A.-M., Sabetkish, N., Khorramirouz, R., Akbarzadeh, A., Seyedian, S. L., Pasalar, P., Orangian, S., Beigi, R. S. H., Aryan, Z., Akbari, H., & Tavangar, S. M. (2015). Whole-organ tissue engineering: Decellularization and recellularization of three-dimensional matrix liver scaffolds. *Journal of Biomedical Materials Research. Part A*, *103*(4), 1498–1508. DOI: 10.1002/jbm.a.35291 PMID: 25045886

Salerno, S., Campana, C., Morelli, S., Drioli, E., & De Bartolo, L. (2011). Human hepatocytes and endothelial cells in organotypic membrane systems. *Biomaterials*, *32*(34), 8848–8859. DOI: 10.1016/j.biomaterials.2011.08.004 PMID: 21871658

Schneider, C. A., Rasband, W. S., & Eliceiri, K. W. (2012). NIH Image to ImageJ: 25 years of image analysis. *Nature Methods*, *9*(7), 671–675. DOI: 10.1038/nmeth.2089 PMID: 22930834

Sysmex Partec. (2014). *CyFlow® Cube 6 instrument operating manual* (CyView™ software version 1.5; Rev. 05). https://www.wolflabs.co.uk/documents/Sysmex-partec_flow-cytometers_cyflow-Cube-6_manual.pdf

Thanapirom, K., Caon, E., Papatheodoridi, M., Frenguelli, L., Al-Akkad, W., Zhen-zhen, Z., Vilia, M. G., Pinzani, M., Mazza, G., & Rombouts, K. (2021). Optimization and validation of a novel three-dimensional co-culture system in decellularized human liver scaffold for the study of liver fibrosis and cancer. *Cancers (Basel)*, *13*(19), 4936. Advance online publication. DOI: 10.3390/cancers13194936 PMID: 34638417

Thermo Fisher Scientific. (2023). *Quant-iT™ dsDNA broad-range assay kit: User guide* (Cat. No. Q33130). https://assets.thermofisher.com/TFS-Assets/LSG/manuals/Quant_iT_dsDNA_BR_Assay_UG.pdf

Chapter 9
Isolation, Characterization, and Anti-Cancer Properties of Microalgal Monogalactosyldiacylglycerol (MGDG)

Baghish Harutyunyan
http://orcid.org/0000-0002-6665-810X
Scientific and Production Center "Armbiotechnology" of the National Academy of Sciences of Armenia, Armenia

Anna Zakoyan
http://orcid.org/0000-0002-2360-3539
Scientific and Production Center "Armbiotechnology" of the National Academy of Sciences of Armenia, Armenia

Ani Aghabekyan
http://orcid.org/0009-0005-4918-5231
Scientific and Production Center "Armbiotechnology" of the National Academy of Sciences of Armenia, Armenia

Angelina Sayadyan
http://orcid.org/0009-0005-4918-5231
Scientific and Production Center "Armbiotechnology" of the National Academy of Sciences of Armenia, Armenia

Elena Kalita
http://orcid.org/0000-0002-8785-2817
CANDLE Synchrotron Research Institute, Institute of Molecular Biology NAS RA, Armenia

Nelly Babayan
http://orcid.org/0000-0001-9205-7693
Faculty of Biology, Institute of Molecular Biology NAS RA, Yerevan State University, Armenia

Iren Shahazizyan
http://orcid.org/0009-0007-1392-6557
Faculty of Biology, Research Institute of Biology, Yerevan State University, Armenia

DOI: 10.4018/979-8-3373-1524-9.ch009

Vigen B. Goginyan
http://orcid.org/0000-0002-7697-6515
Scientific and Production Center "Armbiotechnology" of the National Academy of Sciences of Armenia, Armenia

ABSTRACT

Glycoglycerolipids are amphiphilic compounds found in thylakoid membranes of cyanobacteria, algae, and plants, attracting interest for their therapeutic potential, especially anticancer applications. They interact with biological molecules and show anticancer, antiviral, and anti-inflammatory properties. Among them, monogalactosyldiacylglycerol (MGDG) inhibits cancer cell proliferation through DNA polymerase inhibition, kinesin repression, and T-cell activation. While plant sources are well studied, microalgal MGDGs remain understudied. This work isolated, purified, and characterized MGDGs from the green microalgae Ch. vulgaris and N. oleoabundans and evaluated their anticancer activity. Microalgae were cultured in a photobioreactor, lipids were extracted and MGDGs purified by chromatography. HPLC and NMR confirmed their profiles. In MTT assays, microalgal MGDGs markedly reduced MCF-7 cell viability, with stronger effects when combined with 5-aminolevulinic acid, while parsley MGDG showed moderate activity. These findings highlight the strong anticancer potential of microalgal MGDGs.

INTRODUCTION

Cancer remains one of the main causes of death worldwide, severely impacting human health and healthcare systems of the WHO (WHO, 2024). Despite the fact that continuous progress has been achieved in very early diagnosis, molecular profiling, and many modalities of treatments, many cancers, such as hepatic, pancreatic, breast, and glioblastoma cancers, still show poor prognosis and limited therapeutic options. This was according to (cancer.gov, 2024). Generally, conventional chemotherapeutic agents are limited by systemic toxicity, multi-drug resistance, and low selectivity toward malignant cells, which pressed for urgent demands for new effective and biocompatible therapeutic compounds. In this respect, structurally diverse and biologically potent natural bioactive molecules from microorganisms, algae, and plants have recently emerged as one of the main sources for anticancer compounds (Rojas et al., 2025, and Safe S., 2024).

Among these natural compounds, glycoglycerolipids represent a class of membrane lipids that are distributed widely in cyanobacteria, marine algae, and higher plants (Benning, 1998; Hölzl et al., 2007; Benning, 2008). These amphiphilic molecules play essential roles in maintaining membrane structure, photosynthetic function, and stress adaptation. The most abundant species, monogalactosyldiacylglycerol (MGDG), is known for its wide medicinal benefits, which include anti-inflammation, antivirus, and anticancer properties (Maeda et al., 2011; Kim et al., 1999; Gustafson, 2015). Mechanistically, the reported modes of action of MGDG include selective inhibition of DNA polymerases and centromere-associated kinesin proteins, enzymes that are important in the process of DNA replication and mitotic progression in cancer cells (Hou et al., 2007; Bebenek and Kunkel, 2004; Hübscher et al., 2002; Murakami et al., 2003). Additionally, MGDG can modulate the immune response by activating T lymphocytes and enhancing tumour cell recognition and elimination (Kinjo et al., 2006).

While MGDG derived from higher plants has been extensively investigated, microalgal sources of MGDG, such as *Chlorella vulgaris* and *Neochloris oleoabundans*, remain comparatively underexplored, particularly in the context of anticancer activity. Microalgae possess many advantages for the production of bioactive lipids in that they have rapid growth rates, can be grown in controlled photobioreactor systems, and are capable of accumulating high levels of lipids under conditions of environmental or nutritional stress (Gao et al., 2019). More precisely, nitrogen deprivation is known to induce the biosynthesis of lipids, which could also increase MGDG yield. Hence, the investigation of lipid accumulation dynamics and the optimization of MGDG extraction from microalgae stand for a promising new source for the sustainable production of natural anticancer agents (Bandaru et al., 2025).

Meanwhile, 5-aminolevulinic acid (ALA) has been widely investigated as one of the most powerful photosensitizers in photodynamic therapy (PDT) due to its endogenous presence in the human body as a part of the heme biosynthesis pathway. After administration, ALA metabolism generates PpIX, which is well known for its ability to preferentially accumulate in malignant cells. This compound is photoactivated with specific light, inducing the formation of reactive oxygen species (ROS) capable of causing lethal damage to tumor cells and selective cell destruction through apoptosis and necrosis. ALA-PDT has already found broad applications in dermatology and oncology; nonetheless, its potential could be significantly improved by combination with other bioactive molecules with either direct sensitizing action against cancer cells or those affecting the oxidative stress pathways.

With the complementary mechanisms of MGDG and ALA, their combined application may offer a new and synergistic therapeutic modality. MGDG might potentiate ALA-mediated phototoxicity by modulating membrane permeability, ROS sensitivity, or immune signaling in a way that enhances the overall antitumor response. This study will investigate the optimization of lipid accumulation in the selected microalgal strains under controlled stress; isolate and purify MGDG from microalgal biomass; evaluate the anticancer potency of MGDG alone and in combination with ALA-induced PDT *in vitro*. This work also explores the possibility of developing natural low-cost, renewable, and safe therapeutic agents in cancer treatment, with the integration of microalgal biotechnology and photodynamic therapy, which might offer an environmentally friendly alternative to synthetic chemotherapeutics.

MATERIALS AND METHODS

Microorganisms and Culture Conditions

Green microalgae *Chlorella vulgaris* and *Neochloris oleoabundans* were provided by the Culture Collection of the Laboratory of Alternative Energy Sources, Scientific and Production Center "Armbiotechnology" of National Academy of Sciences of the Republic of Armenia (NAS RA). These strains were cultured in liquid medium at temperatures ranging from 26 to 30 °C in an illuminated incubator with continuous illumination of LEDs (Harutyunyan et al., 2025).

Culture Maintenance

Stock cultures were cultured in sterile 250 mL Erlenmeyer flasks filled with 100 mL each of liquid nutrient media, which was continually illuminated with a light irradiance of 116.4 μmol m^{-2} s^{-1}. Periodically, these stock cultures were subcultured to maintain them in exponential growth.

Batch Culturing, Medium Optimization, Cell Growth

For identifying optimal culture conditions, microalgal strains were cultivated in 500mL Erlenmeyer flasks containing 250mL of Bold's Basal Medium (BBM)(Lloyd et al., 2021) and Ferty II nutrient medium (F2), (Rifai et al., 2022). These culture conditions were optimized in laboratory settings. Culture conditions included temperatures of 28 °C with agitation at 300 rpm, while illumination was continuous with an intensity of 116.4μmol m^{-2} s^{-1}, which included controlled LEDs with efficient photosynthetically active radiation. For reproducibility to be achieved, triplicates were used in all growth experiments (Wolf et al., 2015).

Cultivation in a Photobioreactor

Larger-scale cultivations were conducted in a 3 L laboratory-scale photobioreactor with controlled temperatures and lighting. Illumination was continuous with red-blue LEDs (630nm, 420nm), with an intensity of 116.4 μmols m^{-2} s^{-1}, spectral irradiance analysis having been performed with a spectrophotometer (OHSP-350P, Hopoocolor, China) (Harutyunyan et al.,2025). It was continuously supplied with filtered air (0.2μm in size) with 1–2% v/v CO_2 supplementation, to provide optimal support for photosynthesis. Daily, it was checked for any changes in pH, maintaining $6.8 < pH < 7.2$ (Lecina et al., 2016).

Growth Monitoring

Growth of microalgae was determined by measurement of optical density (OD) at 540 nm with a UV–Vis spectrophotometer (VWR, USA) (Chi et al., 2019). Cell biomass concentrations were also checked periodically by matching OD readings with cell dry weight (CDW) values obtained gravimetrically. The specific growth rate (μ, day^{-1}) was calculated during the exponential phase according to the equation:

$$\mu = \frac{\ln(N_2) - \ln(N_1)}{t_2 - t_1}$$

where N_1 and N_2 are the concentrations of biomass at times t_1 and t_2, respectively.

Microscopic Examination

Cell morphology and division patterns were observed with a phase-contrast light microscope (Boeco, Germany) under 1000× magnification. Samples were examined periodically for cell integrity, cell morphology, and potential microbial contaminants.

Cultivation Under Nitrogen-Limited Conditions

For the stimulation of lipid accumulation, microalgal strains were cultivated under both nitrogen-sufficient (N^+) and nitrogen-deficient (N^-) conditions. The N^+ media served as control cultures and consisted of standard formulations of Bold's Basal Medium (BBM), the composition is as follows (per liter) $NaNO_3$ – 250 mg, $CaCl_2{\cdot}2H_2O$ – 25 mg, $MgSO_4{\cdot}7H_2O$ – 75 mg, K_2HPO_4 – 75 mg, KH_2PO_4 175 mg, NaCl – 25 mg, trace metals – 10 mL of trace stock $ZnSO_4{\cdot}7H_2O$ – 8.82 g, $MnCl_2{\cdot}4H_2O$ – 1.44 g, MoO_3 – 0.71 g, $CuSO_4{\cdot}5H_2O$ – 1.57 g, $Co(NO_3)_2{\cdot}6H_2O$ – 0.49 g, Fe–EDTA – 1 mL of stock. pH – 6.6 to 6.8. The trace stock solution is composed of: $ZnSO_4{\cdot}7H_2O$ – 8.82 g/l, $MnCl_2{\cdot}4H_2O$ – 1.44 g/l, MoO_3 – 0.71 g/l, $CuSO_4{\cdot}5H_2O$

– 1.57 g/l, $Co(NO_3)_2 \cdot 6H_2O$ – 0.49 g/l, Fe–EDTA stock solution is composed of $FeSO_4 \cdot 7H_2O$ – 4.98 g/l, Na_2EDTA – 8.36 g/l. (Lloyd et al., 2021). For nitrogen-limited (N^-) cultivation, the nitrogen sources in their respective media were either excluded or substituted with equimolar NaCl to maintain ionic balance, based on modified versions of Gao et al. (2019) and Restiawaty et al. (2023). Substituting these media successfully imposed nitrogen starvation with less differences in osmo-regulatory responses among treatment groups.

Batch cultures were first allowed to grow in N^+ conditions until mid-exponential phase, then centrifuged (3,000 x g, 10 min) to harvest the cells, which were subsequently resuspended in N^- medium with the same composition but without the nitrogen source. Growth was continued in the same light intensity (116.4 μmol m^{-2} s^{-1}), temperature (28 °C), and shaking rate (300 rpm) for 5-7 days to stimulate lipid production.

Cell growth and physiological condition were determined by measurement of optical density (OD_{540}) and microscopy, while the emergence of nitrogen starvation was indicated by analysis of residual nitrate levels in the medium with Griess reagent method (APHA, 2017). Cell biomass was harvested for lipid analysis at the end of culturing.

Lipid Extraction

Total lipids were extracted by the conventional method described by Bligh and Dyer in 1959, with some modifications. Microalgal cell mass was obtained from late exponential growth phases (A_{640} = 3.0 or A_{540} = 3.0, with about 168 hours of growth). Cell mass aliquots were centrifuged at 15,000 x g for 15 minutes at 4 °C to obtain cell pellets. These pellets were washed twice with water to remove any components from the medium, described in Singh, G., & Patidar, S. K. (2018).

For extraction, a mixture consisting of 1 mL of concentrated cell suspension was blended with 3.75 mL of chloroform: methanol (1:2, v/v); thereafter, it was vigorously mixed for 15 minutes for cell lysis and optimal lipid solubilization. Then, 1.25 mL of chloroform was added, with 1-minute mixing. Finally, after 1.25 mL of distilled water was added, gentle mixing was performed, resulting in the formation of two layers, with the lower layer consisting of total lipids in chloroform being carefully retrieved, followed by removal of solvent under reduced pressure with a rotary evaporator. Then, 1 mL of chloroform was used to dissolve the lipid extract, which was finally preserved at –20 °C for analysis (De Jesus et al., 2019).

Isolation of MGDG by Column Chromatography

MGDG was isolated from total lipid extracts through column chromatography with silica gel, using gradient elution based on the method of Sugawara and Miyazawa (1999). The stationary phase utilized was 10 to 40 µm particle size silica gel (Merck). Elution was carried out with a solvent mixture of increasing polarity consisting of chloroform and acetone.

Fractions were collected sequentially, with analysis via thin-layer chromatography (TLC) in a solvent system consisting of chloroform, methanol, and water with a volumetric ratio of 65:25:4. Bands specifically indicating MGDG were detected after spraying with methanolic-sulfuric acid (10%) and α-naphthol reagent and heating at 110 °C for 5 minutes. Pooled fractions containing MGDG, concentrated under nitrogen gas, were kept at −20 °C for further analysis (Wang et al., 2011). Quantification of MGDG was carried out with the use of high performance liquid chromatography (HPLC) (Waters, USA). HPLC analysis was done based on Foseid et al. (2022), with the method adapted by Abedin & Barua (2021). Chromatographic analysis was carried out on the C18 reverse-phase column with dimensions (250 x 4.6mm, 5µm, with binary solvent systems composed of Chloroform, solvent A, and methanol: water (95:5 The flow rate was kept at 1.0 mL min^{-1}, while the column temperature was maintained at 30 °C. The wavelength setting on the UV detector was 240 nm, with 20-25 µL sample injection per analysis. MGDG was identified and quantified based on its retention time and peak area with those of a standard MGDG (Sigma-Aldrich, USA; purity >99%).

Structural Verification by NMR Analysis

Structural confirmation of MGDG was carried out using ^{1}H nuclear magnetic resonance (^{1}H NMR) spectroscopy. Each purified lipid fraction was dissolved in deuterated chloroform ($CDCl_3$), and spectra were collected and characterized using a 400 MHz NMR spectrometer (Bruker, Germany). Chemical shifts were compared with those of the standard MGDG (Sigma-Aldrich, USA; MW = 752.36) and interpreted using established reference databases (Gunstone & Harwood, 2007). Diagnostic signals corresponding to the galactosyl headgroup (δ = 3.2–4.2 ppm) and acyl chains (δ = 0.8–2.8 ppm) confirmed the presence of MGDG.

In *Vitro* Cytotoxicity Assay

The cytotoxic potential of both isolated MGDG and commercial standard MGDG was evaluated using the MCF-7 human breast adenocarcinoma cell line (Comşa et al., 2015). Cells were cultured in RPMI-1640 or Hybri-Care ATCC 46-X media supplemented with 10% fetal bovine serum (FBS) and 1% penicillin–streptomycin. Cultures were maintained at 37 °C in a humidified condition containing 5% CO_2 (Osborne et al., 1987).

For cytotoxicity testing, approximately 1×10^4 cells were seeded in 96-well plates and allowed to adhere overnight before being treated with a range of MGDG concentrations (0–10,000 ng mL^{-1}) for 48 h. Cell viability was assessed using the MTT assay (Van de Loosdrecht et al., 1994; Kim et al., 2007). MTT reagent (3-(4,5-dimethylthiazol-2-yl)-2,5-diphenyltetrazolium bromide; MW = 414) was added to each well and incubated for 4 h. The formazan crystals were dissolved in DMSO, and the resulting solution's absorbance was measured at 570 nm with a microplate reader. Cell viability (%) was calculated by comparing their results to the untreated controls (Buch et al., 2012).

Combination Treatment of MGDG and ALA With Photodynamic Exposure

To evaluate the synergistic cytotoxic and photodynamic effects of MGDG in combination with 5-aminolevulinic acid (ALA), MCF-7 cells were seeded in 96-well plates (1×10^4 cells per well) and treated with different concentrations of MGDG (extracted or standard) and ALA (0 -10 mg/mL) for 24 h. Following incubation, cells were irradiated with a combination of blue light (λ = 400–410 nm) and red light (λ = 615–635 nm) to activate the photodynamic process, according to Ishizuka et al. (2011).

Cell viability was measured by MTT assay after an additional 24 hours of incubation, as described previously. The potential synergism between MGDG and ALA was analyzed by comparing the combined treatment effects to single-agent controls.

Statistical Analysis

All experiments were carried out with three independent replicates, and the results are given in the form of mean values with their respective standard deviations (SD). Data obtained from growth curve analysis, lipid measurement, and cytotoxicity tests were statistically evaluated for differences among groups being compared. Normality was confirmed with Shapiro-Wilk tests, while homogeneity of variance was determined by Levene's test. Comparison between these two groups was done with t-tests. In other situations, there were more than two groups being compared, such as different concentrations regarding MGDG or ALA, with one varying group in each comparison or different conditions, like whether cultures were nitrogen-sufficient or -deficient, analysis of variance was done, followed by Tukey's test for multiple comparisons. Differences were considered statistically significant at $p < 0.05$, $p < 0.01$, and $p < 0.001$, which have been indicated in figures and tables.

Graphical analyses were performed with GraphPad Prism version 10.0/Origin-Pro 2023, while statistical analyses were done with IBM SPSS Statistics version 27, purchased from IBM Corp., USA (George & Mallery, 2024). All the tests were conducted in at least three independent replicates to verify the reproducibility and consistency of results.

RESULTS AND DISCUSSION

Culture conditions and growth evaluation

Effects of different culture mediums, LED lighting spectra, and pH levels on growth and biomass accumulation in selected microalgal strains were examined, due to important role played by these factors in impacting microbial metabolism, thereby affecting lipid synthesis. During the cultivation, the cell morphology in algal cultures was observed under microscopy to note their structural integrity and physiological condition. Representative images illustrating *C. vulgaris* and *N. oleoabundans* are presented in Figure 1.

Figure 1. Microscopic images of cells of (A) Chlorella vulgaris and (B) Neochloris oleoabundans (Magnification 1000×)

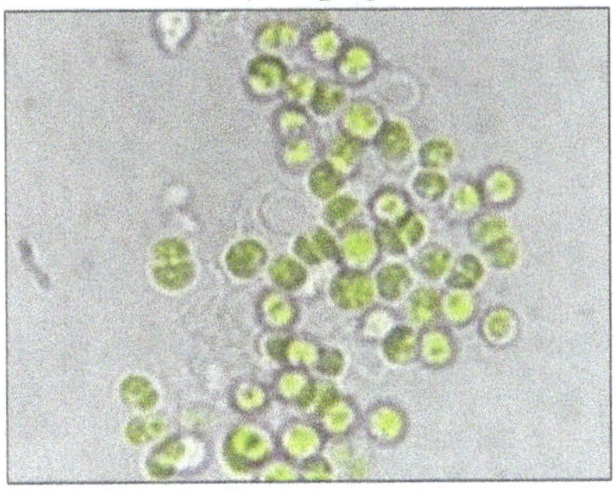

A

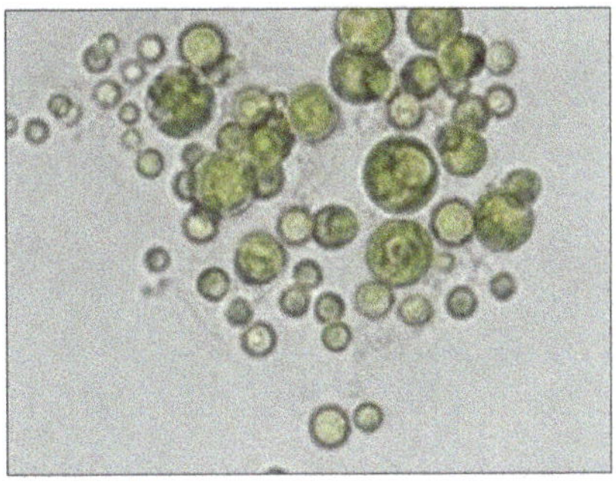

B

The results have shown that both strains exhibited faster growth rates and produced higher concentrations of biomass yields in BBM medium compared to BG-11 and SOT media (Figure 2).

Figure 2. Growth of Neochloris oleoabundans and Chlorella vulgaris in different culture media

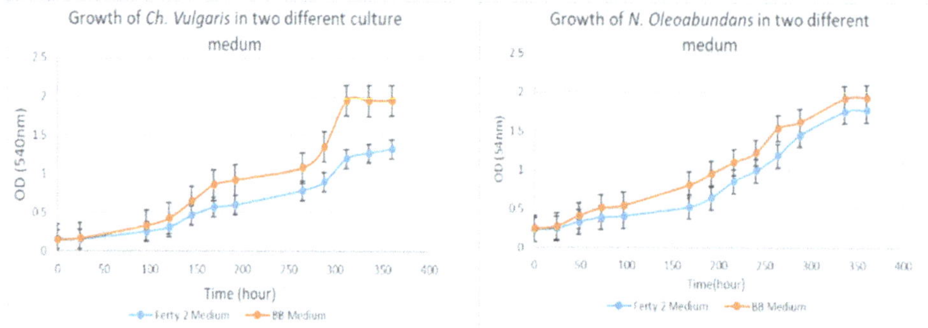

To further evaluate the influence of pH on biomass production, *N. oleoabundans* and *Chlorella vulgaris* were cultivated in BBM medium with different pH values. Optimal growth and biomass synthesis for *Neochloris oleoabundans* were observed at pH 7.5, while *Chlorella. vulgaris* reached maximum biomass production at pH 7.0 (Figure 3).

Figure 3. Growth of Neochloris oleoabundans and Chlorella vulgaris under different pH conditions of the BB media

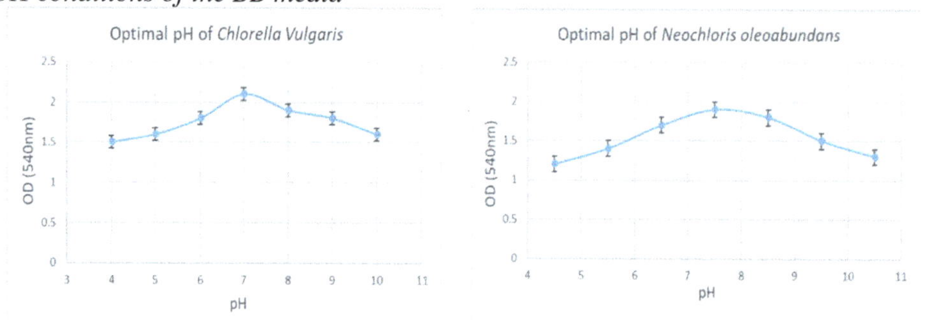

Determining the optimal pH for cultivation is crucial for large-scale applications, as it allows efficient biomass production under stable and optimum physiological conditions. To obtain a sufficient system for lipid extraction and glycoglycerolipids fractionation, both species were cultivated in a 3 L photobioreactor under the controlled blue-red LED illumination (Figure 4). The growth kinetics of the cultures are shown in Figure 5.

Figure 4. Cultivation of Chlorella vulgaris and Neochloris oleoabundans microalgae in a 3 L photobioreactor with blue-red light illumination

Figure 5. Growth dynamics of C Chlorella vulgaris and Neochloris oleoabundans microalgae in the photobioreactor

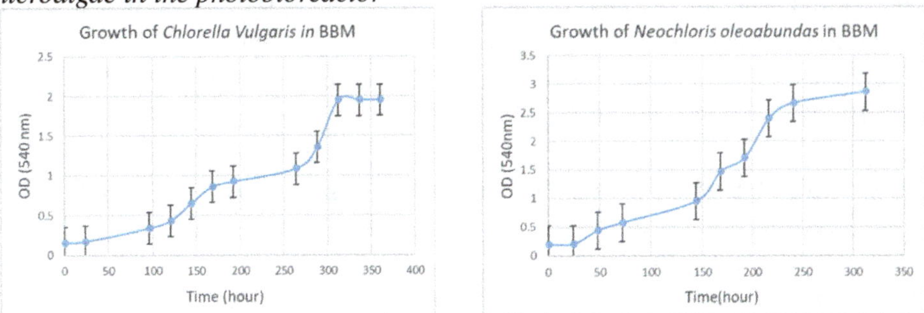

During cultivation, both strains reached their respective exponential growth phases, after which the growth rate (μ) and maximum dry biomass concentration were determined. For *Chlorella vulgaris*, μ and maximum biomass were 0.182 day^{-1} and 1.50 g L^{-1}, respectively, while *Neochloris oleoabundans* demonstrated slightly higher values - 0.201 day^{-1} and 2.21 g L^{-1}.

According to previous reports, *Chlorella vulgaris* tends to accumulate lipids under stress conditions, particularly when nitrogen is limited. Therefore, this strain was cultivated in BB medium under N^+ and N^- conditions. The results demonstrated that in N^+ medium, *Chlorella vulgaris* grew nearly twice as fast and synthesized a larger biomass quantity, whereas in N^- medium, the cells became yellowish due to chlorophyll degradation. Gravimetric analysis of lipids revealed a 2.3-fold increase in total lipid content of biomass cultivated under N^- conditions, proving the induction of lipid biosynthesis in response to nutrient stress, particularly to nitrogen.

Lipid Extraction, Quantitative and Qualitative Analysis of MGDG

The process of extracting total lipids from the biomass of *Chlorella vulgaris* and *Neochloris oleoabundans* followed the method described by Bligh and Dyer (1959). After the separation of the solvents, the crude lipid extracts were concentrated and then processed through silica gel column chromatography for fractionation. We used a gradient elution with chloroform: acetone mixtures that had a gradually increasing concentration of acetone.

Identification and Preliminary Screening by TLC

To identify glycolipids, particularly MGDG, the fractions were screened using TLC on silica gel plates. Visualization with iodine vapor and methanol-sulfuric acid staining showed distinct bands in the 3rd and 4th fractions, matching the Rf value of the standard MGDG (parsley-derived) (0.56 ± 0.02). The bands turned violet after methanol-sulfuric acid treatment and yellow with iodine, confirming the presence of unsaturated glycolipids. These fractions were preserved at -20 °C for further purification and analysis.

Quantitative Determination and Purity Assessment by HPLC

Quantitative and qualitative analysis of MGDG was done using HPLC (Waters, USA) with a C18 reverse-phase column and a UV detector set at 240 nm. Gradient elution was conducted with chloroform (solvent A) and methanol/water (95:5, solvent B) (Stolyhwo & Privett, 1973). The chromatograms showed a clear peak at 5.3 minutes, corresponding to the retention time of the standard MGDG (99%, Sigma-Aldrich) (Figure 6A). Analysis of microalgal samples confirmed the presence of MGDG with retention times identical to the standard (Figure 6B). The MGDG content was quantified as 0.93 mg mL^{-1} in *Chlorella vulgaris* and 0.42 mg mL^{-1} in *Neochloris oleoabundans*. Peak purity analysis (area normalization method) indicated that the isolated MGDG fractions had chromatographic purity levels of 92.4 ± 1.3% for *Chlorella. vulgaris* and 88.7 ± 1.8% for *Neochloris oleoabundans*.

Figure 6. A) Chromatogram of standard MGDG; B) Chromatogram of MGDG isolated from microalgae

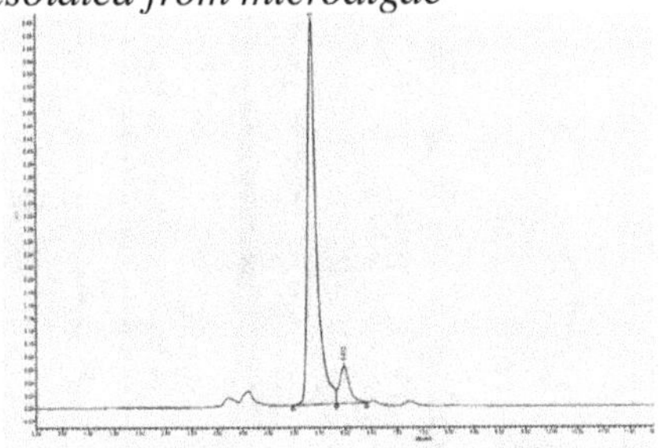

A

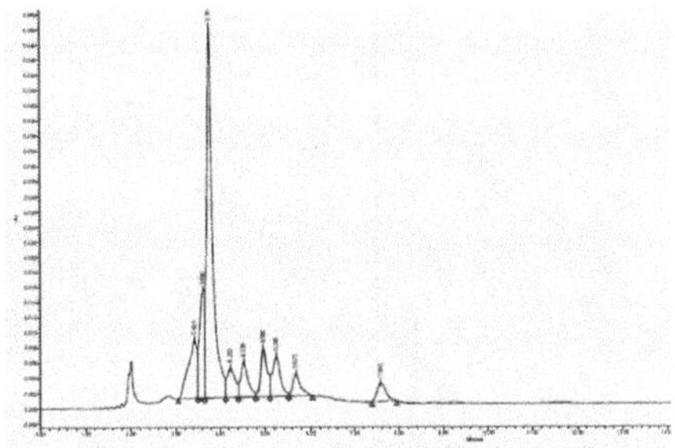

B

Structural Characterization by NMR Spectroscopy and Mass Spectrometry

NMR Spectroscopy

Structural confirmation of the isolated monogalactosyldiacylglycerol (MGDG) was performed using ^1H and ^13C nuclear magnetic resonance (NMR) spectroscopy (Bisht et al., 2021). The NMR spectra were acquired in $CDCl_3$ on a 500 MHz spectrometer and processed with Renova software. Complete signal assignments were made by comparison with reference data (Gunstone & Harwood, 2007) and verified using standard MGDG spectra (Figure 7).

The ^1H NMR spectrum of *C. vulgaris* MGDG exhibited characteristic resonances at δ 3.6–4.3 ppm corresponding to galactosyl protons, δ 5.2–5.4 ppm for olefinic hydrogens, and δ 2.2–2.4 ppm for allylic methylene groups. Notably, downfield shifts near δ 6.0 ppm indicated the presence of conjugated double bonds adjacent to carbonyl carbons, suggesting partial dehydrogenation of the acyl chains.

The ^13C NMR spectrum confirmed carbonyl carbons at δ 173 to 175 ppm and unsaturated carbons at δ 128 to 132 ppm. Additional signals at δ 134 to 136 ppm offered more support for the presence of conjugated double bonds. These findings indicate that *Chlorella vulgaris*-derived MGDG contains modified polyunsaturated fatty acid parts, which differ from the fully saturated or mono-unsaturated acyl chains usually found in plant-derived MGDGs.

Figure 7. ^1H NMR spectrum of MGDG isolated from Chlorella vulgaris microalgae.

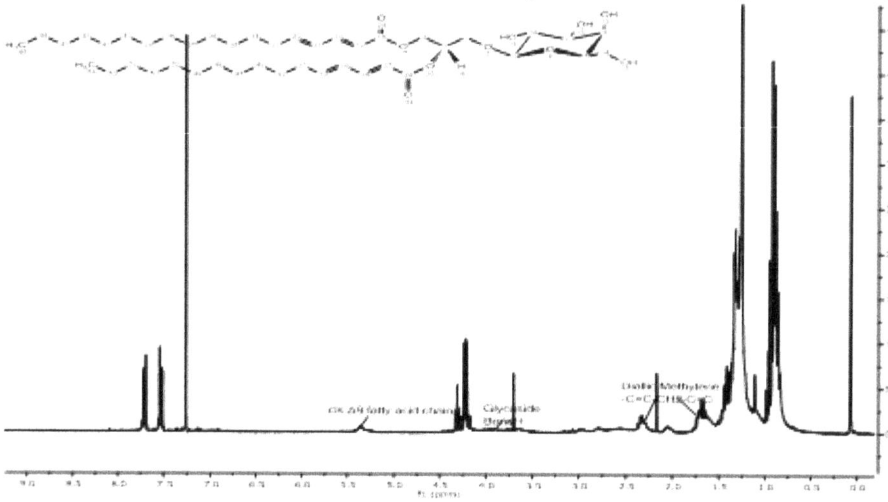

Mass Spectrometry Confirmation

To confirm the presence of conjugated unsaturation and to define the molecular composition, electrospray ionization tandem mass spectrometry (ESI-MS/MS) was performed (Schneiter et al., 1999). The spectra showed a main $[M\text{-}H]^-$ ion at m/z 752.5, which matches the expected molecular mass of MGDG ($C_{41}H_{78}O_{10}$). The fragmentation patterns displayed ions at m/z 495.3 and 253.2, linked to monoacylglycerol-galactosyl and fatty acyl fragments, respectively. The MS/MS data pointed to the dominance of C18:3 (linolenic) and C16:1 acyl chains. Additionally, ions with a mass deficit of 2 Da hinted at partial dehydrogenation, supporting the NMR observations. To check the molecular uniformity, we also conducted LC-MS lipidomic profiling. The analysis identified a single dominant MGDG species that made up 91-94% of the total detected glycolipid mass, confirming both the high purity and structural integrity of the isolated compound.

Interpretation

The combined results from TLC, HPLC, NMR, and MS analyses confirm that MGDG was successfully isolated from *C. vulgaris* and *N. oleoabundans*, and its structure was identified. Chromatographic purity exceeds 90%, and the retention times are consistent. The data from ^1H/^13C NMR and MS/MS further support the molecular identity of the isolated glycolipids. Notably, the presence of conjugated double bonds next to acyl carbonyl groups in *Chlorella vulgaris* MGDG suggests improved electron delocalization. This may help produce photoinduced reactive oxygen species (ROS). These structural features might explain the increased cytotoxic and photodynamic activities seen in the MCF-7 cell assays that followed.

Table 1. Microalgae-derived MGDG's physicochemical and structural characteristics

Parameter	**Standard MGDG**	*C. vulgaris* **MGDG**	*N. oleoabundans* **MGDG**	**Analytical Method**
Retention time (min)	5.30 ± 0.02	5.31 ± 0.03	5.32 ± 0.04	HPLC (C18 column, UV 240 nm)
Concentration (mg mL^{-1})	–	0.93 ± 0.05	0.42 ± 0.03	HPLC quantification
Chromatographic purity (%)	99.0	92.4 ± 1.3	88.7 ± 1.8	HPLC (peak area normalization)
Major acyl chains	C16:0, C18:2	C16:1, C18:3	C16:0, C18:2	ESI–MS/MS
$[M–H]^-$ molecular ion (*m/z*)	752.5	752.5	752.5	ESI–MS
Key fragment ions (*m/z*)	495.3, 253.2	495.3, 253.2, 251.2	495.3, 253.2	ESI–MS/MS
Characteristic ^1H NMR peaks (δ, ppm)	3.6–4.3 (galactose), 5.3 (C=C), 2.3 (–CH_2–)	3.6–4.3, 5.3, 2.3, 6.0 (conjugated C=C)	3.6–4.3, 5.3, 2.3	^1H NMR (500 MHz, $CDCl_3$)
Characteristic ^13C NMR peaks (δ, ppm)	173–175 (C=O), 128–130 (C=C)	173–175, 128–132, 134–136 (conjugated C=C)	173–175, 128–130	^13C NMR (125 MHz, $CDCl_3$)
Structural modification	–	Partial dehydrogenation; conjugated double bonds near carbonyls	Not detected	NMR + HR–MS
Estimated molecular homogeneity (%)	100	91–94	87–90	LC–MS lipidomics

Summary Comment

The combined analytical evidence shows that *Chlorella vulgaris*-derived MGDG has unique structural features. It notably contains conjugated double bonds next to acyl carbonyl groups, which suggests partial dehydrogenation. These patterns of unsaturation, backed by high-resolution NMR and MS data, likely contribute to the improved bioactivity of *Chlorella vulgaris* MGDG seen in the cytotoxicity assay.

Study of the Anticancer Activity of MGDG

The anticancer potential of MGDG from microalgae was tested using the MCF-7 breast cancer cell line. To check for possible combined effects, the study was done to reveal the MGDG's activity with 5-aminolevulinic acid (ALA), a known photosensitizer that boosts the production of ROS. The cytotoxicity of MGDGs was investigated *in vitro,* using the MTT assay. The results showed that MGDGs extracted from *Chlorella vulgaris* and *Neochloris oleoabundans* significantly reduced MCF-7 cell growth, with stronger effects observed at higher concentrations. At 10,000 ng/mL, MGDG from *Chlorella vulgaris* decreased cell viability down to 4.2 ± 2.5%, while MGDG from *N. oleoabundans* brought viability down to 7.3 ± 3.0%. The IC_{50} values were about 500 ng/mL for *C. vulgaris* and 1,000 ng/mL for *N. oleoabundans,* indicating promising cytotoxic ability. Co-treatment with 5-aminolevulinic acid HCl (ALA) at 0.80 mg/ml boosted the cytotoxic response even more, reducing cell viability to 0.4 ± 1.5% and 0.6 ± 2.0% after photodynamic exposure. In comparison, MGDG from parsley had a lower effect at 10,000 ng/mL, showing 16 ± 3.5% cell viability, with an IC_{50} of 1,500 ng/mL.

CONCLUSION

This study showed that cultivation conditions, especially nutrient composition, pH, and light quality, are crucial for optimizing the growth and lipid accumulation of *Chlorella vulgaris* and *Neochloris oleoabundans*. BBM medium was found to be the best option, with optimal pH values of 7.0 for *Chlorella* vulgaris and 7.5 for *Neochloris oleoabundans*. Cultivation in a photobioreactor produced efficient biomass in a comparatively short period of cultivation. Nitrogen limitation in *Chlorella vulgaris* increased lipid accumulation. Lipid extraction and chromatographic analysis identified MGDG as a major glycoglycerolipid, *with Chlorella vulgaris* producing higher concentrations than *Neochloris oleoabundans*. NMR structural characterization revealed dehydrogenation and conjugated double bonds near carbonyl groups in MGDG from *Chlorella vulgaris*, features likely related to its bioactivity. Functional tests confirmed that MGDGs from both species have anticancer effects against MCF-7 breast cancer cells, with increased potency when combined with ALA. Overall, this study shows the potential of microalgal MGDGs as bioactive compounds, especially in combination with ALA, and emphasizes the need for more extensive research to understand their mechanisms of action and possible therapeutic applications.

ACKNOWLEDGMENT

This research was supported by the Higher Education and Science Committee of MESCS RA (Research projects No. 22YR-2I015 and No. 25YR-2I016).

REFERENCES

Abedin, M. R., & Barua, S. (2021). Isolation and purification of glycoglycerolipids to induce apoptosis in breast cancer cells. *Scientific Reports*, *11*(1), 1–13. DOI: 10.1038/s41598-020-80484-x PMID: 33446783

Bandaru, N., Patil, Y. P., Ekghara, S. D., & Bonthu, M. G. (2025). Exploring marine-derived compounds as potential anti-cancer agents: Mechanisms and therapeutic implications. *Cancer Pathogenesis and Therapy*. Advance online publication. DOI: 10.1016/j.cpt.2025.08.004 PMID: 41732213

Bebenek, K., & Kunkel, T. A. (2004). DNA repair and replication. *Advances in Protein Chemistry*, *69*, 137–165. DOI: 10.1016/S0065-3233(04)69005-X PMID: 15588842

Benning, C. (1998). Biosynthesis and function of the sulfolipid sulfoquinovosyl diacylglycerol. *Annual Review of Plant Biology*, *49*(1), 53–75. DOI: 10.1146/annurev.arplant.49.1.53 PMID: 15012227

Benning, C. (2008). A role for lipid trafficking in chloroplast biogenesis. *Progress in Lipid Research*, *47*(5), 381–389. DOI: 10.1016/j.plipres.2008.04.001 PMID: 18440317

Bisht, B., Kumar, V., Gururani, P., Tomar, M. S., Nanda, M., Vlaskin, M. S., Kumar, S., & Kurbatova, A. (2021). The potential of nuclear magnetic resonance (NMR) in metabolomics and lipidomics of microalgae-a review. *Archives of Biochemistry and Biophysics*, *710*, 108987. DOI: 10.1016/j.abb.2021.108987 PMID: 34260946

Blair, M. F., Kokabian, B., & Gude, V. G. (2014). Light and growth medium effect on Chlorella vulgaris biomass production. *Journal of Environmental Chemical Engineering*, *2*(1), 665–674. DOI: 10.1016/j.jece.2013.11.005

Bligh, E. G., & Dyer, W. J. (1959). A rapid method of total lipid extraction and purification. *Canadian Journal of Biochemistry and Physiology*, *37*(8), 911–917. DOI: 10.1139/o59-099 PMID: 13671378

Buch, K., Peters, T., Nawroth, T., Sänger, M., Schmidberger, H., & Langguth, P. (2012). Determination of cell survival after irradiation via clonogenic assay versus multiple MTT Assay-A comparative study. *Radiation Oncology (London, England)*, *7*(1), 1. DOI: 10.1186/1748-717X-7-1 PMID: 22214341

Chi, N. T. L., Duc, P. A., Mathimani, T., & Pugazhendhi, A. (2019). Evaluating the potential of green alga Chlorella sp. for high biomass and lipid production in biodiesel viewpoint. *Biocatalysis and Agricultural Biotechnology*, *17*, 184–188. DOI: 10.1016/j.bcab.2018.11.011

Comşa, Ş., Cimpean, A. M., & Raica, M. (2015). The story of MCF-7 breast cancer cell line: 40 years of experience in research. *Anticancer Research*, *35*(6), 3147–3154. PMID: 26026074

de Jesus, S. S., Ferreira, G. F., Moreira, L. S., Maciel, M. R. W., & Maciel Filho, R. (2019). Comparison of several methods for effective lipid extraction from wet microalgae using green solvents. *Renewable Energy*, *143*, 130–141. DOI: 10.1016/j.renene.2019.04.168

Duse, L., Pinnapireddy, S. R., Strehlow, B., Jedelská, J., & Bakowsky, U. (2018). Low level LED photodynamic therapy using curcumin loaded tetraether liposomes. *European Journal of Pharmaceutics and Biopharmaceutics*, *126*, 233–241. DOI: 10.1016/j.ejpb.2017.10.005 PMID: 29017954

Foseid, L., Devle, H., Naess-Andresen, C. F., & Ekeberg, D. (2022). Laminaria hyperborea as a Source of Valuable Glyceroglycolipids—A Characterization of Galactosyldiacilglycerols in Stipe and Blade by HPLC-MS/MS. *AppliedChem*, *2*(4), 185–198. DOI: 10.3390/appliedchem2040013

Gao, Y., Feng, J., Lv, J., Liu, Q., Nan, F., Liu, X., & Xie, S. (2019). Physiological changes of Parachlorella Kessleri TY02 in lipid accumulation under nitrogen stress. *International Journal of Environmental Research and Public Health*, *16*(7), 1188. DOI: 10.3390/ijerph16071188 PMID: 30987041

George, D., & Mallery, P. (2024). *IBM SPSS statistics 29 step by step: A simple guide and reference*. Routledge. DOI: 10.4324/9781032622156

Gunstone, F. D., & Harwood, J. L. (2007). *The lipid handbook with CD-ROM*. CRC press., DOI: 10.1201/9781420009675

Gustafson, K. R. (2015). Mining the extensive chemical diversity of the NCI natural products repository for new agents that can target HIV. *Planta Medica*, *81*(11), IL1. DOI: 10.1055/s-0035-1556098

Harutyunyan, B., Mühlstädt, G., Seibel, L., Hilpmann, G., Sayadyan, A., Aghabekyan, A., & Goginyan, V. (2025). Algal biomass as a food and feed supplement: Process engineering analysis. *Functional Foods in Health and Disease, 15*(7), 380-395. DOI: 10.31989/ffhd.v15i7.1662

Hölzl, G., & Dörmann, P. (2007). Structure and function of glycoglycerolipids in plants and bacteria. *Progress in Lipid Research, 46*(5), 225–243. DOI: 10.1016/j.plipres.2007.05.001 PMID: 17599463

Hou, C.-C., Chen, Y.-P., Wu, J.-H., Huang, C.-C., Wang, S.-Y., Yang, N.-S., & Shyur, L.-F. (2007). A galactolipid possesses novel cancer chemopreventive effects by suppressing inflammatory mediators and mouse B16 melanoma. *Cancer Research, 67*(14), 6907–6915. DOI: 10.1158/0008-5472.CAN-07-0158 PMID: 17638902

Hübscher, U., Maga, G., & Spadari, S. (2002). Eukaryotic DNA polymerases. *Annual Review of Biochemistry, 71*(1), 133–163. DOI: 10.1146/annurev.biochem.71.090501.150041 PMID: 12045093

Ishizuka, M., Abe, F., Sano, Y., Takahashi, K., Inoue, K., Nakajima, M., Kohda, T., Komatsu, N., Ogura, S., & Tanaka, T. (2011). Novel development of 5-aminolevurinic acid (ALA) in cancer diagnoses and therapy. *International Immunopharmacology, 11*(3), 358–365. DOI: 10.1016/j.intimp.2010.11.029 PMID: 21144919

Jiang, M., Hong, K., Mao, Y., Ma, H., Chen, T., & Wang, Z. (2022). Natural 5-aminolevulinic acid: Sources, biosynthesis, detection, and applications. *Frontiers in Bioengineering and Biotechnology, 10*, 841443. DOI: 10.3389/fbioe.2022.841443 PMID: 35284403

Kim, Y. H., Choi, J. S., Hong, J., Yoo, J. S. & Kim, M. S. (1999). Identification of acylated glycoglycerolipids from a cyanobacterium, Syn-echocystis sp., by tandem mass spectrometry. *Lipids, 34*, 847–853. DOI: 10.1007/s11745-999-0432-2

Kim, Y. H., Kim, E.-H., Lee, C., Kim, M.-H. & Rho, J.-R. (2007). Two new monogalactosyl diacylglycerols from brown alga Sargassum thunbergii. *Lipids, 42*, 395–399. DOI: 10.1007/s11745-007-3035-7

Kinjo, Y., Tupin, E., Wu, D., Fujio, M., Garcia-Navarro, R., Benhnia, M. R.-E.-I., Zajonc, D. M., Ben-Menachem, G., Ainge, G. D., Painter, G. F., Khurana, A., Hoebe, K., Behar, S. M., Beutler, B., Wilson, I. A., Tsuji, M., Sellati, T. J., Wong, C.-H., & Kronenberg, M. (2006). Natural killer T cells recognize diacylglycerol antigens from pathogenic bacteria. *Nature Immunology, 7*(9), 978–986. DOI: 10.1038/ni1380 PMID: 16921381

Lecina, M., Nadal, G., Solà, C., Prat, J., & Cairó, J. J. (2016). Optimization of ferric chloride concentration and pH to improve both cell growth and flocculation in Chlorella vulgaris cultures. Application to medium reuse in an integrated continuous culture bioprocess. *Bioresource Technology*, *216*, 211–218. DOI: 10.1016/j.biortech.2016.05.063 PMID: 27240237

Lisnjak, I. O., Kutsenok, V. V., Polyschuk, L. Z., Gorobets, O. B., & Gamaleia, N. F. (2005). Effect of photodynamic therapy on tumor angiogenesis and metastasis in mice bearing Lewis lung carcinoma. *Experimental Oncology*, *27*(4), 333–335. DOI: 10.1007/s00018-009-0016-4 PMID: 16404357

Lloyd, C., Tan, K. H., Lim, K. L., Valu, V. G., Fun, S. M. Y., Chye, T. R., Mak, H. M., Sim, W. X., Musa, S. L., Ng, J. J. Q., Bte Nordin, N. S., Bte Md Aidzil, N., Eng, Z. Y. W., Manickavasagam, P., & New, J. Y. (2021). Identification of microalgae cultured in Bold's Basal medium from freshwater samples, from a high-rise city. *Scientific Reports*, *11*(1), 4474. DOI: 10.1038/s41598-021-84112-0 PMID: 33627771

Maeda, N., Kokai, Y., Ohtani, S., Hada, T., Yoshida, H., & Mizushina, Y. (2009). Inhibitory effects of preventive and curative orally administered spinach glycoglycerolipid fraction on the tumor growth of sarcoma and colon in mouse graft models. *Food Chemistry*, *112*(1), 205–210. DOI: 10.1016/j.foodchem.2008.05.059 PMID: 23251235

Maeda, N., Matsubara, K., Yoshida, H., & Mizushina, Y. (2011). Anti-cancer effect of spinach glycoglycerolipids as angiogenesis inhibitors based on the selective inhibition of DNA polymerase activity. *Mini-Reviews in Medicinal Chemistry*, *11*(1), 32–38. DOI: 10.2174/138955711793564042 PMID: 21034405

Murakami, C., Kumagai, T., Hada, T., Kanekazu, U., Nakazawa, S., Kamisuki, S., Maeda, N., Xu, X., Yoshida, H., Sugawara, F., Sakaguchi, K., & Mizushina, Y. (2003). Effects of glycolipids from spinach on mammalian DNA polymerases. *Biochemical Pharmacology*, *65*(2), 259–267. DOI: 10.1016/S0006-2952(02)01483-1 PMID: 12504801

Osborne, C. K., Hobbs, K., & Trent, J. M. (1987). Biological differences among MCF-7 human breast cancer cell lines from different laboratories. *Breast Cancer Research and Treatment*, *9*(2), 111–121. DOI: 10.1007/BF01807363 PMID: 3620713

Pignatelli, P., Umme, S., D'Antonio, D. L., Piattelli, A., & Curia, M. C. (2023). Reactive oxygen species produced by 5-aminolevulinic acid photodynamic therapy in the treatment of cancer. *International Journal of Molecular Sciences*, *24*(10), 8964. DOI: 10.3390/ijms24108964 PMID: 37240309

Restiawaty, E., Marwani, E., Steven, S., Mega Rahayu, G., Hanif, F., & Prakoso, T. (2023). Cultivation of Chlorella vulgaris in mediums with varying nitrogen sources and concentrations to induce the lipid yield. *Indian Chemical Engineer, 65*(4), 369–380. DOI: 10.1080/00194506.2022.2164525

Rifai, A., Jamilah, S., Admirasari, R., & Prayitno, J. (2022, June). Alternative Source of Nutrients for Microalgae Cultivation in a Photobioreactor System. In *IOP Conference Series: Earth and Environmental Science* (Vol. 1034, No. 1, p. 012005). IOP Publishing. DOI: 10.1088/1755-1315/1034/1/012005

Rojas, A., González, I., & Morales, M. A. (2025). Natural products and cancer: The urgent need to bridge the gap between preclinical and clinical research. *World Journal of Gastrointestinal Oncology, 17*(4). Advance online publication. DOI: 10.4251/wjgo.v17.i4.100484 PMID: 40235887

Safe, S. (2024). Natural products as anticancer agents and enhancing their efficacy by a mechanism-based precision approach. *Exploration of Drug Science, 2*(4), 408–427. DOI: 10.37349/eds.2024.00054

Schneiter, R., Brügger, B., Sandhoff, R., Zellnig, G., Leber, A., Lampl, M., Athenstaedt, K., Hrastnik, C., Eder, S., Daum, G., Paltauf, F., Wieland, F. T., & Kohlwein, S. D. (1999). Electrospray ionization tandem mass spectrometry (ESI-MS/MS) analysis of the lipid molecular species composition of yeast subcellular membranes reveals acyl chain-based sorting/remodeling of distinct molecular species en route to the plasma membrane. *The Journal of Cell Biology, 146*(4), 741–754. DOI: 10.1083/jcb.146.4.741 PMID: 10459010

Singh, G., & Patidar, S. K. (2018). Microalgae harvesting techniques: A review. *Journal of Environmental Management, 217*, 499–508. DOI: 10.1016/j.jenvman.2018.04.010 PMID: 29631239

Stolyhwo, A., & Privett, O. S. (1973). Studies on the analysis of lipid classes by gradient elution adsorption chromatography. *Journal of Chromatographic Science, 11*(1), 20–25. DOI: 10.1093/chromsci/11.1.20

Stummer, W., Pichlmeier, U., Meinel, T., Wiestler, O. D., Zanella, F., & Reulen, H. J.ALA-Glioma Study Group. (2006). Fluorescence-guided surgery with 5-aminolevulinic acid for resection of malignant glioma: A randomised controlled multicentre phase III trial. *The Lancet. Oncology, 7*(5), 392–401. DOI: 10.1016/S1470-2045(06)70665-9 PMID: 16648043

Sugawara, T., & Miyazawa, T. (1999). Separation and determination of glycolipids from edible plant sources by high-performance liquid chromatography and evaporative light-scattering detection. *Lipids*, *34*(11), 1231–1237. DOI: 10.1007/s11745-999-0476-3 PMID: 10606047

Van de Loosdrecht, A. A., Beelen, R. H. J., Ossenkoppele, G., Broekhoven, M. G., & Langenhuijsen, M. M. A. C. (1994). A tetrazolium-based colorimetric MTT assay to quantitate human monocyte mediated cytotoxicity against leukemic cells from cell lines and patients with acute myeloid leukemia. *Journal of Immunological Methods*, *174*(1-2), 311–320. DOI: 10.1016/0022-1759(94)90034-5 PMID: 8083535

Wang, Z., & Benning, C. (2011). Arabidopsis thaliana polar glycerolipid profiling by thin layer chromatography (TLC) coupled with gas-liquid chromatography (GLC). *Journal of Visualized Experiments*, *e2518*(49). DOI: 10.3791/2518-v PMID: 21445048

Wolf, J., Ross, I. L., Radzun, K. A., Jakob, G., Stephens, E., & Hankamer, B. (2015). High-throughput screen for high performance microalgae strain selection and integrated media design. *Algal Research*, *11*, 313–325. DOI: 10.1016/j.algal.2015.07.005

Chapter 10
Ecological Resources and Environmental Solutions for Sustainable Development in Togo and Sierra Leone:
Challenges of Innovative Management in a Multicultural and Globalized African Context

Sèmèvo Magloire Assogba
http://orcid.org/0009-0009-3825-8836
Baikal State University, Russia

Malame Yakinambe
Baikal State University, Russia

Agossou Elie Adona
Baikal State University, Russia

Victor Yakovlevich Kuzevanov
Baikal State University, Russia

Shirley Koroma
Baikal State University, Russia

DOI: 10.4018/979-8-3373-1524-9.ch010

ABSTRACT

The nations of West Africa, notably Togo and Sierra Leone, face complex ecological and managerial challenges in a globalized and culturally diverse environment. This study focuses on issues related to small and medium-sized enterprises (SMEs), logistics, e-commerce, and intercultural management, which are key factors in achieving the sustainable development of these countries. SMEs in these two nations face limited access to finance, poor infrastructure, and insufficient managerial skills. Simultaneously, the rapid-growing e-commerce sector encounters significant challenges such as last-mile delivery problems, lack of adequate addressing systems, and high transport costs. Integrating sustainable development principles reveals a direct link to adapting management practices to local cultural contexts. The difficulties in adapting management practices, particularly in commercial interactions and international cooperation with partners such as Russia, China, the United States, and the European Union, have a directly affect the effectiveness of collaborations. This chapter proposes integrated strategies for sustainable development, including digitalization, change management approaches, promotion of green technologies, and the integration of local and regional supply chains. These strategies align the Sustainable Development Goals (SDGs) of the 2030 Agenda, particularly SDG 8 (Decent Work), SDG 9 (Industry, Innovation and Infrastructure), SDG 11 (Sustainable Cities and Communities, Eco-Cities), and SDG 17 (Partnerships to Achieve the Goals). Geospatial solutions will significantly advance e-commerce development.

INTRODUCTION

Since their adoption on 15 September 2015, the United Nations Sustainable Development Goals (SDGs) have provided a global framework for addressing the world's pressing environmental, social, and economic challenges by 2030 (Arora N et al., 2019). For many African nations, particularly in West Africa, achieving these goals depends on the sustainable management of their ecological resources. In countries like Togo and Sierra Leone, however, this management faces threats from rapid demographic growth, globalization, and intense urbanization (Kombate A et al., 2022, Stevens C M et al., 2025). The SDGs are not merely aspirations but form an interdependent network; failure in one area, such as climate action (SDG 13), hinders progress on others, including health (SDG 3) and poverty eradication (SDG 1). This highlights how achieving the entire 2030 Agenda relies on the sustainable stewardship of ecological resources.

Sierra Leone and Togo, nations blessed with tremendous ecological diversity and abundant mineral wealth, stand in profound vulnerability. Despite their natural capital, they face the triple perils of land degradation, climate change, and energy insecurity. This reveals a poignant paradox: the very abundance of resources that should guarantee prosperity instead fuels a cycle of unsustainable development. This cycle stems deep-seated systemic governance failures.

Further exacerbating this crisis, both nations confront a devastating coastal degradation epidemic. This scourge is rapidly spreading across sub-Saharan Africa. In a response, Togo collaborates with initiatives like the West Africa Coastal Areas Management (WACA) program and the Reducing Emissions from Deforestation and Forest Degradation (REDD+) committee. These initiatives have launched ambitious projects to increasing forest cover over the next three decades (Kombate A et al., 2022).

National development trajectories face intricate cross-cultural dynamics, pervasive institutional weaknesses, and a crippling dependence on formal economic structures (Stevens C M et al., 2025). The informal sector, often beyond the effective governance, lacks capacity for environmental compliance. It also faces challenges in accessing financing for green technologies, and remains difficult for governments to support or regulate.

Africa's e-commerce sector is burgeoning. It is poised to reach 500 million users by 2025, with a striking annual growth rate of 17% (ITA, 2025). However, this progress generates severe environmental costs, including packaging waste and carbon-intensive transportation emissions. This research examines the potential for innovative management practices to mitigate these pressing challenges within Africa's unique multicultural context. It focuses on optimizing resource use, implementing robust environmental policies, and harnessing international cooperation. The goal is to advance SDGs 7, 11, 13, and 17.

Figure 1. Location of Togo and Sierra Leone in West Africa. The map illustrates their geographical position on the Atlantic coast, highlighting coastal vulnerability to climate change and regional cooperation within the West African coastal zone (e.g., the WACA initiative). Source: BBC News (1999), modified for research purposes.

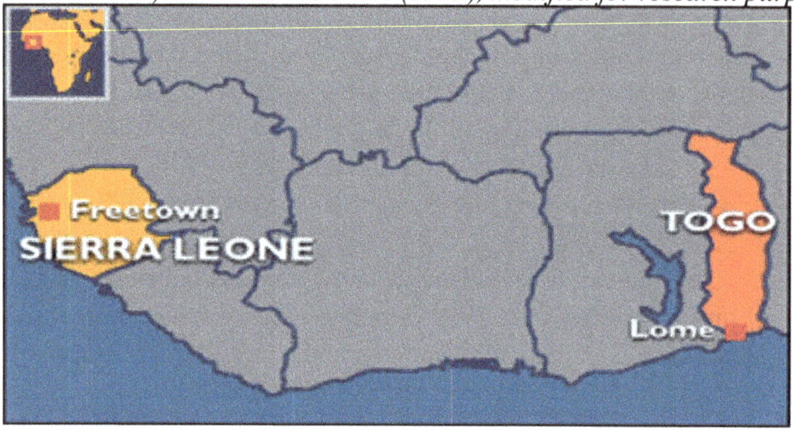

Togo and Sierra Leone illustrate the inherent tension between development aspirations and ecological preservation. Ecology is one of the key factors in development. Socio-cultural resistance to change" is a crucial, often underestimated factor. It suggests that policy failures are not only due to a lack of political will or resources, but also stem from deeply rooted cultural norms and practices.

In Togo, more than 80% of households rely wood and charcoal, derived from the uncontrolled felling of trees. This practice leads to forests loss of more than 1.5% per year (Ehlui K et al., 2024). To develop adaptable solutions, stakeholders must address energy access based on principles that preserve the environment, and avoid negative health impacts.

In Sierra Leone, mining activities, particularly titanium and diamond extraction, destroy mangroves in critical coastal areas. This threatens carbon sequestration and fisheries (Stevens C M et al., 2025). As Daniel Tang's demonstrates, soil chemical changes and, alterations in microbial flora contribute significantly to greenhouse gas emissions (Muthmainna S K et al., 2022). despite commitments under Nationally Determined Contributions (NDCs) to promote renewable energy public investment in clean technologies remains below fossil fuel subsidies (Nsabimana A., 2022). This mismatch between political intention and practical execution stems from, cross-cultural differences in decision-making. The aim of this study is to assess current environmental governance mechanisms and propose integrated solutions. These solutions will leverage innovation, cross-border collaboration and local knowledge systems to support long-term ecological resilience and inclusive development.

Sustainable development in West Africa has been widely studied through the lenses of political economy, environmental science, and international aid. Researchers emphasize the importance of institutional coherence and participatory governance in achieving the SDGs. UNEP research highlights that weak enforcement of environmental regulations often result from overlapping mandates and corruption in resource-rich countries (Nsabimana A., 2022). The recurring theme of "weak enforcement" and "corruption" indicate that institutional capacity and integrity are as crucial as policy design.

This implies that even well designed policies and significant international assistance can be ineffective if the implementing institutions are compromised or lack of capacity. Therefore, governance reforms and anti-corruption measures as prerequisites for effective environmental management.

The Uppsala model of internationalization offers insights into how small businesses in Sierra Leone struggle to penetrate global markets due to logistical and cultural barriers (Nsabimana A., 2022). Meanwhile, Hofstede's cultural dimensions theory reveals that strong hierarchical distance and collectivist values in Togo influence hierarchical management styles, limiting adaptive responses to ecological crises (Hofstede G et al., 2009).

Agro ecology and precision agriculture have shown promise for improving food security while reducing environmental impact (Klerkx, L., et al.,2024), contributing to MDGs 2 and 12. However, a notable gap in studies that combine geospatial analysis with socio-economic field data to evaluate sustainability interventions in multi-ethnic and low-income contexts. This study fills this gap by proposing three key objectives: (1) to map trends in ecological degradation using remote sensing; (2) to analyze socio-political factors affecting the adoption of environmental policies; and (3) to develop context-specific recommendations for green innovation and cross-border cooperation.

MATERIALS AND METHODS

This study relies on a meticulous examination of desk research and survey data, utilizing quantitative information sourced from reliable outlets. The methodological approach unfolded in two primary phases: first, content analysis of key strategic documents, notably Togo's National Strategy for Sustainable Development (SNDD) and Sierra Leone's National Policy on Climate Change. Second, examination of geospatialdata and bilateral agreements between Togo and Sierra Leone. Cultural dimensions were analyzed the Human Development Index (HDI) and Geert Hofstede's Cultural Dimensions framework, visualized via Canva. Field surveys were impracticable due to physical constraints; consequently, online quantitative data was deemed unreliable. A quantitative research design ensures data integrity. Ecological resources are defined using Constanza's model (Constanza R et al., 1997), which integrates ecological resources and economic processes across nine commodities and processes. Notably, it distinguishes between two 'economic' processes (urban and agricultural) and seven 'ecological' ones, encompassing both terrestrial and marine systems.

Our analysis targeted strategic sectors including agriculture, fisheries, ecological management, and e-commerce. This focused approach, further enriched by a thorough review of relevant academic studies and sectoral statistics, was instrumental in enabling a rigorous and insightful examination.

RESULTS AND DISCUSSION

Due to their geographical position within Africa, Togo and Sierra Leone face environmental challenges that plague much of sub-Saharan Africa; these challenges trigger economic and managerial difficulties. Analysis reveals a profound interconnection between natural resources, institutional, and management cultures. Intervening in one-area ripples through others.Thus, a piecemeal approach fails. Integrated action. Such a holistic strategy is imperative to reconcile the socio-ecological system, ensuring policy coherence and stakeholder's coordination.

Despite mining wealth, both nations face a daunting paradox: they are rich in resources yet rank among the lowest on the Human Development Index (Aboudala S I et al., 2018). This troubling contradiction is compounded by critical infrastructure deficits and a heavy unsustainable dependence on traditional energy sources. Together, these formidable barriers are severely hindering progress towards crucial development goals, specifically SDG 7 (Affordable and Clean Energy), SDG 8 (Decent Work and Economic Growth), and SDG 13 (Climate Action). In direct response to these intersecting challenges, Togo has strategically prioritized SDG 3 (Good Health and Well-being) in its allocation of development aid, recognizing that a healthy population is a fundamental prerequisite for overcoming broader developmental obstacles.

The reliance on firewood and charcoal for cooking extends to over 80% of households, driving a severe crisis of forest ecosystem degradation. This is particularly devastating, as large trees, crucial for attracting rainfall are increasingly felled by charcoal producers and ravaged by seasonal bushfires. The consequences of this deforestation create a harmful cascading effect throughout the entire environmental chain.

A similar damaging pattern is observed in Sierra Leone, where traditional slash-and-burn agriculture remains the predominant method for clearing land. This practice persistently leads to extensive soil erosion and a tragic loss of biodiversity. A stark report by the FAO in 2021 confirms that this method accounts for 60% of agricultural techniques in the country's rural areas, resulting in an alarming annual forest cover loss of 2.5% and significantly worsened soil erosion.

Figure 2. Distribution of Official Development Assistance (ODA) by SDG in Togo and Sierra Leone. Modified from the source (Nsabimana A., 2022). The chart shows that the largest share of aid is directed towards SDG 3 (Good Health and Well-being), followed by SDG 9 (Industry, Innovation and Infrastructure), indicating donor priorities and the potential underfunding of environmental programs

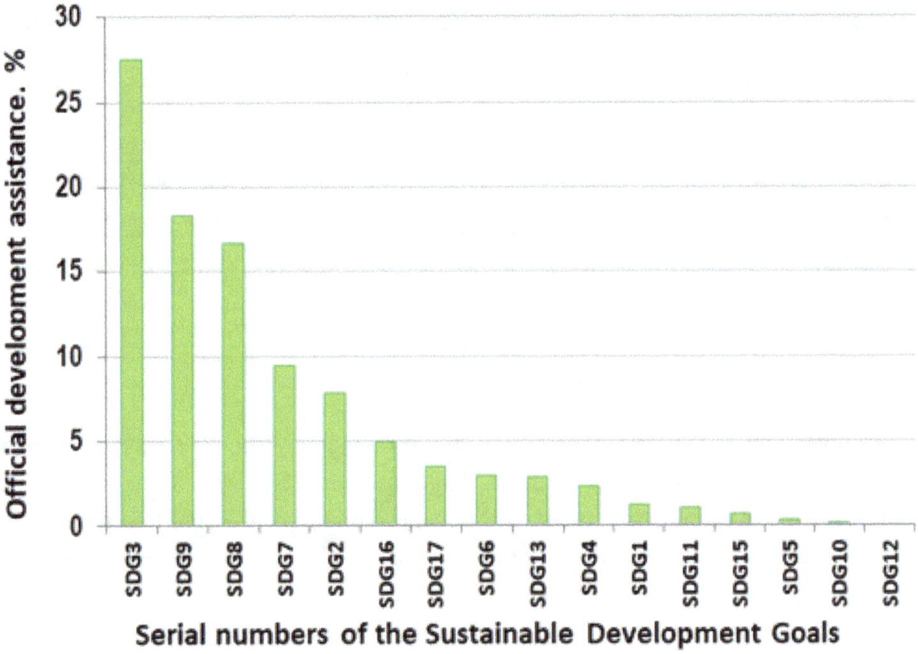

Analysis suggests that these environmentally destructive behaviors are, to a significant extent, deeply embedded cultural attitudes. This makes them exceptionally difficult to manage, as they are connected with long-standing local values and traditional practices.

Our own upbringing in Togolese villages provided a firsthand perspective on a devastating economic reality: communities, driven by necessity for cooking fuel and income, are often forced to target large trees for charcoal production. This practice, while providing a vital source of domestic energy and meager income, comes at an immense environmental cost. Regrettably, analysis confirms that this has led to substantial environmental degradation in both Togo and Sierra Leone. Alarming data shows that Togo's forest cover diminished by 18% between 2015 and 2025, a loss primarily driven by agricultural expansion and relentless firewood collection (Kombate A et al., 2022). The situation is equally dire in Sierra Leone, where mangrove loss has reached 22% in mining-affected regions, severely compromising natural coastal defenses and critical biodiversity (Stevens C M et al., 2025). The scale of these interconnected challenges clearly illustrate by key indicators of environmental degradation (Table1).

Table 1. Key indicators of environmental degradation in Togo and Sierra Leone (2015–2025). The table summarizes data on biomass use, tree cover loss, slash-and-burn farming practices, and energy financing. The indicators highlight the scale of environmental challenges and the need for comprehensive solutions

Indicators	Togo	Sierra Leone
Percentage of households using wood/coal for energy	80%	Similar (traditional practices)
Annual rate of forest loss due to wood/charcoal production	1.5%	2.5% (due to slash-and-burn agriculture)
Decline in forest cover (2015–2025)	18%	N/A (focus on mangroves)
Loss of mangroves in mining regions	N/A	22%
Percentage of slash-and-burn agriculture in rural areas	N/A	60%
Diesel subsidies vs. renewable energy budget	N/A	$50 million vs. $8 million

Modified from the source (Ehlui K et al., 2024)

Environmental degradation can also be explained by the loss of forest cover observed over the period 2001 to 2024. Togo, as shown in Figure 3, lost 82.1 Kilo hectares (Kha) of forest cover during this decade, equivalent to 82100 hectares, while Sierra Leone lost 2.17 Mega hectares (Mha), equivalent to 217000 Kilo hectares, representing a major loss of forest cover.

Figure 3. The dynamics of the annual tree cover loss in Togo and Sierra Leone. Between 2001 and 2024, Togo lost 82.100 hectares of tree cover, which has a link to unsustainable fuelwood collection and agricultural expansion (2017 recorded the highest loss.). Sierra Leone reached its highest deforestation levels in 2015 and 2017 due to the uncontrolled cutting of trees. Modified from the source (Global Forest Watch., 2025) based on Landsat remote sensing data

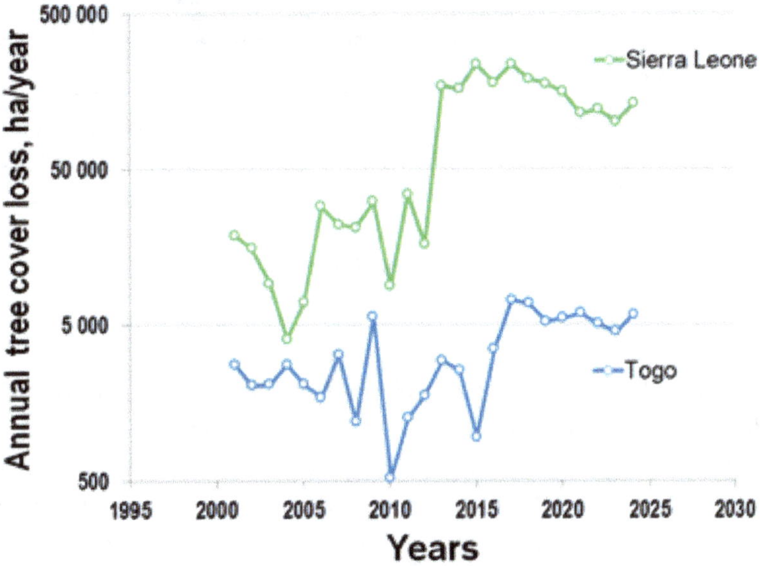

The intercultural context of management plays a particularly crucial role in shaping environmental governance. A comparative analysis using Hofstede's cultural dimensions reveals striking similarities between Togo, Sierra Leone, and Russia — all characterized by high power distance (PDI = 77 and 93 respectively) and low individualism (IDV = 20 and 46) (Trompenaars F., 2012). These cultural traits foster highly centralized management styles and inherent skepticism toward new initiatives proposed by civil society or international partners. The high power distance and collectivist orientation concentrate decision-making authority within hierarchical structures, leaving little room for individual initiative or bottom-up innovation. This creates a significant bottleneck for environmental programs promoted by external donors, which often face resistance or passive opposition at the local level.

Furthermore, the strong uncertainty avoidance observed in these societies (UAI = 54 in Togo versus 95 in Russia) reinforces risk-averse behavior among local entrepreneurs and government officials. This cultural characteristic substantially slows adaptation to climate change and impedes the adoption of green technologies, which are often perceived as uncertain or untested. The causal relationship is compelling: deeply embedded cultural values directly influence institutional responses to environmental challenges and technological innovation.

Table 2. Hofstede's Cultural Dimensions for Selected Countries (PDI, IDV, MAS, UAI, LVSO)

Parameters	Countries			
		Armenia	Togo and Sierra Leone	East African region
Hierarchical Distance Index (PDI)	93	85	77	64
Individualism vs. Collectivism (IDV)	46	17	20	27
Masculinity vs. Femininity (MAS)	36	50	46	41
Uncertainty Avoidance Index (UAI)	95	88	54	52
Long-Term vs. Short-Term Orientation (LVSO)	58	38	9	32

Modified from the sources [Hofstede Comparison Tool, Hofstede's survey file 2015]

Figure 4. Culture Dimensions of Geert Hofstede's Cultural Dimensions: A Comparison of Togo, Sierra Leone, Russia, and Armenia. The chart compares six cultural dimensions, including Power Distance (PDI), Individualism (IDV), and Uncertainty Avoidance (UAI). High PDI scores in Togo and Russia indicate centralized governance, which affects the perception of top-down environmental initiatives. Modified from the source (Hofstede G et al., 2010)

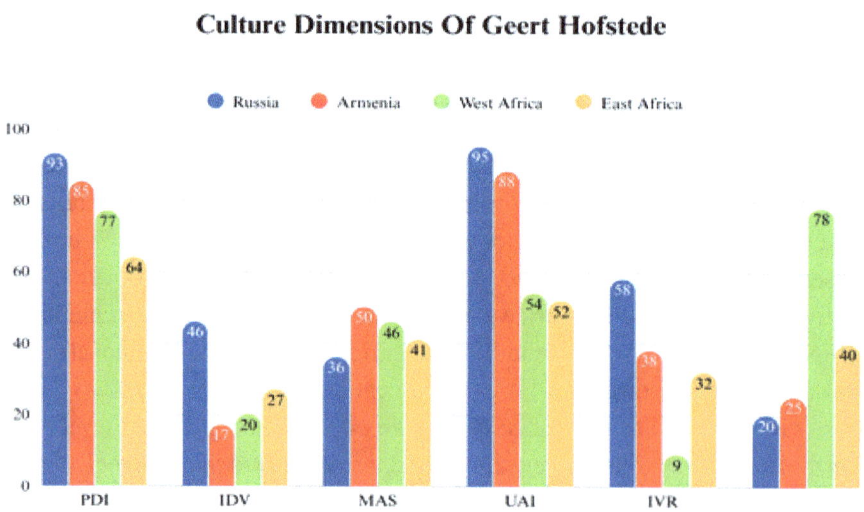

The effectiveness of international cooperation is affected by cultural differences. Considering intercultural and geopolitical dimensions into account is therefore essential if sustainable development initiatives are to be strengthened. As *Sizykh* and *Assogba (2025)* point out. An analysis framed by Trompenaars' model reveals that while countries like Russia, Armenia, Togo, and Sierra Leone show relative alignment on dimensions such as universalism/particularism (26 vs. 32) and affectivity/neutrality (6 vs. 51), a significant cultural chasm emerges when these nations are compared to the United States (*Table 3*). The U.S. scores exceptionally high (92) on universalism, indicating a societal preference for rules, contracts, and objective criteria. This divergence creates a critical mismatch in international partnerships. Western partners, who strongly adhere to Universalist principles, prioritize formal agreements and standardized procedures. In contrast, partners in Togo and Sierra Leone, where particularism is dominant, prioritize personal relationships, context, and trust above formal rules. This fundamental clash in expectations regarding decision-making, accountability, and implementation is a primary source of misunderstanding and a leading cause of project failure.

Table 3. Trompenaars' Cultural Dimensions for Togo, Sierra Leone, Russia and Armenia

Parameters	Countries			
	Russia	Armenia	Togo	Sierra Leone
Particularism vs Universalism	26	30	32	32
Individualism vs Communitarianism (Collectivism)	92	32	28	28
Broadcast vs Specific	86	33	27	27
Affective vs Neutral	6	51	43	43
Ascription Vs Achievement	16	30	30	30

Adopted and summarized from (Trompenaars F., 2012)

Therefore, for any sustainable development initiative to succeed, it is not merely beneficial but essential to integrate a comprehensive intercultural training component for all involved parties. This is crucial for bridging the gap between Universalist and particularistic approaches, fostering mutual understanding, and aligning expectations from the outset. Companies in Sierra Leone's commercial sector face a complex array of obstacles, including a pronounced lack of product standardization, difficulties in obtaining certifications, persistent logistical challenges, and insufficient training for future managers. In this challenging context, digitalization is poised to play a pivotal, yet complicated, role.

The development of e-commerce, though still nascent in Togo, presents promising new opportunities. However, its environmental integration faces a significant hurdle: the last mile of delivery can constitute up to 40% of total logistics costs, representing a major sustainability challenge. This is particularly critical given the structure of Togo's economy, where small and medium-sized enterprises (SMEs) account for approximately 90% of all economic entities and provide nearly 75% of private sector jobs (Togo First., 2024).

Figure 5. Trompenaars Cultural Dimensions: A Comparison of West African Countries and Their International Partners. A comparison based on dimensions such as Universalism/Particularism and Affective/Neutral relationships reveals cultural gaps between Western partners and African countries, which can hinder effective cooperation in sustainable development. Adopted and modified from the source (Trompenaars F., 2012)

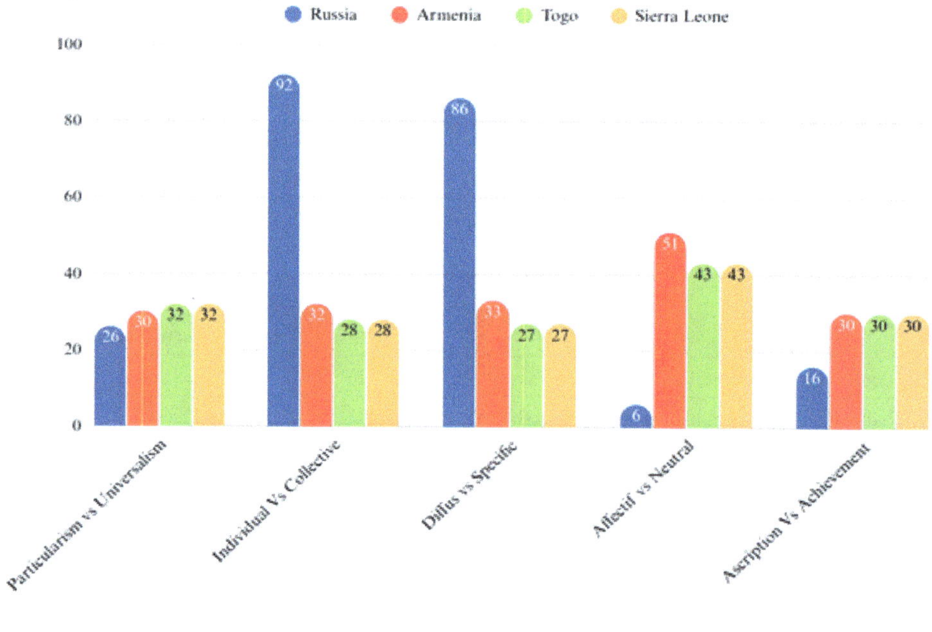

While digitalization is often hailed as a transformative solution, its promise is tempered by stark realities such as the lack of a clear addressing system, low financial inclusion, and high last-mile costs.

These issues reveal a critical disconnection. Without first addressing these fundamental infrastructural and financial gaps, the potential of e-commerce to drive sustainable development remains severely limited. In fact, inefficient logistics could lead to increased emissions. Nevertheless,we can overcome these challenges through innovative geospatial solutions, including the implementation of GPS tagging and the comprehensive mapping of informal settlements. In Togo, small and medium-sized enterprises (SMEs) constitute approximately 90% of all economic entities and serve as a critical anchor for employment. Paradoxically, however, their contribution to the national GDP remains disproportionately low, primarily due to constrained access to credit, cumbersome administrative barriers, and a pervasive shortage of innovative solutions 1(Togo First., 2024) 8. This discrepancy is widely illustrated by data from the Centre de Formalités des Entreprises (CFE= Centre de Formalités des Entreprises) in Lomé, which reported over 15,000 new business registrations in 2024, yet only a minimal number managed to formalize and expand beyond the informal sector. The fact that SMEs dominate the economic landscape yet contribute minimally to GDP underscores a fundamental development challenge. This structural limitation is largely driven by limited financial access — exacerbated by low banking penetration — complex regulatory hurdles, and insufficient innovation. This dynamic creates a vicious cycle: informal enterprises cannot obtain the resources necessary for growth, thereby constraining their ability to adopt sustainable practices or integrate meaningfully into the formal economy. In response, one promising avenue may be the establishment of business incubators specifically focused on green technologies, such as bio fertilizer production, waste recycling, or decentralized solar mini-grids for rural electrification. Despite these challenges, several encouraging developments are emerging. Pilot projects introducing solar-powered irrigation and block chain-enabled mineral tracking are already demonstrating potential. Such digital tools have enhanced supply chain transparency, thereby supporting progress toward SDG 9 (Industry, Innovation, and Infrastructure) and SDG 16 (Peace, Justice, and Strong Institutions). Similarly, community-led reforestation initiatives have shown markedly higher success rates when designed in alignment with traditional land-use customs (Nsabimana A., 2022). This stands in direct contrast to the distrust often encountered by top-down conservation programs, highlighting a crucial insight: effectiveness depends not only on the solution itself, but also on the approach to implementation. Bottom-up, culturally resonant strategies prove substantially more effective, reaffirming the vital importance of intercultural management (Hofstede Get al., 2010) .

These findings collectively emphasize the need for governance reforms that are both culturally sensitive and technologically smart — prioritizing equity, local context, and ecological integrity in the pursuit of sustainable development.

The demographic landscape of Togo and Sierra Leone is undergoing significant transformation. The general population in both countries continues to grow steadily, yet the most transformative trend lies in rapid urbanization. Currently, a majority of the population in both nations still resides in rural areas. However, projections indicate that around 2035, a critical demographic turning point will occur: the urban population will reach parity with the rural population and subsequently become dominant. This shift will inevitably have a profound impact on all aspects of life, including energy consumption patterns, pressure on urban infrastructure, education, healthcare, and the implementation of sustainable development goals related to the economy and ecology. Regarding well-being, over the past 30 years, its dynamics — as measured by the Human Development Index (HDI) — have indeed shown volatility, particularly during periods of political instability. However, recent decades have seen a confident positive trend. Looking at the evolution of the Human Development Index, Togo and Sierra Leone experienced constant growth from 1990 to 2023, but according to the classifications, Togo has an average HDI and Sierra Leone a low HDI. The growth rates are as follows: Togo, 1990–2023: 0.412–0.571; Sierra Leone, 1990–2023: 0.313–0.467.

The rise in HDI in both countries can be explained by a combination of several key factors:

1) Improved access to education: expansion of school enrolment, especially among girls and young women.
2) Progress in healthcare: reduction in child and maternal mortality, combating epidemics, and improved access to basic medical care.
3) Gradual economic development: despite persistent challenges, there has been slow but steady economic growth, partly supported by improved management in industry and trade, as well as the sustainable use of ecological resources.
4) Urbanization:, moving to cities often provides populations with access to better social services, income opportunities, and education, which directly contributes to improvements in the health and welfare metrics accounted for in the HDI.

Thus, although both countries face significant challenges, the steady growth in HDI indicates slow but consistent progress in human development.

Figure 6. Evolution of the Human Development Index (HDI) as well as total, urban and rural populations in Sierra Leone (upper plot) and Togo (lower plot). Both countries show positive HDI dynamics; however, Sierra Leone remains in the low human development category. HDI growth is link to improvements in education and healthcare but requires the integration of environmental factors for long-term sustainability. Adopted and summarized from (Herre B et al., 2023)

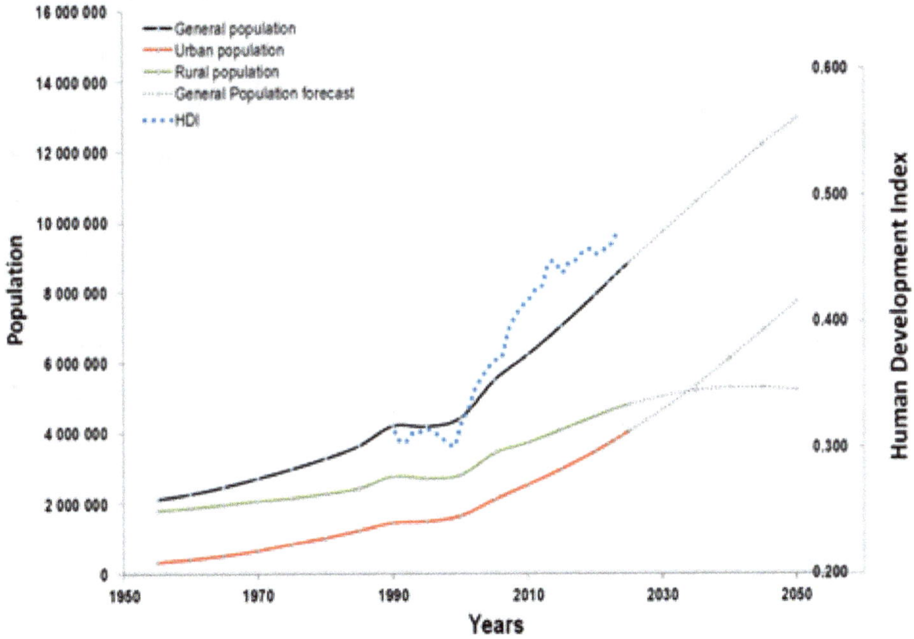

Figure 7. Evolution of the Human Development Index (HDI) as well as total, urban and rural populations in Sierra Leone (upper plot) and Togo (lower plot). Both countries show positive HDI dynamics; however, Sierra Leone remains in the low human development category. HDI growth is link to improvements in education and healthcare but requires the integration of environmental factors for long-term sustainability. Adopted and summarized from (Herre B et al., 2023)

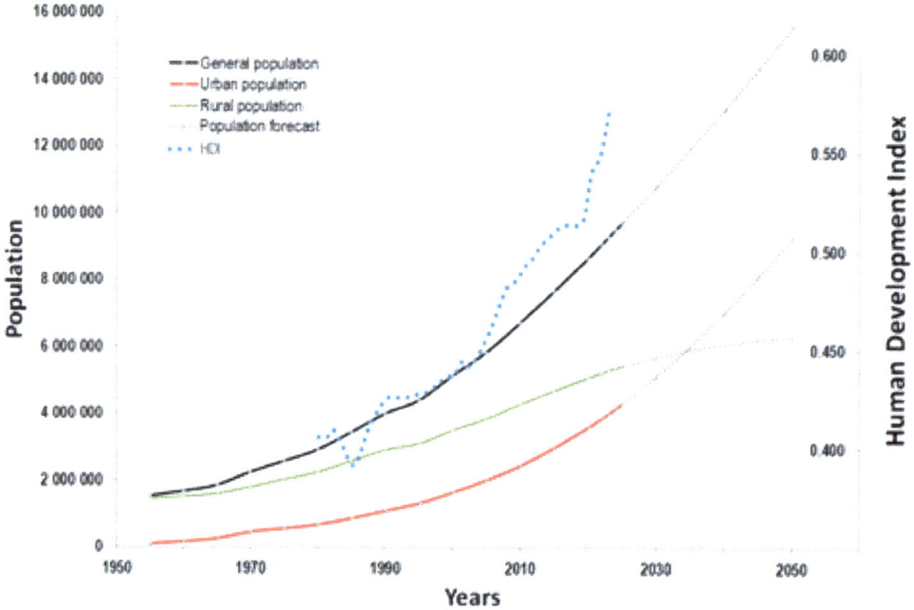

CONCLUSION

Togo and Sierra Leone must not limit themselves to mere political declarations in their pursuit of sustainable development. Profound and structural action and reforms must encompass all sectors and not be isolated measures. Environmental, economic, and social challenges are interlinked and must be addressed holistically, as they take into account chain reactions. Climate policy takes into account employment, education, culture, and other factors. A multidimensional approach to these challenges is necessary for both countries, taking into account the concept of balanced sustainable development and aligning policies across all sectors. The fundamental levers for achieving the objectives are decentralized renewable energies, such as the CIZO projects in Togo, which contribute to the achievement of the SDGs (African Development Bank Group., 2019) 19, aligning with community management of resources (forests, fisheries, agricultural land). Policies on climate goals and renewable energy enable progress to be made on several SDGs.

Environmental challenges are deeply connected with social, educational, and economic spheres, particularly in developing nation. . For Sierra Leone and Togo, addressing these interconnected issues demands a multi-faceted approach: enacting robust legislation, advancing environmental education, and fostering ecological and fair-trade markets. The ultimate aim of these measures is to ensure the sustainable protection of natural resources while simultaneously boosting employability and fostering inclusive economic growth in both countries. Within this framework, science, biodiversity conservation, education, and sustainable economic practices (including free markets and community initiatives) function as circular catalysts, each reinforcing the other in a cycle of positive feedback. To operationalize this, urban centers must harness sophisticated management systems to positively influence local ecology, thereby facilitating the integration of green energy solutions into our terrestrial ecosystems and built environments.

REFERENCES

Aboudala, S. I., Minkilabe, D., & Thiou, T. K. T. (2018). Impacts socio-économiques de l'exploitation minière sur les populations riveraines. Cas du gisement de calcaires de Tabligbo dans la préfecture de Yoto au Togo. *Revue ANYASA, 8,* 152-159. https://hal.science/hal-02265636v1

African Development Bank Group. (2019). Projet d'électrification rurale CIZO – Résumé CPR. African Development Bank. https://golnk.ru/ZB2E4

Arora, N. K., & Mishra, I. (2019). United Nations Sustainable Development Goals 2030 and environmental sustainability: race against time. *Environmental Sustainability, 2,* 339 342. DOI: 10.1007/s42398-019-00092-y

BBC News. (1999). Massacre Sierra Leone (Afrique). BBC News. http://news.bbc.co.uk/2/hi/africa/325227.stm

Costanza, R., d'Arge, R., de Groot, R., Farber, S., Grasso, M., Hannon, B., Limburg, K., Naeem, S., O'Neill, R. V., Paruelo, J., Raskin, R. G., Sutton, P., & van den Belt, M. (1997). the value of the world's ecosystem services and natural capital. *Nature, 387*(6630), 253–260. DOI: 10.1038/387253a0

Ehlui, K. S., Atakpama, W., von Wehrden, H., Bah, A., Kola, E., Anthony-Krueger, C., Egbelou, H., Kokou, K. B., & Boukpessi, T. (2024). Anthropogenic Threats to Degraded Forest Land in the Savannahs' Region of Togo from 1984 to 2020, West Africa. *Journal of Geoscience and Environment Protection, 12*(1), 164–179. DOI: 10.4236/gep.2024.121011

Global Forest Watch. (2025). Togo Dashboard. https://golnk.ru/gr325

Herre, B., & Arriagada, P. (2023). The Human Development Index and related indices. *Published online at OurWorldinData.org*. https://golnk.ru/MvWEZ

Hofstede, G., Hofstede, G. J., & Minkov, M. (2010). Table data (Russia, Armenia, Togo, Sierra Leone and East Africa). Cultures and Organizations: Software of the Mind (3rd ed.). McGraw-Hill. https://golnk.ru/851O5

ITA. (2025). Mobile-Driven Africa. https://www.trade.gov/rise-ecommerce-africa

Johanson, J., & Vahlne, J.-E. (2009). The Uppsala internationalization process model revisited. *Journal of International Business Studies, 40*(9), 1411–1431. DOI: 10.1057/jibs.2009.24

Klerkx, L. (2024, December 18). The evolution of precision agriculture and food safety: A bibliometric review. *Frontiers in Sustainable Food Systems*, *8*, 1475602. DOI: 10.3389/fsufs.2024.1475602

Kombate, A., Folega, F., Atakpama, W., Dourma, M., Wala, K., & Goïta, K. (2022). Characterization of Land-Cover Changes and Forest-Cover Dynamics in Togo between 1985 and 2020 from Landsat Images Using Google Earth Engine. *Land (Basel)*, *11*(11), 1889. DOI: 10.3390/land11111889

Muthmainna, S. K. (2020). A Systematic Review of Slash-and-Burn Agriculture as an Obstacle to Future-Proofing Climate Change. *International Journal of Research in Agricultural Sciences*, *4*(1), 1–8. https://golnk.ru/5DArv

Nsabimana, A. (2022). *Évaluation et stratégie du financement et développement du Togo*. PNUD, Ministère de l'Economie et des Finances (MEF) et du Ministère de la Planification du Développement et de la Coopération (MPDC). https://inff.org/assets/resource/efd_pnud-togo-dfa-final.pdf

Robitaille, P. (2004). Les diamants en Sierra Leone, de la mine à la mort. https://ceim.uqam.ca/db/spip.php?page=article-cirdis&id_article=2065

Stevens, C. M. B., Bojang, O. A., Amara, E., & Msuya, A. S. (2025). Comprehensive Overview of Deforestation Dynamics and Sustainable Forest Management in Sierra Leone. *European Journal of Ecology. Biology and Agriculture*, *2*(2), 18–29. DOI: 10.59324/ejeba.2025.2(2).03

Togo First. (2024). Près de 75% des emplois dans le secteur privé créés par les TPME. https://golnk.ru/BdewE

Trompenaars, F. (2012). Culture Xplore. *Cultural Dimensions Tool*. https://golnk.ru/KkyEO

Chapter 11
Assessment of Heavy Metals Contamination in Soils of Mining Area of Tumanyan Region, Armenia: Distribution and Environmental Risks

Hasmik Vardumyan
http://orcid.org/0009-0009-6571-8471
Faculty of Biology, Yerevan State University, Armenia

Abhishek Singh
http://orcid.org/0000-0002-0049-6910
Faculty of Biology, Yerevan State University, Armenia

Rupesh Kumar Singh
http://orcid.org/0000-0002-2536-1967
Centre for the Research and Technology of Agro-Environmental and Biological Sciences (CITAB), Inov4Agro, University of Trás-os-Montes and Alto Douro, Portugal

João Ricardo Sousa
http://orcid.org/0000-0003-4365-9086
Centre for the Research and Technology of Agro-Environmental and Biological Sciences (CITAB), Inov4Agro, University of Trás-os-Montes and Alto Douro, Portugal

DOI: 10.4018/979-8-3373-1524-9.ch011

Karen Ghazaryan
http://orcid.org/0000-0002-8973-2595
Faculty of Biology, Yerevan State University, Armenia

ABSTRACT

Heavy metal pollution of soils has brought potential risk to the environment and human health. The chapter is a long-term study evaluating the extent of heavy metal pollution in soils of one of Armenia's most important mining regions. Soil samples were taken from various distances (0–20 cm) of the surface of both active and abandoned mining sites. The sampling sites were selected based on accessibility and their distance from contaminant sources. Three other baseline soil properties, pH, texture, and organic matter content, were also taken to aid interpretation. Analytical methods for ecological risk assessment indices like potential ecological risk index (PERI), pollution index (PI), Nemerow integrated pollution index (NIPI) were useful for this study. The findings revealed that the concentration levels of some heavy metals were considerably higher than the local and world set standards, in close proximity to tailings and ore processing areas. Spatial distribution maps indicate a spatial association of heavy metals uptake with the distance from mining activities. According to the ecological risk assessment, there is high potential of phytotoxicity and wide contamination of food chain in treated sites. These results highlight the necessity for frequent monitoring and the application of reparation techniques such as phytoremediation to reduce heavy-metal risks in mining-affected areas of Armenia. The work adds to the limited literature on post-mining land use planning and environmental health in South Caucasus.

INTRODUCTION

Soil acts as a sink for several environmental pollutants such as heavy metals (HMs) (Xia et al., 2025). Earth soil pollution has attracted global attention in recent decades due to its possible effects on human health (Brevik et al., 2020). Preventing, or mitigating soil pollution is also supportive of the Unite Nations SDGs with goal 2 "zero hunger", since HM soil pollution decreases food quality and yield, goal 3 "good health and well-being", and goal 15 "life on land" (Briassoulis, 2023; Lal et al., 2021; Rajan Selin & Shanmugasundram, 2024). The soil is a complex and dynamic, constantly changing part of the earth ecosystem. It is necessary for human kind to survive and civilization to advance. Soil quality impacts food safety, crop quality, and ultimately human health (Rajput et al., 2024; Taghavi et al., 2024). The impact of pollutants in the environment is increasingly becoming more due to a series of human activities such as excessive industrialization, urban construction, resource exploitation by mining, pesticide application, and the exhaust emissions from motor vehicles (figure 1) (Taghavi et al., 2024). Degradation of soils by different pollutants is turning out to be a major environmental problem for many countries, including Armenia (Rajput et al., 2024). Particularly, the heavy metals pollutions and potential human health risks in crop-growing and agriculture area can become a serious issue (Vasilachi et al., 2023). Heavy metals include iron (Fe), aluminium (Al), cadmium (Cd), nickel (Ni), copper (Cu), lead (Pb), zinc (Zn) and mercury (Hg) (Vasilachi et al., 2023). Heavy metals can occur in the soil from both natural geological process and anthropogenic activities (Angon et al., 2024a). The impact of human activities can cause significant changes in soil properties, increasing the content of heavy metals found in larger levels than those contributed by natural sources (Angon et al., 2024a). Emissions of increasing industrial areas, chemical fertilizers, organic manures, untreated wastewater sludge, pesticides (including acaricides, insecticides, fungicides, rodenticides and herbicides), wastewater for irrigation purposes and atmospheric pathways would be the main sources of metal accumulation in the soil of several regions (Vácha, 2021; Wuana & Okieimen, 2011). Therefore, heavy metals are hazardous to human health and the transfer of metals into food chain through soil application become accumulated in crops for human consumption (Angon et al., 2024a).

Figure 1. HMs contamination and accumulation that associated risks in environmental and ecological food chains. (A) explain the various type of source responsible for HMs. (B-C) bioaccumulation of HMs at the terrestrial and aquatic levels

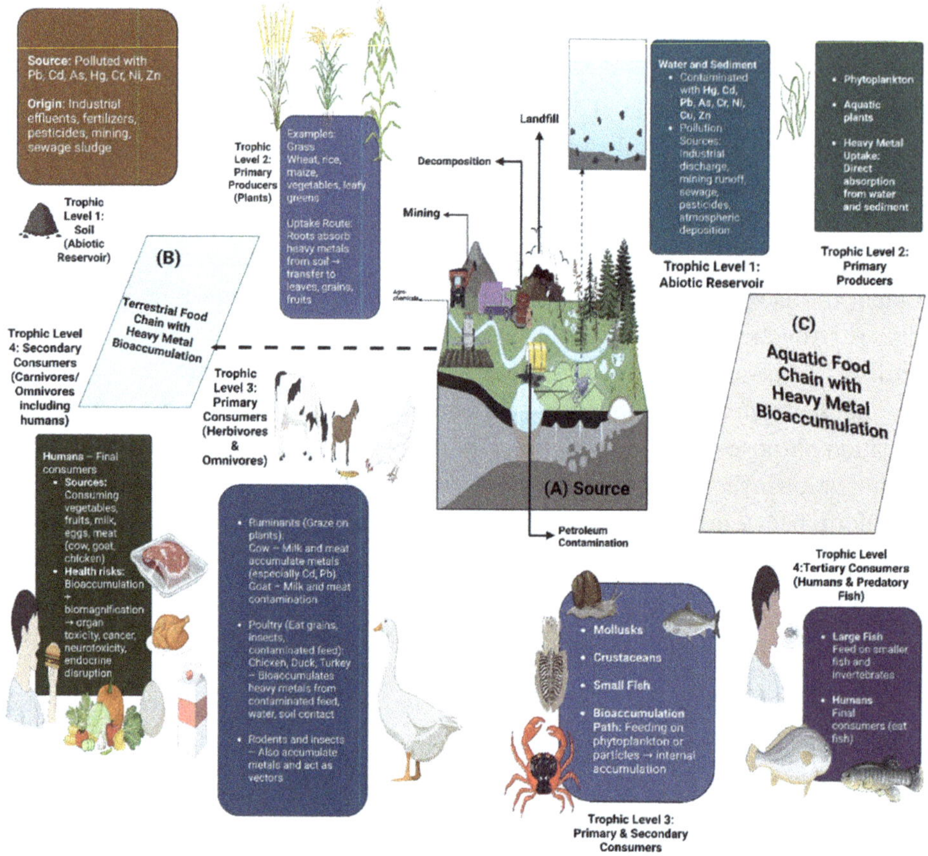

The mining sector is an essential component of the world's economic development providing crucial materials for various industries (Soltani et al., 2017). Recently, the issue of soil contamination and other environmental impacts caused by mining has become a significant environmental concern that demands urgent attention (Haghighizadeh et al., 2024a).

The level of hazard posed by HMs soil pollution can be assessed through various single and composite indices, including the *pollution index (PI), contamination factor (CF), enrichment factor (EF), geoaccumulation index (Igeo), integrated pollution index (IPI),* and *pollution load index (PLI).* These indices are derived from the estimation of HMs in analyzed samples relative to background concentrations (Weissmannová & Pavlovský, 2017).Ecological risks associated with soil heavy metals can be assessed through various ecological indicators, including the single *ecological risk index (Er), potential ecological risk index (PERI), modified ecological risk index (MRI), mean ERM Quotient (MERMQ)*, and *contamination severity index (CSI)* (Weissmannová & Pavlovský, 2017). Environmental researchers often use PI, CF, EF, Igeo, Er, and PERI to keep attention on HMs pollution and the risks that come with this pollutant into the soil (Adewumi & Ogundele, 2024; Ali et al., 2017; Kachoueiyan et al., 2024; Laudiño et al., 2024; Liu et al., 2020; Oyebamiji et al., 2024; Sonkar et al., 2024; Sulieman et al., 2017; Tian et al., 2020). The significance and quantity of research on ecological risk assessment have risen in recent years (Kalipci et al., 2023; Töre et al., 2021).

The aim of the study was to evaluate ecological and agricultural risk assessment by analysis of hazardous metal concentrations and distributions that could help to explain patterns of soil contamination in the mining region in Tumanyan of Armenia which situated in Lori Province of northern Armenia. The level of contamination in the soil was assessed by the *Potential ecological risk index (PERI), Pollution index (PI), Nemerow integrated pollution index (NIPI).* This work will provide basic information on the status of distribution and accumulation of toxic metals in soils, thus facilitating the development of a pollution prevention plan to control the migration and draining of pollutants in areas with mining activities.

MATERIALS AND METHODS

The soils in the vicinity of Tumanyan region were investigated in october 2022. The investigated soils belong to the type of mountain cambisol. Observation sites were selected around Akhtala Mining and Processing Enterprise CJSC, TEGHOUT CJSC, "Nahatak" Tailings Dump, Shamlugh and Alaverdi copper mining area. For different regions, 3 different control sites were selected: *Background N1, Background N2, Background N3* (Figure 2-4).

The soil samples were obtained from a depth of 0–20 cm and transferred into well labeled polyethylene bags for storage and laboratory analysis. The collected samples were air-dried at room temperature. The dried samples were grounded into powder by a laboratory mortar and pestle, sieved with 1 mm mesh and stored in an air tight container prior to analysis. The soil samples were digested by the Aqua Regia (conc. *HCl* and conc. HNO_3, 3:1) digestion method (D & MA, 2012). An atomic absorption spectrometer (*AAS, PG990, PG Instruments Ltd, UK*) was used to measure the amount of heavy metals (*Standard Methods for the Examination of Water and Wastewater, 20th Edition - Google Books*, 1998).

Figure 2. Sampling area N 1.

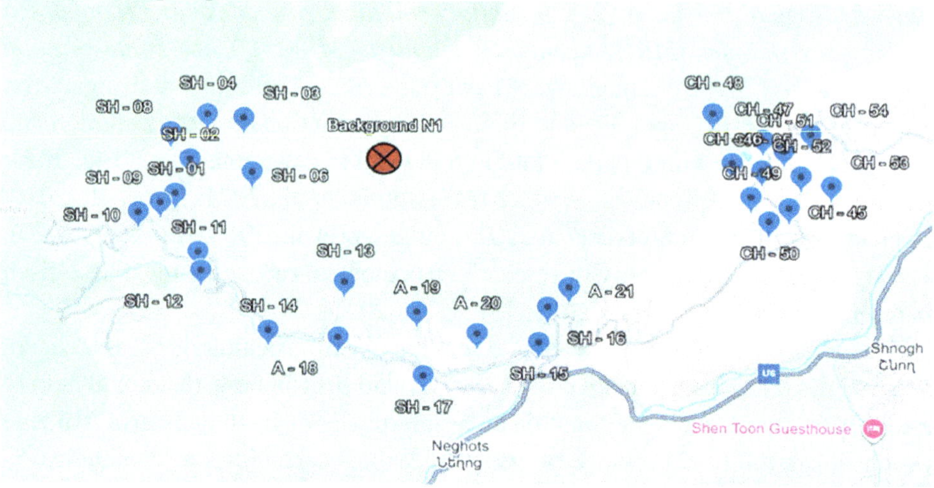

Figure 3. Sampling area N 2.

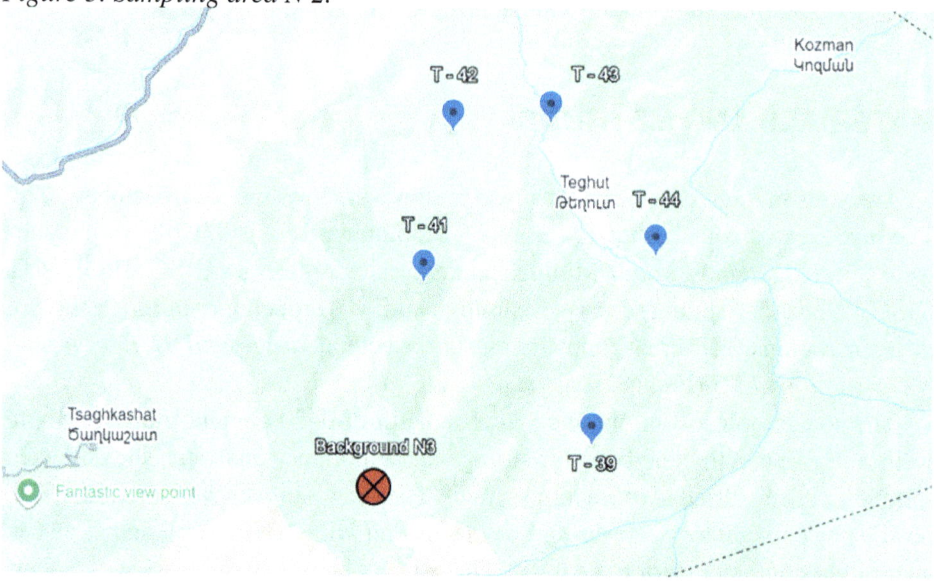

Figure 4. Sampling area N 3.

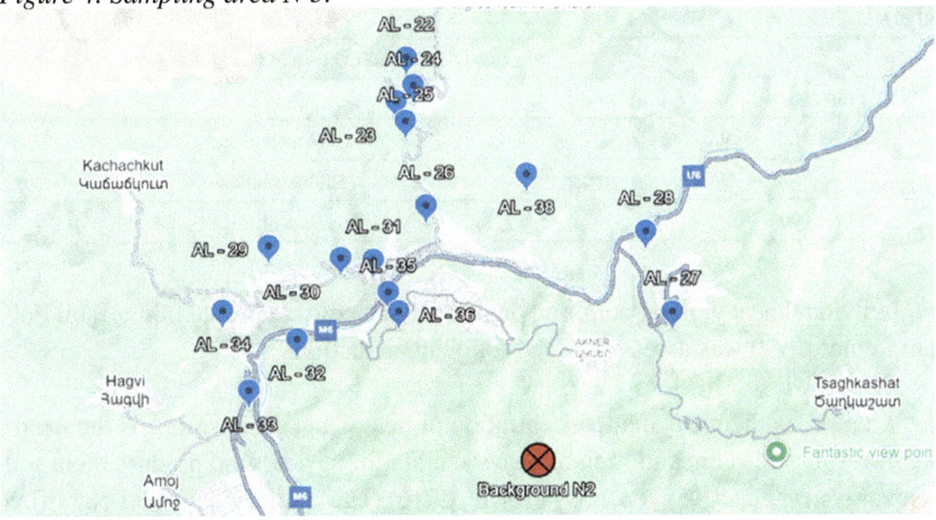

The ecological risks of heavy metals in the soils were assessed according to the Potential ecological risk index (*PERI*) method calculated as follows (Hakanson, 1980):

$C_r^i = C_s^i / C_n^i$, $E_r^i = C_r^i T_r^i$, $RI = \sum E_r^I$,

where C_r^i, C_s^i, C_n^i are the pollution factor; the measured concentration and the background concentration of a single element in soil, respectively; E_r^i is the *PERI* of a single element;

T_r^i is the toxic response factor for a single element; *RI* is the comprehensive *PERI*. The classification of the E_r^i and *PERI* values are presented in Table. 1 and 2, respectively (Su et al., 2014).

Table 1. Ecological risk level classification of single element

E_r^i	**Ecological risk level**
$E_r^i < 40$	Low
$40 \leq E_r^i < 80$	Moderate
$80 \leq E_r^i < 160$	Considerable
$160 \leq E_r^i < 320$	High
$E_r^i \geq 320$	Serious

Table 2. The adjusted grading standard of potential ecological risk of heavy metals in soil.

PERI	**Potential Ecological Risk Index level**
PERI **<150**	Low
$150 \le PERI < 300$	Moderate
$300 \le PERI < 600$	High potential
$PERI \ge 600$	Significantly high

Individual heavy metal pollution degree for agricultural production on soil Pollution index (*PI*) was assessed by the following equation:

$$PI = C_i/S_i,$$

where C_i is the measured concentration of heavy metal *i* in soil; S_i is the maximum permissible concentration of heavy metal *i* for agricultural production in soil (Gevorgyan et al., 2015; Tepanosyan et al., 2016). The *PI* of each metal is classified into five pollution categories presented in Table 3 (Tepanosyan et al., 2016).

Table 3. Pollution categories of Pollution index of each metal

PI **value**	**Pollution category**
$PI < 1$	Non-pollution
$1 \le PI < 2$	Low level of pollution
$2 \le PI < 3$	Moderate level of pollution
$3 \le PI < 5$	Strong level of pollution
$PI > 5$	Very strong level of pollution

Integrated heavy metal pollution degree for agricultural production on soil was evaluated by the Nemerow integrated pollution index (*NIPI*):

$$NIPI = \sqrt{PI_{avg}^2 + PI_{max}^2/2},$$

where PI_{avg} is the average value of the single pollution indices of all heavy metals; PI_{max} is the maximum value of the single pollution indices of all heavy metals (Tepanosyan et al., 2016; Wieczorek et al., 2018a).

The *NIPI* classification is presented in Table 4 (Tepanosyan et al., 2016).

Table 4. Pollution categories of Nemerow integrated pollution index (NIPI)

NIPI value	Pollution category
***NIPI* < 0,7**	No pollution
0.7≤ *NIPI* < 1	Pollution alarm/ warning
1 ≤ *NIPI* < 2	Low pollution
2 ≤ *NIPI* < 3	Moderate pollution
***NIPI* > 3**	High pollution

RESULTS AND DISCUSSION

The results of this study reveal significant spatial variation in heavy metal concentrations across the mining-affected soils of the Tumanyan region, Armenia. Elevated levels of key pollutants such as *Cu, Cd, Pb* and *Zn* were detected in areas surrounding active and legacy mining sites, indicating substantial environmental impact from mining, ore processing and transport activities (Angon et al., 2024).

Table 5 presents baseline concentrations of heavy metals in unpolluted reference soils of the Tumanyan region, serving as a comparative standard for evaluating contamination levels in mining-affected areas.

Table 5. Heavy metal concentrations (mg/ kg) in the backgrounds.

Background N	*Zn*	*Cu*	*Pb*	*Cd*	*Cr*	*Ni*
1	93,76	67,76	7,94	0,26	135,93	69,74
2	71,99	125,18	10,61	0,55	48,59	22,85
3	48,34	21,77	4,94	0,19	45,22	17,54

Tables 6, 7 and 8 present the concentrations of heavy metals in soils collected from different mining-influenced areas of the Tumanyan region, Armenia. These tables serve as a basis for understanding the spatial distribution of metal contamination in the region and reveal clear signs of industrial impact, particularly from mining and ore transportation activities.

Table 6. Heavy metal concentrations (mg/ kg) in the soils around Alaverdi copper mining area (Al 22 – Al 38).

Sample	*Zn*	*Cu*	*Pb*	*Cd*	*Cr*	*Ni*
max	1 030,1	3 562,8	274,1	7,6	89,0	47,5
min	68,7	75,4	8,1	0,3	35,9	4,6
average	269,2	818,2	70,1	2,4	57,4	25,5

Table 7. Heavy metal concentrations (mg/ kg) in the soils around Akhtala Mining and Processing Enterprise CJSC, "Nahatak" Tailings Dump, Shamlugh (SH 2 – SH 17, A 19 – 21, CH 45 – 54)

Sample	*Zn*	*Cu*	*Pb*	*Cd*	*Cr*	*Ni*
max	2 689,0	2 969,3	538,2	13,9	90,9	40,4
min	32,7	21,1	3,4	0,1	2,8	6,4
average	362,0	390,9	54,9	1,7	56,5	25,8

Table 8. Heavy metal concentrations (mg/ kg) in the soils around TEGHOUT CJSC (T 39, T 41 – T 44)

Sample	*Zn*	*Cu*	*Pb*	*Cd*	*Cr*	*Ni*
max	61,6	323,5	12,0	0,3	85,3	41,9
min	26,2	60,9	0,7	0,0	35,1	13,3
average	37,6	122,4	6,0	0,1	60,1	26,2

For assessing the biological health risks of heavy metal pollution in the investigated soils, the PERI values, which shows the sensitivity of various biological communities to harmful elements and illustrate the potentioal ecological risk, were applied. The *PERI* analysis demonstrated that soil samples collected from the Shamlugh mining area (*SH 9 – SH 13*) exhibit moderate to significantly high ecological risk levels (Qi et al., 2022; Haghighizadeh et al., 2024) . Similarly, areas along the ore transportation corridor from Shamlugh to Akhtala (*SH 15 – SH 16*) and in proximity to Akhtala Mining and Processing Enterprise CJSC (*A 19 – A 21*) show high to significantly high *PERI* values. These elevated indices likely reflect the cumulative effect of emissions from mining, smelting and transportation activities, which are known to result in atmospheric deposition and runoff of heavy metals into surrounding soils (Table 9).

Notably, cadmium presents a particularly serious ecological risk in several areas (*SH 15, A 19 – A 21, Al 26, Al 33*), with high levels also found in *SH 10 – SH 11, SH 16, A 18, Al 22* and *Al 24*. Cd is well-documented for its high toxicity and bioavailability and its persistence in the soil poses a significant long-term threat to terrestrial ecosystems. Similarly, lead also contributes to moderate to serious ecological risk levels, especially in samples *A 20 – A 21* (moderate), *SH 15, Al 22, Al 26* and *Al 33* (considerable) and *A 19* (high) (Yi & Huang, 2019).

These results align with similar findings in other post-mining regions, where *Cd* and *Pb* are typically identified as the dominant pollutants due to their historical and ongoing use in ore processing and smelting operations (Wieczorek et al., 2018; Xing et al., 2019). The accumulation patterns observed here suggest not only localized contamination but also regional transport of pollutants, possibly facilitated by wind and surface runoff.

Table 9. Ecological risk (E_r^i) level of single element and RI.

Sample	$E_r^i Zn$	$E_r^i Cu$	$E_r^i Pb$	$E_r^i Cd$	$E_r^i Cr$	$E_r^i Ni$	RI
SH - 01	0,59	2,17	4,14	22,38	0,65	1,39	31,31
SH - 02	1,54	12,03	9,76	78,59	0,64	0,82	103,38
SH - 03	0,71	2,88	6,28	34,55	0,89	2,57	47,89
SH - 04	0,79	1,95	6,80	33,70	0,82	2,50	46,56
SH - 05	0,79	1,56	10,69	33,11	0,56	1,20	47,90
SH - 06	0,60	2,69	11,98	49,98	0,40	0,60	66,26
SH - 08	0,66	2,13	3,72	29,33	0,83	2,44	39,09
SH - 09	2,27	16,55	26,40	128,93	0,42	1,05	175,61
SH - 10	9,33	45,55	339,02	288,79	0,93	1,59	685,23
SH - 11	8,69	32,48	33,11	197,58	0,04	0,46	272,37
SH - 12	1,84	16,36	35,03	159,65	0,36	1,69	214,93
SH - 13	3,92	12,59	6,26	99,66	1,03	1,81	125,27
SH - 14	1,42	15,23	6,94	61,82	1,34	2,01	88,76
SH - 15	9,61	140,96	97,93	487,85	1,13	1,48	738,95
SH - 16	15,80	169,71	36,59	314,76	0,84	1,86	539,55
SH - 17	1,15	8,64	10,59	86,65	0,69	2,34	110,07
A - 18	1,98	43,34	22,84	216,91	0,82	2,79	288,68
A - 19	11,02	27,94	163,88	780,47	0,84	1,78	985,94
A - 20	6,07	28,29	61,66	572,67	0,89	2,65	672,24
A - 21	28,68	219,09	76,69	1 601,61	0,63	1,80	1 928,51

continued on following page

Table 9. Continued

Sample	$E^i_r Zn$	$E^i_r Cu$	$E^i_r Pb$	$E^i_r Cd$	$E^i_r Cr$	$E^i_r Ni$	*RI*
CH - 45	0,96	4,43	6,50	66,74	0,87	2,51	82,00
CH - 46	1,05	4,44	6,54	72,33	0,86	2,48	87,71
CH - 47	1,28	4,57	5,74	106,03	1,19	2,89	121,70
CH - 48	0,77	3,62	3,82	44,89	1,32	2,60	57,01
CH - 49	1,06	11,21	11,85	97,57	0,75	2,27	124,71
CH - 50	0,82	8,17	9,12	72,14	0,82	2,44	93,51
CH - 51	0,35	2,83	2,14	15,47	1,20	0,91	22,90
CH - 52	0,82	8,83	7,73	85,13	0,95	1,82	105,26
CH - 53	0,82	11,64	9,95	97,32	0,87	1,69	122,28
CH - 54	0,47	3,31	3,20	26,47	1,33	1,06	35,84
AL - 22	10,23	142,31	111,04	307,86	1,62	2,82	575,88
AL - 23	2,35	11,97	13,97	57,06	1,48	1,58	88,41
AL - 24	9,09	48,11	25,48	200,49	2,08	1,73	286,97
AL - 25	1,30	17,74	8,50	43,41	2,21	1,00	74,17
AL - 26	6,95	77,63	129,19	417,32	1,93	4,97	637,98
AL - 27	1,39	9,34	8,46	54,77	3,17	6,11	83,24
AL - 28	1,68	28,61	20,46	107,72	2,07	9,15	169,69
AL - 29	2,03	41,56	22,89	133,73	2,36	7,76	210,32
AL - 30	1,36	15,00	12,63	78,49	1,87	6,40	115,74
AL - 31	1,76	26,21	27,23	114,86	2,29	6,61	178,96
AL - 32	1,94	21,15	22,77	109,42	2,47	6,78	164,54
AL - 33	14,31	59,84	94,92	320,75	3,60	6,41	499,81
AL - 34	0,95	4,73	3,80	20,42	3,66	10,39	43,95
AL - 35	1,62	9,22	11,48	51,14	2,82	7,13	83,42
AL - 36	1,86	6,52	9,39	58,90	2,23	4,95	83,84
AL - 38	1,02	3,01	6,83	17,38	1,97	5,58	35,79
T - 39	0,70	14,12	12,12	15,75	1,55	11,94	56,18
T - 41	0,54	14,00	11,14	14,01	2,95	6,49	49,12
T - 42	1,27	20,73	5,08	40,72	3,10	10,77	81,68
T - 43	0,74	74,30	1,13	20,31	1,91	4,32	102,71
T - 44	0,64	17,45	0,69	7,58	3,77	3,79	33,92

Beyond ecological risks, the findings demonstrate a considerable threat to agricultural soil productivity and food safety. The Pollution Index (*PI*) and the Nemerow Integrated Pollution Index (*NIPI*) clearly indicate that the highest risk to agriculture is located near ore transport routes (*SH 15 – SH 16*), Akhtala Mining and Processing Enterprise CJSC (*A 21*) and Alaverdi copper mining area (*Al 22, Al 26, Al 33*). These zones also correspond to areas of elevated *PERI* values, indicating a spatial overlap of biological and agricultural risk (Table 10) (Rashid et al., 2023).

Specifically, copper (*Cu*) was identified as a major contributor to high *PI* values in these areas. This is expected in copper mining regions, but the extent of contamination especially where high *PI* of *Cu* is accompanied by only moderate or low *NIPI* values (*Al 24, Al 25, Al 28, Al 29, Al 31*) suggests that *Cu* alone can significantly degrade soil health even in the absence of multi-element contamination.

Even though Zn is considered one of the metals that are necessary for human and animal growth, its presence in high concentration can become poisonous to living things. It has been demonstrated by several research works that Zn may penetrate the soil from vehicle brake wear, car tire wear, vehicle engine lubrication oil leakages, and vehicle engine cylinder head gaskets. Chemical fertilizers, decaying matter and pesticides are other important sources of Zn pollution. Zn is also a constituent used in the production of paints, fungicides, batteries and several alloys like brass and bronze (Sayadi & Sayyed, 2011). Zn and Cd also emerged as significant contributors to soil degradation (Kubier et al., 2019; Padoan et al., 2020). In this study high *PI* of *Zn* was evident in *SH 16, A 21* and moderate level in *SH 15, A 19, Al 22, Al 33*. while *PI* of *Cd* reached high values in *A 21* and moderate levels in *A 19* and *Al 26* (Table 10).

Pietrzykowski et al. (2018)examined Zn content in contaminated soil from southern Poland. The concentration of Zn was 10638 mg Kg^{-1}, higher than the allowable limits, signifying tottered damage to plants (crops) and humans by this kind soil. These data are in good agreement with Ciarkowska et al. (2016) also substantiated the high content of Zn in the studied soil, which detrimentally affects the activities of soil enzymes. Wyszkowska et al. (2013)studied the adverse effect of heavy metal toxicity on soil properties, microbes, and enzymes in Poland. According to these authors, high pH increased the bioavailable Zn in their study. High contents of Zn and the other metals disturb soil homeostasis. Further, inhibition of microbial enzymatic activity also observed.

Cd is toxic to the phenotype (e.g., plant weight, root length, and shoot length), cytotoxicity (chlorophyll content decrease, photosynthesis performance inhibition), and metabolic compounds such as chlorosis and cell damage (Hayat et al., 2019). Seregin and Kozhevnikova, (2008)found that xylem parenchyma could largely account for the translocation of heavy metals to conducting tissue, although xylem-limited Cd transportation also occurred. Gratão et al. (2009)reported that Cd toxicity was able to reduce the dry mass and length of roots. A recent study of Vardhan et al. (2019)*Lactuca sativa* seeds were used as the experimental material to study Cd ions impact on seedling development through *Pichia* sp. and *Azotobacter chroococcum*.

It was found that Cd (II) suppressed seed germination, decreased biomass, and impaired root growth and development. Furthermore, the Cd (II) acted negatively on microbial growth. The toxicological properties of Zn, although less severe than Cd, can still adversely impact microbial activity, soil respiration, and crop yield, particularly in soils already compromised by other metals (Ćwieląg-Drabek et al., 2020; Lan et al., 2020).

Table 10. PI and NIPI values for agricultural production of the soils.

Sample	PI_{Zn}	PI_{Cu}	PI_{Pb}	PI_{Cd}	PI_{Cr}	PI_{Ni}	*NIPI*
SH - 01	0,12	0,14	0,02	0,06	0,29	0,17	0,23
SH - 02	0,32	0,78	0,05	0,23	0,29	0,10	0,59
SH - 03	0,15	0,19	0,03	0,10	0,40	0,32	0,32
SH - 04	0,25	0,19	0,04	0,10	0,37	0,46	0,37
SH - 05	0,25	0,15	0,06	0,10	0,25	0,22	0,22
SH - 06	0,19	0,26	0,06	0,14	0,18	0,11	0,22
SH - 08	0,14	0,14	0,02	0,08	0,37	0,30	0,29
SH - 09	0,47	1,07	0,14	0,37	0,19	0,13	0,81
SH - 10	1,94	2,94	1,79	0,84	0,42	0,20	2,29
SH - 11	1,81	2,10	0,18	0,57	0,02	0,06	1,58
SH - 12	0,58	1,58	0,19	0,46	0,17	0,31	1,19
SH - 13	0,82	0,81	0,03	0,29	0,47	0,22	0,66
SH - 14	0,45	1,47	0,04	0,18	0,61	0,37	1,11
SH - 15	2,00	9,10	0,52	1,41	0,51	0,18	6,63
SH - 16	3,29	10,95	0,19	0,91	0,38	0,23	7,97
SH - 17	0,24	0,56	0,06	0,25	0,31	0,29	0,44
A - 18	0,41	2,80	0,12	0,63	0,37	0,35	2,05
A - 19	2,30	1,80	0,87	2,26	0,38	0,22	1,87
A - 20	1,26	1,83	0,33	1,66	0,40	0,33	1,46
A - 21	5,98	14,14	0,41	4,64	0,28	0,22	10,45
CH - 45	0,20	0,29	0,03	0,19	0,39	0,31	0,33
CH - 46	0,22	0,29	0,03	0,21	0,39	0,31	0,32
CH - 47	0,27	0,29	0,03	0,31	0,54	0,36	0,44
CH - 48	0,16	0,23	0,02	0,13	0,60	0,32	0,46
CH - 49	0,22	0,72	0,06	0,28	0,34	0,28	0,56

continued on following page

Table 10. Continued

Sample	PI_{Zn}	PI_{Cu}	PI_{Pb}	PI_{Cd}	PI_{Cr}	PI_{Ni}	*NIPI*
CH - 50	0,17	0,53	0,05	0,21	0,37	0,30	0,42
CH - 51	0,07	0,18	0,01	0,04	0,54	0,11	0,40
CH - 52	0,17	0,57	0,04	0,25	0,43	0,23	0,45
CH - 53	0,17	0,75	0,05	0,28	0,39	0,21	0,57
CH - 54	0,10	0,21	0,02	0,08	0,60	0,13	0,45
AL - 22	2,46	25,45	0,79	1,87	0,26	0,17	18,36
AL - 23	0,56	2,14	0,10	0,35	0,24	0,10	1,57
AL - 24	1,45	5,74	0,18	1,22	0,34	0,07	4,19
AL - 25	0,31	3,17	0,06	0,26	0,36	0,06	2,30
AL - 26	1,11	9,25	0,91	2,54	0,31	0,20	6,76
AL - 27	0,22	1,11	0,06	0,33	0,51	0,25	0,84
AL - 28	0,40	5,12	0,14	0,65	0,33	0,56	3,72
AL - 29	0,32	4,95	0,16	0,81	0,38	0,32	3,60
AL - 30	0,22	1,79	0,09	0,48	0,30	0,26	1,32
AL - 31	0,28	3,12	0,19	0,70	0,37	0,27	2,28
AL - 32	0,31	2,52	0,16	0,67	0,40	0,28	1,85
AL - 33	2,29	7,13	0,67	1,95	0,58	0,26	5,27
AL - 34	0,15	0,56	0,03	0,12	0,59	0,42	0,47
AL - 35	0,26	1,10	0,08	0,31	0,46	0,29	0,83
AL - 36	0,45	1,17	0,07	0,36	0,36	0,30	0,88
AL - 38	0,16	0,36	0,05	0,11	0,32	0,23	0,29
T - 39	0,11	0,44	0,04	0,03	0,23	0,56	0,43
T - 41	0,06	0,29	0,04	0,03	0,44	0,20	0,34
T - 42	0,14	0,43	0,02	0,09	0,47	0,34	0,37
T - 43	0,08	1,54	0,00	0,04	0,29	0,13	1,12
T - 44	0,10	0,54	0,00	0,02	0,57	0,18	0,43

The *NIPI* results further support this assessment. High pollution levels were found near ore transportation routes (*SH 15 – SH 16*), Akhtala (*A 21*) and Alaverdi (*Al 22, Al 26, Al 33*) indicating these areas as pollution hotspots. Moderately polluted zones were identified in *Al 24, Al 28* and *Al 29*, while low pollution was observed near Shamlugh (*SH 10*), Akhtala (*A 18*) and Alaverdi (*Al 25, Al 31*) suggesting a gradient of contamination extending from core industrial sites outward.

CONCLUSION AND FUTURE DIRECTIONS

The present study evaluates heavy metal contamination and ecological and agricultural risks in the mining-mediated soils of the Tumanyan region, Armenia. Heavy metal content in soils shows unequivocal spatial variation with the highest concentration of Cu, Zn, Pb and Cd observed near active and historical mining sites, mineral processing plants and network of transport infrastructure. The background soil information shows that the high metal concentrations are substantially higher than geochemical natural backgrounds, and demonstrate a major input by long-life mining, smelting and transportation activities. The Potential Ecological Risk Index (PERI) and Pollution Indices (PI and NIPI) method reliably same a few important air pollution hotspots especially close to Akhtala Mining and Processing Enterprise CJSC (A 21), the ore transport corridor (SH 15–SH 16), and Alaverdi Copper Mining Area (Al 22, Al 26, Al 33). High to extreme ecological and agriculture risks were observed in these areas, which are mainly attributed to the toxic and bioavailability of Cd and Pb, as well as the excessive accumulation of Cu and Zn. The effects of these metals on the environment are accumulated and persistent, which will have long-term impacts on soil, biodiversity and food safety. Specifically, Cd was the most harmful element with predominant contribution to both ecological and agricultural risk indices because of its strong soil-plant transportability. The results are consistent with the worldwide trends of heavy metal migration documented in other post-mining regions, and highlight the combined role of historical contamination, atmospheric deposition and surface runoff in transport of metals outside proximal mining areas. The detected trend in the industrial cores to peripheral areas contamination gradient further emphasizes the necessity for ongoing monitoring of these environments, as well as remediation.

Future Directions

There are following future directions (Figure 5):

(A) Phytoremediation and Biochar-Based Remediation

Further studies should consider the utilization of hyperaccumulator plants and biochar, or nano-biochar amendments for immobilization or phytoextraction of heavy metals. "These nature-based technologies could Improves soil fertility and lower bioavailability of toxic elements.

(B) Long-Term Monitoring and Risk Mapping

The establishment of a network to monitor the region is needed using spatial analysis such as GIS and remote sensing to monitor changes of contaminated patterns and locations, new hotspots, evaluate effectiveness for the mitigation process in future.

(C) Soil-to-Plant Transfer and Risk Assessment for the Food Chain

Therefore, it is important to determine the transfer factors of Cd, Pb, Cu and Zn from contaminated soil to locally cultivated crops in order to quantify human exposure through dietary ingestion. This needs to be considered within the context of a more comprehensive food safety risk assessment.

(D) Microbial and Enzymatic Health Indicators

Future studies should investigate the effects of heavy metals on soil microbial diversity, enzymatic activity and nutrient cycling processes in detail, to gain a broader understanding of how contamination affects soil ecosystem functioning and resilience.

(E) Policy and Management Implications

These results will contribute to establish specific soil quality standards, guidelines for post-mining land reclamation and environmentally-friendly mining. Improved regulation of the sector and local engagement in monitoring activities could provide long-term sustainability.

(F) Integration With Climate and Land Use Models

Since heavy metals may become mobile under changing climate conditions (e.g., more precipitation or erosion) in the future, soil contamination data can be incorporated into climate-adaption and land-use planning models to avoid secondary threats of pollution.

Figure 5. Future direction and suggestions.

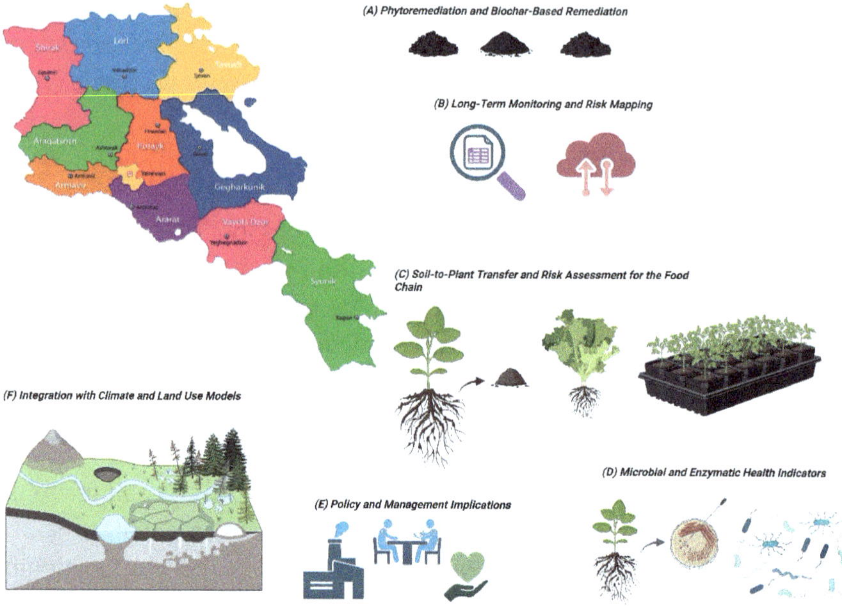

ACKNOWLEDGEMENT

AS is supported by the internal grant program of Yerevan State University (2024).

REFERENCES

Adewumi, A. J., & Ogundele, O. D. (2024). Hidden hazards in urban soils: A meta-analysis review of global heavy metal contamination (2010-2022), sources and its Ecological and health consequences. *Sustainable Environment, 10*(1), 2293239. Advance online publication. DOI: 10.1080/27658511.2023.2293239

Ali, M., Elhagwa, A., Elfaki, J., & Sulieman, M. (2017). Influence of the artisanal gold mining on soil contamination with heavy metals: A case study from Dar-Mali locality, North of Atbara, River Nile State, Sudan. *Eurasian Journal of Soil Science, 6*(1), 28–36. DOI: 10.18393/ejss.284261

Angon, P. B., Islam, M. S., Kc, S., Das, A., Anjum, N., Poudel, A., & Suchi, S. A. (2024a). Sources, effects and present perspectives of heavy metals contamination: Soil, plants and human food chain. *Heliyon, 10*(7), e28357. DOI: 10.1016/j.heliyon.2024.e28357 PMID: 38590838

Angon, P. B., Islam, M. S., Kc, S., Das, A., Anjum, N., Poudel, A., & Suchi, S. A. (2024b). Sources, effects and present perspectives of heavy metals contamination: Soil, plants and human food chain. *Heliyon, 10*(7), e28357. DOI: 10.1016/j.heliyon.2024.e28357 PMID: 38590838

Brevik, E. C., Slaughter, L., Singh, B. R., Steffan, J. J., Collier, D., Barnhart, P., & Pereira, P. (2020). Soil and Human Health: Current Status and Future Needs. *Air, Soil and Water Research, 13*. Advance online publication. DOI: 10.1177/1178622120934441

Briassoulis, D. (2023). Agricultural plastics as a potential threat to food security, health, and environment through soil pollution by microplastics: Problem definition. *The Science of the Total Environment, 892*, 164533. DOI: 10.1016/j.scitotenv.2023.164533 PMID: 37285997

Ciarkowska, K., Gargiulo, L., & Mele, G. (2016). Natural restoration of soils on mine heaps with similar technogenic parent material: A case study of long-term soil evolution in Silesian-Krakow Upland Poland. *Geoderma, 261*, 141–150. DOI: 10.1016/j.geoderma.2015.07.018

Ćwieląg-Drabek, M., Piekut, A., Gut, K., & Grabowski, M. (2020). Risk of cadmium, lead and zinc exposure from consumption of vegetables produced in areas with mining and smelting past. *Scientific Reports, 10*(1), 3363. DOI: 10.1038/s41598-020-60386-8 PMID: 32099081

D, S. T., & MA, S. (2012). Calculating Integrated Pollution Indices for Heavy Metals in Ecological Geochemistry Assessment Near Sugar Mill. *Journal of Research in Biology, 2*(5), 489–498. https://ojs.jresearchbiology.com/index.php/jrb/article/view/241

Gevorgyan, G. A., Movsesyan, H. S., Grigoryan, K. V., & Ghazaryan, K. A. (2015). Environmental risks of heavy metal pollution of the soils around Kajaran Town, Armenia. *Proceedings of the YSU B: Chemical and Biological Sciences, 49*(2 (237)), 50–55. DOI: 10.46991/PYSU:B/2015.49.2.050

Gratão, P. L., Monteiro, C. C., Rossi, M. L., Martinelli, A. P., Peres, L. E. P., Medici, L. O., Lea, P. J., & Azevedo, R. A. (2009). Differential ultrastructural changes in tomato hormonal mutants exposed to cadmium. *Environmental and Experimental Botany*, *67*(2), 387–394. DOI: 10.1016/j.envexpbot.2009.06.017

Haghighizadeh, A., Rajabi, O., Nezarat, A., Hajyani, Z., Haghmohammadi, M., Hedayatikhah, S., Asl, S. D., & Aghababai Beni, A. (2024a). Comprehensive analysis of heavy metal soil contamination in mining Environments: Impacts, monitoring Techniques, and remediation strategies. *Arabian Journal of Chemistry, 17*(6), 105777. DOI: 10.1016/j.arabjc.2024.105777

Haghighizadeh, A., Rajabi, O., Nezarat, A., Hajyani, Z., Haghmohammadi, M., Hedayatikhah, S., Asl, S. D., & Aghababai Beni, A. (2024b). Comprehensive analysis of heavy metal soil contamination in mining Environments: Impacts, monitoring Techniques, and remediation strategies. *Arabian Journal of Chemistry, 17*(6), 105777. Advance online publication. DOI: 10.1016/j.arabjc.2024.105777

Hakanson, L. (1980). An ecological risk index for aquatic pollution control.a sedimentological approach. *Water Research, 14*(8), 975–1001. DOI: 10.1016/0043-1354(80)90143-8

Hayat, M. T., Nauman, M., Nazir, N., Ali, S., & Bangash, N. (2019). Environmental Hazards of Cadmium: Past, Present, and Future. *Cadmium Toxicity and Tolerance in Plants: From Physiology to Remediation*, 163–183. DOI: 10.1016/B978-0-12-814864-8.00007-3

Kachoueiyan, F., Karbassi, A., Nasrabadi, T., Rashidiyan, M., & De-la-Torre, G. E. (2024). Speciation characteristics, ecological risk assessment, and source apportionment of heavy metals in the surface sediments of the Gomishan wetland. *Marine Pollution Bulletin, 198*, 115835. DOI: 10.1016/j.marpolbul.2023.115835 PMID: 38039575

Kalipci, E., Cüce, H., Ustaoğlu, F., Dereli, M. A., & Türkmen, M. (2023). Toxicological health risk analysis of hazardous trace elements accumulation in the edible fish species of the Black Sea in Türkiye using multivariate statistical and spatial assessment. *Environmental Toxicology and Pharmacology, 97*, 104028. DOI: 10.1016/j.etap.2022.104028 PMID: 36455837

Kubier, A., Wilkin, R. T., & Pichler, T. (2019). Cadmium in soils and groundwater: A review. *Applied Geochemistry, 108*, 104388. DOI: 10.1016/j.apgeochem.2019.104388 PMID: 32280158

Lal, R., Bouma, J., Brevik, E., Dawson, L., Field, D. J., Glaser, B., Hatano, R., Hartemink, A. E., Kosaki, T., Lascelles, B., Monger, C., Muggler, C., Ndzana, G. M., Norra, S., Pan, X., Paradelo, R., Reyes-Sánchez, L. B., Sandén, T., Singh, B. R., & Zhang, J. (2021). Soils and sustainable development goals of the United Nations: An International Union of Soil Sciences perspective. *Geoderma Regional, 25*, e00398. DOI: 10.1016/j.geodrs.2021.e00398

Lan, M. M., Liu, C., Liu, S. J., Qiu, R. L., & Tang, Y. T. (2020). Phytostabilization of Cd and Pb in Highly Polluted Farmland Soils Using Ramie and Amendments. *International Journal of Environmental Research and Public Health, 17*(5), 1661. DOI: 10.3390/ijerph17051661

Laudiño, F. A. R., Agtong, R. J. M., Jumawan, J. C., Fukuyama, M., & Elvira, M. V. (2024). Assessment of contamination and potential ecological risks of heavy metals in the bottom sediments of Lake Mainit, Philippines. *Journal of Hazardous Materials Advances, 15*, 100443. DOI: 10.1016/j.hazadv.2024.100443

Liu, P., Hu, W., Tian, K., Huang, B., Zhao, Y., Wang, X., Zhou, Y., Shi, B., Kwon, B. O., Choi, K., Ryu, J., Chen, Y., Wang, T., & Khim, J. S. (2020). Accumulation and ecological risk of heavy metals in soils along the coastal areas of the Bohai Sea and the Yellow Sea: A comparative study of China and South Korea. *Environment International, 137*, 105519. DOI: 10.1016/j.envint.2020.105519 PMID: 32014790

Oyebamiji, A. O., Olaolorun, O. A., Popoola, O. J. P., & Zafar, T. (2024). Assessment of heavy metal pollution in soils of Jebba Area, Nigeria: Concentrations, source analysis and implications for ecological and human health risks. *The Science of the Total Environment, 945*, 173860. DOI: 10.1016/j.scitotenv.2024.173860 PMID: 38871321

Padoan, E., Hernandez Kath, A., Vahl, L. C., & Ajmone-Marsan, F. (2020). Potential Release of Zinc and Cadmium From Mine-Affected Soils Under Flooding, a Mesocosm Study. *Archives of Environmental Contamination and Toxicology*, *79*(4), 421–434. DOI: 10.1007/s00244-020-00777-0 PMID: 33175188

Pietrzykowski, M., Antonkiewicz, J., Gruba, P., & Pajak, M. (2018). Content of Zn, Cd and Pb in purple moor-grass in soils heavily contaminated with heavy metals around a zinc and lead ore tailing landfill. *Open Chemistry*, *16*(1), 1143–1152. DOI: 10.1515/chem-2018-0129

Qi, Y., Wei, X., Zhao, M. J., Pan, W., Jiang, C., Wu, J., & Li, W. C. (2022). Heavy metal pollution characteristics and potential ecological risk assessment of soils around three typical antimony mining areas and watersheds in China. *Frontiers in Environmental Science*, *10*, 913293. DOI: 10.3389/fenvs.2022.913293

Rajan Selin, A., & Shanmugasundram, S. (2024). Traditional approaches of agriculture soil remediation: boon or bane for global food security. *Bioremediation of Emerging Contaminants from Soils: Soil Health Conservation for Improved Ecology and Food Security*, 43–62. DOI: 10.1016/B978-0-443-13993-2.00003-7

Rajput, P., Singh, A., Agrawal, S., Ghazaryan, K., Rajput, V. D., Movsesyan, H., Mandzhieva, S., Minkina, T., & Alexiou, A. (2024). Effects of environmental metal and metalloid pollutants on plants and human health: Exploring nano-remediation approach. *Stress Biology*, *4*(1), 1–25. DOI: 10.1007/s44154-024-00156-y PMID: 38777953

Rashid, A., Schutte, B. J., Ulery, A., Deyholos, M. K., Sanogo, S., Lehnhoff, E. A., & Beck, L. (2023). Heavy Metal Contamination in Agricultural Soil: Environmental Pollutants Affecting Crop Health. *Agronomy (Basel)*, *13*(6), 1521. DOI: 10.3390/agronomy13061521

Sayadi, M. H., & Sayyed, M. R. G. (2011). Comparative assessment of baseline concentration of the heavy metals in the soils of Tehran (Iran) with the comprisable reference data. *Environmental Earth Sciences*, *63*(6), 1179–1188. DOI: 10.1007/s12665-010-0792-z

Seregin, I. V., & Kozhevnikova, A. D. (2008). Roles of root and shoot tissues in transport and accumulation of cadmium, lead, nickel, and strontium. *Russian Journal of Plant Physiology: a Comprehensive Russian Journal on Modern Phytophysiology*, *55*(1), 1–22. DOI: 10.1134/S1021443708010019

Soltani, N., Keshavarzi, B., Moore, F., Sorooshian, A., & Ahmadi, M. R. (2017). Distribution of potentially toxic elements (PTEs) in tailings, soils, and plants around Gol-E-Gohar iron mine, a case study in Iran. *Environmental Science and Pollution Research International, 24*(23), 18798–18816. DOI: 10.1007/s11356-017-9342-5 PMID: 28620857

Sonkar, V., Jaswal, V., Chenlak, S., & Nandabalan, Y. K. (2024). Pollution status and health risk assessment of heavy metals in the soil of Sahibzada Ajit Singh (SAS) Nagar district of Punjab, India and its source apportionment. *Journal of Geochemical Exploration, 261*, 107453. DOI: 10.1016/j.gexplo.2024.107453

Standard Methods for the Examination of Water and Wastewater, 20th Edition - Google Books. (1998). APHA American Public Health Association. https://books.google.am/books/about/Standard_Methods_for_the_Examination_of.html?id=2BcoYAAACAAJ&redir_esc=y

Su, C., Jiang, L., & Zhang, W. (2014). *A review on heavy metal contamination in the soil worldwide: Situation, impact and remediation techniques. The International Academy of Ecology and Environmental Sciences*. IAEES.

Sulieman, M. M., Elfaki, J. T., Adam, M. M., Dafalla, M. S., Ali, S. H., Ahmed, H. M., & Ali, M. E. (2017). Assessment of heavy metals contamination in the Nile River water and adjacent sediments: A case study from Khartoum City and Nile River State, Sudan. *Eurasian Journal of Soil Science, 6*(3), 285–294. DOI: 10.18393/ejss.298949

Taghavi, M., Bakhshi, K., Zarei, A., Hoseinzadeh, E., & Gholizadeh, A. (2024). Soil pollution indices and health risk assessment of metal(loid)s in the agricultural soil of pistachio orchards. *Scientific Reports, 14*(1), 8971. Advance online publication. DOI: 10.1038/s41598-024-59450-4 PMID: 38637594

Tepanosyan, G., Sahakyan, L., Belyaeva, O., & Saghatelyan, A. (2016). Origin identification and potential ecological risk assessment of potentially toxic inorganic elements in the topsoil of the city of Yerevan, Armenia. *Journal of Geochemical Exploration, 167*, 1–11. DOI: 10.1016/j.gexplo.2016.04.006

Tian, K., Wu, Q., Liu, P., Hu, W., Huang, B., Shi, B., Zhou, Y., Kwon, B. O., Choi, K., Ryu, J., Seong Khim, J., & Wang, T. (2020). Ecological risk assessment of heavy metals in sediments and water from the coastal areas of the Bohai Sea and the Yellow Sea. *Environment International, 136*, 105512. DOI: 10.1016/j.envint.2020.105512 PMID: 31999973

Töre, Y., Ustaoğlu, F., Tepe, Y., & Kalipci, E. (2021). Levels of toxic metals in edible fish species of the Tigris River (Turkey); Threat to public health. *Ecological Indicators, 123*, 107361. DOI: 10.1016/j.ecolind.2021.107361

Vácha, R. (2021). Heavy metal pollution and its effects on agriculture. *Agronomy (Basel), 11*(9), 1719. Advance online publication. DOI: 10.3390/agronomy11091719

Vardhan, K. H., Kumar, P. S., & Panda, R. C. (2019). A review on heavy metal pollution, toxicity and remedial measures: Current trends and future perspectives. *Journal of Molecular Liquids, 290*, 111197. DOI: 10.1016/j.molliq.2019.111197

Vasilachi, I. C., Stoleru, V., & Gavrilescu, M. (2023). Analysis of Heavy Metal Impacts on Cereal Crop Growth and Development in Contaminated Soils. *Agriculture, 13*(10), 1983. DOI: 10.3390/agriculture13101983

Weissmannová, H. D., & Pavlovský, J. (2017). Indices of soil contamination by heavy metals – methodology of calculation for pollution assessment (minireview). *Environmental Monitoring and Assessment, 189*(12), 1–25. DOI: 10.1007/s10661-017-6340-5 PMID: 29116419

Wieczorek, J., Baran, A., Urbański, K., Mazurek, R., & Klimowicz-Pawlas, A. (2018). Assessment of the pollution and ecological risk of lead and cadmium in soils. *Environmental Geochemistry and Health, 40*(6), 2325–2342. DOI: 10.1007/s10653-018-0100-5 PMID: 29589150

Wuana, R. A., & Okieimen, F. E. (2011). Heavy Metals in Contaminated Soils: A Review of Sources, Chemistry, Risks and Best Available Strategies for Remediation. *ISRN Ecology, 2011*(1), 402647. DOI: 10.5402/2011/402647

Wyszkowska, J., Borowik, A., Kucharski, M., & Kucharski, J. (2013). Oddzia£ywanie kadmu, miedzi i cynku na roœliny, drobnoustroje i enzymy glebowe. *Journal of Elementology, 18*(4), 769–796. DOI: 10.5601/jelem.2013.18.4.455

Xia, Y., Gao, T., Liu, Y., Qi, M., Zhu, J. M., Tong, H., Lv, Y., & Liu, C. (2025). Cd/Pb behavior during combustion in a coal-fired power plant and their spatiotemporal impacts on soils: New insights from Cd/Pb isotopes. *Journal of Environmental Sciences (China), 150*, 582–593. DOI: 10.1016/j.jes.2024.03.002 PMID: 39306431

Xing, W., Zheng, Y., Scheckel, K. G., Luo, Y., & Li, L. (2019). Spatial Distribution of Smelter Emission Heavy Metals on Farmland Soil. *Environmental Monitoring and Assessment, 191*(2), 115. DOI: 10.1007/s10661-019-7254-1 PMID: 30697651

Yi, Q., & Huang, C. (2019). Review of heavy metal pollution by mining. *E3S Web of Conferences, 118*, 04028. DOI: 10.1051/e3sconf/201911804028

Chapter 12

Assessing the Environmental Impacts of Pesticides: A Case Study From Armenia

Meri Arzumanyan
http://orcid.org/0000-0002-9777-7830
Faculty of Biology, Research Institute of Biology, Armenia

Iren Shahazizyan
http://orcid.org/0009-0007-1392-6557
Faculty of Biology, Research Institute of Biology, Armenia

Lusine Margaryan
http://orcid.org/0000-0002-3520-6100
Faculty of Biology, Research Institute of Biology, Armenia

Anzhela Sargsyan
http://orcid.org/0000-0003-2720-0355
Faculty of Biology, Research Institute of Biology, Armenia

Narine Zakaryan
http://orcid.org/0009-0000-3597-6271
Faculty of Biology, Research Institute of Biology, Armenia

Anahit Atoyants
http://orcid.org/0000-0001-5076-5675
Research Institute of Biology, Armenia

Siranush Nanagulyan
http://orcid.org/0000-0003-0993-7249
Faculty of Biology, Research Institute of Biology, Armenia

Ruzanna Matevosyan
http://orcid.org/0000-0001-9903-0805
Faculty of Biology, Research Institute of Biology, Armenia

Inessa Eloyan
http://orcid.org/0009-0006-3016-7743
Faculty of Biology, Research Institute of Biology, Armenia

DOI: 10.4018/979-8-3373-1524-9.ch012

ABSTRACT

This research investigates the impacts of pesticides use on the soil-plant-fungi-animal chain in the mountain regions of Armenia, using Tradescantia plants for mutation detections, micromycetes, molluscs richness for secondary impact of pesticides on the non-target biodiversity. Data sampling has been conducted from 2022 to 2023 in the potato fields, during spring, summer, and autumn seasons on three different locations: Gogaran, Arevshat, and Lichk mountain-steppe regions of Armenia. Soil chemical analysis revealed that experimental areas contained four dangerous active components of pesticides including Arsenous. Using two model tests of Tradescantia, Trad-SHM (stamen hair mutations of Tradescantia) and Trad-MN (micronuclei in tetrads of plant microspores), a significant increase in the level of mutation events was shown in all three experimental soil variants compared to their background ones. A total 95 species of micromycetes was recorded, out of which the diversity noticeably declined. As a result, some beneficial species of Mucoromycetes were destroyed, which led to the emergence of a resistant pathogenic fungal species Agroathelia rolfsii. As for molluscs species richness remained constant throughout the experiment. However, molluscs species abundance and mutation levels appeared varying in experimental and control locations. The authors observed genotoxic effects of pesticides on Georginapaeous hohenackeri with high levels of DNA damage in pesticide used plots. These findings highlight the importance of adopting sustainable agricultural practices to mitigate negative impacts on non-target species and preserve biodiversity, thereby ensuring the integrity of ecosystems and underscoring the urgent need for stringent pesticide regulation and monitoring the usage in Armenia.

INTRODUCTION

Armenia is a mountainous country known for its rich biodiversity which is facing decline due to several reasons including growth of agriculture. Pesticides are easily accessible and effective tools against harmful pests which results not only soil and water ecosystem pollutions but also massive non-targeted biodiversity lost (Köhler and Triebskorn, 2013; Convention on Biological Diversity of RA, 2014; Fayvush, 2023). However, pesticides impact on the biodiversity of Armenia has been poorly studied. There are no effective methods to detect and prevent harmful components use in agriculture of Armenia. There are several bioindicator organisms including soil living fungi, terrestrial molluscs which can play crucial role to understand impact of harmful components on the biodiversity (Druart et al., 2011; Gall and Tooker, 2017; Warnasuriya al., 2023). Indicator plants, particularly Tradescantia can help determine the level of mutations caused by pesticides. Molluscs also use to assess the genotoxic impact of pesticides using comet assay (Petridis et al., 2009; Martins and Costa, 2015).

Understanding the intricate dynamics of pesticide use and its effects on the soil-plant-fungi-animal chain is essential for environmentally friendly agricultural practices and ecosystems health. In Armenia such type of research has not been performed so far to understand the impact of pesticides on different biological species and on different levels of live organisations: starting from populations and moving to level of DNA damage. Therefore, this research aims to investigate the impacts of pesticides on soil-plant-fungi-animal chain in mountain regions of Armenia, providing valuable insights into the environmental consequences of pesticide use in the mountain farming.

MATERIALS AND METHODS

Study Area and Sampling Design

Three locations in the mountain steps were selected for sampling in the northern and the central regions of Armenia: Gogaran in Lori Province (GPS: 40.87N, 44.18E), Arevshat in Shirak Province (GPS: 40.65N, 44.03E), and Lichk in Gegharkunik Province (GPS: 40.15N, 45.22E) (Fig. 1). Sampling activities were conducted in the following order: 1st - spring months - before applying fertilizers; 2nd - summer months after applying fertilizers; and 3rd - autumn months after harvesting. The sampling has been performed in 2022 and 2023.

Lichk, Gogaran, and Arevshat are mountain chernozems of moderately moist steppes, carbonate and typical, in combination with shallow, poorly developed chernozems. Experimental soils are cultivated irrigated soils of mountain steppes.

Sampling locations were potato farming areas in the mountain steppes, encompassing a 5x5 m^2 plot at each 3 farming locations. Soil samples, totalling 1.5 kg, were collected from 4 places around the potato fields. These sampling points were designated as experimental locations.

Control sites were defined outside the designated agricultural fields, at a distance of 100 m, and were characterized by the absence of any agricultural activity, serving as reference sites.

Figure 1. Sampling location in Armenia. 1 - Gogaran, 2 - Arevshat, 3 - Lichk.

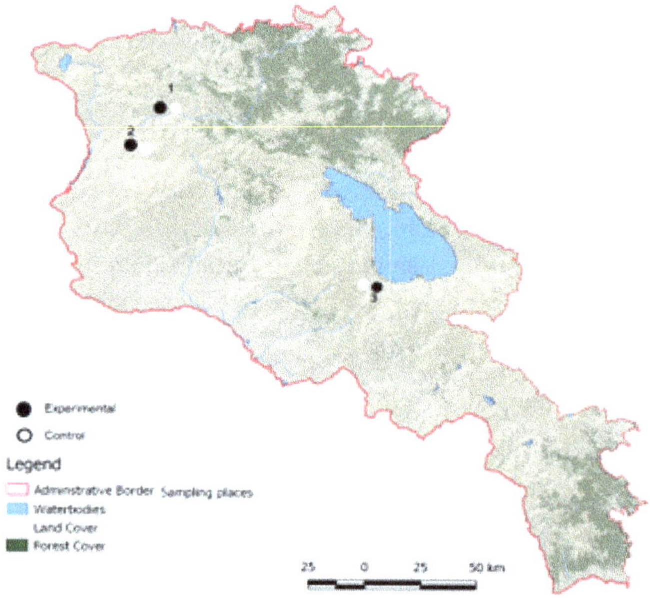

Molluscs Sampling

Terrestrial snails were manually sampled. Sampled specimens have become part of the Yerevan State University Zoological collection (Collection Numbers are: Mol_P2022-Mol_P2023). Only several living individuals of *Georginapaeus hohenackeri* (L. Pfeiffer, 1848) species from the sampled locations were selected for genotoxicological testing to assess DNA damage levels in their hepatopancreas' cells using COMET method.

Molluscs species identification has been conducted using Akramowskii's species identification book (Akramowskii, 1976). Molluscs species abundance was calculated adding up all the individuals' numbers, and richness was calucalted by adding up all the species number.

Tradescantia Growth

Tradescantia (clone 02) has been used as a model plant for biotesting, which is a natural interspecific hybrid between *Tradescantia occidentalis* Britton Rudb. and *T. ohiensis* Raf. It is a heterozygous plant in terms of flower coloration: blue coloration is dominant, while pink is a recessive trait. This clone has been selected as it has several advantages for mutagen screening and bioindication: The main marker tests for Tradescantia clone 02 during the analysis where: stamen hair mutations of Tradescantia (Trad-SHM) to identify somatic mutations, morphological changes in stamen hair (genotoxic effect - Trad-SHM test), and the micronucleus detection test to determine the scale of microsporogenesis development with the formation of micronuclei (MN) in tetrads of microspores (clastogenic effect - Trad-MN test). Both biotests: Trad- SHM and Trad-MN, are part of the International Program on Plant Bioassays (IPPB) under the auspices of the United Nations Environment Programme (UNEP) (Ma, 1999).

Additional to our samples from Gogaran, Arevshat and Lichk we used an overall control check using samples of plants grown in Yerevan State University greenhouse conditions. Tradescantia plants (clone 02) were cultivated in a greenhouse under controlled conditions: 18/6 h day/night cycle, 50-60% humidity, and temperatures ranging from 20 to 26°C, using soil from sampled locations surface (0-10 cm deep). Each soil sample was tested in duplicate, with each vessel containing 500 g of soil. For each soil sample, two sets of vegetation vessels were prepared, each containing 5-7 Tradescantia plants planted during the leaf phase of organogenesis. Irrigation was carried out with tap water. Within 2-3 weeks after the appearance of young buds, samples for using the Trad-MN test were fixed with 10-15 flower buds in aceta-alcohol (3:1). The cells were then stained with acetocarmine, temporary slides were prepared, and analysed using a light microscope (magnification 10x20 or 10x40, Motic Images Swift M10L) according to the generally accepted method (Ma et al., 1994b). The frequency of micronuclei occurrence was calculated per 100 tetrads (i.e. percent of MCN). In each version, 3000 tetrads were reviewed. The remaining flower buds were allowed to mature and flower for subsequent analysis using the Trad-SHM test. To analyze somatic mutations, staminate hairs in a flower were carefully separated from the base with tweezers, carefully straightened, and placed on a glass slide. The preparations were viewed under a binocular loupe (magnification 40x, Motic SMZ-171 or magnification 32x MBS-9 binocular loupe). The hairs on each stamen were counted and recessive pink mutation events (PME) and genetically undefined colourless mutational events (CME). The morphological changes in the stunted stamen hair (Trad-SH) and hair branching (Trad-BH) were recorded as well. In addition, some morphological changes in the number of stamens and petals in the flower, or absence of an anther were recorded. In each variant, an average of 12-15 thousand hairs were analysed. The calculation was carried out per 1000 hairs in accordance with the generally accepted method (Ma et al., 1994a). This approach allowed for a comprehensive assessment of mutagenic effects in the soil-plant system using standardized growth conditions and specific testing protocols developed to assess genetic damage to Tradescantia plants.

Fungi Growth Methodology

The mycological investigations of micromycetes were conducted based on the dilution assay, entails the preparation of an aqueous suspension followed by its inoculation onto a wort agar medium, adjusted to pH 4.5 to inhibit bacterial growth.

For the preparation of suspension cultures, 10 grams of soil were mixed with 90 mL of water, after which the mixture was further diluted with distilled water at dilution ratios of 1:100, 1:1000, and 1:10000. Then 1 ml of the suspension from each dilution was distributed into individual Petri dishes and, using a sterile spatula, spread over the entire surface of the agar plate. The experiment was carried out in triplicate. The fungal cultivation was conducted at a temperature range of 23–28°C, with the duration of cultivation spanning from 4 to 14 days (Kurakov & Kokaeva, 2023).

Species identification has been conducted using identification book (Pidoplichko and Milko, 1971; Koval et al., 2016).

Genotoxic Measurements-Comet Assay

Georginapaeous hohenackeri snails from all sampling locations were used to assess DNA damage in their digestive gland cells. The standard alkaline comet assay (Tice et al., 2000) was modified (Dailianis et al., 2005) for snails. Concisely, 20 μl of digestive gland cells mixed with 100 μl of 0.5% low melting point agarose were placed on slides coated with 1% normal melting point agarose. Slides were submerged in a lysis solution (2.5 M NaCl, 100 mM Na 2 EDTA, 10 mM Tris, 1% Triton X-100, 10% DMSO, pH 10.0) for 1 hour at 4°C, followed by alkaline electrophoresis buffer (300 mM NaOH, 1 mM Na 2 EDTA, pH 13) for 20 minutes at 4°C, and the electrophoresis was run at 300 mA and 1 V/cm for 15 minutes. After neutralization in 0.4 M Tris buffer (pH 7.5) for 10 minutes, slides were stained with 20 μg/ml ethidium bromide. DNA damage was assessed using the Comet Assay IV imaging system, with 150 cells analysed from three replicates per snail. DNA damage was quantified by the percentage of DNA in the comet tail. Comet data are presented as mean ± standard error (SE). Normality of comet data was tested using the Shapiro-Wilk test, and then the data were compared using the nonparametric Mann-Whitney U test. The level of statistical significance was $p < 0.05$.

Pesticides Measurement

The qualitative analysis of pesticides in soil has been performed by gas chromatography using PerkinElmer Clarus 680 GC instruments at Institute of Chemical Physics after A.B. Nalbandyan of NAS RA (ISO 10382: 2002).

Statistical Analysis

R studio 3.6.3 (2013) was used for statistical analysis. Pearson correlation coefficient has been used to analyse relationship between pesticides and fungi richness, molluscs richness, and genotoxic impact.

For statistical analysis of Tradescantia species and the comet assay data the software STATGRAPHICS Centurion 16.2 (StatPoint Technologies, Inc. USA; Warrenton, VA) were applied. Student t-test was used to determine the differences in levels of MN, PME and SSH among various groups.

All data are presented as mean ± standard error (SE). Standard comet assay data were compared using the nonparametric Mann-Whitney U test. The difference of $p < 0.05$ was considered statistically significant.

RESULTS AND DISCUSSION

This study aimed to investigate the impact of pesticides on the interconnected soil-plant-fungi-animal chain, with a focus on populations, species richness, abundance and genotoxicology.

Pesticides

The following pesticides were found at the studied locations (Table 1). The safety level of the pesticides has been checked based on the NCBI database (NCBI, 2023).

1-Dodecanol is a common component of insect pheromones and plant oils and used predominantly in orchards. It is also used as a biopesticide and as a surfactant. Substance is used to control *Cydia pomella*, *Adoxophyes orana* and *Pandemis heparana*.

2,4-Di-tert-butylphenol is also known as 2,4-DTBP is reported to have an impact on different organisms (Zhao et al., 2020).

Arsenous compounds are known with their impact on the soil and aquatic ecosystems (Shankar et al., 2014; Ravenscroft, et al., 2009).

Salicylaldehyde is used for fungicidal and herbicidal control (Kim & Chan, 2022).

All the recorded 4 pesticides have been detected from the Gogaran's sampled soil. In Arvshat's soil samples 2,4-DTBP and Salicylaldehyde have been reported. In Lichk the only reported pesticide was 2,4-DTBP.

Table 1. The recorded pesticides with their safety based on the NCBI database.

Chemical	Safety based on NCBI
1-Dodecanol $C_{12}H_{26}O$	Dangerous: irritant, very toxic to aquatic life with long lasting effects
2,4-Di-tert-butylphenol $C_{14}H_{22}O$	Dangerous: irritant, very toxic to aquatic life with long lasting effects
Arsenous acid AsH_3O_3	Dangerous: Highly toxic and carcinogenic
Salicylaldehyde $C_7H_6O_2$	Dangerous: irritant, poisonous, health damage causing genetic defects

Tradescantia

The analysis of somatic mutations in Tradescantia reveled a significant increase in PME frequency across all soil sample from study sites in 2022 compared with the control. Farming plots demonstrated a higher genotoxicity than their corresponding control plots. Notably, the highest mutation frequency was observed in plants grown in soil samples from Gogaran and Arevshat, with levels 3.9 to 5.0 times higher than the control. The frequency of PMEs in the Lichk's soil samples were from 1.5 to 3.4 times higher than the control. Spring samples taken before tillage were the cleanest, while the dirtiest samples were found during the summer months.

Additionally, various types of morphological changes were reported in Tradescantia SHM assays. Dwarf filamentous hairs containing 2 to 10 cells were more common in plants exposed to the soil from experimental plots in Gogaran, indicative of teratogenic effects.

According to the Trad-MN (Micronucleus) test, all studied soil samples induced a high level of MN formation in tetrads. The maximum MN induction was observed in soil samples from Gogaran and Arevshat, exceeding the control by 2.0 to 3.6 times, indicating that the experimental plots were more genotoxic than their controls (Table2).

Table 2. Induction of genotoxic effects in soil samples arable lands of the territory of Armenia in Tradescantia (clone 02)

Test	Sampling site	Season	2022		2023	
			Test	Control	Test	Control
Somatic Mutation Test (Trad-PME Test)	Arevshat	spring	1.59±0.50*	1.04±0.39	1.56±0.42*	1.01±0.32
		summer	1.83±0.51**	1.20±0.42	1.58±0.50*	0.93±0.33
		autumn	1.58±0.52*	0.98±0.44	1.54±0.49*	0.68±0.34
	Gogaran	spring	2.05±0.45***	1.21±0.35	1.64±0.55**	1.09±0.63
		summer	2.22±0.57***	1.75±0.50**	2.03±0.43***	1.26±0.45
		autumn	1.66±0.53**	0.96±0.34	1.60±0.41*	0.89±0.36
	Lichk	spring	1.48±0.38*	0.89±0.28	1.29±0.36*	0.84±0.28
		summer	1.63±0.44*	1.08±0.36	1.47±0.42*	0.70±0.35
		autumn	1.23±0.41	0.73±0.33	1.06±0.38	0.59±0.30
	greenhouse		0.40±0.16		0.46±0.21	
Morphological changes Trad-SHM assays /1000±m	Arevshat	spring	4,81±0,67	2,32±0,56	3,52±0,59	2,12±0,46
		summer	8,03±1,12	4,81±0,67	7,47±1,16	5,24±0,71
		autumn	5,16±0,72	1,63±0,49	2,58±0,61	1,67±0,53
	Gogaran	spring	3,45±0,65	1,58±0,53	3,39±0,58	1,92±0,44
		summer	17,59±2,52***	7,47±1,16	5,24±0,71	2,08 ±0,50
		autumn	3,64±0,76	1,96±0,62	2,92±0,67	1,64±0,49
	Lichk	spring	2,32±0,56	1,45±0,46	2,45±0,53	2,17±0,51
		summer	2,92±0,67	1,67±0,53	3,23±0,59	2,95±0,66
		autumn	2,06±0,57	1,64±0,49	2,33±0,63	1,63±0,49
	greenhouse		0,53±0,27		0,8±0,30	
Micronucleus Test (Tetr-MN Test)	Arevshat	spring	5.1±0.40***	3.9±0.35***	5.1±0.40***	3.5±0.34***
		summer	6.8±0.46***	4.9±0.39***	5.4±0.41***	3.8±0.35***
		autumn	4.3±0.37***	3.5±0.34***	4.1±0.36***	3.2±0.32**
	Gogaran	spring	5.8±0.43***	4,9±0.39***	5.6±0.42***	4.7±0.39***
		summer	8.0±0.46***	5.1±0.40***	6.8±0.46***	4.0±0.36***
		autumn	5.8±0.43***	4.7±0.39***	5.6±0.42***	3.9±0.35***
	Lichk	spring	4.9±0.39***	3.5±0.34***	4.3±0.37***	2.9±0.31
		summer	6.7±0.46***	4.8±0.39***	4.9±0.39***	3.2±0.32*
		autumn	3.8±0.35***	2.9±0.31	2.9±0.31	2.7±0.30
	greenhouse		2.2±0.26		2.3±0.27	

Statistically significant difference compare to the control is indicated at *p < 0.05, **p < 0.01, *** p < 0.001

In the second year of the study, the results were largely consistent with those of the first year. Cultivated plots continued to exhibit higher genotoxicity compared to controls. The frequency of observed mutations was higher in the summer months compared to the spring months. As in the previous year, the most contaminated samples were from the cultivated lands in Gogaran, exceeding background levels by 3.7 to 4.4 times.

Our investigation into Tradescantia highlighted significant genotoxicity in soil samples. The highest mutation rates were found in Gogaran, with frequencies up to 5 times higher than controls. The recurrent high contamination in Gogaran points to significant environmental impacts from agricultural activities.

Fungi

From 2022 to 2023, our research recorded a total 95 species of micromycetes from 3 locations. Identified species belong to 36 genera, 27 families, 20 orders, 11 classes, and 4 phyla. We have analysed the diversity of soil fungi in potato fields both before and after applying pesticides. The Pearson correlation coefficient indicated negative correlation between fungi richness and pesticides ($r = -0.07$, $P < 0.05$).

For all locations the farming (experimental) plots consistently had fewer species than the control plots in both years. In the first year, from Gogaran control plots were recorded 25 species and 21 species on Gogaran farming plots. This plot had the highest richness of micromycetes. On the next year the experimental plot showed a sharper decline in species diversity - from 21 to 16 species (fig. 2).

Figure 2. Micromycetes species diversity in control and experimental plots across three locations (Gogaran, Arevshat, and Lichk) over two years.

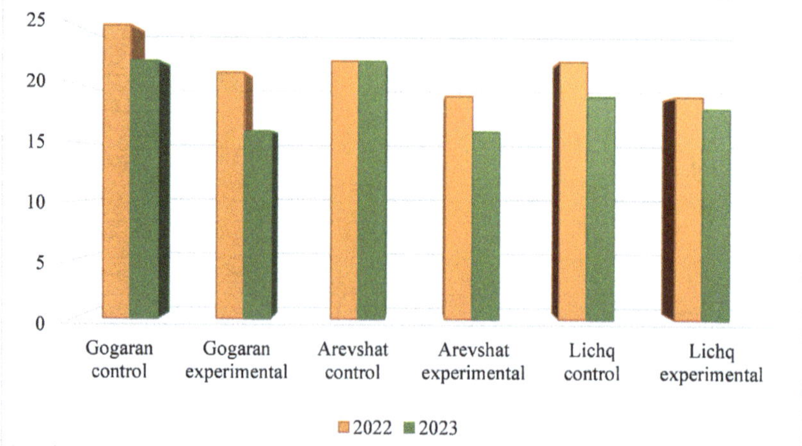

The Arevshat control plot maintained constant species diversity (22 species) over the two years. The experimental plot showed a decrease in species diversity from 19 to 16 species.

Twenty-two species were recorded from Lichk control location in 2022 and 19 species in 2023. The experimental plot showed a slight decrease in species diversity from 19 to 18 species.

We have separated 15 fungal species that caused damage to potatoes (Table 3). As indicated in the table, most identified micromycetes were observed during the spring and summer months of the first year of the study. Most of the detected species disappeared after the first application of pesticides. After applying pesticides in 2022 autumn and in the spring of the subsequent year, we detected only two species: *Alternaria alternata* and *Rhizoctonia solani*.

Table 3. List of micromycetes affecting potatoes.

Observation year	**2022**			**2023**		
Observation season	**spring**	**summer**	**autumn**	**spring**	**summer**	**autumn**
Species of micromycetes						
Agroathelia rolfsii	-	-	-	-	-	+
Alternaria alternata	+	+	+	+	-	-
Alternaria brassicae	+	+	-	-	-	-
Alternaria brassicicola	+	+	-	-	-	-
Alternaria consortialis	+	+	-	-	-	-
Alternaria malvae	+	+	-	-	-	-
Alternaria humicola	+	-	-	-	-	-
Fusarium javanicum	+	+	-	-	-	-
Fusarium verticillioides	+	+	-	-	-	-
Fusarium oxysporum	+	-	-	-	-	-
Fusarium sporotrichioides	+	+	-	-	-	-
Neocosmospora solani	+	+	-	-	-	-
Papulaspora irregularis	+	+	-	-	-	-
Rhizoctonia solani	+	+	+	+	-	-

The first applying pesticides suppressed the growth species of the *Alternaria* and *Fusarium* genera, both known for their phytopathogenic characteristics (Liu et al., 2022; Steglińska et al., 2023). This indicates a significant impact of pesticide application on soil fungal biodiversity and highlights potential shifts in microbial community dynamics within the potato field ecosystem.

R. solani and *A. alternata* were detected in the spring of 2023 and after pesticides application to the soil, the fungi were not detected.

In the autumn of 2023 on the experimental plot of Gogaran, the new pathogenic fungus *Agroathelia rolfsii* was found, which causes potato root and stem rot. The rotting of underground tubers caused by this fungus can reduce the quality and quantity of the potato crop. Above-ground symptoms include yellowing or browning of the leaves, and white powdery mildew appears on the stem in wet weather (Kator et al., 2015).

Regarding micromycetes species richness it has been shown that pesticides use can impact both harmful and useful species. Across all locations, control plots had more species of micromycetes than experimental plots in both years. With the experimental plots showing a more significant decline of species diversity from 2022 to 2023. Species existed before applying disappeared and they were replaced by one new pathogenic species resistant to applied pesticides: *A. rolfsii*. Pesticide use in potato cultivation, directly and indirectly, affects soil microorganisms, including Mucoromycetes. Although pesticides are primarily designed to control pests and phytopathogens, they also affect non-target organisms, including beneficial soil fungi such as Mucomycetes, which play an important role in nutrient cycling in the soil (Dzurendova et al., 2021; Devi et al., 2022).

Mollusca

From 2022 to 2023, our research recorded a total of 12 species of terrestrial molluscs across control and experimental plots in three studied localities. Several species were found in both experimental and control locations: *Georginapaeus hohenackeri*, *Chondrula tridens* (O. F. Müller, 1774), *Pupilla inops* (Reinhardt, 1877), *Xeropicta derbentina* (Krynicki, 1836), *Harmozica selecta* (Klika, 1894), *Lauria cylindracea* (Da Costa, 1778), *Sphyradium doliolum* (Bruguière, 1792), *Truncatellina costulata* (Nilsson, 1823), *Vallonia costata* (O. F. Müller, 1774), *Orcula doliolum* (Bruguière, 1792). Two species were recorded only from Arevshat's control location: *Harmozica pisiformis* (L. Pfeiffer, 1846), and *Truncatellina cylindrica* (A. Férussac, 1807). Molluscs recorded richness in the studied localities within the examined two years are presented in the figure below.

Based on the Figure 3 slight richness change can be fixed in studied experimental and control locations. However, this change is not that significant ($p>0.05$). In contrast of species richness species abundance has shown significant difference in control and experimental plots of studied locations ($p<0.05$) (Figure 4). Sampled molluscs species abundance is presented in the below figure.

Figure 3. Molluscs species richness within 2022-2023 sampling from 3 locations.

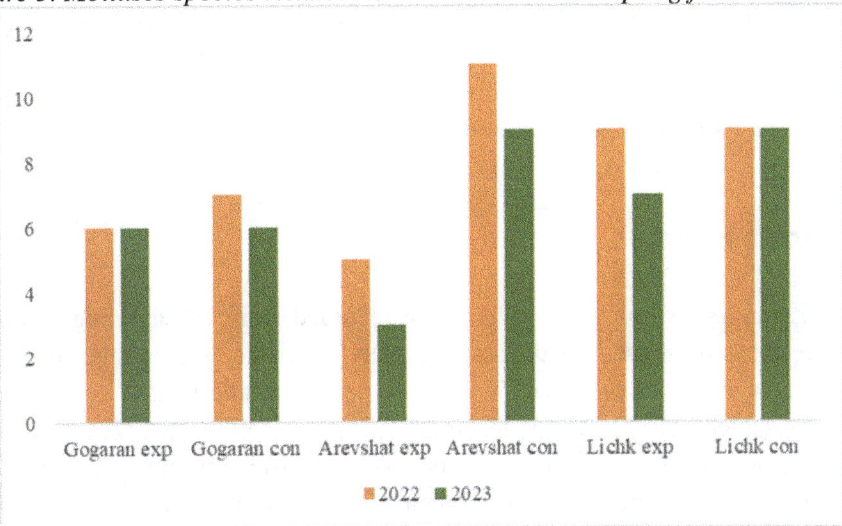

The comparison between experimental and control locations for molluscs species richness has not shown a significant change ($p > 0.05$), suggesting that pesticide application did not impact the number of mollusc species. Further, the Pearson correlation coefficient indicated a negative correlation between mollusc richness and pesticide levels ($r = -0.36$, $P < 0.05$), hinting on the impact of pesticides on biodiversity.

Figure 4. Molluscs species abundance within 2022-2023 sampling from 3 locations.

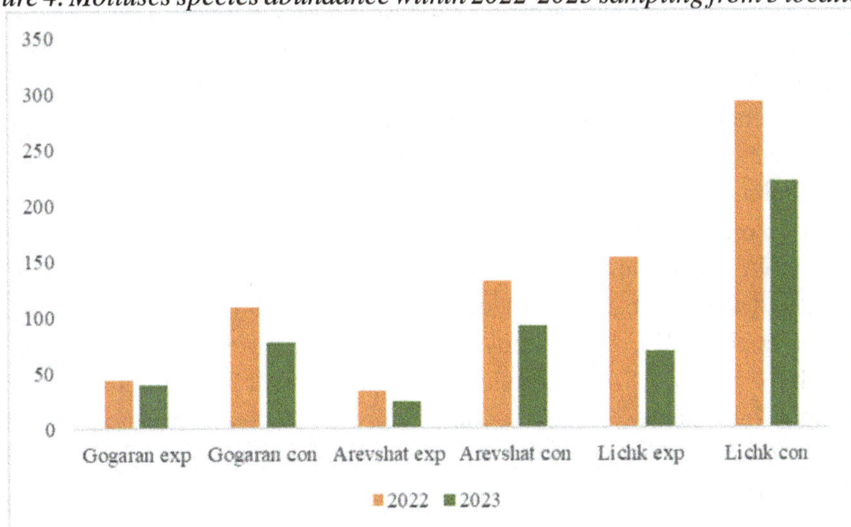

This decrease in abundance shows that pollutants can reduce population sizes of non-target organisms, including molluscs, due to their toxic effects (Druart et al., 2011; Sánchez-Bayo 2011; Radwan et al., 2020). This correlation underscores the need for careful management of pesticide use to mitigate its detrimental effects on terrestrial mollusc populations and maintain ecological balance (Isenring 2010; Dudley et al. 2017; Sor et al. 2020).

Genotoxic Effects

Georginapaeous hohenackeri has been recorded from all sampling locations therefore it has been used for the assessing DNA damage level in digestive gland cells. The mean values (± SE) of % DNA in the comet tail of snails *G. hohenackeri* are presented in Table 4. The levels of DNA damage in snails *G. hohenackeri* sampled from sites treated with pesticides were significantly higher compared to snails from the corresponding control areas. Moreover, the highest level of DNA damage was observed in snails living location of Gogaran (41.04 ± 1.56% DNA in comet tail), where farming field was treated with 4 pesticides. Similar trends were observed in Arevshat and Lichk, where pesticide exposure resulted in increased DNA damage in snails. The Pearson correlation coefficient indicated positive correlation between DNA % in comet tail and pesticides ($r = 0.92$, $P < 0.05$), this explains that the higher is the pollution more is the DNA damage.

Living organisms are typically exposed to mixtures of pollutants, whose combined effects may be synergistic or antagonistic (Chen et al., 2015; Singh et al., 2017). Moreover, Chen et al. (2015) demonstrated that the nature of these interactions shifts depending on the effect level, and that the relevance of synergistic effects increases with the complexity of the pollutant mixture. Consequently, the more complex mixture detected in Gogaran (all the recorded 4 pesticides have been detected from the Gogaran's sampled soil) compared with the other sampling sites may explain the higher genotoxic potential of the soil from this location.

Universal threshold values for population-level impairment in terrestrial gastropods, including *G. hohenackeri*, have not yet been formally defined. Previous study on *Helix lucorum*, collected from an unpolluted area in Armenia, demonstrated baseline DNA damage levels of approximately 24% in digestive gland cells, which is considered a normal background level for species inhabiting clean environments (Sargsyan et al., 2022). In our study, *G. hohenackeri* from each pesticide-treated site exhibited higher DNA damage levels compared to snails captured from unpolluted areas. Importantly, the mollusc abundance was lower in polluted areas than in non-polluted sites, suggesting that chronic genotoxic stress may negatively affect on population viability.

Table 4. Levels of DNA damage (mean ± SE) in digestive gland cells of snails G. hohenackeri inhabiting Gogaran, Arevshat and Lichk.

Sampling Season	% DNA in comet tail					
	Gogaran		Arevshat		Lichk	
	Con	Exp	Con	Exp	Con	Exp
Spring 1st sampling	**17.03 ± 0.97**	**40.51 ± 1.52**	**15.89 ± 1.36**	**36.54 ± 1.78**	**11.89 ± 0.58**	**27.56 ± 1.56**
Summer 1st sampling	17.61 ± 0.98	39.99 ± 1.53	16.66 ± 1.09	33.68 ± 1.66	10.78 ± 1.03	31.02 ± 1.78
Autumn 1st sampling	19.85 ± 0.86	41.58 ± 1.77	15.12 ± 1.26	35.74 ± 1.57	13.69 ± 0.65	29.78 ± 1.03
Spring 2nd sampling	18.44 ± 1.09	44.43 ± 1.51	13.92 ± 1.69	38.12 ± 1.89	12.58 ± 0.97	30.25 ± 1.24
Summer 2nd sampling	18.28 ± 1.02	38.60 ± 1.40	15.21 ± 0.85	37.45 ± 1.56	14.35 ± 1.12	28.56 ± 1.47
Autumn 2nd sampling	17.98 ± 1.15	41.15±1.65	14.61 ± 1.23	32.82 ± 1.44	12.14 ± 1.01	29.13 ± 1.15
Average	**18.20 ± 1.01**	**41.04 ± 1.56 ***	**15.24 ± 1.25**	**35.73 ± 1.65 ***	**12.57 ± 0.89**	**29.38 ± 1.37 ***

Con – Control, Exp – Experimental, *$p < 0.05$—significant difference compared to the corresponding control group.

Comparison of DNA damage of different locations summed for 2 years is also presented in the below Figure 5.

Figure 5. Comparison between DNA damage (mean ± SE of % DNA in comet tail) observed in digestive gland cells of G. hohenackeri at each sampling site. [a] *$p < 0.05$—significant difference compared to the control group of Gogaran;* [b] *$p < 0.05$—significant difference compared to the control group of Arevshat;* [c] *$p < 0.05$—significant difference compared to the control group of Lichk.*

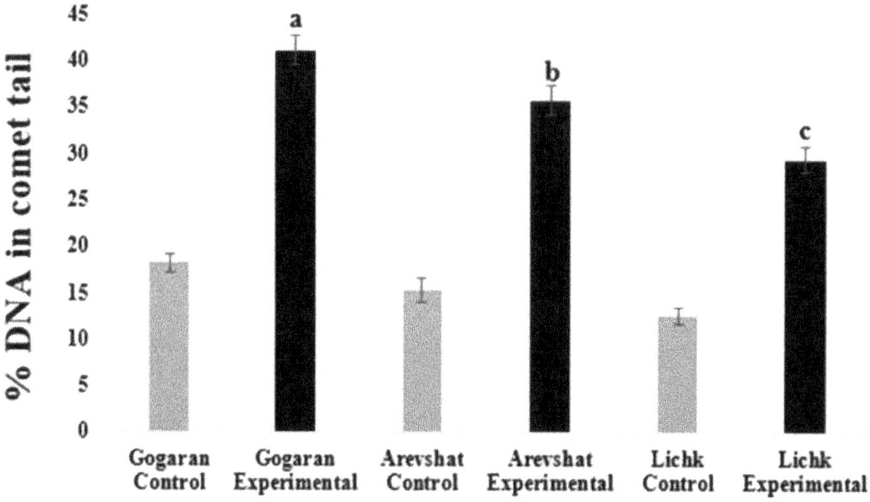

Our investigation into the genotoxic effects of pesticide exposure on the terrestrial mollusc *G. hohenackeri* revealed significant DNA damage in digestive gland cells, as indicated by the comet assay. The strong positive correlation found between the percentage of DNA in the comet tail and pesticide levels indicates a clear relationship between pesticide pollution and DNA damage, supporting previous findings that environmental contaminants can induce significant genetic damage in non-target organisms. These results align with other studies demonstrating the genotoxic effects of pesticides on various invertebrate species. For instance, studies on earthworms and aquatic snails have shown similar DNA damage responses to pesticide exposure, reinforcing the notion that pesticides pose a substantial risk to the genetic health of terrestrial and aquatic invertebrates. Various agrochemical components have mutagenic properties, causing mutations, chromosomal changes or DNA damage (Bolognesi, 2003; Kapeleka et al., 2019). Genotoxic effects of various pesticides were evaluated by the comet assay mainly using non-target aquatic animals (Jiang et al., 2023). An increase of DNA damage in haemocytes of freshwater snail Lymnea leuteola exposed to organophosphate insecticide profenofos was observed using the comet assay (Ali et al., 2018). Genotoxic effects of insecticide Match 5%EC (its active ingredient is lufenuron 5% EC) (Ibrahim et al., 2018) and herbicides butralin, glyphosate-isopropylammonium and pendimethalin (Ibrahim et al., 2019)

were assessed in haemocytes of freshwater snail Biomphalaria alexandrina by the comet assay. Our results are in agreement with reports that indicate the possibility of using the snails as bioindicators for the assessment of genotoxicity of pesticides. The consistent findings across different seasons and locations suggest that pesticide contamination in agricultural soils is a persistent issue that can lead to long-term genetic damage in mollusc populations. This has broader ecological implications, as genetic damage can affect population viability and ecosystem health.

CONCLUSION

This study reveals significant ecological and genotoxic impacts of agricultural pollution on high mountain steppe ecosystems of Armenia, focusing on species richness, abundance and genotoxicity. The experimental sites in Gogaran, Arevshat and Lichk showed significant contamination with organic matter (e.g. dodecanol), resulting in reduced biodiversity, increased genotoxicity and significant DNA damage.

Among the comparative methods for assessing agricultural soil pollution in the high mountain steppe regions of Armenia, the comet assay proved to be the most sensitive and efficient, providing rapid detection of snail DNA damage. This makes the comet assay an effective tool for large-scale monitoring of soil pollution in agricultural areas. Stamen hair mutations of Tradescantiat (Trad-SHM) test and comet assay have proven to be effective in detecting genotoxic effects, while the Trad-MN test represents a cost-effective method. Micromycete diversity and mollusc abundance also served as reliable bioindicators, showing significant declines in abundance across the experimental plots, particularly in summer when pesticide exposure was most intense.

A probably reason why Gogaran showed stronger genotoxic effects than Lichk may lie in the specific agricultural pressures and environmental characteristics of the site. Agricultural activity in Gogaran tends to be more intensive, which could mean the use of a broader spectrum or greater quantities of pesticides, some of which may possess higher genotoxic potential. Moreover, differences in soil parameters such as the level of organic matter, pH balance, or soil structure can affect how long pesticides remain active in the environment and how readily they are taken up by organisms, ultimately amplifying their biological impact.

This study shows that using several types of bioindicators together—Tradescantia assays (Trad-SHM and Trad-MN), micromycete communities, and mollusc populations—creates a much more complete and reliable picture of soil pollution and its ecological effects. Each of these indicators responds to environmental stress in its own way, and their combination allows for a deeper and more nuanced assessment of both ecological disturbances and genotoxic damage.

When rapid genotoxicity methods, such as the comet assay and Tradescantia tests, are complemented by broader ecological indicators like micromycetes and molluscs, the resulting system becomes especially valuable for environmental managers and policymakers. It provides a solid scientific basis for building monitoring programs, assessing pesticide-related risks, and planning more sustainable agricultural practices.

Overall, the results highlight the need for such comprehensive evaluations in Armenia, where the impact of pesticides on biodiversity remains poorly understood. They also emphasize the importance of adopting pest-management approaches that are less harmful to the fragile ecosystems of the high mountain steppe.

This study demonstrates the effectiveness of integrating bioindicators such as micromycetes and molluscs to create a comprehensive assessment system for soil pollution and its cascade effects. The combined use of these bioindicators provides valuable information to policymakers and conservationists, serving as a reliable tool for monitoring pesticide pollution and developing conservation strategies.

The results of this research show first the importance of performing such kind of evaluations in Armenia where pesticides impact on biodiversity is poorly studied and second, it's important to develop environmentally friendly options fighting against agricultural pests which will have low impact on the biodiversity in general.

ACKNOWLEDGMENT

The work was supported by RA MoESCS Higher Education and Science Committee, research project N° 21T-1F281.

REFERENCES

Akramowskii, N. N. (1976) Fauna of the Armenian SSR. Molluscs. Academy of Sciences of the Armenian SSR.

Ali, D., Ali, H., Alifiri, S., Alkahtani, S., Alkahtane, A. A., & Huasain, S. A. (2018). Detection of oxidative stress and DNA damage in freshwater snail Lymnea leuteola exposed to profenofos. *Frontiers of Environmental Science & Engineering*, *12*(5), 1. DOI: 10.1007/s11783-018-1039-6

Bolognesi, C. (2003). Genotoxicity of pesticides: A review of human biomonitoring studies. *Mutation Research/Reviews in Mutation Research*, *543*(3), 251–272. DOI: 10.1016/S1383-5742(03)00015-2 PMID: 12787816

Chen, C., Wang, Y., Qian, Y., Zhao, X., & Wang, Q. (2015). The synergistic toxicity of the multiple chemical mixtures: Implications for risk assessment in the terrestrial environment. *Environment International*, *77*, 95–105. DOI: 10.1016/j.envint.2015.01.014 PMID: 25667058

Convention on Biological Diversity of RA. (2014). 5th National Report.

Dailianis, S., Piperakis, S. M., & Kaloyianni, M. (2005). Cadmium effects on ROS production and DNA damage via adrenergic receptors stimulation: Role of Na+/H+ exchanger and PKC. *Free Radical Research*, *39*(10), 1059–1070. DOI: 10.1080/10715760500243765 PMID: 16298731

Devi, R., Kaur, T., Kour, D., Yadav, A., Yadav, A.N., Suman, A., Ahluwalia, A.S., & Saxena, A.K. (2022). Minerals solubilizing and mobilizing microbiomes: A sustainable approach for managing minerals' deficiency in agricultural soil. *J Appl Microbiol.*, *133*(3), 1245-1272. DOI: 10.1111/jam.15627

Druart, C., Millet, M., Scheifler, R., Delhomme, O., Raeppel, C., & de Vaufleury, A. (2011). Snails as indicators of pesticide drift, deposit, transfer and effects in the vineyard. *The Science of the Total Environment*, *409*(20), 4280–4288. DOI: 10.1016/j.scitotenv.2011.07.006 PMID: 21784506

Dudley, N., Attwood, S. J., Goulson, D., Jarvis, D., Bharucha, Z. P., & Pretty, J. (2017). How should conservationists respond to pesticides as a driver of biodiversity loss in agroecosystems? *Biological Conservation, 209*, 449–453. DOI: 10.1016/j.biocon.2017.03.012

Dzurendova, S., Zimmermann, B., Kohler, A., Reitzel, K., Nielsen, U. G., Dupuy—Galet, B. X., Leivers, S., Horn, S. J., & Shapaval, V. (2021). Calcium affects polyphosphate and lipid accumulation in Mucoromycota fungi. *Journal of Fungi (Basel, Switzerland), 7*(4), 300. DOI: 10.3390/jof7040300 PMID: 33920847

Fayvush, G. (2023). *Biodiversity of Armenia.* Springer International Publishing. DOI: 10.1007/978-3-031-34332-2

Ibrahim, A. M., Ahmed, A. K., Bakry, F. A., & Abdel-Ghaffar, F. (2018). Hematological, physiological and genotoxicological effects of Match 5% EC insecticide on *Biomphalaria alexandrina* snails. *Ecotoxicology and Environmental Safety, 147*, 1017–1022. DOI: 10.1016/j.ecoenv.2017.09.059 PMID: 29976004

Ibrahim, A. M., Ahmed, A. K., Bakry, F. A., Rabei, I., & Abdel-Ghaffar, F. (2019). Toxicological impact of butralin, glyphosate-isopropylammonium and pendimethalin herbicides on physiological parameters of Biomphalaria alexandrina snails. *Molluscan Research, 39*(3), 224–233. DOI: 10.1080/13235818.2019.1592296

Isenring, R. (2010). *Pesticides and the loss of biodiversity.* Pesticide Action Network Europe.

Jiang, N., Naz, S., Ma, Y., Ullah, Q., Khan, M. Z., Wang, J., Lu, X., Luosang, D.-Z., Tabassum, S., Chatha, A. M. M., & Basang, W. D. (2023). An overview of comet assay application for detecting DNA damage in aquatic animals. *Agriculture, 13*(3), 623. DOI: 10.3390/agriculture13030623

Kapeleka, J.A., Sauli, E., & Ndakidemi, P.A. (2021) Pesticide exposure and genotoxic effects as measured by DNA damage and human monitoring biomarkers. *Int J Environ Health Res., 31*(7), 805-822. DOI: 10.1080/09603123.2019.1690132

Kator, L., Hosea, Z. Y., & Oche, O. D. (2015). Sclerotium rolfsii: Causative organism of southern blight, stem rot, white mold and sclerotia rot disease. *Annals of Biological Research, 6*(11), 78–89.

Kim, J. H., & Chan, K. L. (2022). Natural salicylaldehyde for fungal and pre-and post-emergent weed control. *Applied Sciences (Basel, Switzerland), 12*(8), 3749.

Köhler, H. R., & Triebskorn, R. (2013). Wildlife Ecotoxicology of Pesticides: Can We Track Effects to the Population Level and Beyond? *Science, 341*(6147), 759–765. DOI: 10.1126/science.1237591 PMID: 23950533

Koval, E. Z., Rudenko, A. V., & Voloshuk, N. M. (2016). *Penicillium*. NNIRXV.

Kurakov, A. V., & Kokaeva, L. Yu. (2023). *Methods for studying the diversity of fungi in terrestrial and aquatic ecosystems*. Public National Academy of Mycology.

Le Gall, M., & Tooker, J. F. (2017). Developing ecologically based pest management programs for terrestrial molluscs in field and forage crops. *Journal of Pest Science*, *90*(3), 825–838. DOI: 10.1007/s10340-017-0858-8

Liu, J., Sun, Z., Zou, Y., Li, W., He, F., Huang, X., Lin, C., Cai, Q., Wisniewski, M., & Wu, X. (2022). Pre- and postharvest measures used to control decay and mycotoxigenic fungi in potato (*Solanum tuberosum* L.) during storage. *Critical Reviews in Food Science and Nutrition*, *62*(2), 415–428. DOI: 10.1080/10408398.2020.1818688 PMID: 32924541

Ma, T. H. (1999). The international program on plant bioassays and the report of the follow-up study after the hands-on workshop in China. *Mutation Research. Fundamental and Molecular Mechanisms of Mutagenesis*, *426*(2), 103–106. DOI: 10.1016/S0027-5107(99)00049-4 PMID: 10350579

Ma, T. H., Cabrera, G. L., Cebulska-Wasilevska, A., Chen, R., Loarca, A. I., Vandenberg, A. L., & Salamone, M. F. (1994a). Tradescantia stamen hair mutation bioassay. *Mutation Research*, *310*(2), 211–220. DOI: 10.1016/0027-5107(94)90114-7 PMID: 7523892

Ma, T. H., Cabrera, G. L., Chen, R., Gill, B. S., Sandhu, S. S., Vandenberg, A. L., & Salamone, M. F. (1994b). Tradescantia micronucleus bioassay. *Mutation Research*, *310*(2), 220–230. DOI: 10.1016/0027-5107(94)90115-5 PMID: 7523893

Martins, M., & Costa, P. M. (2015). The comet assay in Environmental Risk Assessment of marine pollutants: Applications, assets and handicaps of surveying genotoxicity in non-model organisms. *Mutagenesis*, *30*(1), 89–106. DOI: 10.1093/mutage/geu037 PMID: 25527732

NCBI. (2023). https://www.ncbi.nlm.nih.gov/

Petridis, P., Jha, A. N., & Langston, W. J. (2009). Measurements of the genotoxic potential of (xeno-)oestrogens in the bivalve mollusc Scrobicularia plana, using the Comet assay. *Aquatic Toxicology (Amsterdam, Netherlands)*, *94*(1), 8–15. DOI: 10.1016/j.aquatox.2009.05.010 PMID: 19540004

Pidoplichko, N. M., & Milko, A. A. (1971). *Atlas of Mucoral Fungi*. Naukova Dumka.

Radwan, M. A., El-Gendy, K. S., & Gad, A. F. (2020). Biomarker responses in terrestrial gastropods exposed to pollutants: A comprehensive review. *Chemosphere*, *257*, 127218. DOI: 10.1016/j.chemosphere.2020.127218 PMID: 32497833

Ravenscroft, P., Brammer, H., & Richards, K. (2011). *Arsenic pollution: a global synthesis*. John Wiley & Sons.

Sánchez-Bayo, F. (2011). Impacts of agricultural pesticides on terrestrial ecosystems. *Ecological impacts of toxic chemicals*, 63-87. .DOI: 10.2174/978160805121210063

Sargsyan, A., Hovhannisyan, G., Simonyan, A., Arakelyan, M., Arzumanyan, M., & Aroutiounian, R. (2022). Application of land snail Helix lucorum for evaluation of genotoxicity of soil pollution. *Mutation Research/Genetic Toxicology and Environmental Mutagenesis*, *878*, 503500. DOI: 10.1016/j.mrgentox.2022.503500 PMID: 35649673

Shankar, S., & Shanker, U., & Shikha. (2014). Arsenic contamination of groundwater: A review of sources, prevalence, health risks, and strategies for mitigation. *TheScientificWorldJournal*, •••, 304524.

Singh, N., Gupta, V. K., Kumar, A., & Sharma, B. (2017). Synergistic effects of heavy metals and pesticides in living systems. *Frontiers in Chemistry*, *5*, 70. DOI: 10.3389/fchem.2017.00070 PMID: 29075624

Sor, R., Ngor, P. B., Boets, P., Goethals, P. L., Lek, S., Hogan, Z. S., & Park, Y. S. (2020). Patterns of mekong mollusc biodiversity: Identification of emerging threats and importance to management and livelihoods in a region of globally significant biodiversity and endemism. *Water (Basel)*, *12*(9), 2619. DOI: 10.3390/w12092619

Steglińska, A., Sulyok, M., Janas, R., Grzesik, M., Liszkowska, W., Kręgiel, D., & Gutarowska, B. (2023). Metabolite formation by fungal pathogens of potatoes (*Solanum tuberosum* L.) in the presence of bioprotective agents. *International Journal of Environmental Research and Public Health*, *20*(6), 5221. DOI: 10.3390/ijerph20065221 PMID: 36982130

Tice, R. R., Agurell, E., Anderson, D., Burlinson, B., Hartmann, A., Kobayashi, H., Miyamae, Y., Rojas, E., Ryu, J. C., & Sasaki, Y. F. (2000). Single cell gel/comet assay: Guidelines for in vitro and in vivo genetic toxicology testing. *Environmental and Molecular Mutagenesis*, *35*(3), 206–221. DOI: 10.1002/(SICI)1098-2280(2000)35:3<206::AID-EM8>3.0.CO;2-J PMID: 10737956

Warnasuriya, S. D., Udayanga, D., Manamgoda, D. S., & Biles, C. (2023). Fungi as environmental bioindicators. *The Science of the Total Environment*, *892*, 164583. DOI: 10.1016/j.scitotenv.2023.164583 PMID: 37277042

Chapter 13 Antioxidant Enzyme Activities in Yeast Candida guilliermondii NP-4 Exposed to Millimeter Waves

Anna Muradyan
http://orcid.org/0009-0000-9472-4672
Yerevan State University, Armenia

Syuzan Marutyan
Yerevan State University, Armenia

Hasmik Karapetyan
Yerevan State University, Armenia

Meri Hayrapetyan
Yerevan State University, Armenia

Seda Marutyan
http://orcid.org/0000-0002-8019-9093
Yerevan State University, Armenia

ABSTRACT

Modern telecommunication technologies use the high frequency microwave radiation, the application of which continues to expand rapidly. Exposure to non-ionizing radiation can induce oxidative stress in cells. The biochemical processes involved

DOI: 10.4018/979-8-3373-1524-9.ch013

in overcoming this stress are highly energy-consuming and require additional energy expenditure which consequently enhances the ATP hydrolysis. The indicator of oxidative stress is the intensification of lipid peroxidation, with malondialdehyde (MDA) serving as a key biomarker. In eukaryotic cells, various bioactive dietary supplements, including royal jelly (RJ), may exert beneficial effects on mitigating oxidative stress due to their antioxidant and other biologically active components. Yeasts represent a convenient model for studying the development of oxidative stress and the mechanisms of its overcoming in eukaryotic cells, because they are similar to higher eukaryotes by their structural, reproductive, and functional peculiarities. This study aimed to investigate the effect of 51.8 GHz microwave radiation on ATPase activity, lipid peroxidation, and antioxidant enzyme activity in homogenates and mitochondria of Candida guilliermondii NP-4, as well as to assess the potential role of RJ in migtigating of oxidative stress in yeast cells. The results demonstrated that the exposure to millimeter waves of 51.8 GHz and the following repair period led to intensification of lipid peroxidation, evidenced by an increase in MDA level by 75%. In repaired cells, the MDA content was reduced by ≈27%. Microwave radiation also stimulated catalase and superoxide dismutase (SOD) activity by 65% and 30%, respectively, in repaired cells the SOD activity increased by 13%, while catalase activity - by 25%.

INTRODUCTION

The mitochondrial ATPsynthase is a rotary molecular machine composed of multiple subunits (Xu et al., 2015). ATPsynthase plays a key role as a molecular target for drug development and can be used in medical practiceas antimicrobial and anticancer agent. The non-ionizing radiation induces different structural damages that must be repaired before cells can resume normal growth and proliferatio (Faccenda et al., 2012). Eukaryotic cells possess different mechanisms for repairing of radiation-induced damages. These mechanisms are highly energy-intensive, and increase the cellular demand for ATP. From this point of view, the ATP production in repairing cells, as well as its hydrolysis as a universal energy source, critically increases (Hovnanyan et al., 2020). Radiation exposure also induces oxidative stress in living cells (Lei, 2016), becausex the overproduction of reactive oxygen species (ROS) leads to increase of lipid peroxidation and antioxidant enzymes activity, such as superoxide dismutase (SOD) and catalase (Lei, 2016). Some non-enzymatic defense mechanisms are also activated, among which the amino acid proline plays a significant role. Proline functions as a stress-adaptive molecule and can neutralize ROS. It can also serve as an additional source of energy and nitrogen, participate in protein folding and stabilization, and thus contribute to the maintenance of cellular functions (Mat Nanyan et al., 2020).

Oxidative stress mitigation and cell survival can also be enhanced by use of biologically active compounds, such as royal jelly (RJ). RJ traditionally used in medicine, functional nutrition, and cosmetics worldwide. It is known to reduce lipid peroxidation levels and enhance the activity of antioxidant enzymes such as SOD and catalase (Yang et al., 2017). The biological activity of RJ depends on its content of bioactive constituents (Miyashita et al., 2016) including proteins, lipids, carbohydrates, vitamins, flavonoids, polyphenols, and other biologically active substances (Ramanathan et al., 2016). RJ can have a radioprotective effect on eukaryotic cells, and may be integrated into strategies for protecting living systems from various types of radiation exposure.

RJ also acts as a natural antibiotic and contains antioxidants with the potential to reduce the risk of cancer, hypertension, diabetes, and cardiovascular diseases (Bok et al., 2019).

The present study aimed to investigate the effects of 51.8 GHz microwave radiation on ATPase activity, lipid peroxidation, and antioxidant enzyme function in homogenates and mitochondria of *Candida guilliermondii* NP-4, in the presence and absence of RJ, to elucidate its potential protective effects against non-ionizing radiation in yeast cells. In this study, yeasts were used as a model of higher eukaryotes (Marutyan, 2021), as the structure and functions of yeast genetic material, as well as the basic metabolic pathways of yeasts are similar with those of higher eukaryotic organisms (Salazar et al, 2016). The results obtained will allow to elucidate the molecular mechanisms of the effects of microwave radiation on eukaryotic organisms, as well as contribute to the development of protective mechanisms for living organisms against such harmful external factors.

MATERIALS AND METHODS

Object of the Study

The experimental object was *Candida guilliermondii* NP-4 yeast (Navasardyan et al., 2017), cultivated in tubes containing 2% sucrose agar and grown for 48 h at 30 °C.

Yeast Growth Medium

Yeast cells were cultivated in liquid synthetic medium containing: $(NH)_2HPO_4$ – 3.8 mM, $NH_4H_2PO_4$ – 17.5 mM, K_2SO_4 – 1 mM, $MgSO_4 \times 7H_2O$ – 1.6 mM, and D-glucose – 100 mM (pH 5.5). The medium was sterilized at 1 atm for 20 min.

Preparation of Yeast Biomass

To the growth medium, lyophilized RJ (Arpimed LLC, Kotayk region, RA) was added to a final concentration of 50 μg mL^{-1}, along with biotin at 3×10^6 g L^{-1}. This amount of RJ was the effective concentration for the studied yeasts. Yeast cells were incubated for 24 h in a thermostatic orbital shaker-incubator (INFORS HT Multitron Standard, Switzerland) at 180 rpm and 30 °C. The biomass was harvested by centrifugation at 3000 × g for 10 min, washed with distilled water, and the wet weight was recorded spectrophotometrically (GENESIS 10S UV-VIS, Thermo Fisher Scientific Inc., USA) (Marutyan et al., 2021).

Exposure of Yeast Cells to Microwaves

A portion of the wet biomass was exposed to millimeter-waves (MMW) at the Faculty of Radiophysics, Yerevan State University, under the following conditions: λ=6mm, ν=51.8 GHz, power density – 10mW cm^{-2}), exposure time – 1 hour. The distance from antenna was determined by equation $r > \frac{2D^2}{\lambda}$ (Capozzoli et. al, 2024), where D – diameter of sphere containing yeasts, cm; and r – distance of exposed object from antenna, cm. The temperature remained constant during exposure: $\Delta t=0.1^{\circ}C$.

Post-Exposure Repair

A portion of the MMW-exposed biomass was incubated under repair conditions (30 °C, 100 mM glucose, 50 μg mL^{-1} RJ) in the same growth medium used for initial cultivation.

Preparation of Yeast Homogenates

Homogenates were prepared using a high-pressure homogenizer (Avestin EmulsiFlex-C3). Each sample was disrupted under 15,000-18,000 Pa for 5 min. The cell-free extract was suspended in 0.05 M Tris-HCl buffer (pH 8.5).

Isolation of Mitochondria

Mitochondria were isolated from the yeast homogenate by differential centrifugation in 0.25 M sucrose medium at 0°C, as described by Karapetyan et al. (2024). Differential centrifugation was performed in the following steps: 600 x g, supernatant 2000 x g, then supernatant 9500 x g (Guėrin, et al, 1979) The resulting sediment, representing the mitochondrial fraction, was homogenized with a glass homogenizer in 0.05 M Tris-HCl buffer (pH 8.5) to disrupt mitochondrial membranes.

Determination of Protein Amount

Protein content in homogenates and mitochondrial fractions was determined by the Lowry method (Lowry et al., 1951).

ATPase Activity Assay

ATPase activity was determined by measuring the amount of inorganic phosphate released. Yeast homogenates and mitochondrial fractions were incubated with 16 μM ATP in 0.05 M Tris-HCl buffer, followed by the addition of 0.5 mL 8 mM $MgSO_4 \times 7H_2O$ (in 0.05 M Tris-HCl buffer). The mixtures were incubated at 37 °C for 45 min. Reactions were stopped by adding 6 mL of 6% perchloric acid (PCA). From the supernatant, 1 mL was taken and mixed sequentially with 0.08 mL 20% KOH (pH 4), 7 mL acetate buffer (pH 4), 1 mL 1% ascorbic acid in 1 mM $CuSO_4 \times 5H_2O$, and 1 mL 1% ammonium molybdate in 0.05 N H_2SO_4. Absorbance was measured at $\lambda = 670$ nm using spectrophotometer (Genesis 10 UV-VIS), and phosphate concentration was calculated from a standard curve (Fig. 1a). Specific ATPase activity was expressed as μg Pi/mg protein. P_i was determined spectrophotometrically (Genesis UV-VIS 10, USA) as described by Blbultyan and Trchounian (2015). N, N′-dicyclohexylcarbodiimide (DCCD) was used as an FoF_1 inhibitor of H^+-ATPase. The samples were treated with DCCD for 5 min before assay. The application of DCCD to yeast cells significantly reduces H^+ efflux and prevents the formation of a transmembrane potential (Serrano, 1983). Additionally, DCCD binds to the proton transport subunits of the enzyme, resulting in irreversible inhibition (Bowman et al., 1988). Therefore, the use of DCCD for studying yeast H^+-ATPase activityis fully justified (Hong & Peterson, 2008) and complies with internationally accepted experimental protocols.

Determination of MDA Amount

MDA content in yeast homogenates was measured spectrophotometrically (Fig.1b) by assessing the absorbance of thiobarbituric acid-reactive substances (TBARS) as described by Mejía-Barajas et al. (2017). This method of MDA determination is fully sufficient to assess the presence of oxidative stress in exposed cells (Ito et al., 2019; Kenneth et al, 2023). The MDA concentration was calculated using the following formula: MDA concentration (μM mL-1)=A/εl, where: *A* absorbance was measured at 532 nm, ε- molar extinction coefficient of the MDA–TBA complex (typically 1.56×10^5 $M^{-1} \cdot cm^{-1}$), l - optical path.

Enzyme Activity Assays

Catalase: Catalase activity was assayed by incubating the enzyme sample in 1 mL substrate (65 μmol L^{-1} H_2O_2 in 60 μmol L^{-1} sodium-potassium phosphate buffer, pH 7.4) at 37 °C for 3 min. The reaction was stopped with ammonium molybdate, and the absorbance of the yellow molybdate–H_2O_2 complex was measured at 374 nm (Mahmoud, 2016). Activity was calculated using the formula:

$$KU = (2.303/t) \times [\log(S^0/(S-M))] \times V \ / V ,$$

where S^0 = absorbance of standard, S = absorbance of sample, M = absorbance of sample buffer, V = total sample volume, and V = enzyme sample volume.

Superoxide dismutase (SOD). SOD activity was determined by the inhibition of adrenaline auto-oxidation at 480 nm (Carbone et al., 2003). The maximum rate of oxidation (0.025 A min^{-1}) was considered 0% inhibition; 50% inhibition corresponded to one unit of SOD activity, expressed as U/mg protein. Protein concentration was determined by the Lowry method.

Enzymes of Proline biosynthesis. Total activity of proline biosynthesis enzymes was determined as described by Karapetyan et al. (2024), based on the amount of proline formed (μmol proline mg^{-1} protein), estimated by chemical assay.

Statistical Analysis

Data were analyzed and the graphs were constructed using GraphPad Prism 8 (GraphPad Software Inc., San Diego, CA, USA). Results are presented as mean ± standard error of the mean (SEM). Group differences were evaluated using two-way ANOVA. The number of biological replicates (n) and data of significance (p-values) are given in the legend of each graph. A p-value < 0.05 was considered statistically significant.

RESULTS AND DISCUSSION

ATPase Activity in *Candida guilliermondii* NP-4 Homogenates and Mitochondria in the Presence and Absence of RJ

In yeast cells *C. guilliermondii* NP-4, total ATPase activity in homogenates and mitochondria was approximately fourfold higher than H^+-ATPase activity (Fig. 1a). Moreover, both total and H^+-ATPase activities were significantly higher in mitochondria compared to homogenates (Fig. 1b).

Exposure to millimeter-waves (MMW) and the following repair period led to an increase in ATPase activity in both homogenates and mitochondria: the total

ATPase activity in homogenates of MMW-exposed yeast increased by ~30% in comparison to native controls.

In repaired cells, total ATPase activity increased by ~36% compared to MMW-exposed cells. The similar effect was observed for H^+-ATPase activity. After MMW exposure, it increased approximately twofold, and after repair period, it rose by an additional ~17% (Fig. 1a).

In MMW-exposed mitochondria, total ATPase activity increased more than twofold (Fig. 1b), while in repaired cells, it was 1.6 times higher than in the MMW-exposed yeasts mitochondria. DCCD-sensitive ATPase activity in mitochondria increased by 52.8% and 71.3% in MMW-exposed and repaired cells, respectively.

Figure 1. ATPase activities in yeast C. guilliermondii NP-4 exposed to miocriwaves and after repair: a - standard curve for phosphate, b- homogenate, c – mitochondria (n=3, p<0.05, p<0.01, p<0.001).

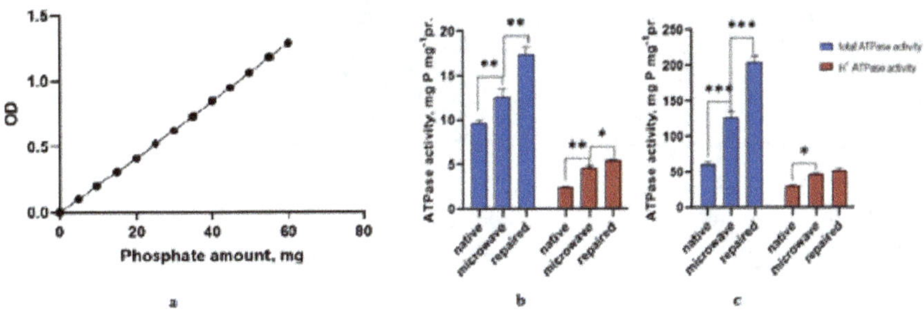

An increase in ATPase activity was also observed when 50 µg mL-1 RJ was added to the growth medium of yeasts (Fig. 2). In homogenates of yeast grown in the presence of RJ, MMW exposure led to a threefold increase in total ATPase activity compared to native yeasts. In repaired cells homogenate, when the incubation was realised in the presence of RJ, the total ATPase activity increased by 24.6% in comparison with MMW-exposed cells. H^+-ATPase activity in yeast homogenates in the presence of RJ increased ˜6-fold in MMW-exposed and by 14.6% in repaired cells.

In yeast mitochondria grown in the presence of RJ (Fig. 2b), the total ATPase activity increased by ˜37% after MMW exposure, and by 80% in repaired cells in comparison of MMW-exposed yeast mitochondria. The H^+-ATPase activity increased by 27% in MMW-exposed yeast mitochondria and by 35% in repaired yeast mitochondria compared to the MMW-exposed yeasts.

The total ATPase activity in living cells is determined by ATP-dependent ion transport systems, which are driven by various membrane-bound ATPases (Serrano et al., 1986). The plasma membrane ATPase of yeast facilitates active nutrient uptake via an H^+-symport mechanism, generating an asymmetric distribution of Na^+ and K^+ ions between the cell and its environment. This ionic gradient supports essential metabolic processes. The higher total ATPase activity compared to H^+-ATPase activity in homogenates is likely due to the contribution of Na^+/K^+-ATPase (Bhattacharya and Basak, 2017).

Figure 2. ATPase activities in yeast C. guilliermondii NP-4 exposed to miocrowaves and after repair in the presence of RJ: a - homogenate, b – mitochondria (n=3, $p<0.05$, $p<0.01$, $p<0.001$).

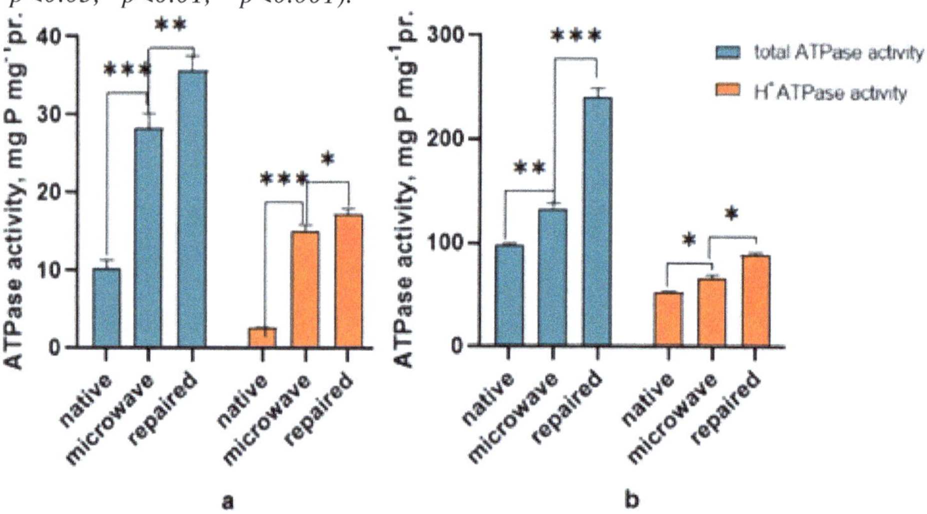

The observed increase in H^+-ATPase activity in MMW-exposed and repaired cells may be explained by the highly energy-intensive processes occurring in yeast cells to repair MMW-induced damage. These processes require enhanced ATP hydrolysis, which is reflected in the elevated H^+-ATPase activity, particularly in yeast mitochondria.

Effect of Millimeter Waves on Lipid Peroxidation Processes and the Antioxidant System in *C. guilliermondii* NP-4

Microwave exposure stimulated lipid peroxidation processes in yeast cells, resulting in the accumulation of malondialdehyde (MDA) as the end product (Cordiano et al., 2023). In MMW-exposed yeast, MDA content increased by 75% compared to native cells (Fig. 3). In repaired cells, the MDA level decreased by ~27% in comparison to MMW-exposed cells, although it remained ~30% higher than in native yeast. This phenomenon is probably related to changes in the activity of the antioxidant system.

Figure 3. Effect of millimeter waves on the MDA amount (μM mg⁻¹ protein.) in C. guilliermondii NP-4 yeasts grown in the absence and presence of RJ (n=3, p<0.05, p<0.01, p<0.001)

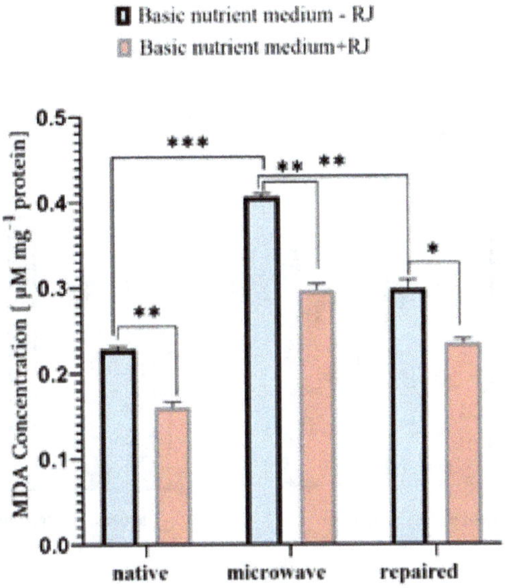

Specifically, MMW exposure enhanced catalase (Fig. 4a) and SOD (Fig. 4b) activities by 65% and 30%, respectively. In repaired cells, catalase and SOD activities were further elevated by 25% and 13%, respectively.

Figure 4. Effect of millimeter waves on the activity of antioxidant enzymes in C. guilliermondii NP-4 yeasts grown in the absence and in the presence of RJ (n=3, p<0.01, p<0.001) a- catalase (U mg^{-1} protein); b- SOD (U mg^{-1} protein)

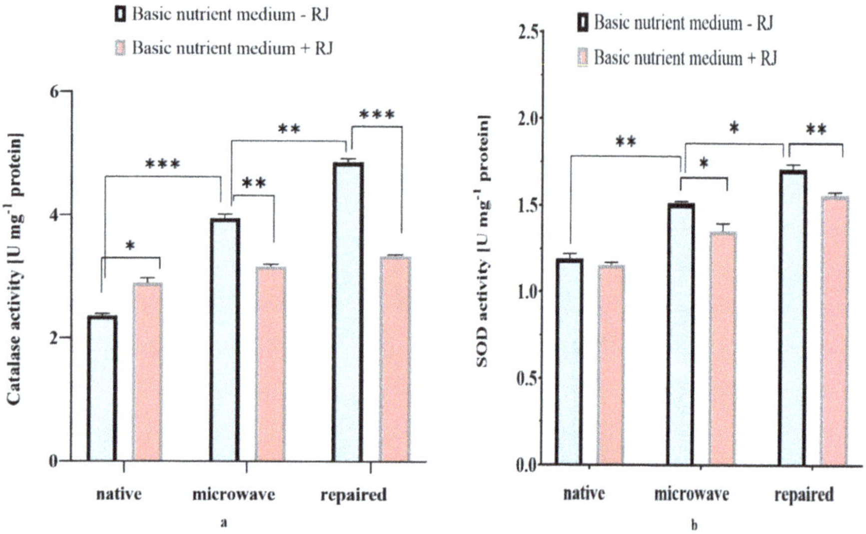

Cultivation of yeast in the presence of RJ markedly reduced lipid peroxidation levels: in MMW-exposed cells, MDA content decreased by ~27% in compared to yeast grown without RJ, while in repaired cells the reduction was ~18% (Fig. 3). These findings indicate that RJ can mitigate MMW-induced oxidative stress in yeast by lowering lipid peroxidation. Additionally, in native yeast grown in the presence of RJ, the catalase activity increased by ~22% (Fig. 4a), whereas no significant changes were detected in SOD activity (Fig. 4b). This suggests that RJ primarily enhances hydrogen peroxide detoxification rather than superoxide anion removal.

Millimeter waves also influenced the non-enzymatic antioxidant defense system, particularly by increasing proline content. Free proline levels increased by 31% in MMW-exposed yeast compared to native yeasts, and by an additional 12% in repaired cells in comparison to MMW-exposed cells (Fig. 5a).

The increase in proline content was associated with elevated activities of its biosynthetic enzymes - ornithine transaminase (OT) and pyrroline-5-carboxylate reductase (P5CR). No significant differences in OT and P5CR activity were observed between MMW-exposed and repaired cells; however, in repaired cells, proline biosynthesis was ~36% higher compared to native yeasts (Fig. 5b).

In yeasts grown in the presence of RJ, free proline content decreased by ~16% in unexposed cells and by 10% in both MMW-exposed and repaired cells compared to yeasts grown without RJ. Proline biosynthetic enzyme activities (OT and P5CR) did not show significant changes in RJ-containing medium, suggesting that RJ may stabilize proline metabolism without enhancing its activity under MMW-exposure conditions. Overall, these data demonstrate that in MMW-exposed *C. guilliermondii* NP-4 cells, cultivation with RJ leads to a relatively weak oxidative stress response. At the same time, the antioxidant system - both enzymatic and non-enzymatic components - remains largely unaffected.

Figure 5. Effect of millimeter waves on the proline amount (μM mg^{-1} protein) (a) and total activity of proline biosynthesis enzymes (OT and P5KR) (U mg^{-1} protein) (b) in C. guilliermondii NP-4 yeasts grown in the absence and in the presence of RJ (n=3, $p<0.05$, $p<0.01$)

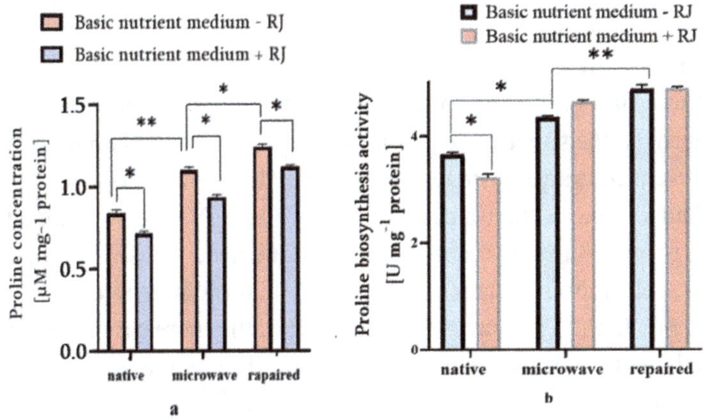

This indicates a potential protective or modulatory role of RJ in maintaining cellular homeostasis under radiation stress. It is likely that the short-chain (8–12 carbon atoms) fatty acids present in RJ possess strong antioxidant activity (Hora et al., 2018), thereby reducing the level of oxidative stress induced by MMW exposure and contributing to yeast survival.

CONCLUSION

Microwave exposure induces oxidative stress in *C. guilliermondii* NP-4 yeasts, associated with enhanced lipid peroxidation processes and increased reactive oxygen species (ROS) levels. MDA is a stable and highly sensitive end product of lipid peroxidation, which is widely accepted as a primary and reliable indicator in studies of UV and millimeter wave exposure to cells (Ito et al., 2019; Kenneth et al, 2023). The accumulation of malondialdehyde (MDA) observed in exposed yeast directly reflects the intensification of membrane polyunsaturated fatty acid peroxidation (Zheng et al., 2024).

Under the influence of 51.8 GHz millimeter waves, changes in the ratios of individual enzymes are recorded in the antioxidant enzyme system, which is characterized by a significant increase in the catalase/SOD ratio (CAT/SOD ratio). This is consistent with the kinetic properties of these enzymes. The reaction rate constant of SOD is $\approx 10^8$–10^9 M^{-1} s^{-1}, due to which H_2O_2 is formed from the rapid dismutation of superoxide, while catalase, being an enzyme with a high kcat ($\approx 10^6$ s^{-1}), ensures the rapid decomposition of H_2O_2, preventing the formation of hydroxyl radicals through the Fenton reaction (Rynkowska et al., 2020). Thus, the increase in the CAT/SOD ratio under millimeter-wave-induced stress conditions reflects a shift in the balance towards more efficient neutralization of H_2O_2 during cell adaptation to reduce secondary ROS damage.

Furthermore, irradiation increased the catalase/SOD activity ratio, which may represent an additional adaptive factor. This shift is advantageous because the product of SOD activity, the superoxide anion radical, is less toxic than hydrogen peroxide, whose high concentrations can lead to the formation of the more dangerous hydroxyl radical. Thus, higher catalase activity relative to SOD may enhance the ability of yeast cells to overcome stress consequences (Cui et al., 2023).Under oxidative stress conditions, cells mobilize their energetic and material resources to survive extreme environments. In the absence of royal jelly (RJ), the increase in the CAT/SOD ratio in exposed cells clearly indicates that the cell directs the antioxidant response to the activation of the catalase-dependent neutralization pathway. In the presence of RJ, however, a significantly weaker oxidative stress is observed, since a 27–30% decrease in MDA levels and a moderate ($\approx$20%) activation of catalase without a significant change in SOD lead to a more balanced CAT/SOD ratio. This indicates that RJ reduces ROS formation under millimeter wave exposure and maintains a stable regulation of the antioxidant enzyme system. Thus, the analysis of the CAT/SOD ratio serves not only as a biochemical indicator for the assessment of microwave-induced oxidative stress, but also confirms the protective action of RJ by providing controlled detoxification of H_2O_2 and preventing secondary ROS damage.

The marked increase in ATPase activity observed in the mitochondria of MMW-exposed and repaired yeast, compared to native cells, indicates an elevated energy demand. This is likely due to morphological alterations and structural and functional disruptions in macromolecules caused by the thermal and resonance effects of millimeter waves (Marutyan et al., 2021), which must be repaired to restore normal cell function.

The stress-adaptive amino acid proline also plays a direct role in defense and recovery processes under MMW-exposure. Its chemical structure enables interaction with certain ROS species (Suzuki, 2015), facilitating their neutralization.

Additionally, similar to molecular chaperones, proline can prevent stress-induced protein aggregation and denaturation (Kolupaev and Karpets, 2019) and stimulate the expression of SOD, catalase, and ascorbate peroxidase (Mat Nanyan et al., 2020). Supplementation of the growth medium with RJ promotes adaptive responses aimed at repairing MMW-induced damage. The biologically active compounds in RJ protect cells from oxidative stress by reducing ROS levels and inhibiting lipid peroxidation, in part through hydroxyl radical scavenging and hydrogen peroxide acceptance (Kumar et al., 2024). Thus, MRJPs and RJ-derived peptides possess documented ROS-scavenging activity, while phenolics such as quercetin, kaempferol and chlorogenic acid are known to suppress oxidative stress by neutralizing superoxide and hydroxyl radicals. These components provide a biochemical basis for the reduction of MDA levels observed in RJ-treated microwave-exposed cells (Seker et al., 2025). Besides direct radical scavenging, RJ may also act by stabilizing membrane structure through its medium-chain fatty acids (notably a unique fatty acid found exclusively in RJ *trans*-10-hydroxy-2-decenoic acid or 10-HAD (Spanidi, et al, 2022), thereby limiting lipid peroxidation, and by supporting mitochondrial homeostasis. The 10-HAD unsaturated fatty acid enhances the proyton gradient and thereby increases the mitochondrial membrane potential, and improves the efficiency of the ETC. As a result, mitochondrial functional activity increases and leads to increase of ATP synthesis (Yuan et al., 2024).

Taken together, these findings demonstrate that the presence of RJ in the post-exposure repair medium protects yeast cells against MMW-induced damage. The radiation-protective properties of RJ could be used to develop protective means against different external stresses for higher animals and humans, using the yeast cells as a model system in these studies.

FUNDING

This work was supported by the Higher Education and Science Committee of MESCS of RA in the frames of Research Project N° 25RG-1F111.

DATA AVAILABILITY

The data presented in this study are available on request from the corresponding author on reasonable request.

CONFLICTS OF INTERESTS

The authors declare no relevant financial or non-financial interests to disclose.

The authors declare that there are no conflicts of interest regarding the publication of this manuscript. All authors have participated in the work and have reviewed and approved the final manuscript. Furthermore, there are no financial, personal, or professional relationships with other individuals or organizations that could inappropriately influence (bias) this study.

ACKNOWLEDGMENT

The authors are grateful to V. Ghazarian, Director of Arpimed LLC, for providing lyophilized samples of RJ. The authors also are grateful to V. Khalantaryan (Faculty of Radiophysics, YSU) for exposure of yeast samples to MMW.

REFERENCES

Ali, A., Rashid, M. A., Huang, Q. Y., & Lei, C. L. (2016). Effect of UV-A radiation as an environmental stress on the development, longevity, and reproduction of the oriental armyworm, Mythimna separata (Lepidoptera: Noctuidae). *Environmental Science and Pollution Research International*, *23*(17), 17002–17007. DOI: 10.1007/s11356-016-6865-0 PMID: 27206749

Bhattacharya, M., & Basak, T. (2017). A comprehensive analysis on the effect of shape on the microwave heating dynamics of food materials. *Innovative Food Science & Emerging Technologies*, *39*, 247–266. DOI: 10.1016/j.ifset.2016.12.002

Bok, E., Jo, M., Lee, S., Lee, B. R., Kim, J., & Kim, H. J. (2019). Dietary Restriction and Neuroinflammation: A Potential Mechanistic Link. *International Journal of Molecular Sciences, 20*(3), 464. DOI: 10.3390/ijms20030464 PMID: 30678217

Bowman, E. J., Siebers, A., & Altendorf, K. (1988). Inhibitors of the ATP-dependent proton pump in yeast plasma membranes. *Biochemistry, 27*(9), 3089–3095.

Capozzoli, A., Curcio, C., D'Agostino, F., & Liseno, A. (2024). A Review of the Antenna Field Regions. *Electronics (Basel), 13*(11), 2194. DOI: 10.3390/electronics13112194

Carbone, M. C., Tatone, C., Monache, S. D., Marci, R., Caserta, D., Colonna, R., & Amicarelli, F. (2003). Antioxidant enzymatic defences in human follicular fluid: Characterization and age-dependent changes. *Molecular Human Reproduction, 9*(11), 639–643. DOI: 10.1093/molehr/gag090 PMID: 14561807

Cordiano, R., Di Gioacchino, M., Mangifesta, R., Panzera, C., Gangemi, S., & Minciullo, P. L. (2023). Malondialdehyde as a Potential Oxidative Stress Marker for Allergy-Oriented Diseases: An Update. *Molecules (Basel, Switzerland), 28*(16), 5979. DOI: 10.3390/molecules28165979 PMID: 37630231

Cui, Z., He, F., Li, X., Li, Y., Huo, C., Wang, H., Qi, Y., Tian, G., Zong, W., & Liu, R. (2023). Response pathways of superoxide dismutase and catalase under the regulation of triclocarban-triggered oxidative stress in Eisenia foetida: Comprehensive mechanism analysis based on citotoxicity and binding model. *The Science of the Total Environment, 854*, 158821. DOI: 10.1016/j.scitotenv.2022.158821 PMID: 36116645

Devenish, R.J., Prescott, M., & Rodgers, A.J.W. (2008). The structure and function of mitochondrial F1F0-ATP synthase. *Int Rev Cell Mol Biol,* 267, 1–58. DOI: 10.1016/S1937-6448(08)00601-1

Faccenda, D., & Campanella, M. (2012). Molecular regulation of the mitochondrial F1F0-ATP synthase: Physiological and pathological significance of the inhibitory factor 1 (IF1*). International Journal of Cell Biology, 1*, 1–12. DOI: 10.1155/2012/367934 PMID: 22966230

Fratini, F., Cilia, G., Mancini, S., & Felicioli, A. (2016). Royal Jelly: An ancient remedy with remarkable antibacterial properties. *Microbiological Research, 192*, 130–141. DOI: 10.1016/j.micres.2016.06.007 PMID: 27664731

Guérin, B., Labbe, P., & Solmo, P. (1979). Preparation of Yeast Mitochondria (*Saccharomyces cerevisiae*) with good P/O and respiratory control Ratios. *Methods in Enzymology, 55*, 149–159. DOI: 10.1016/0076-6879(79)55021-6 PMID: 379498

Hong, C. C., Kume, T., & Peterson, R. T. (2008). Role of crosstalk between phosphatidylinositol 3-kinase and extracellular signal-regulated kinase/mitogen-activated protein kinase pathways in artery–vein specification. *Circulation Research, 103*(6), 573–579. DOI: 10.1161/CIRCRESAHA.108.180745 PMID: 18796644

Hong, S., & Pederson, P. L. (2008). ATP Synthase and the Actions of Inhibitors Utilized To Study Its Roles in Human Health, Disease, and Other Scientific Areas. *Microbiology and Molecular Biology Reviews : MMBR, 72*(4), 590–641. DOI: 10.1128/MMBR.00016-08 PMID: 19052322

Hovnanyan, K., Marutyan, S., Marutyan, S., Hovnanyan, M., Navasardyan, L., & Trchounian, A. (2020). Ultrastructural investigation of acidocalcisomes and ATPase activity in yeast Candida guilliermondii NP-4 as 'complementary'stress-targets. *Letters in Applied Microbiology, 71*(4), 413-419. DOI: 10.1016/j.jff.2024.106505

Hovnanyan, K., Marutyan, S., Marutyan, S. A., Hovnanyan, M., Navasardyan, L., & Trchounian, A. (2020). Ultrastructural investigation of acidocalcisomes and ATPase activity in yeast *Candida guilliermondii* NP-4 as "complementary" stress-targets. *Letters in Applied Microbiology, 71*, 413–419. DOI: 10.1111/lam.13348 PMID: 32623751

Ito, F., Sono, Y., & Ito, T. (2019). Measurement and Clinical Significance of Lipid Peroxidation as a Biomarker of Oxidative Stress: Oxidative Stress in Diabetes, Atherosclerosis, and Chronic Inflammation. *Antioxidants, 8*(3), 72. DOI: 10.3390/antiox8030072 PMID: 30934586

Kamakura, Y., Yamanashi, K., & Sato, A. (2012). Freeze-dried royal jelly maintains its developmental and physiological bioactivity in *Drosophila melanogaster. Bioscience, Biotechnology, and Biochemistry, 76*(11), 2107–2111. DOI: 10.1271/bbb.120496 PMID: 23132575

Karapetyan, H., Marutyan, S., Muradyan, A., Badalyan, H., Marutyan, S. V., & Trchounian, K. (2024, April). Changes in ATPase activity, antioxidant enzymes and proline biosynthesis in yeast *Candida guilliermondii* NP-4 under X-irradiation. *Journal of Bioenergetics and Biomembranes, 56*(2), 141–148. DOI: 10.1007/s10863-024-10003-4 PMID: 38308068

Kenneth, S. K., Nneka, O. F., Kalu, A. K., & Joseph, A. O. (2023). Radioprotective Potencies of Allium Cepa Extract (ACE) against Radiation-Induced Hepatoxicity in Wistar Rats. *International Journal of Medical Physics, Clinical Engineering and Radiation Oncology, 12*(3), 59–83. DOI: 10.4236/ijmpcero.2023.123007

Kumar, R., Thakur, A., Kumar, S., & Hajam, Y. A. (2024). Royal jelly a promising therapeutic intervention and functional food supplement. *A systematic review. Heliyon*, *10*(17), e37138. DOI: 10.1016/j.heliyon.2024.e37138 PMID: 39296128

Lowry, O. H., Rosebrough, H. J., Farr, A. L., & Randall, R. J. (1951). Protein Measurement with the Folin Phenol Reagent. *The Journal of Biological Chemistry*, *193*(1), 265–278. DOI: 10.1016/S0021-9258(19)52451-6 PMID: 14907713

Mahmoud, H., & Hussein, N. (2016). Data supporting the spectrophotometric method for the estimation of catalase activity. *Data in Brief*, *6*, 194–199. DOI: 10.1016/j.dib.2015.12.012 PMID: 26862558

Marutyan, S. V., Marutyan, S. A., Navasardyan, L. H., Hovnanayan, K. O., & Trchounian, A. H. (2021). Changes in growth kinetic parameters, morphology and mitotic activity of yeasts Candida guilliermondii exposed to the low-intensity waves of 51.8-GHz frequency. *Archives of Microbiology*, *203*(6), 3707–3714. DOI: 10.1007/s00203-021-02336-0 PMID: 33938972

Mejía-Barajas, J., Montoya-Perez, R., Salgado-Garciglia, R., Aguilera-Aguirre, L., Cortés-Rojo, C., Mejía-Zepeda, R., Arellano-Plaza, M., & Saavedra-Molina, A. (2017). Oxidative stress and antioxidant response in a thermotolerant yeast. *Brazilian Journal of Microbiology : Publication of the Brazilian Society for Microbiology*, *48*(2), 326–332. DOI: 10.1016/j.bjm.2016.11.005 PMID: 28094115

Miyashita, A., Kizaki, H., Sekimizu, K., & Kaito, C. (2016). Body-enlarging effect of royal jelly in a non-holometabolous insect species. *Gryllus bimaculatus. Biology Open*, *5*(6), 770–776. DOI: 10.1242/bio.019190 PMID: 27185266

Nanyan, N. S. B. M., & Takagi, H. (2020). Proline homeostasis in *Saccharomyces cerevisiae*: How does the stress-responsive transcription. factor Msn2 play a role? *Frontiers in Genetics*, *11*, 438. DOI: 10.3389/fgene.2020.00438 PMID: 32411186

Navasardyan, L., Marutyan, S., Hovnanyan, K., & Trchounian, A. (2017). Survival and changes in morphology, mitotic and metabolic activity of yeast *Candida guilliermondii* exposed to X-irradiation. *Indian Journal of Biochemistry & Biophysics*, *54*, 273–280. http://nopr.niscair.res.in/handle/123456789/43366

Obrador, E., & Montoro, A. (2023). Ionizing Radiation, Antioxidant Response and Oxidative Damage: Radiomodulators. *Antioxidants*, *12*(6), 1219. DOI: 10.3390/antiox12061219 PMID: 37371949

Ramanathan, A. N. K. G., Nair, A. J., & Sugunan, V. S. (2016). A review on Royal Jelly proteins and peptides. *Journal of Functional Foods*, *44*, 255–264. DOI: 10.1016/j.jff.2018.03.008

Ramón, S., Kielland-Brandt, M. C., & Fink, G. R. (1986). Yeast plasma membrane ATPase is essential for growth and has homology with ($Na^+ + K^+$), K^+- and Ca^{2+}-ATPases. *Nature*, *319*(6055), 689–693. DOI: 10.1038/319689a0 PMID: 3005867

Rynkowska, A., Stępniak, J., & Karbownik-Lewińska, M. (2020). Fenton Reaction-Induced Oxidative Damage to Membrane Lipids and Protective Effects of 17β-Estradiol in Porcine Ovary and Thyroid Homogenates. *International Journal of Environmental Research and Public Health*, *17*(18), 6841. DOI: 10.3390/ijerph17186841 PMID: 32962175

Salazar, H. G. M., Salem, A. Z. M., Kholif, A. E., Monroy, H., Perez, L. S., Zamora, J. L., & Gutiérez, A. (2016). Yeast: Desciption and structure. In *Yeast additive and animal production, India*. PubMed Central Research Publishing Services. DOI: 10.1016/0014-5793(83)80237-3

Seker, I., Karlidag, S., Akyol, A., Yil, G., Koseman, A., Uyumlu, A. B., Edrdogan, S., & Yilmaztekin, M. (2025). Volatile organic compounds of royal jellies obtained from honeybee (Apis mellifera L.) colonies fed with different industrial sugars at different locations. *Journal of Food Measurement and Characterization*, *19*(3), 2019–2034. DOI: 10.1007/s11694-025-03097-w

Spanidi, E., Athanasopoulou, S., Liakopoulou, A., Chaidou, A., Hatziantoniou, S., & Gardikis, K. (2022). Royal jelly components encapsulation in a controlled release system - skin functionality and biochemical activity for skin applications. *Pharmaceuticals (Basel, Switzerland)*, *15*(8), 907. DOI: 10.3390/ph15080907 PMID: 35893731

Srivastava, A. P., Luo, M., Zhou, W., Symersky, J., Bai, D., Chambers, M. G., Faraldo-Gómez, J. D., Liao, M., & Mueller, D. M. (2018). High-resolution cryo-EM analysis of the yeast ATP synthase in a lipid membrane. *Science*, *360*(6389), 1–8. DOI: 10.1126/science.aas9699 PMID: 29650704

Suzuki, N. (2015) ROS as Key Players of Abiotic Stress Responses in Plants. *Reactive Oxygen Species and Oxidative Damage in Plants Under Stress,* 57–82.

Wei, Y., Zhang, Y. J., Cai, Y., & Xu, M. H. (2015). The role of mitochondria in mTOR-regulated longevity. *Biological Reviews of the Cambridge Philosophical Society*, *90*(1), 167–181. DOI: 10.1111/brv.12103 PMID: 24673778

Yang, W., Tian, Y., Han, M., & Miao, X. (2017). Longevity extension of worker honey bees (Apis mellifera) by royal jelly. Optimal dose and active ingredient. *PeerJ*, *5*, e3118. DOI: 10.7717/peerj.3118 PMID: 28367370

Yuan, C. Y., Yang, S. Z., Gao, S. M., Bai, T. M., Han, T. T., Jiang, S. S., Liu, S.-L., Gao, K.-K., Mao, X., Jiang, H., & Li, Y. (2024). Biologically synthesized 10-Hydroxy-2-Decenoic acid activates cornified envelope formation via PPARα signaling, and regulates oxidative respiratory chain to improve mitochondrial function. *Journal of Functional Foods*, *122*, 106505. DOI: 10.1016/j.jff.2024.106505

Zheng, Y., Sun, J., Luo, Z., & Huang, Y. (2024). Emerging mechanisms of lipid peroxidation in regulated cell death and its physiological implications. *Cell Death & Disease*, *15*(11), 859. DOI: 10.1038/s41419-024-07244-x PMID: 39587094

Chapter 14
Antibiotic Activity of Plant Metabolites and Possibilities of Their Application

Aram Aramyan
http://orcid.org/0009-0005-4546-6976
Yerevan State University, Armenia

Silvard Tadevosyan
http://orcid.org/0009-0005-3818-7274
Research Institute of Biology, Armenia

Alvard Minasyan
http://orcid.org/0009-0009-2837-0146
Research Institute of Biology, Armenia

Anush Babayan
http://orcid.org/0009-0008-5300-6170
Research Institute of Biology, Armenia

Naira Sahakyan
http://orcid.org/0000-0002-7370-1380
Research Institute of Biology, Armenia

ABSTRACT

The growing threat of antibiotic resistance demands the search for novel alternatives to existing antimicrobial agents. This study aimed to investigate the antibacterial potential of four natural flavonoids—catechin, gallic acid, rutin, and quercetin—

DOI: 10.4018/979-8-3373-1524-9.ch014

and one sugar alcohol, myo-inositol, all abundant in plants used in Armenian folk medicine. Tested bacterial strains included both Gram-negative and Gram-positive species. Myo-inositol demonstrated growth inhibition zones of 7.0–9.0 mm for E. coli K12, 11.0–18.0 mm for B. subtilis, and 12.0–21.0 mm for S. aureus. Rutin showed 12.0–15.0 mm zones for B. subtilis and 15.0–16.0 mm for S. aureus. MTT assays revealed that quercetin (0.25 mg/mL) reduced the MIC of kanamycin fourfold for E. coli pARG25 and halved the MIC of ampicillin for E. coli DH5α-pUC18. Gallic acid and rutin reduced the kanamycin MIC twofold, while catechin halved the ampicillin MIC. These findings indicate the potential of flavonoid-based agents in combating antibiotic resistance, encouraging further investigation into their mechanisms of action and clinical relevance.

INTRODUCTION

The rise of the problem of antimicrobial resistance (AMR) has become one of the most urgent global health challenges of the 21st century, diminishing or even neutralising the efficacy of many conventional antibiotics and complicating the treatment of common infections (Interagency Coordination Group on Antimicrobial Resistance, 2019). A huge number of pathogenic bacteria among which some species such as *Escherichia coli*, *Staphylococcus aureus*, and *Salmonella typhimurium* increasingly getting the resistance to many first-line conventional antibiotics like ampicillin and kanamycin, prompting the search for alternative strategies to either replace or potentiate existing antimicrobial agents (Abreu et al., 2023; Melek, 1994). Among the emerging approaches, natural products — especially plant-derived polyphenolic substances — have garnered growing attention due to their structural diversity, low toxicity, and multiple biological activities, including the desired antibacterial, antioxidant effects, as well as the potential synergistic properties with numerous conventional antibiotics (Ginovyan et al., 2025; Sahakyan et al., 2020).

Flavonoids, a broad class of polyphenolic compounds abundant in fruits, vegetables, and medicinal plants, have demonstrated antimicrobial activities in several studies (Sharma et al., 2025). In plants, flavonoid substances generally synthesizing against different stressors and as signalling molecules for symbiotic bacteria. They play a crucial role in plant organism (Mathesius, 2018). In addition to their unreplaceable roles for plants these active metabolites are essential also for different areas of human activity (Ullah et al., 2020). Compounds such as quercetin, rutin, catechin, and gallic acid have been investigated for their capacity to disrupt bacterial membranes, chelate metal ions, or inhibit essential enzymes, although results have been variable and often species-specific (Jomova et al., 2025; Tarabishi et al., 2024). While some flavonoids exhibit direct bactericidal activity against different bacterial species, others act indirectly by modulating membrane permeability or altering efflux pump activity, thus sensitizing bacteria to antibiotics by this way (Joseph et al., 2025). These compounds have shown particular promise as antibiotic adjuvants, restoring the efficacy of conventional antibiotics against resistant strains (Donadio et al., 2021).

Myo-inositol, a polyol also present in many plant tissues, is less studied in the context of antibacterial activity but is known for its role in cell signalling and osmoregulation (Al-Mushhin et al., 2021). Preliminary evidence suggests it may influence microbial growth or metabolism (Jessen & Bui, 2025), yet its antibacterial properties remain poorly understood.

Medicinal plants used in Armenian folk medicine form a vast and largely untapped treasury of biologically active compounds, many of which hold promising antimicrobial potential (Ginovyan et al., 2022). For centuries, rural communities across Armenia's diverse landscapes - stretching from high alpine meadows to sun-warmed valleys - have relied on these plants as frontline defences against infections, inflammation, and various ailments. This enduring tradition suggests a deep reservoir of empirical knowledge, accumulated through generations of observation and practice. However, despite their widespread use and cultural significance, the scientific investigation of these species remains surprisingly limited. A substantial gap persists between traditional understanding and modern biochemical evidence (Sahakyan et al., 2019).

Yet for many species used in Armenian ethnomedicine, the precise identity and abundance of these constituents are still unknown. Even fewer studies have mapped how these molecules interact with bacterial or fungal cells, leaving the mechanisms governing their antimicrobial activity largely speculative. This lack of documentation presents both a scientific challenge and an extraordinary opportunity. Modern antimicrobial research is searching for new solutions at a time when antibiotic resistance continues to rise globally, outpacing the development of novel drugs. Many natural metabolites disrupt membrane integrity, interfere with protein or nucleic acid synthesis, modulate quorum sensing, or induce oxidative stress - yet these processes require careful characterization that is currently missing for most Armenian medicinal species (Tadevosyan, 2025; Tadevosyan et al., 2025; Tadevosyan & Sahakyan, 2024; Tadevosyan et al., 2023).

Bridging this gap calls for interdisciplinary research that integrates ethnobotany, analytical chemistry, microbiology, and molecular biology. Documenting traditional uses provides the starting compass, guiding researchers toward species of greatest therapeutic relevance. High-resolution chromatographic and spectrometric techniques can then unravel the chemical architecture of essential oils, extracts, and isolated metabolites. In parallel, modern microbiological assays can quantify antibacterial and antifungal activities, revealing minimum inhibitory concentrations, bactericidal effects, and potential synergies among compounds. Finally, biophysical and molecular approaches, such as membrane permeability studies, enzymatic inhibition assays, and omics-based analyses, can illuminate the precise mechanisms by which these natural products exert their therapeutic effects.

This study investigates the antibacterial potential of five plant-derived metabolites — quercetin, rutin, catechin, gallic acid, and myo-inositol — against both susceptible and antibiotic-resistant strains of *E. coli*, *B. subtilis*, *S. aureus*, and *S. typhimurium*. By combining disc-diffusion assays, MTT-based MIC determination, and proton flux analysis, we evaluate both direct antibacterial effects and potential synergism with antibiotics. The findings aim to clarify the contribution of selected phytochemicals to bacterial growth inhibition and resistance modulation, highlighting their potential as alternative or adjunctive agents in combating AMR.

MATERIALS AND METHODS

Bacterial Strains and Growth Conditions

The following Gram positive and Gram negative bacterial strains were used in the experiments: *E. coli* K-12, *E. coli* BW 25113, ampicillin-resistant *E. coli* DH5a-pUC18, kanamycin-resistant *E. coli* pARG25, *Salmonella typhimurium* MDC 1754, *Bacillus subtilis* WT-A1 and *Staphylococcus aureus* MDC 5233. The strains were cultivated at 37°C, during the 18-22 hours in liquid peptone media (MP) of the following composition: 5 g L^{-1} NaCl, 2 g L^{-1} glucose, 2 g L^{-1} K_2HPO_4, 20 g L^{-1} peptone. pH was adjusted to 7.5 by 0.1 M NaOH or 0.1 M HCl. For aerobic growth, the cells were cultivated in rotary shaking with 150 rpm conditions.

Investigation of Antimicrobial Activity

The initial evaluation of catechin, gallic acid, rutin, quercetin and myo-inositol antibacterial activity was carried out by disc diffusion assay against the selected bacterial strains listed above (see "Bacterial strains and growth conditions" sub-title) (Moghrovyan et al., 2019). In this assay, bacteria were cultured on solid MP medium, and small filter-paper disks containing the known concentrations of antimicrobial compounds (in our case – the selected flavonoids) are placed on the surface. Once a 6-mm flavonoid-impregnated disk contacts the agar, moisture from the medium is immediately drawn into the disk, allowing the antimicrobial substance to begin diffusing outward into the surrounding agar. Because diffusion through the agar occurs more slowly than the initial release of the compound from the disk, the highest antimicrobial concentration remains closest to the disk, with the concentration gradually decreasing in a logarithmic pattern as the distance from the disk increases. Cellular growth is suppressed in regions where the local antimicrobial concentration exceeds the organism's inhibitory threshold. Growth becomes detectable only when the bacterial population reaches a density at which the antimicrobial concentration is insufficient to impede essential metabolic and replicative processes. This interplay between antimicrobial diffusion kinetics and microbial growth dynamics results in the development of discrete zones of inhibited growth, which serve as quantitative indicators of the organism's susceptibility profile. Despite the drawback of this assay, connected with the diffusion specificities, this method is suitable for the preliminary screening of the antibacterial activity of many substances.

Flavonoids and myo-inositol were tested across a concentration range of 31.25–1000 μg/mL. Kanamycin (50 μg/mL) and ampicillin (50 μg/mL) were used as positive controls, while ethanol (96%) served as the negative control.

Bacterial suspensions, adjusted to 10^8 cells/mL, were dispensed into 96-well plates and incubated at 37 °C for 24 hours. After the incubation, bacterial viability in response to the mentioned metabolites' treatment was assessed using also the MTT assay. The results were reported in terms of minimal inhibitory concentration (MIC) values (Ginovyan et al., 2025).

Determination of Antibiotic Modulatory Activity of Plant Extracts

Antibiotic modulatory activity of test metavbolites was explored by determining the minimal inhibitory concentrations of antibiotics in the presence and absence of plant crude extracts at non-inhibitory concentrations (Ginovyan & Trchounian, 2019). Broth microdilution assay was used for the determination of MIC values, as described previously (Ginovyan & Trchounian, 2019). The decrease in MICs of antibiotics (ampicillin and kanamicyn) in the presence of plant crude extracts indicated antibiotic modulatory activity of the extract. To calculate the modulation factor (MF), which indicates synergistic or modulating interactions, the following formula was used: $MF = MIC_{antibiotic}/MIC_{antibiotic+extract}$. $MF \geq 2$ indicated the occurrence of synergistic interactions, n = 1 meant no impact. Ampicillin (100 μg/mL) and kanamicyn (50 μg/mL) alone were used, as positive controls (https://www.addgene.org/173107/).

Determination of Proton Fluxes

Cells used for proton flux investigations were cultivated for 18-22 hours under aerobic conditions and were harvested by centrifugation (Sorvall LYNX 6000 Superspeed Centrifuge, Thermo Scientific, USA) at 3500g for 15 min, then washed with distilled water twice. Centrifugation was repeated twice for 10m and final pellet was washed in 150 mM Tris–HCl (pH = 7.5) phosphate buffer containing 0.4 mM

$MgSO_4$, 1 mM NaCl, and 1 mM KCl. The proton flux rates (J_{H+}) were determined using potentiometric assay, by pH/mV/ISE meter (HI5222, HANNA instruments, Portugal) equipped with a selective H^+ electrode (HI1131), by registering the pH value changes, upon the addition of glucose (2 g L^{-1}) and above-mentioned metabolites (50 μg mL^{-1}). J_{H+} was calculated as the negative logarithm of proton concentration in millimolar units. Results were expressed in mmol min^{-1} 10^8 $cells^{-1}$ in 1 unit of volume (mL) (Tadevosyan et al., 2025).

In proton flux analysis, the measurement of pH variation following substrate addition reflects the dynamics of proton translocation across the cell membrane and the metabolic activity associated with proton-coupled processes. Because glucose metabolism rapidly activates proton-exporting mechanisms, the observed acidification or alkalinization of the medium can be directly linked to the functionality of membrane-bound transport systems, respiratory chain activity, or proton-ATPase performance. The use of a selective H^+ electrode allows real-time detection of subtle shifts in extracellular proton concentration, thereby enabling highly sensitive quantification of JH^+ under defined physiological conditions. Monitoring proton flux under these controlled settings provides insight into the integrity of membrane permeability, the efficiency of energy-dependent transport, and potential alterations in proton motive force. Furthermore, comparing flux rates in the presence of different metabolites or environmental modifiers can help characterize the regulatory role of specific metabolic pathways, stress responses, or inhibitory compounds on proton-related transport mechanisms (Tadevosyan et al., 2025).

Statistical Analysis

All experiments were independently repeated at least three times. The results are presented as means ± SD. The statistical significance of the results was examined using one-way ANOVA followed by Dunnett's test or two-way ANOVA and Tukey's multiple comparisons tests. The statistical analyses were performed using GraphPad Prism 8 software (GraphPad Software, Inc., San Diego, CA, USA) and Microsoft Excel 10 programs. *p*-value of less than 0.05 was considered significant. Graph presentations were carried out by GraphPad Prism.

RESULTS AND DISCUSSION

In our previous research we demonstrated the abundance of phenolic origin secondary metabolites in alcohol extracts of different plants harvested in Armenia. Among the representatives of the main components of these extracts five compounds had been manifested to be catechin, gallic acid, rutin, quercetin and myo-inositol (Ginovyan et al., 2023; Ginovyan et al., 2025).

As already mentioned above, the initial antibacterial activity of these five natural compounds was evaluated using the disc-diffusion method against a diverse panel of bacterial strains. These included *E. coli* K12, *B. subtilis WT-A1*, *S. aureus MDC 5223*, *S. typhimurium* MDC 1754, ampicillin-resistant *E. coli* DH5α-pUC18, and kanamycin-resistant *E. coli* pARG25. This kind of selection of bacterial strains allowed assessment of the compounds across both Gram-positive and Gram-negative groups, as well as against bacteria exhibiting established resistance mechanisms against antibiotics.

Among the tested substances, only the *inositol* and rutin demonstrated measurable antibacterial activity, primarily against the Gram-positive bacteria, suggesting a possible specificity related to cell wall composition or permeability characteristics.

Inositol showed inhibition zones with the diameters of *7.3 ± 0.5 mm to 9.0 ± 0.0 mm* for *E. coli* K12, 13.0 ± 1.1 mm to *17.0 ± 1.7 mm* for *B. subtilis*, and *13.6 ± 0.5 mm* to *16.3 ± 0.5 mm* for *S. aureus*. Rutin exhibited no activity against *E. coli* K12 but inhibited the growth of *B. subtilis* and *S. aureus*, with formed inhibition zones diameters of which ranging from *10.0 ± 4.2 mm* to 15.0 ± 4.2 mm and 13.6 ± 3.2 mm to 15.6 ± 2.5 mm, respectively. These results suggest that the antibacterial properties of rutin may be more pronounced in organisms lacking an outer membrane barrier, which is characteristic of Gram-positive bacteria.

In contrast, gallic acid, catechin, and quercetin showed no detectable antibacterial effect on any of the tested strains under the experimental conditions. Their lack of activity could be attributed to low intrinsic antibacterial potency or potential instability under assay conditions. The insufficient diffusion in agar also could play some role. However, further mechanistic studies would be required to clarify these possibilities.

Additionally, none of the five compounds exhibited any inhibitory effect on the antibiotic-resistant bacterial strains (*E. coli* DH5α-pUC18 and *E. coli* pARG25) or *S. typhimurium* MDC 1754 at any concentration tested.

To evaluate the potential of flavonoids in the process of possible enhancing of antibiotic efficacy, the minimum inhibitory concentrations (MICs) of kanamycin were assessed in combination with sub-inhibitory concentrations of gallic acid, quercetin, and rutin using the MTT viability assay in *kanamycin-resistant E. coli pARG25*.

In the presence of *gallic acid (at concentration of 0.25 mg/mL)*, the MIC of kanamycin was reduced twice - from *5 μg/mL to approximately 2.5 μg/mL*. The lower concentration (0.125 mg/mL) of gallic acid did not produce a significant effect (Figure 1 a).

Quercetin (0.25 mg/mL) showed the strongest potentiation effect, reducing the MIC of kanamycin from *5 μg/mL to approximately 1.25 μg/mL*—a *fourfold decrease*.

Again, the lower concentration (0.125 mg/mL) failed to enhance the kanamycin efficacy (Figure 1b).

Similarly, *rutin (0.25 mg/mL)* reduced the MIC of kanamycin twofold, from 5 µg/mL to 2.5 µg/mL, while the 0.125 mg/mL concentration showed no significant change (Figure 1c).

Statistical analysis (****p < 0.0001) confirmed the significance of the observed synergistic effects at 0.25 mg/mL, while comparisons involving 0.125 mg/mL were not statistically significant (ns) (Figure 1).

These findings suggest that gallic acid, rutin, and especially quercetin can act as modulators of the activity of antibiotic and are able to overcome the resistance of kanamycin-resistant *E. coli* pARG25 when combined with kanamycin at specific concentrations.

*Figure 1. Modulatory effect of Gallic Acid (G) (a), Rutin (R) (b) and Quercetin (Q) (c) on the minimum inhibitory concentrations (MICs) of kanamycin (K) against kanamycin-resistant E. coli pARG25. G, R and Q reduced MIC values of the tested antibiotic at 0.25 mg/mL concentrations, respectively in a statistically significant manner (****: p<0.0001). Data represent means ± SD from three independent experiments performed in triplicate. Experimental conditions: K+ (treated with kanamycin), G+K+, R+K+ and Q+K+ (treated with gallic acid + kanamycin, rutin + kanamycin, quercetin + kanamycin), (K - kanamycin).*

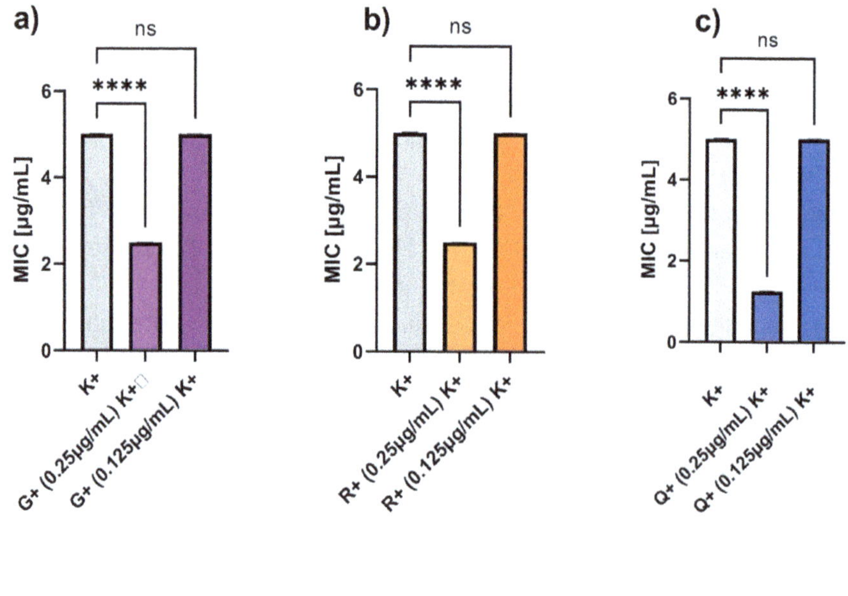

The effect of the investigated flavonoids on the susceptibility of ampicillin-resistant *E. coli* DH5α-pUC18 was assessed by determining changes in the MIC values of ampicillin in the presence of *catechin* and *quercetin* using the MTT assay.

When *catechin (at concentration of 0.25 mg/mL)* was combined with ampicillin, the MIC of this conventional antibiotic was reduced from *24 μg/mL to the approximately value reaching 12 μg/mL*, indicating a *twofold decrease* in resistance. However, a lower concentration of catechin (0.125 mg/mL) did not result in any significant change, and the MIC remained at the baseline levels. The reduction observed at 0.25 mg/mL was statistically significant (****$p < 0.0001$), while, according to our calculations, in case of applying the 0.125 mg/mL dose the reduction was not statistically significant (ns).

Similarly, *quercetin (at concentration of 0.25 mg/mL)* significantly potentiated the activity of ampicillin, lowering the MIC of antibiotic from *24 μg/mL to approximately 12 μg/mL*. There is observed a *twofold reduction again*. The lower dose of quercetin (0.125 mg/mL) had no measurable effect on MIC values.

These results demonstrate that both catechin and quercetin, at sufficient concentrations, can partially restore the susceptibility of resistant *E. coli* strains to ampicillin, highlighting their potential as antibiotic adjuvants.

*Figure 2. Modulatory effect of Quercetin (Q) (a) and Catechin (C) (b) on the minimum inhibitory concentrations (MICs) of ampicillin (A) against ampicillin-resistant E. coli DH5α-pUC18. Q and C reduced MIC values at 0.25 mg/mL concentrations, respectively, in a statistically significant manner (**** p<0.0001). Data represent means ± SD from three independent experiments performed in triplicate. Experimental conditions: A+ (treated with ampicillin), Q+A+ and C+A+ (quercetin + ampicillin, catechin + ampicillin), (A = ampicillin).*

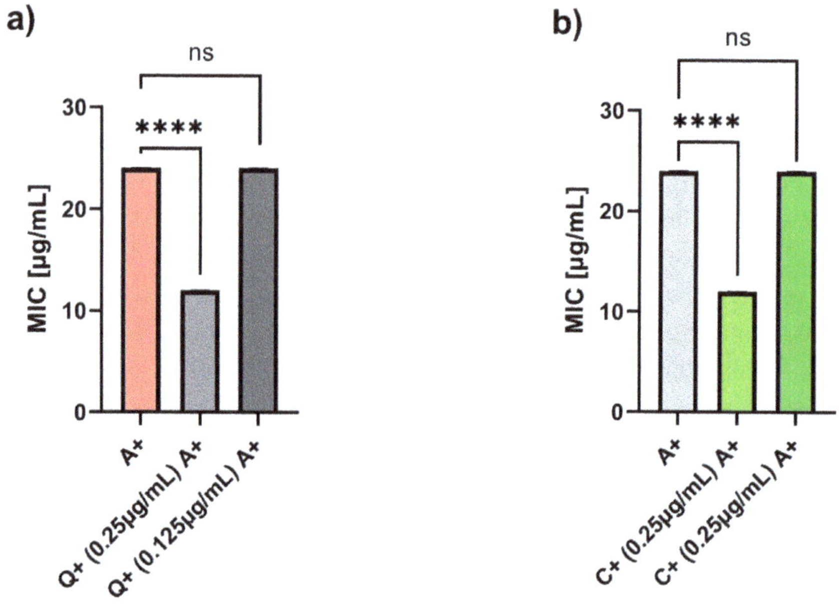

To assess whether flavonoids affect membrane permeability, the proton fluxes (J_H^+) were evaluated in wild-type *E. coli* BW25113 (as a reference strain) and kanamycin-resistant *E. coli* pARG25 cell membrane following the treatment with kanamycin (K) alone or in combination with rutin (R), quercetin (Q), or gallic acid (G).

In wild type *E. coli* strain, both K^-G^+ and K^+G^+ caused a reduction in proton flux. The combined treatment with kanamycin and gallic acid (K^+G^+) decreased efflux from ~4.0 to ~2.0 mmol/min per 10^8 cells (****$p < 0.0001$). In the resistant *E. coli* pARG25, proton flux remained low across all treatments (~1.3 mmol/min), with no significant differences between treated and untreated conditions (Figure 2 a).

In wild-type *E. coli* strain, rutin alone (K^-R^+) or kanamycin alone (K^+R^-) did not significantly affect proton flux compared to the untreated control (K^-R^-). However, the combined treatment with kanamycin and rutin (K^+R^+) resulted in some decrease (around 30%) in proton efflux from ~1.0 to ~0.7 mmol/min per 10^8 cells (p <

0.01). In *E. coli* pARG25, the same combination caused a more pronounced decrease in this parameter - from ~1.1 to ~0.5 mmol/min per 10^8 cells (****p < 0.0001), while individual treatments showed no visible effects (Figure 2 b).

The kanamycin and quercetin simultaneous (K^+Q^+) treatment decreased the flux up to ~*2.5 mmol/min per 10^8 cells* compared to the control where this parameter value was recorded to be ~3.5 mmol/min (K^-Q^-) (****p < 0.0001). In the kanamycin-resistant *E. coli* pARG25, again the individual treatment with quercetin produced no significant change, with proton flux value remaining around *2.0 mmol/min per 10^8 cells*. In contrast, the K^+Q^+ combination led to a *substantial reduction*, decreasing the flux value to ~*1.2 mmol/min per 10^8 cells* (*p < 0.01*) (Figure 2 c).

*Figure 3. The impact of gallic acid (G) (a), rutin (R) (b) and quercetin (Q) (c) on total H^+-fluxes on the wild-type E. coli BW 25113 and kanamycin-resistant E. coli pARG-25 bacterial membranes under the treatment with 25 μg mL^{-1} concentration of flavonoids. K^-G^-, K^-R^-, and K^-Q^- - are the control cells without the treatment with flavonoids, K-G+, K-R+, and K-Q+ are the cells treated only with flavonoids, K^+G^-, K^+R^-, and K^+Q^- are the cells treated only with kanamycin, and K+G+, K+R+, and K+Q+ are the cells treated with both flavonoids and kanamycin. The results are means ± SD of three independent experiments (ns – not significant; **p<0.01; ***p<0.001; ****p<0.0001).*

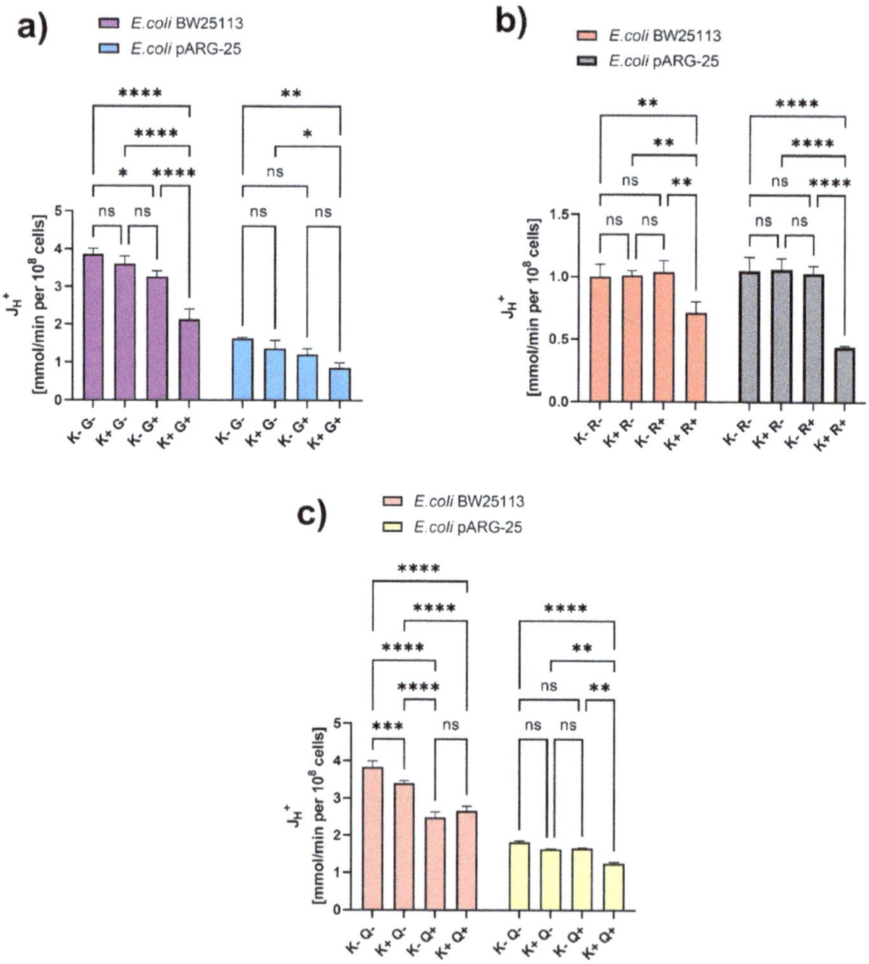

To further explore flavonoid- and antibiotic interactions, proton fluxes (J_H^+) were measured in *E. coli* BW25113 and *ampicillin-resistant E. coli DH5α-pUC18*

strains, following the treatment with *catechin* or *quercetin*, alone or in combination with *ampicillin (A)*.

In case of using *catechin*, wild-type *E. coli* displayed a significant reduction in proton fluxes when treated with both ampicillin *and catechin* (A^+C^+), compared to the control (A^-C^-) and single-agent treatments (A^+C^- and A^-C^+) (****$p < 0.0001$). The flux rate decreased from approximately *3.5 to 2.8 mmol/min per 10^8 cells*. In contrast, no significant changes in proton fluxes were observed in the ampicillin-resistant *E coli* strain in case of any treatment condition, indicating that catechin did not alter membrane proton transport in this strain (Figure 3a).

In case of applying the *quercetin as an acting agent*, proton fluxes were reduced in both strains upon both ampicillin and quercetin (A^+Q^+) simultaneous treatment. In *E. coli* BW25113, the flux rate value decreased from ~2.0 to *~1.5 mmol/min per 10^8 cells* (***$p < 0.001$), while in the resistant strain the value dropped from ~3.0 to *~2.5 mmol/min per 10^8 cells* (***$p < 0.001$). Neither ampicillin nor quercetin alone produced a significant effect (Figure 4a).

*Figure 4. The impact of Catechin (C) (a) and Quercetin (Q) (b) on total H^+-fluxes through the E. coli BW 25113 and ampicillin-resistant E. coli DH5α-pUC18 membranes under the treatment with 25 μg mL^{-1} concentration of flavonoids (aerobic conditions). A-C-, and A-Q- are the control cells without the treatment with flavonoids, A-C+, and A-Q+ are the cells treated only with flavonoids, A+C-, and A+Q- are the cells treated only with ampicillin, and A+C+, and A+Q+ are the cells treated with both flavonoids and ampicillin. The results are means ± SD of three independent experiments (ns – not significant; *p<0.05; ***p<0.001; ****p<0.0001).*

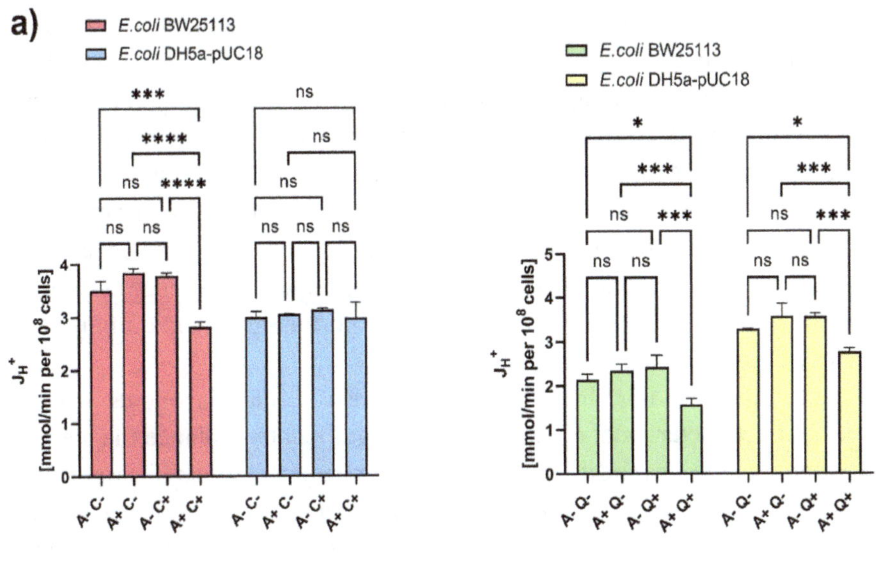

The obtained results suggest that quercetin enhances the activity of ampicillin in both wild-type and resistant E. coli strains by modulating the proton fluxes.

CONCLUSION

This study provides novel insights into the antibacterial potential of five plant-derived compounds—inositol, rutin, gallic acid, catechin, and quercetin—against both antibiotic-sensitive and resistant bacterial strains. While inositol and rutin exhibited measurable direct antibacterial activity, particularly against Gram positive *B. subtilis* and *S. aureus*, the other flavonoids tested (gallic acid, catechin, quercetin) showed no significant effects under the tested conditions. Nevertheless, the absence of direct inhibition does not preclude biological relevance. Many phytochemicals exert indirect or modulatory effects rather than classical bactericidal activity, and this trend became evident when their interaction with conventional antibiotics was assessed. However, their potential as antibiotic adjuvants proved more promising.

When co-administered with kanamycin or ampicillin, gallic acid, rutin, catechin, and especially quercetin significantly reduced the MIC values of these antibiotics against resistant *E. coli* strains, suggesting a capacity to restore partial susceptibility. The magnitude of these synergistic responses varied in a concentration-dependant manner, indicating that the enhancement of antibiotic activity may follow saturable interaction kinetics. These synergistic effects were concentration-dependent. Such concentration-driven behavior is consistent with mechanisms involving partial inhibition of efflux pumps, increased membrane permeability, or metabolic sensitization, all of which can facilitate greater intracellular accumulation of antibiotics.

Further mechanistic insights revealed that combined treatment with flavonoids and antibiotics altered membrane-associated proton fluxes in bacterial cells. The most notable reductions were observed in the resistant strains upon co-treatment with quercetin or rutin and the respective antibiotics. This suggests that disruption of membrane energetics may underlie the observed potentiation of antibiotic efficacy. Given that the proton motive force is essential for ATP synthesis, nutrient transport, and maintenance of membrane homeostasis, its disruption can compromise multiple physiological processes simultaneously. This multi-target interference may sensitize bacteria to antibiotics by weakening active transport systems and diminishing the efficiency of resistance-associated pathways.

Together, these findings highlight the dual role of certain flavonoids: while they may lack direct bactericidal effects, they can significantly enhance the activity of conventional antibiotics and disrupt bacterial membrane functions. This underscores their potential application as natural resistance-modifying agents, particularly in the context of multidrug-resistant Gram-negative pathogens. The broader implication of these results is that plant – derived metabolites may serve as valuable scaffolds for the development of adjunct therapies capable of revitalizing older antibiotics whose effectiveness is declining. Further investigations into their mechanisms of action, pharmacodynamics, and potential clinical use are warranted to harness their full therapeutic potential in antimicrobial strategies.

FUNDING

The research was supported by the Higher Education and Science Committee of MESCS RA, in the frames of the research projects N° 24WS-1F003.

COMPETING INTERESTS

The authors have declared that no competing interests exist.

REFERENCES

Abreu, R., Semedo-Lemsaddek, T., Cunha, E., Tavares, L., & Oliveira, M. (2023). Antimicrobial Drug Resistance in Poultry Production: Current Status and Innovative Strategies for Bacterial Control. *Microorganisms, 11*(4), 953. DOI: 10.3390/microorganisms11040953 PMID: 37110376

Al-Mushhin, A., Qari, S., Fakhr, M., Alnusairi, G., Alnusaire, T., ALrashidi, A., Latef, A., Ali, O., Khan, A., & Soliman, M. (2021). Exogenous Myo-Inositol Alleviates Salt Stress by Enhancing Antioxidants and Membrane Stability via the Upregulation of Stress Responsive Genes in Chenopodium quinoa L. *Plants, 10*(11), 2416. DOI: 10.3390/plants10112416 PMID: 34834781

Donadio, G., Mensitieri, F., Santoro, V., Parisi, V., Bellone, M. L., De Tommasi, N., Izzo, V., & Dal Piaz, F. (2021). Interactions with Microbial Proteins Driving the Antibacterial Activity of Flavonoids. *Pharmaceutics, 13*(5), 660. DOI: 10.3390/pharmaceutics13050660 PMID: 34062983

Ginovyan, M., Hovhannisyan, S., Javrushyan, H., Sevoyan, G., Karabekian, Z., Zakaryan, N., Sahakyan, N., & Avtandilyan, N. (2022). Screening revealed the strong cytotoxic activity of *Alchemilla smirnovii* and *Hypericum alpestre* ethanol extracts on different cancer cell lines. *AIMS Biophysics, 10*(1), 12–22. DOI: 10.3934/biophy.2023002

Ginovyan, M., Javrushyan, H., Petrosyan, G., Kusznierewicz, B., Koss-Mikołajczyk, I., Koziara, Z., Kuczyńska, M., Jakubek, P., Karapetyan, A., Sahakyan, N., Maloyan, A., Bartoszek, A., & Avtandilyan, N. (2023). Anti-cancer effect of Rumex obtusifolius in combination with arginase/nitric oxide synthase inhibitors via downregulation of oxidative stress, inflammation, and polyamine synthesis. *The International Journal of Biochemistry & Cell Biology, 158*, 158. DOI: 10.1016/j.biocel.2023.106396 PMID: 36918141

Ginovyan, M., Tadevosyan, S., Shirvanyan, A., Babayan, A., Kusznierewicz, B., Koss-Mikołajczyk, I., Mróz, M., Bartoszek, A., & Sahakyan, N. (2025). The potential of blackcurrant, fig, and grape leaf extracts in the development of new preparations for overcoming antibiotic resistance and enhancing the efficacy of chemotherapeutic agents. *BMC Complementary Medicine and Therapies*, *25*(1), 148. DOI: 10.1186/s12906-025-04859-1 PMID: 40269785

Ginovyan, M., & Trchounian, A. (2019). Novel approach to combat antibiotic resistance : Evaluation of some Armenian herb crude extracts for their antibiotic modulatory and antiviral properties. *Journal of Applied Microbiology*, *127*(2), 472–480. DOI: 10.1111/jam.14335 PMID: 31136046

Ginovyan, M., & Trchounian, A. (2019). Novel approach to combat antibiotic resistance: Evaluation of some Armenian herb crude extracts for their antibiotic modulatory and antiviral properties. *Journal of Applied Microbiology*, *127*(2), 472–480. DOI: 10.1111/jam.14335 PMID: 31136046

Interagency Coordination Group on Antimicrobial Resistance. (2019). *No Time to Wait: Securing the Future from Drug-Resistant Infections. Report to the Secretary General of the United Nations.* https://www.who.int/publications/i/item/no-time-to-wait-securing-the-future-from-drugresistant-infe

Jessen, H., & Bui, T. P. N. (2025). Metabolism of inositol derivatives by the gut microbiome. *Gut Microbes*, *17*(1), 17. DOI: 10.1080/19490976.2025.2564765 PMID: 41021306

Jomova, K., Alomar, S. Y., Valko, R., Liska, J., Nepovimova, E., Kuca, K., & Valko, M. (2025). Flavonoids and their role in oxidative stress, inflammation, and human diseases. *Chemico-Biological Interactions*, *413*, 111489. DOI: 10.1016/j.cbi.2025.111489 PMID: 40147618

Joseph, J., Boby, S., Mooyottu, S. & Muyyarikkandy, M. S. (2025). Antibiotic potentiators as a promising strategy for combating antibiotic resistance. *NPJ Antimicrobials and Resistance, 3*, 53.

Mathesius, U. (2018). Flavonoid Functions in Plants and Their Interactions with Other Organisms. *Plants*, *7*(2), 30. DOI: 10.3390/plants7020030 PMID: 29614017

Melek, S. M. (1994). How much will coverage for universal access to comprehensive behavioral healthcare really cost? *Behavioral Healthcare Tomorrow*, *3*, 31–38. PMID: 10141167

Moghrovyan, A., Sahakyan, N., Babayan, A., Chichoyan, N., Petrosyan, M., & Trchounian, A. (2019). Essential Oil and Ethanol Extract of Oregano (Origanum vulgare L.) from Armenian Flora as a Natural Source of Terpenes, Flavonoids and other Phytochemicals with Antiradical, Antioxidant, Metal Chelating, Tyrosinase Inhibitory and Antibacterial Activity. *Current Pharmaceutical Design, 25*(16), 1809–1816. DOI: 10.2174/1381612825666190702095612 PMID: 31267860

Sahakyan, N., Bartoszek, A., Jacob, C., Petrosyan, M., & Trchounian, A. (2020). *Bioavailability of Tannins and Other Oligomeric Polyphenols: a Still to Be Studied Phenomenon. Current Pharmacology Reports* (Vol. 6). DOI: 10.1007/s40495-020-00217-6

Sahakyan, N., Petrosyan, M., Koss-Mikołajczyk, I., Bartoszek, A., Sad, T. G., Nasim, M. J., Vanidze, M., Kalandia, A., Jacob, C., & Trchounian, A. (2019). The Caucasian flora: A still-to-be-discovered rich source of antioxidants. *Free Radical Research, 53*(sup1), 1153–1162. DOI: 10.1080/10715762.2019.1648799 PMID: 31510813

Sharma, V., Sharma, D., Saini, M., Jain, A., Chaudhary, J., Kaur, N., Sahoo, S., Singh, S. K., Goyal, K., Rekha, A., Ali, H., Gupta, S., Hussain, M. S., & Gupta, G. (2025). Flavonoids as Antimicrobial Agents: A Comprehensive Review of Mechanisms and Therapeutic Potential. *Current Pharmaceutical Biotechnology, 26*, 26. DOI: 10.2174/0113892010384002250628163849 PMID: 40619652

Tadevosyan, S. (2025). B. A. S. N. Antioxidant and Antibiotic-modulatory Potential of Teucrium polium Ethanolic Extract Against Genetically-modified Antibiotic-resistant E. coli Strains. *Iran J Med Microbiol 19*(2).

Tadevosyan, S., Babayan, A., & Sahakyan, N. (2025). Antioxidant and Antibiotic-modulatory Potential of Teucrium polium Ethanolic Extract Against Genetically-modified Antibiotic-resistant E. coli Strains. *Majallah-i Mikrub/Shinasi-i Pizishki-i Iran, 19*(2), 107–117. DOI: 10.30699/ijmm.19.2.107

Tadevosyan, S., & Sahakyan, N. (2024). Influence of menthol on membrane-associated properties of tetracycline-resistant *Escherichia coli. AIMS Biophysics, 11*(3), 329–339. DOI: 10.3934/biophy.2024018

Tadevosyan, S. K., Shirvanyan, A. H., Markosyan, A. A., Petrosyan, M. T., & Sahakyan, N. Z. (2023). Chemical composition and antibacterial activity of essential oil of $mentha~arvensis$ l. harvested at high altitude Armenian flora. *Proceedings of the YSU B: Chemical and Biological Sciences, 57*, 230–237. DOI: 10.46991/PYSU:B/2023.57.3.230

Tarabishi, A. A., Mashhoud, J., & Tahan, Z. S. (2024). Quercetin and rutin as a dual approach to antibacterial and anti-biofilm activity via iron chelation mechanism. *Discover Food*, *4*(1), 189. DOI: 10.1007/s44187-024-00265-7

Ullah, A., Munir, S., Badshah, S. L., Khan, N., Ghani, L., Poulson, B. G., Emwas, A.-H., & Jaremko, M. (2020). Important Flavonoids and Their Role as a Therapeutic Agent. *Molecules (Basel, Switzerland)*, *25*(22), 5243. DOI: 10.3390/molecules25225243 PMID: 33187049

Chapter 15 Adaptation of Ixodid Ticks (Ixodidae) to Parasitism on Cattle in Northeastern Armenia

Martin Yakov Rukhkyan
http://orcid.org/0000-0002-9559-9719
Scientific Center of Zoology and Hydroecology, Armenia

Oleg Valery Shcherbakov
http://orcid.org/0000-0001-7533-1670
Institute of Zoology, Armenia

Ruzan Levon Hovhannisyan
http://orcid.org/0000-0003-4277-7785
Institute of Zoology, Armenia

ABSTRACT

The ecology, abundance, and behavioral characteristics of ixodid ticks, parasitizing on the body of farm animals and having epidemiological significance, were studied. A total of 175 cows and 130 sheep were examined. Manual collection of ticks from animals did not cause them any discomfort or stress. Species diversity of ixodid ticks in the mountain-forest zone of Armenia is rather high. The following tick species were detected: Ixodes ricinus, Ixodes redikorzevi, Ixodes trianguliceps, Rhipicephslus sanguineus, Rhipicephslus annulatus, Rhipicephslus bursa, Hyalomma marginatum, H. asiaticum, and Dermacentor marginatus. Typically, ticks will first cling to the animal forelimbs, head, and neck that is facilitated by the grazing posture of the animal. The distribution of ticks over the animal's body also depends on the host's

DOI: 10.4018/979-8-3373-1524-9.ch015

ability to self-clean and "self-defense". The research data on seasonal dynamics of ticks in the region can be used by agricultural enterprises in planning the timing and conditions of tick control measures in Dilijan National Park.

INTRODUCTION

Ixodid ticks (Ixodidae) widespread in various natural areas of Armenia, have an important medical and veterinary significance as carriers of pathogens of a number of dangerous diseases of humans and farm animals. The constant interest in the study of ixodid ticks and a large number of publications devoted to various aspects of their morphology, physiology, biology, study of life cycles, etc., is explained by their medical and parasitological significance.

Research on ecology, morphology, taxonomy, medical and parasitological aspects of ixodid ticks are known in Armenia. The authors touched upon various topics, i.e. the transmission of infectious agents - spread by ixodid ticks (Zilfyan and Mnatsakanyan, 1964), epidemic rickettsiosis (Kotsinyan, 1959), hemosporidiosis of farm animals (Marutyan, 1979; Hakobyan et al., 2003; 2004; Vardanyan et al., 2005; Naghashyan et al., 2013; Gevorgyan et al., 2014), transmission of arboviruses and circulation of the causative agent of tick-borne borreliosis in Armenia (Gigen and Harutyunyan, 1991; Alekseev and Dubinina, 2000; Rukhkyan et al., 2011), hibernation of ticks in mountain areas of Armenia (Rukhkyan, 1986), olfactometric preferendum for ixodid ticks (Rukhkyan, 2012), the phenomenon of dead-end development, and dispersal of the pasture tick *Dermacentor marginatus* across biotopes (Rukhkyan and Alekseev, 1991), the presence of the Crimean-Congo hemorrhagic fever virus in ixodid ticks of Armenia (Gevorgyan et al., 2019).

Fauna of ticks in Ijevan Region of Armenia involving Dilijan National Park, was studied about 50 to 60 years ago (Marutyan, 1964; Marutyan and Oganesyan, 1974; Marutyan et al., 1974).

Tick species detected were as follows: *I. ricinus, Rh. annulatus (mentioned as Boophiluis calcaratus), Rh. sanguineus, Rh. turanicus, Rh. bursa, H. punctata, H. sulcata, H. inermis, D. pictus (synonym: D. reticulatus), D. marginatus, H. scupense, H. detritum, H. asiatisum, H. anatolicum.*

The choice of parasitized sites by ixodid ticks on the host body, as well as the number and species background of ticks in the northeastern region of Armenia depends on numerous factors. These include the season, animal husbandry conditions, climatic features of the pastures, behavioral adaptation of the ticks, solar radiation, which significantly affects the microclimate of the skin surface, etc. (Babenko and Leonovich, 1985; Grech-Angelini et al., 2016; Bedouhene et al., 2022; Baňas et al., 2023).

It is also obvious that the peculiarity of skin structure of the mammalians, and the species-specific anatomical and physiological structure of the barbed sucking apparatus of ixodid ticks are essential prerequisites for the selection of host body sites for attachment and the initiation of bloodsucking. The pattern of tick infestation of the mammalian body depends directly on the thickness of the epidermis, the depth of blood vessels, hair follicles, and sweat glands (Lee et al., 2024).

The study of the behavior of ixodid ticks when attacking the host and their parasitism on certain parts of the body will make it possible to develop more effective tick control measures to reduce the economic damage caused by ixodid ticks to livestock in certain natural areas of Armenia.

The goal of the research was to study the ecology of ixodid ticks, as well as areas of the body of farm animals that are most affected by ixodid ticks in a specific mountain forest biotope of Armenia. The following tasks were set to reach the mentioned goal: to monitor ixodid ticks, study their biodiversity of ixodid ticks, and identify factors influencing their mosaic distribution on the body of animals.

MATERIALS AND METHODS

The study was conducted from 2019 to 2021 in the mountain forest biotopes of Dilijan National Park. Livestock were examined at field stations to identify body areas most affected by ixodid ticks.

Tick infestation in cattle was monitored every ten days throughout the growing season in high-mountain pastures and along livestock passes in the Taush and Gegharkunik regions. Ticks were also collected manually from animals in the pens of agricultural communities in the villages of Khachardzan (coordinates: 40.7184° N, 45.0528° E), Kalavan (coordinates: 40.6492° N, 45.1081° E), and the village of Aygut (coordinates: 40.6830° N, 45.1742° E) during morning and evening milking. This procedure did not cause any negative emotions or discomfort in the animals, as it was combined with the normal milking process.

A total of 175 cows and 130 sheep were examined.

The data obtained significantly correlated with parameters such as average ambient temperature, seasonality, tick species, and key body areas. The collected ixodid tick material was processed and identified in the General Helminthology and Parasitology Laboratory of the Scientific Center of Zoology and Hydroecology of the National Academy of Sciences of the Republic of Armenia, Yerevan, Armenia. Using the key (Walker et al., 2004).

Statistical processing of the obtained data was performed using vassarstats.net online statistical tool (t-test; Confidence interval 95%, $P < 0.05$).

RESULTS AND DISCUSSION

As for our observation, the species diversity of ixodid ticks in the mountain-forest zone of Armenia is rather high. This statement was also demonstrated by our previous research (Rukhkyan and Hovhannisyan, 2012).

Over 3 years of research, 3237 specimens of ticks were collected, of which 1797 specimens were from cows and 1440 specimens from sheep, and the difference in calendars was presented: *I. ricinus* Linnaeus, 1758 (n=680), *I. redikorzevi* Olenev, 1927 (n=227), *I. trianguliceps* Birula, 1895 (n=194), *Rh. sanguineus* Latreille, 1806 (n=485), *Rh. annulatus* Say, 1821 (synonyms: *Boophilus annulatus, B. calcaratus*) (n=259), *Rh. bursa* Canestrini et Fanzago, 1878 (n=227), *H. marginatum* Koch, 1844 (synonym: *H. plumbeum*) (n=420), *H. asiaticum* Schulze et Schlottke, 1929 (n=193), *D. marginatus* Sulzer, 1776 (n=550).

Only adult ticks, both males and females, have been collected and identified. Almost all the detected ticks are dixenic and trixenic (excepting *Rh. annulatus*), and only adult stages of Ixodids parasitize on sheep and cattle.

According to our research, tick fauna in the examined region has been significantly enriched. Thus, more than 30 years ago the most common species of ixodid ticks for the broad-leaved forests of the north-eastern region of Armenia in the studied biotopes of the recreational areas of the "Dilijan" National Park were as follows: *I. ricinus, Rh. sanguineus, H. marginatum, D. marginatus.* (Oganjanyan et al., 1990).

Unfed sexually mature ticks in nature passively wait for a potential host located at the top of herbaceous and shrubby vegetation (the phenomenon of negative geotaxis), and actively react to the approaching animal, bulging forelimbs (Sonenshine and Roe, 2013).

Typically, ticks will first cling to the animal forelimbs, head and neck that is facilitated by the grazing posture of the animal. Once on the body of the prey animal, the ticks begin to actively probe the animal's body to search for an area to attach and pierce the tissue, and sink the hypostome into the skin for subsequent bloodsucking.

The main conditions when ticks choose host body sites for sucking are the peculiarities of animal cover structure, epidermis thickness, depth of blood vessels, presence of hair follicles and sweat glands of animals, as well as species-specific anatomical and physiological structure of the barbed-sucking apparatus of Ixodids. It is noted that when attacking sheep, ticks avoid attaching to parts of the body with thick hair, which is a dead end for further development (Rukhkyan and Alekseev 1991).

When examining animals, group sucking of ticks of both sexes to a certain body part was often observed, when unfed individuals of other species begin to actively attach around the fed ticks of one species.

Orientation to the source of the pheromone, and subsequently, the attachment of percipients near this source, serves as an additional guarantee for fertilization: females that prefer to suckle to the sites on which males parasitize, where copulation takes place. These pheromones are largely species-specific and stimulate the ticks to suckle in the area of the necrotic focus (Leonovich 2019).

As a rule, in places where ticks accumulate on the animal's body, an open wound with dried blood or inflammatory-purulent exudate forms causing painful suffering to the animal and resulting to a secondary infection. Usually, there are areas with soft tissues, and superficial blood vessels, as well as areas affected by erythema migrans, where tick accumulation is greater.

The distribution of ticks on an animal's body also depends on the host's ability to self-clean and "self-defense." Cattle actively react to the presence of crawling and sucking ticks and attempt to get rid of them in various ways: by vigorously rubbing against hard objects (fences, posts, tree trunks, rock outcrops, etc.), licking ticks with their tongues, knocking them off with tail lashes, and dislodging them with sharp twitches of the skin.

The intensity and effectiveness of these reactions depend on the animal's constitution and temperament.

Sheep show virtually no reaction to ticks freely crawling on their fur, but they intensively scratch at the sites of their mass accumulation on their chests where a wound has formed. Animals often succeed in crushing "swollen females entangled in wool" with kicks. Thus, due to their anatomical features, sheep's ability to self-cleanse from attacking ticks is very limited, whereas this process is more effective in cattle.

In this regard, it is clear that cattle are the primary hosts of ixodid ticks in forest habitats in spring, while sheep play a lesser role in the ticks' diet.

The average numbers of registered ixodid tick species *I. ricinus*, *Rh. turanicus*, *Rh. sanguineus*, and *D. marginatus* served as indicators characterizing the areas of the cattle body most infested by ixodid ticks in a given habitat (Table 1). Calculation of the average number of ticks on sheep before shearing (June-July) showed that the bulk of ticks are localized on the chin (11%), groin (17%) and in the belly (10%) (Table 2).

Table 1. Distribution of adult ticks over the cattle body (M±m)

Average number of adult ticks						
Dewlap N = 175	**Chin N = 175**	**Withers N = 175**	**Groin N = 175**	**Belly N = 175**	**Tail N = 175**	**Front Legs N = 175**
31.11±2.19	21.17±1.36	15.28±1.52	16.11±1.47	7.94±0.55	7.33±0.81	5.94±0.76
CI 95% P = 0.000	CI 95% P = 0.000	CI 95% P = 0.000	CI 95% P = 0.000	CI 95% P = 0.038	CI 95% P = 0.217	

Table 2. Distribution of adult ticks over the sheep body (M±m)

Average number of adult ticks						
Dewlap N = 130	**Chin N = 130**	**Withers N = 130**	**Groin N = 130**	**Belly N = 130**	**Tail N = 130**	**Front Legs N = 130**
9.29±0.56	20.93±2.10	8.71±0.83	17.29±1.36	8.14±0.98	13.00±1.48	4.29±0.70
CI 95% P = 0.000	CI 95% P = 0.000	CI 95% P = 0.000	CI 95% P = 0.000	CI 95% P = 0.003	CI 95% P = 0.000	CI 95% P = 0.000

Shearing of sheep scheduled for summer, imposes certain corrections in tick distribution on the body of sheep: so, localization of ticks on the torso in the first days of grazing after shearing, becomes less mosaic, and covers larger body areas. Data reliably correlate with relatively high values of tick population index in shorn sheep, when ticks are more evenly distributed over the "bare body" withers (25%), back (15%), belly (19%), and tail area (14%).

At the same time, difference between the mean indices of tick abundance in cattle and in sheep is not statistically significant (CI 95%, P = 0.429).

D. marginatus ticks were more frequently recorded withers (35%), chin (22%) and tail (15%), and less frequently on the chest in cattle in spring. During the hot season, *I. ricinus* ticks in cattle were found mainly in the groin area (29%). These places are not exposed to insolation, and the higher daily skin layer temperatures on the open parts of the body on average do not exceed +35 to +37.5°C.

In the fall season with high infestation of cattle with ticks *D. marginatus, Rh. turanicus, Rh. sanguineus, I. ricinus,* the accumulation of sucking ticks is observed on the upper parts of the body (29%), including the head and ears, while in late spring and early summer the maximal number of ticks were registered in the interscapular and groin areas (42%).

In early spring tick collections, the species *D. marginatus* on cows was more often observed at the withers (12%), and on the thorax (8%), as well as at the root of the tail (18%), In the hot season, ticks of the species *I. ricinus* were found mainly in the groin (14%), and on the belly (18%). In late autumn collections of ticks *Rh. sanguineus* were found in small quantities on the back of the head (6%).

The results of seasonal statistics of the body parts of cattle and small ruminants that are the mostly infested by ixodid ticks in spring and fall periods, vary with the indicators of summer collections. This phenomenon can be explained by diapause in ixodid ticks in the summer (Belozerov 1981, Gray et al 2016, Ogden et al 2021, Bedouhene et al 2022).

The index of seasonal abundance of tick infestation in livestock is statistically correlated with the choice of preferred body sites for blood sucking by ticks. It was noted that in sheep, when calculating the index of tick infestation abundance in autumn, the main mass of ticks is localized on the pectoralis, udder, and axillary fold. At relatively low tick abundance index values in early spring, and after sheep shearing, ticks are more evenly distributed over the body, including the tip of the tail, and the rump. In cows with high tick infestation in the fall, there is an accumulation of sucking ticks also on the upper parts of the torso, including the head, neck and ear flaps, and in spring and summer, maximal number of ticks was observed in the interscapular and inguinal areas.

In addition, it can be stated that the reduction of the number of parasitic ticks on cows is favored by the procedure of their mechanical removal during morning and evening milking, which is practiced by milkmaids, cleaning their crotch and body from sucking ticks. Thus, it can be concluded that this procedure also affects the number of ectoparasites on the animal, and therefore, the general background of tick distribution in biotopes.

CONCLUSION

The primary hosts of adult ixodid ticks of these species are almost exclusively farm animals, with preimaginal stages of ticks being found only sporadically.

For the first time, on the territory of the recreational zones of the National Park "Dilijan" in the mountain-forest biotope of Armenia, the ecology, abundance, and behavioral characteristics of ixodid ticks, parasitizing on the body of livestock and having epidemiological and epizootological significance, were studied. Research on the biodiversity of ixodid ticks is very relevant for the recreational areas of the National Park "Dilijan" and forest areas, which are often visited by tourists and the local population. Research data on seasonal livestock ticking in the region can be used by farms when planning the timing and conditions of tick control measures in Dilijan National Park.

REFERENCES

Alekseev, A. N., & Dubinina, H. V. (2000). Abiotic parameters and diel and seasonal activity of Borrelia infected and uninfected *Ixodes persulcatus* (Acarina, Ixodidae). *Journal of Medical Entomology*, *37*(1), 9–15. DOI: 10.1603/0022-2585-37.1.9 PMID: 15218900

Babenko, L. V., & Leonovich, S. A. (1985). Ecology. Behavior upon contact with the host and attachment sites. In Filippova, N. A. (Ed.), *Taiga tick Ixodes persulcatus Schulze (Acarina, Ixodidae). Morphology, systematics, ecology, medical significance* (pp. 249–251). Nauka.

Baňas, M., Šofranková, L., Kurimský, J., Pavlík, M., Pikalík, M., Majláthová, V., Cimbala, R., Pipová, N., Wurfl, L., & Majláth, I. (2023). Interspecific differences in the behavioral response of ticks exposed to radiofrequency electromagnetic radiation. *Experimental & Applied Acarology*, *91*(3), 477–485. DOI: 10.1007/s10493-023-00847-7 PMID: 37819593

Bedouhene, A., Kelanemer, R., Medrouh, B., Kernif, T., Saidi, F., Tail, G., & Ziam, H. (2022). Seasonal Dynamics and Predilection Sites of Ticks (Acari: Ixodidae) Feeding in Cows in the Western Parts of the Djurdjura, Algeria. *Frontiers in Tropical Diseases*, *3*, 856179. DOI: 10.3389/fitd.2022.856179

Belozerov, V. N. (1981). Ecological rhythms in Ixodid ticks and their regulation. *Parassitologia*, *30*, 22–35.

Gevorgyan, H., Grigoryan, G., Atoyan, H., Rukhkyan, M., & Hakobyan, A. (2019). Evidence of Crimean - Congo Haemorrhagic Fever Virus Occurrence in Ixodidae Ticks of Armenia.

Gevorgyan, H., Seitzer, U., & Beer, D. (2014). Implementation or molecular epidemiological approaches as a tool for characterization of tick-borne diseases in Armenia. *Proceedings of the International Conference "Biological Diversity and Conservation Problems of the Fauna of the Caucasus - 2". Yerevan,* 135-138.

Gigen, S., & Harutyunyan, E. S. (1991). An attempt to detect *Borrelia burgdorferi* in Armenian Ixodid ticks. *Biological Journal of Armenia*, *44*(4), 304–307.

Gray, J. S., Kahl, O., Lane, R. S., Levin, M. L., & Tsao, J. I. (2016). Diapause in ticks of the medically important *Ixodes ricinus* species complex. *Ticks and Tick-Borne Diseases*, *7*(5), 992–1003. DOI: 10.1016/j.ttbdis.2016.05.006 PMID: 27263092

Grech-Angelini, S., Stachurski, F., Lancelot, R., Boissier, J., Allienne, J.-F., Marco, S., Maestrini, O., & Uilenberg, G. (2016). Ticks (Acari: Ixodidae) infesting cattle and some other domestic and wild hosts on the French Mediterranean island of Corsica. *Parasites & Vectors, 582*(9), 1–11. DOI: 10.1186/s13071-016-1876-8 PMID: 27842608

Hakobyan, V. J., Boyakhchyan, G. H., Shcherbakov, O. V., & Vardanyan, M. V. (2004). Fauna of pasture ticks of Armavir Marz, Armenia. *Proceedings for Conference dedicated to the 100th Anniversary of S.M. Yablokov-Khndzoryan,* 17-18.

Hakobyan, V. J., Boyakhchyan, G. H., Vardanyan, M. V., & Shcherbakov, O. V. (2003). Some aspects of epidemiology of the bovine piroplasmosis and its control. *Proceedings for Conference "Research and Conservation of Fauna in Southern Caucasus"*, 24-26.

Kotsinyan, M. E. (1959). Epidemic rickettsiosis in the Armenian SSR. *Tenth Meeting on Parasitological Problems and Natural Focal Diseases,* 87–88.

Lee, S.-J., Kim, K.-Y., Kim, G., Moon, S., Park, Y.-C., Cho, H.-S., & Oh, Y. (2024). Potential Tick Defense Associated with Skin and Hair Characteristics in Korean Water Deer (Hydropotes inermis argyropus). *Animals (Basel), 14*(2), 185–192. DOI: 10.3390/ani14020185 PMID: 38254354

Leonovich, S. A. (2019). Comparative analysis of sexual behavior in ixodid ticks (Ixodidae). *The Journal of Parasitology, 53*(4), 294–306. DOI: 10.1134/S0031184719040033

Marutyan, E. M. (1979). Ticks - carriers as a factor in the preservation of foci of some babesiosis in nature. *All-Union conference on natural foci of diseases. Alma-Ata., 1*, 127–128.

Marutyan, Ye. M. (1964). *Haemosporidiosis situation in Ijevan Region of Armenian SSR*. Proceedings for Armenian Scientific Research Institute of Animal Husbandry and Veterinary.

Marutyan, Ye. M., Markosyan, V. M., & Oganesyan, V. V. (1974). Some biological and ecological features of Ixodid ticks, and their role in transmission of piroplasmidoses in mountain forest area of Armenia. *Proceedings for Armenian Scientific Research Institute of Animal Husbandry and Veterinary., 12*, 649–659.

Marutyan, Ye. M., & Oganesyan, V. V. (1974). Epidemiological features of the bovine piroplasmidosis in mountain forest area of Armenian SSR. *Proceedings for Armenian Scientific Research Institute of Animal Husbandry and Veterinary., 12*, 639–647.

Naghashyan, H. Z., Shcherbakov, O. V., & Movsisyan, L. A. (2013). Hard ticks as vectors of Haemosporidia in Goris Region of Armenia. *Proceedings for International Conference "Fundamental and Applied Aspects for Study of Parasitic Arthropodes in XXI Century"*, 109-110.

Oganjanyan, A. M., Harutyunyan, E. S., & Rukhkyan, M. Ya. (1990). Distribution of Ixodid ticks in the Armenian SSR. *Advances of Medical Entomology and Acarology in the USSR, X Congress of the All-Union Entomological Society, Leningrad, Russia, 11-15 September 1989,* 118-119.

Ogden, N.H., Ben Beard, C., Ginsberg, H.S., & Jean I. Tsao, J.I. (2021). Possible Effects of Climate Change on Ixodid Ticks and the Pathogens They Transmit: Predictions and Observations *Journal of Medical Entomology, 58*(4), 1536–1545. DOI: 10.1093/jme/tjaa220

Rukhkyan, M.Ya. (1986). Hibernation of *Dermacentor marginatus* under the conditions.

Rukhkyan, M. Ya. (2012). Olfactometry phyto preferendum for Ixodid ticks in the steppe landscape of Armenia. Proceedings for the XI European Multicolloquium of Parasitology (XI EMOP), 447–448.

Rukhkyan, M. Ya., & Alekseev, A. N. (1991). One of developmental deadlines or the way of habitation of the pasture tick *Dermacentor marginatus. Journal of Medical Parasitology and Parasitic Diseases., 2*, 55–57.

Rukhkyan, M. Ya., Dubinina, E. V., & Alekseev, A. N. (2011). On the possibility of circulation of the causative agent of tick-borne borreliosis in the territory of Armenia. *International Conference "Biological Diversity and Problems of Conservation of the Fauna of the Caucasus"*, 267–271.

Rukhkyan, M. Ya., & Hovhannisyan, R. L. (2012). Evaluation of biodiversity of Armnia ixodids. *Proceedings for XIV International Conference "Biological Diversity of Caucasus",* 12-15.

Sonenshine, D. E., & Roe, R. M. (2014). External and internal anatomy of ticks. In Sonenshine, D. E., & Roe, R. M. (Eds.), *Biology of ticks* (Vol. 2, pp. 74–98). Oxford University Press.

Vardanyan, M. V., Hakobyan, V. J., Shcherbakov, O. V., & Boyakhchyan, G. H. (2005). The course of the bovine piroplasmosis in foothill area of Armenia, and comparative efficaciousness of its control. *Bulletin of Armenian Agricultural Academy., 4*, 53–56.

Walker, A. R., Bouattour, A., Camicas, J.-L., & Estrada-Peña, A. (2003). Ticks of Domestic Animals in the Mediterranean Region. A Guide to Identification of Species. University of Zaragoza.

Zilfyan, V. N., & Mnatsakanyan, A. G. (1964). Geographic distribution and some features of natural tularemia foci in the Armenian SSR. *Journal of Microbiology. Epidemiology and Immunology*, *4*(2), 14.

Chapter 16
Global Scientific Trends in Biochar Applications and Research for Salinity and Drought Stresses Management: A Bibliometric Perspective

Sakshi Singh
Faculty of Biology, Yerevan State University, Armenia

Abhishek Singh
Faculty of Biology, Yerevan State University, Armenia

Rupesh Kumar Singh
http://orcid.org/0000-0002-2536-1967
Centro de Investigação e Tecnologias Agroambientais e Biológicas, Universidade de Trás-os-Montes e Alto Douro, Vila Real, Portugal

João Ricardo Sousa
http://orcid.org/0000-0003-4365-9086
Centro de Investigação e Tecnologias Agroambientais e Biológicas, Universidade de Trás-os-Montes e Alto Douro, Vila Real, Portugal

Karen Ghazaryan
http://orcid.org/0000-0002-8973-2595
Faculty of Biology, Yerevan State University, Armenia

DOI: 10.4018/979-8-3373-1524-9.ch016

ABSTRACT

This bibliometric analysis offers a global overview of published research on use of biochar for combating soil salinity and drought in agriculture during 2013-2026. A total of 110 related articles were reviewed, with 4243 citations indicating high academic attention. Journal articles were the most represented, followed by reviews and book chapters, in line with the trend of synthesizing knowledge and conceptual frameworks. Keyword mapping, word clouds, treemaps, and BERT topic modelling found biochar-mediated soil amendment, crop stress tolerance, sustainable agriculture, and emerging cross-sections with nanotechnology as the core topics. Most of the leading contributions were from China, Pakistan, and India, assisted mainly by large publishers and funding agencies. Open access publications have increased in visibility, and network analysis has revealed solid international collaborations and under-explored research frontiers. The results provide insights into research patterns, knowledge gaps, and future prospects for biochar-based approaches targeting improvement of soil health and crop resilience against abiotic stresses.

1. INTRODUCTION

Global climate change is lead to higher temperatures and more uneven precipitation, which can give rise to serious drought, heat wave and soil salt in arid and semi-arid regions (Burke et al., 2006; Trenberth et al., 2013). Furthermore, improper irrigation management reduces water-use efficiency and drought intensity. About 30% of the world's land area is projected to be under severe drought by the 2090s (Cruz De Carvalho, 2008a). Drought is the most severe of the abiotic stresses affecting crop growth (Fig 1 A) (Gran et al., 2011). Moreover, increased evaporation leads to water loss and enhances the accumulation of salt (Dehaan et al., 2002). Soil salinity can also be caused by non-optimal irrigation practices leading to decreased crop yield (Suzuki et al., 2014). The area affected by salinity is nearly 1.2 billion hectares of land as per a recent survey (Zhu, 2016). Drought stress, which is often accompanied by soil salinity, are one of the most common and often occur simultaneously with other abiotic stresses being major threats to global food security, especially in arid and semi-arid regions worldwide (Bargaz et al., 2016; Vincent, 2017; Yang et al., 2016). Reduction in land productivity through salinity has led to estimations of an annual economic loss of USD 27.2 billion due to crop losses in irrigated agriculture (Qadir et al., 2014). It is projected that revenue loss will increase by 69% without halting land degradation. Moreover, the increased release of carbon (C) from degraded lands increases the cost of restoration. Excess soluble salts adversely affect the physicochemical and biological properties of soils through different mechanisms (Fig 1 A). Appropriate crop production depends on the relationship with soil moisture condition and salinity (Maas et al., 1999). The main impact of salinity is the reduction of biomass production, which results in less C input, and soil decline (Wong et al., 2009). Drought stress is also an important environmental factor leading to the decrease of growth and plant yield (Bodner et al., 2015; L. Wang et al., 2014). In recent years, drought has severely affected terrestrial agriculture. For example, Cd uptake by peanuts cultivated in a Cd-contaminated calcareous soil was enhanced due to drought stress (Xia et al., 2015). Plants are among the victims of both drought and salt stresses resulting in decreased water contents of leaves, nutrient acquisition, photosynthesis, growth, and ultimate yield (Noman et al., 2015; Siddiqui et al., 2015). Moreover, drought and salt stress also produce oxidative stress in plants via generation of ROS (reactive oxygen species) (Abbasi et al., 2015; Cruz De Carvalho, 2008b). Therefore, overcoming drought and salt stress is a major endeavor to achieve global food security. Biochar is attracting great interest as a soil amendment due to its potential to improve soil quality. Biochar is a black, highly porous carbon material produced by heating woody materials (biomass) in the absence of air (anaerobic or oxygen-limited conditions) (Fig 1 B). Since a news piece, named "Black the new green" (Marris, 2006), unveiled for the first time in print that large quantities of black char had been produced by ancient tribes and that this was responsible for mysterious of Terra Preta its fertility all along since 2006, the carbon store potentiality of biochar has been shown through extensive data analyses (Lehmann, 2007). Furthermore, in relation to the economic incentives of agricultural residues management (Sheer et al., 2024), pyrolyzed biomass addition to farmland was even promoted, that could increase crop yield on an interim basis. Since then, researchers have conducted a number of studies to determine how various biochar return strategies can modify crop yields, even in adverse environments. Many superior physicochemical properties are owned by biochar, which has rich pore structure with high specific surface area, plenty of ash content (as crops nutrient), alkaline (most biochar is alkaline; while only very little percentage of biochar is in acidic State), and highly stable form (Sheer et al., 2024). Due to its stability, peculiar physical configurations and chemical composition, it has been a powerful tool in study to enhance stress resistance and productivity of crops. Biochar is well known to have large surface functional groups and mesopores for the removal of heavy metals and organic pollutants in soils through adsorption or microbial degradation. Its microphysical qualities also enhance the stability of soil aggregates and WHC, which buffer against drought (Dengxiao et al., 2024; Park et al., 2023; Raza et al., 2023) and makes up the moisture shortage of saline soil (Ran et al., 2020; J. Wang et al., 2023), thereby defending the roots (Hazman et al., 2023). Moreover, biochar is high in nitrogen (N), phosphorus (P), potassium (K), calcium (Ca), magnesium (Mg) and other mineral nutrients (Okebalama et al., 2023). Due to its high value of cation exchange capacity (CEC) and alkaline in nature, biochar has been shown to assist in nutrient holding and pH balancing of soil under conditions like salinity (Duan et al., 2023) and drought (Park et al., 2023). Therefore, biochar addition may increase the soil microorganism activity and nutrient cycling availability and uptake, which enhances the soil quality, cultivation root characteristic, crop yield and grains quality (Anwar, Shehzadi, et al., 2023; Dengxiao et al., 2024b; Irfan et al., 2022). The broad application of biochar into the soil could potentially contribute toward building a healthy and productive agriculture, while increasing food security and sustainability.

Figure 1. The current illustration depicts (A) the process of creating biochar, (B) the effects of biochar on drought and salinity, and (C) a meta-analysis based on bibliometric data.

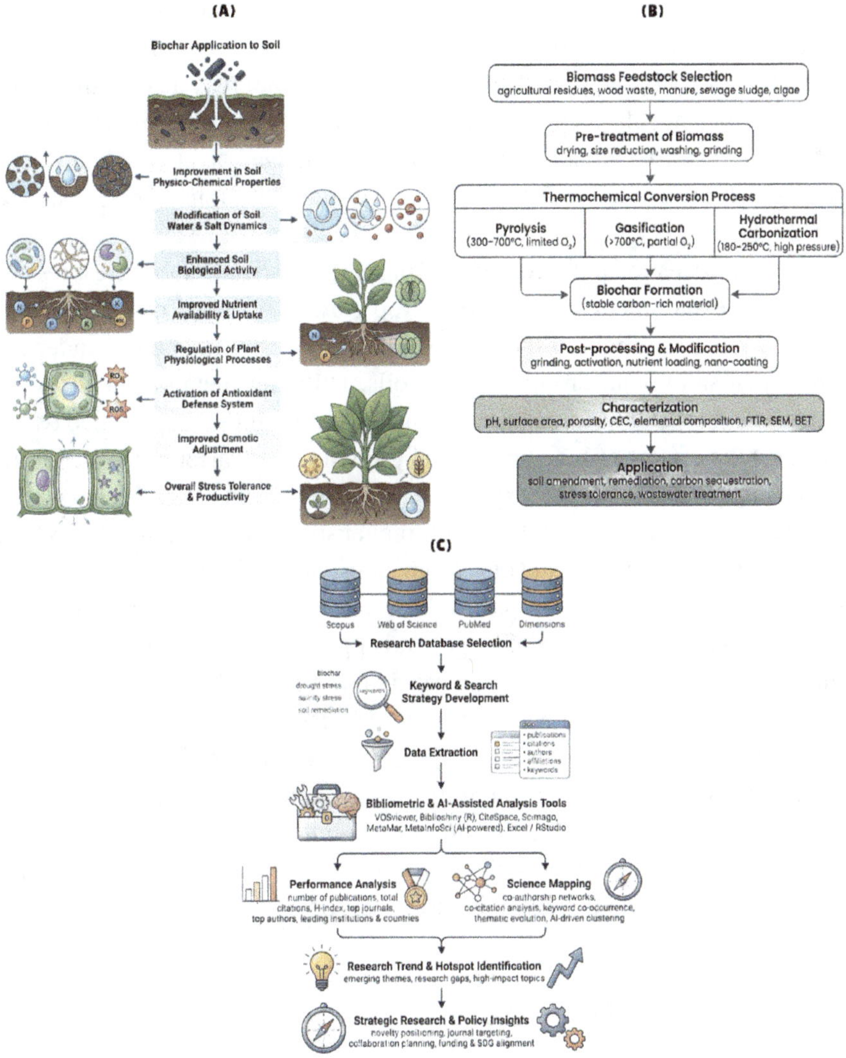

Bibliometric study is an essential tool for bringing forth and consolidating the fast-expanding scientific knowledge through structured data extraction of various databases (Scopus, Web of Science, PubMed and Dimensions) with specific keywords as well as search strategy (Fig 1 C) (Singh et al., 2025). The bibliographic information retrieval related to scientific publications, citations, authors, authorship, affiliations and keywords is considered using traditional search options and AI pertinent software as VOSviewer, Biblioshiny(R), CiteSpace (Scimago) MetaMar or MetaInfoSci (Kirby, 2023; Sharmaa et al., 2025). These utilities allow the performance evaluation (publication tendency, citation impact, leading authors, journals, institutions and countries) as well as science mapping by co-authorship,co-citation, keyword co-occurrence and thematic evolution (Akbari et al., 2020; Kirby, 2023). Incorporation of AI-based platforms such as MetaInfoSci may also further improve information filtering, clustering and trend analysis; thus, supporting the identification of research trends, emerging topics, and knowledge gaps towards influencing future research directions, collaborations and policy decisions (Corsi et al., 2022; Stopar, 2016).

Biochar improves soil physicochemical properties, enhances soil stress resistance and crop root development, and promotes crop growth. A review of 110 papers from the past 14 years highlighted the role of biochar in crop resilience under stress, particularly in saline and drought-prone soils. Utilizing metainfosci software for bibliometric analysis of biochar-based salinity and drought stress revealed two main findings: (a) the effects of biochar on crop growth under various stress conditions, and (b) its influence on soil properties such as nutrients and organic matter. Despite extensive research, there is a need for a comprehensive understanding of how biochar improves stress tolerance and root development in plants. This study aims to integrate current findings to clarify the impact of biochar on soil attributes and crop growth under stress, addressing future challenges and opportunities in its application for soil enhancement, including i) mapping keyword trends, research themes, and emerging topics related to biochar, soil health, and crop stress resilience; ii) profiling publication patterns, document types, and citation impacts across journals, publishers, and open-access categories; iii) examining research networks and collaborations at multiple levels, including authors, institutions, countries, and funding agencies; and iv) identifying knowledge gaps and future research directions to guide sustainable soil plant management and stress mitigation strategies.

2. MATERIAL AND METHOD

2.1 Data Collection

All the data are extracted from Scopus databased from 2013 to 2026 timeline by giving some search keywords query for first step data identification: "*salinity AND drought AND biochar*" with selection of search category are: *Article, Abstract, Keywords* (Luo et al., 2025; Quintarelli et al., 2024; Stopar, 2016). The first identification stage of data extraction provided 110 documents in the Scopus database. After the data identification process, the data were screened based on "*subject area*

(Environmental Science and Agricultural and Biological Sciences)"; "Document type (Articles, Reviews, Book chapters"; "Source type (Journals, Book)"; "Language (English)"; "Publication stage (Final)". In the final stage of inclusion, we selected 110 more appropriate scientific documents for further analysis.

2.2 Data visualization Tools

We utilized the MetaInfoSci bibliometric platform (https://metainfosci.bmu.edu.in) as our main tool for bibliometric analysis (Fig 2). Our research approach involves processing all gathered data through MetaInfoSci, a comprehensive platform that generates the bibliographic information necessary for subsequent analyses (Sharmaa et al., 2025).

Figure 2. Scopus database summary report on biochar and heavy metals generated using MetaInfoSci.

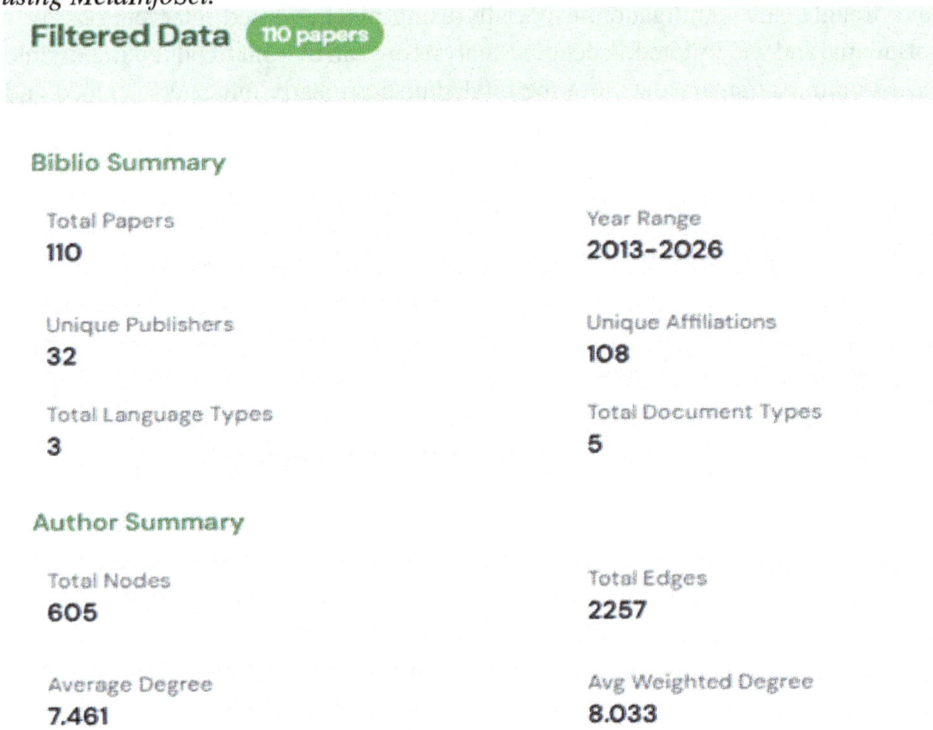

3. RESULTS AND DISCUSSION

3.1 Performance Analysis

(A) Annual Scientific Publication Citation Analysis

This study considered a long-time span that extended from 2013 to 2026 and concentrated in particular on studies concerning salinity stress, drought stress, and biochar application in agricultural and environmental systems. A total of 110 published works were generated during this period, and they provide further evidence of the enduring and ongoing pattern of productivity in these important stress-related areas. These 110 papers have accumulated an impressive 4243 citations, indicating excellent academic impact and adoption of the work in the research community (Fig 3). The high average citation rate (38.57 citations per paper) reveals the excellent impact of the individual studies, suggesting that research on salinity and drought stress mitigation, especially using biochar-based interventions, was influential and widely cited. A detailed analysis of year-by-year trends is impossible based on the summary data, but the total statistics clearly indicate sustained and strong productivity and contribution to knowledge about stress resilience, habitat remediation, and sustainable agriculture over 14 years.

Figure 3. The figure presents yearly data on scientific publications and citations concerning the use of biochar to address both salinity and drought stress.

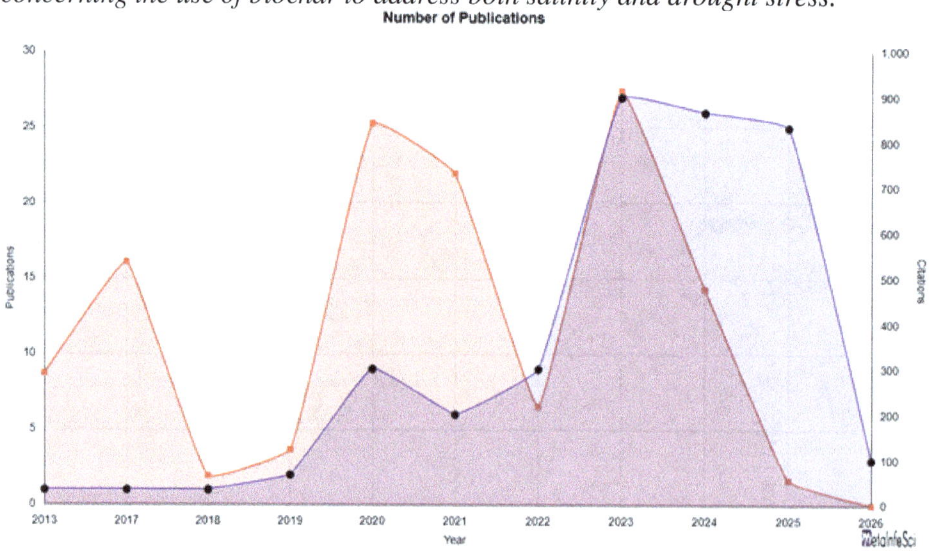

(B) Articles Types

This dataset comprises documents published from 2013 to 2026, which represent research on salinity stress, drought stress, and biochar-related interventions. It has a highly concentrated distribution among the five examined types, which tends to be skewed towards traditional scholarly formats. Journal articles far outweigh the entity of other outputs, with 51 publications (almost one-half of a total of 110) and reveal considerable focus on original research related to stress physiology, soil degradation, and biochar solutions for their mitigation. Reviews are the second most frequent category (33 papers), which demonstrate considerable attempts to integrate and critically appraise available knowledge on salinity- and drought-stress management, and the multifunctional roles of biochar in sustainable agriculture (Fig 4). Book chapters are also a major source (24 entries), which implies the potential to deepen our understanding of longer, comprehensive discussions on stress adaptation mechanisms, soil amendment technologies, and biochar application. Conversely, for conference-related outputs, very little can be found with the identification of only one conference paper and one conference review. Such a pronounced imbalance in the first preference reflects an intentional choice of peer-reviewed journals and edited books to disseminate information on salinity stress, drought tolerance, and biochar application. The lack of conference papers might be due to the focus or the criteria used for indexing or collecting data of that dataset, rather than no academic activity in the conferences themselves. In general, the publication tendency reflects the criteria of depth, strictness, and lasting academic effect in developing stress mitigation studies and viable soil–plant strategies (Zhang et al., 2023).

Figure 4. The figure illustrates the categories of scientific articles published from 2013 to 2026 on biochar, focusing on addressing both salinity and drought stress simultaneously.

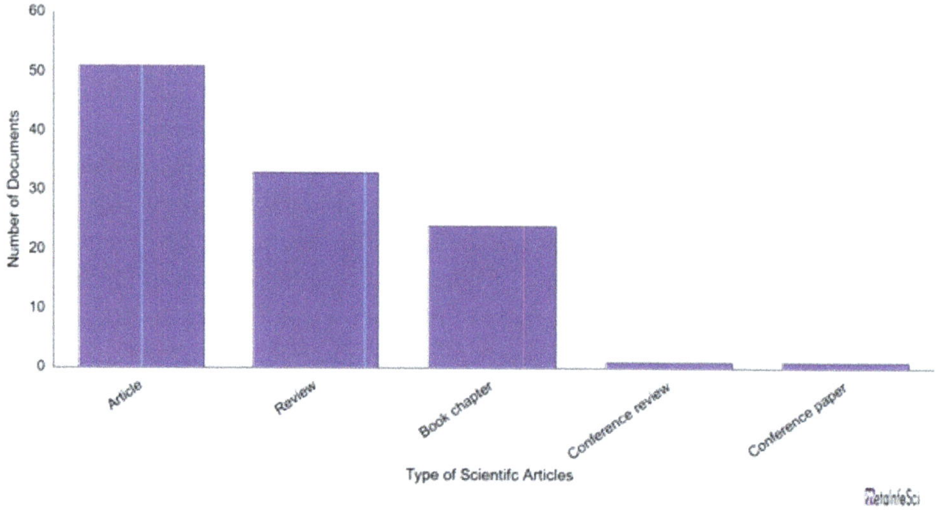

(C) Articles per Decade

The document type trend of these decades is prominently increasing and doddery change for salinity stress, drought stress, and biochar applications. In the previous period (2013-2019) research activity was minimal, with only five journal article publications and no contributions in other document types (Fig 5). This is indicative of the first (or bewildered) phase of research dissemination, limited to articles and original papers only. However, more recent times (2020–2026) show a remarkable increase in the volume of publications, as well as diversity and variety. Journal publications increased substantially from 5 to 46 (blue bar), the highest absolute growth among the metrics analyzed, indicating research acceleration on salinity and drought stress alleviation as well as biochar-based interventions. Of particular importance, reviews (33 documents violate bar) and book chapters (24 documents, red bar) completely nonexistent in the previous time frame came to the fore as important publication types suggestive of some level of maturity, as well as a trend toward knowledge synthesis, conceptual frameworks, and extensive discussions on stress resilience and sustainable soil–plant systems. Conference papers (green bar) and conference reviews (brown bar) also featured for the first time in this period, although only once each, indicating some but limited interaction with disseminating work via conferences. Overall, the findings indicate accelerated growth in scholarly productivity and efforts to diversify publication formats into different types of sources of information by strategic dissemination focusing on the last decade, which reflects increasing scientific relevance and academic consolidation of studies concerning salinity stress, drought stress, and biochar (Murtaza et al., 2023).

Figure 5. The illustration presents the classification of scientific articles published per decade, concentrating on biochar and its role in addressing the dual challenges of salinity and drought stress.

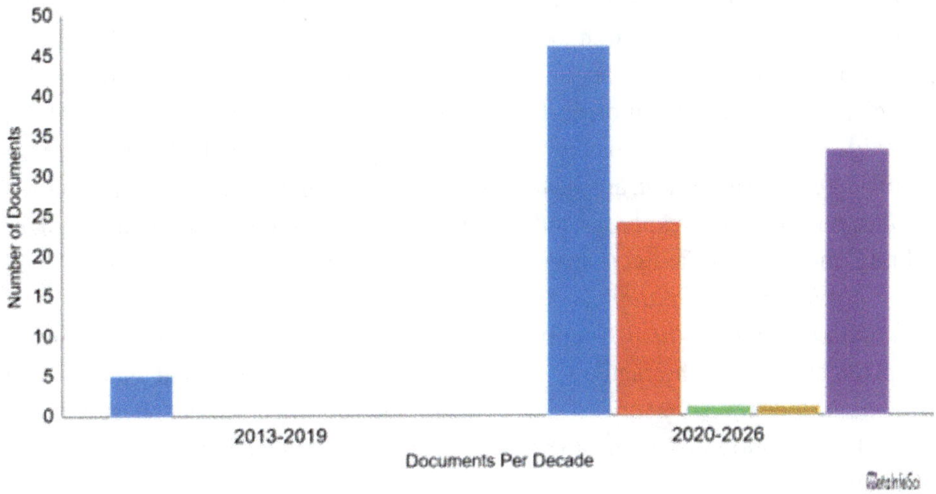

(D) Top Journals Analysis by Number of Publications

This review of top journal publications reflects a prominent theme in plant science and agricultural sustainable development from to 2013-2026. Agronomy, Biochar, and Science of the Total Environment followed with five papers each, illustrating a large interest in general agricultural science principles, biochar application to enhance soil fertility and biochar, and interwoven environmental challenges (Fig 6). Plants and Journal of Plant Nutrition were the second and third most frequent journals for general plant biology and nutrient meta-analysis, respectively. Journals outside that range but still having a clear botanical or agricultural focus (*Journal of Soil Science and Plant Nutrition, Frontiers in Plant Science, BMC Plant Biology, and Agricultural Water Management*) also support this trend with research that performs work in specific areas of plant–soil interactions, plant biology, and sustainable resource use. In summary, the profile underscores strong research activity, largely focusing on improving plant health and quality in an environmentally friendly manner. Key themes also include a significant focus on nutrient cycling, minimizing agricultural practices, and the wider environmental impacts of plant-based systems. Given the regularity with which plant-centered journals appear, their importance in communicating this research is clear. The database query, which covered a five-year period from 2013 to 2026, allowed for the observation at high resolution of trends in top journal distribution with regard to two key themes: crop production, environmental sustainability, salinity stress, drought stress, and biochar. Thus, this type of analysis would show a strong and coherent thematic focus. *Agronomy, Biochar, and Science of the Total Environment* topped the list with an equal number of five papers each, indicating a significant research focus on agronomic practices for stress alleviation, potential functions of biochar for soil quality improvement, nutrient-conserving capacities and C sequestration abilities, salinity and drought amelioration effects, and wider environmental and ecosystem-level significance. Plants, followed by *the Journal of Plant Nutrition*, are the next pair of journals that indicate a high emphasis on plant physiological responses, nutrient dynamics, and adaptive mechanisms under salinity and water-deficit conditions. The presence of *the Journal of Soil Science* and *Plant Nutrition*, *Frontiers in Plant Science*, *BMC Plant Biology*, and *Agricultural Water Management* also supports this tendency, with these journals focusing on plant–soil–water interactions, stress physiology, molecular and biochemical responses, and strategies for sustainable water and soil management. Together, the journal distribution highlights a strong and interdisciplinary research community dedicated to improving plant resilience, nutrient cycling, and productivity in saline and drought-prone systems using environmentally sustainable practices, with biochar as an agent for soil amendment and climate-resistant agriculture (GĄSIOR et al., 2017).

Figure 6. The figure categorizes leading journals based on the number of scientific articles published, focusing on biochar and its significance in addressing the combined issues of salinity and drought stress.

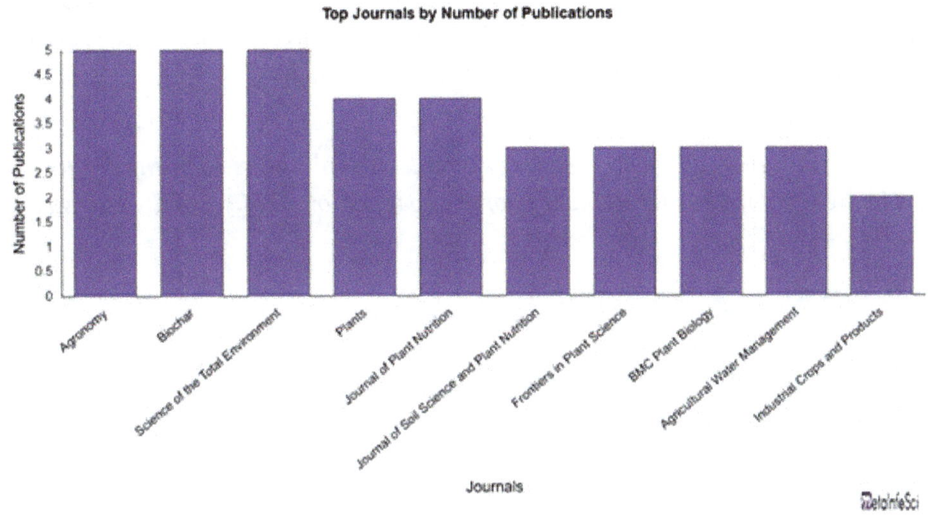

(E) Top Publisher Analysis

This publisher distribution analysis, covering the period of 2013–2026 and involving 110 publications, highlights an uneven dissemination of science applied to biochar applications and management of salinity or drought stress (Fig. 7). The top two publishers are Elsevier (24 journals) and Springer (22 journals), which contribute 41.8% of the total output, indicating their strong positions in environmental science, plant physiology, soil science, and sustainable agriculture (Fig 7). When MDPI (14 papers) is considered, the first three publishers account for more than 54.5% of all publications with high impact in stress-resilient agriculture, soil amendment technologies, and biochar-based mitigation strategies. Although 32 distinct publishers are represented, effective diversity among prolific contributors is quite low. A typical long-tail pattern is apparent, with 22 publishers (almost 69% of unique publishers) having only one publication. This may indicate that despite biochar and stress-type research being published in various sources, the leading core work is found in a small set of dominant international publishers. This level of focus is common for mature research areas, such as salinity and drought stress physiology, soil reclamation, and climate-smart agricultural systems. There are also a number of oddities in the data, such as redundant entries for the same publisher with slightly anomalous name or contact information (e.g., Springer, Taylor and

Francis, BioMed Central, Nova Science Zenodo and PTIE) and one record for a "blank" publisher indicating that care must be taken when cleaning the dataset. In general, the publisher distribution reflects that the topic of biochar-mediated relief from salinity and drought stress is predominantly supported by major international publishing houses, with a few scattered contributions from specialized, regional, or institutional publishers (Anwar, Munwwar, et al., 2023).

Figure 7. The figure categorizes leading journals and publishers, focusing on biochar and its significance in addressing the combined issues of salinity and drought stress.

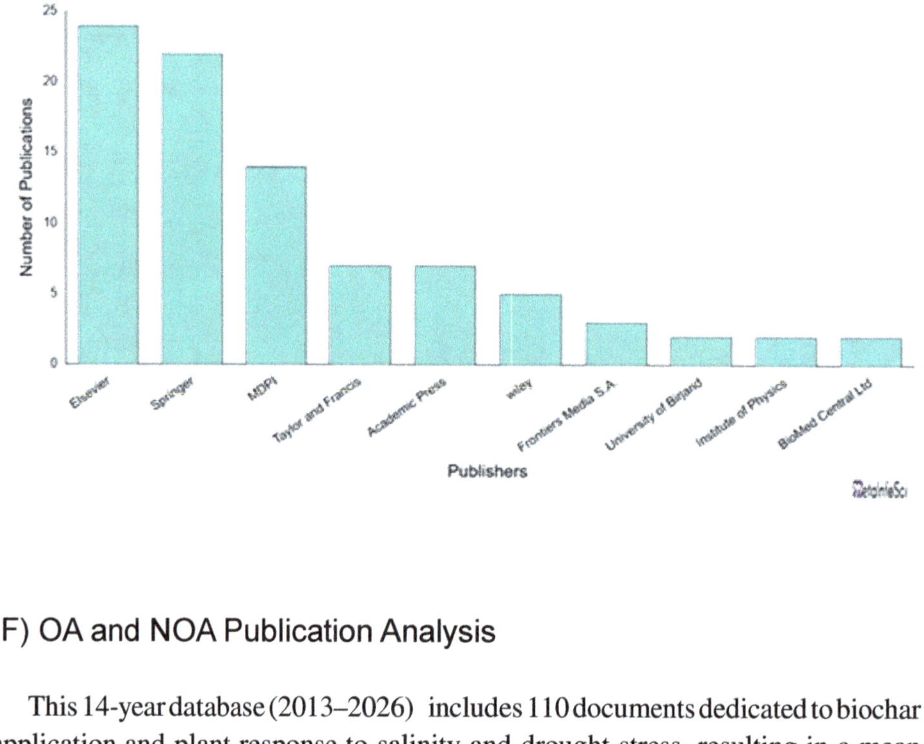

(F) OA and NOA Publication Analysis

This 14-year database (2013–2026) includes 110 documents dedicated to biochar application and plant response to salinity and drought stress, resulting in a mean publication growth of five documents per year for the entire period (Fig 8). The research space within the first period considered (2013–2019) was sparse, at a rate of 1–2 documents per year overall Easy Access, most Not Open Access, articulating an immature state of research dissemination in this particular field. There was a spike in productivity in 2020 with 9 publications, followed by 27 papers in 2023, which was the "Top Year". This growth was largely attributed to open-access papers (17 OA versus 10 not Open Access), which reflects improved availability and dissemination of research on biochar-based mitigation strategies for salinity and drought stress. Open-access publications, nearly non-existent in the prior period, grew more common as of 2020 and predominated in output from 2023 to 2024 (15 OA vs. 11 Not Open Access), providing evidence of burgeoning interest in "free to read" research on stress physiology, soil remediation, and sustainable agriculture. 2025 By 2025, the trend is reversed, with NOA documents leading (16 NOA vs. 9 OA). The data foresee a huge spike in 2026 to only three items (one OA and two NOA). In general, the pattern indicates a period of heavy growth in scientific output from 2020 to 2025, including a rise in research on saline and drought stress, biochar intervention measures, and a prominent, though less stable, ascent of Open Access within this salient field.

Figure 8. The figure categorizes publications into OA and NOA, highlighting the importance of biochar in mitigating the combined challenges posed by salinity and drought stress.

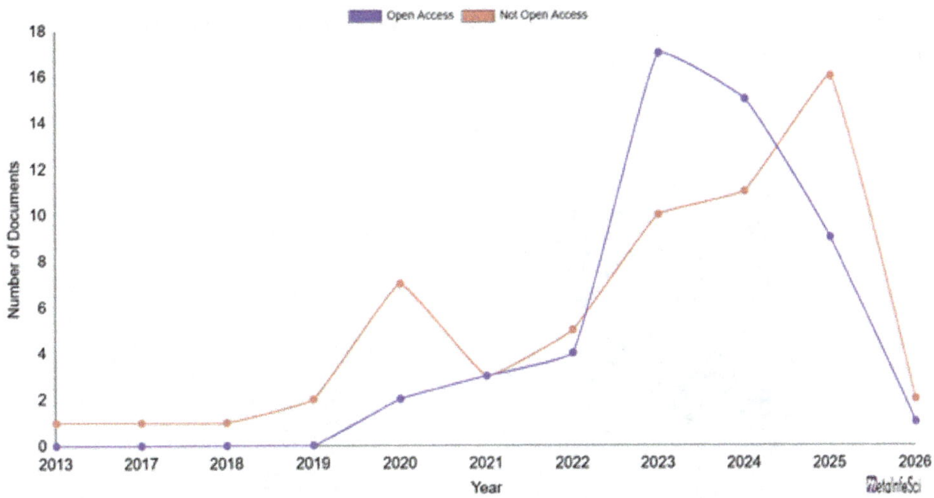

(G) Open Access Citation Analysis

This dataset includes 4243 citations from 2013 to 2026 and provides an analysis of the distribution of the cited work on subject categories (biochar applications and plant responses to salinity and drought stress) across 'Not Open Access' and 'Open Access' content (Fig 9). The results show that NOA has the most cited publications (2338), and the number of citations for OA materials is large at 1,905. Event-based burst detection reveals that over this period, traditional non-open-access publications still received a somewhat higher proportion of academic engagement (in the form of citations) in salinity and drought stress abatement and biochar-based soil/plant interventions. Notably, the gap between the two groups was not very large. With a citation difference of 433, Open Access publications show high competitiveness and visibility, indicating the increasing relevance of publishing works focused on research articles on stress physiology, soil amendment strategies, and sustainable agriculture practices. Although publishing according to the traditional models continues to have its influence, the data indicate a rapidly growing adoption and impact of open-access publications, especially regarding the broader availability and readership of papers focused on biochar-mediated salinity and drought stress mitigation studies. This again reflects two aspects: the continued strength of citations for established NOA publishing trucks and an evident increase in awareness and use by research communities in this sensitive area.

Figure 9. The diagram in the figure classifies publication citations into OA, emphasizing the significance of biochar in addressing the dual challenges of salinity and drought stress in agriculture.

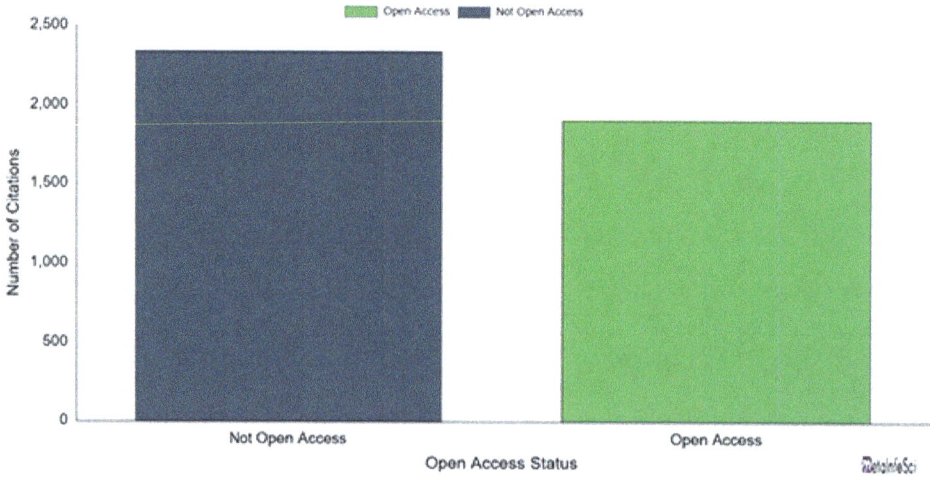

(H) Scientific Production Trends in Leading Countries

The review of 133 scientific papers published between 2013 and 2026 covering biochar applications and plant responses to salinity and drought stress in 10 countries demonstrated a median skew distribution (Fig 10). China ranks highest with 34 papers, an impressive lead over the average, as well as runners-up. Pakistan (24) and India (21) are the second and third, respectively, which clearly demonstrates a very strong regional focus in South Asia, as these three top Asian countries contribute almost 60% of the total output (79 publications). This reflects the strong research emphasis on strategies to minimize salinity and drought stress and the use of biochar for better soil management in such countries. An interesting trend is the large representation of countries from the Middle Eastern region, with Saudi Arabia (9), Egypt (9), Iran (8), and Turkey appearing in the top-10 countries, reflecting the interest in the region for understanding and addressing arid-semi-arid agricultural issues through biochar and stress-resilient crops. However, some traditional research giants, such as the United States (8) and Australia (9), are ranked lower with fewer publications than China, Pakistan, and India in this discipline. This distribution indicates a geographic focus of research, especially in Asia and the Middle East, on the biochar-induced reduction of salinity- and drought stress in soil, owing to its strategic relevance for these regions (Fig 10) and reflects an increasing regionalization along affected regions by abiotic stress compared to previous traditional global research patterns.

Figure 10. Trends in scientific output from leading nations regarding the use of biochar to tackle the combined issues of salinity and drought stress.

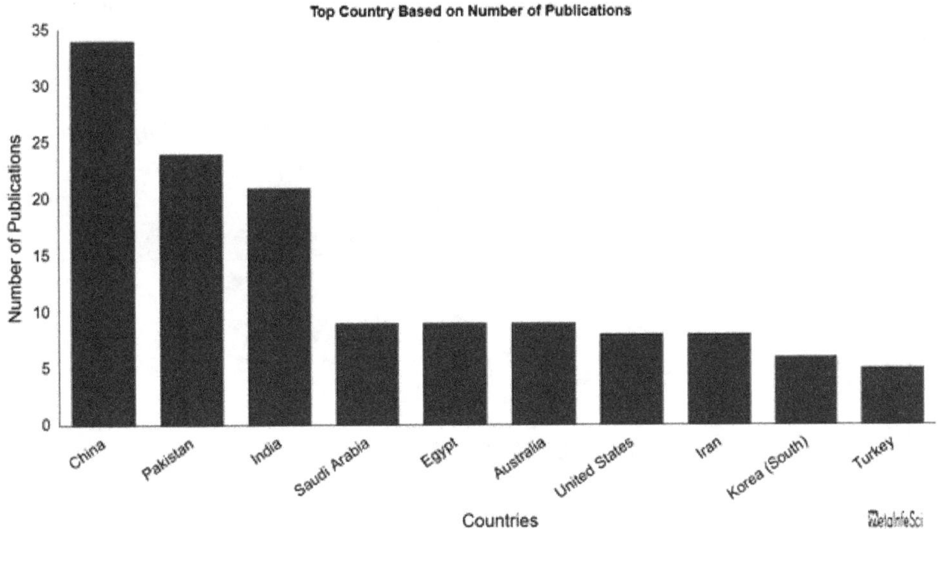

(I) Funding Agencies

The funding analysis of studies on the use of biochar and plant responses under salinity and drought indicates the involvement of 30 agencies, which also present a highly skewed pattern with few contributing agencies (Fig 11). The first two sources of funding were Chinese governmental programs: the National Key Research and Development Program of China (n=5) and the National Natural Science Foundation of China (n=4) (Fig 11). This highlights the strong influence of Chinese national policies on emerging research on salinity and drought stress alleviation and biochar development for soil management. Institutional support is similarly key, even outside national programs. Københavns Universitet and 3 by King Saud University each Individual with more than one publication either received substantial funding from a university source or were managed through an external grant. It is also noteworthy that specialized research institutions play a significant role in promoting studies on biochar intervention for improving plant resilience under abiotic stress, as indicated by two publications from the Deanship of Scientific Research. In total, these top five agencies were responsible for 17 publications, demonstrating the large contribution to this work made by a small group of the most active organizations. The distribution is what is called a 'power law' with the few major funders giving large support; however, each of the 25 other agencies individually gives less support. It is again noteworthy that these topmost authors are from China, Denmark, and Saudi Arabia, indicating the international dimensions of funding in promoting the research aspects of biochar-mediated amelioration of salinity and drought stress along with sustainable agricultural practices in stress-prone areas.

Figure 11. Trends in scientific output from funding agencies regarding the use of biochar to tackle the combined issues of salinity and drought stress.

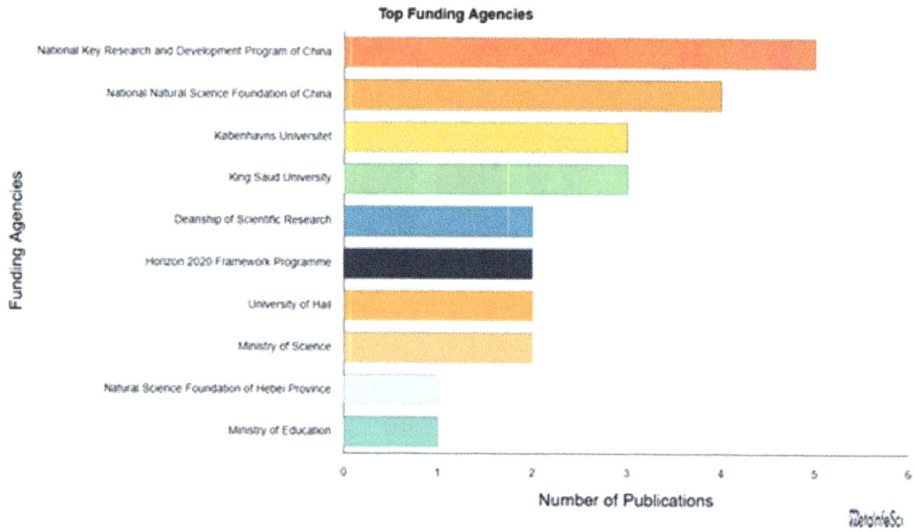

3.2 Science Mapping

(A) Keyword Analysis

I. Keyword Mapping

This network, comprising 50 nodes and 840 edges, demonstrates a high degree of interconnectedness among the displayed concepts (Fig 12 A). It is organized into three distinct communities, with Community 2 being the largest and most influential (20 nodes, total weight: 257). "Biochar" is by far the most dominant node, anchored in Community 2, suggesting that it is a central theme. This community also contains related concepts such as "soil amendment" and "sustainable agriculture," indicating a focus on biochar applications. Prominent environmental stressors, such as "drought," "salinity," "drought stress," and "abiotic stress" appeared across all communities, highlighting the network's emphasis on stress mitigation. Community 1 specifically links "drought stress," "salt stress," and "plant growth," while Community 0 broadly covers "abiotic stress" and "charcoal." Overall, the network explores the role of biochar and similar materials in addressing environmental stress to foster sustainable agriculture (Singh et al., 2025).

Figure 12. Most frequent keywords linked to biochar in drought and salinity.

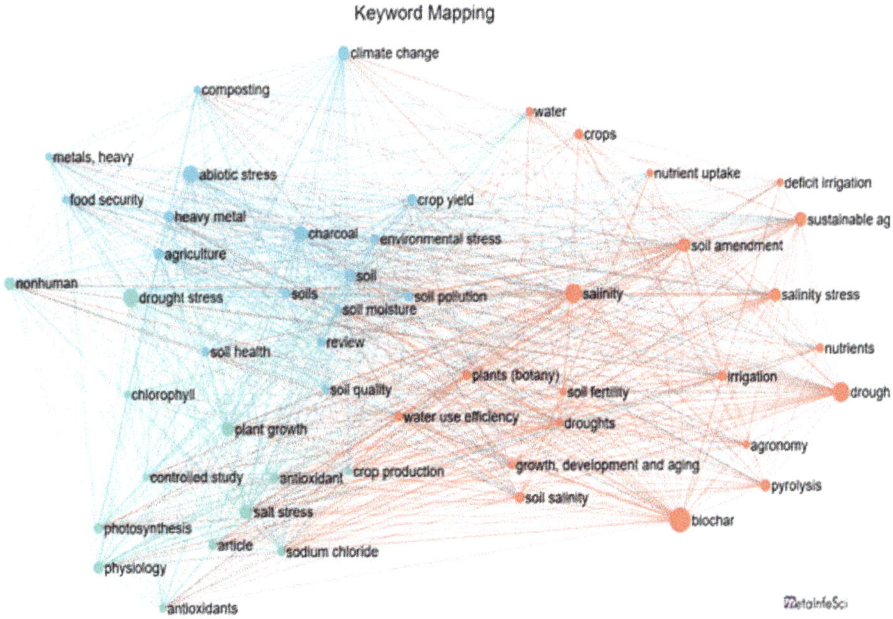

II. Word Cloud

This TF-IDF (Term Frequency–Inverse Document Frequency)-based word cloud provides an obvious and concentrated research focus on biochar, possessing a maximum TF-IDF score (9.718) and accumulated frequency (54), indicating it as the core theme of studies in this dataset (Fig 13). The word cloud mainly represents stress challenges in the environment and agriculture, such as "drought", "abiotic stress", "salinity", "drought stress", "salinity stress, "and "climate change". Keywords effectively emphasize the challenges faced by agricultural systems under unfavorable conditions. Crucially, the study also indicates that biochar is being explored in response to these stresses as a vital mitigation means for enhanced soil qualities, plant tolerance, and crop productivity. Supportive words like "sustainable agriculture," "soil amendment," and "pyrolysis also indicate biochar's double identity: as a technical product from biomass and also simply as a practical solution for enhancing crop stress tolerance. Collectively, the word cloud reflects a hypothetical discourse around biochar production, use, and benefits for salinity and drought stress management by highlighting its potential to enhance a resilient and sustainable agricultural economy under adverse environmental conditions.

Figure 13. Word cloud related to biochar in drought and salinity.

III. Tree Map

A treemap is a visualization method widely used in bibliometric analysis to present hierarchical data with nested rectangles (Fig 14). It is useful for visual identification and visual differentiation of key themes, keywords, sources, authors or countries by their frequency and significance with in a particular literature body (Chandran et al., 2024; Ismail et al., 2025).

Figure 14. The tree map illustrates the connections between biochar in drought and salinity, highlighting their contributions to soil management.

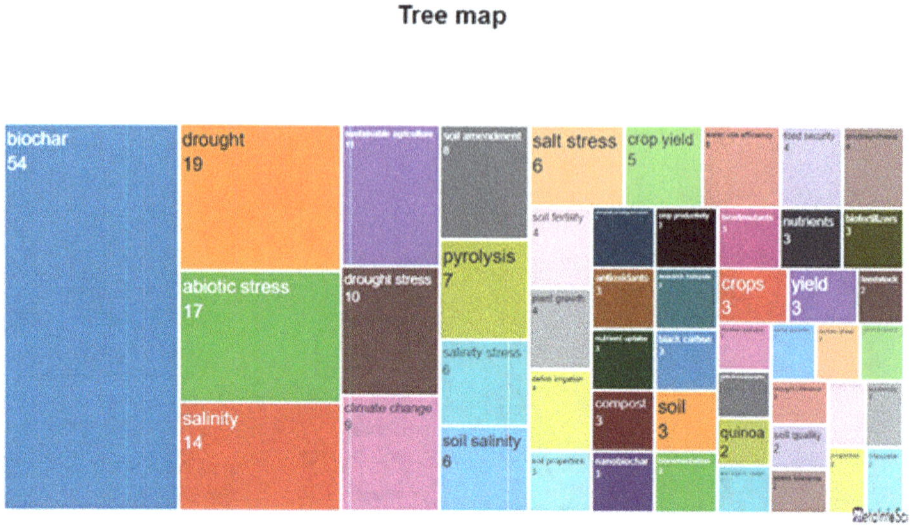

IV. BERT Topic Modelling

BERT topic modeling from 2013 to 2026 (using nine topics) reveals a very focused research backdrop. Perhaps unsurprisingly, given the imposition of such strong environmental stress, the top theme on biochar-stress-soil drought occurred 76 times (Fig 15). This demonstrates continued and intense research interest in the application of biochar, particularly with respect to environmental issues such as drought and soil health. In contrast, the other eight topics are very minor, only appearing 2-4 times. These are notably "*rhizobacteria-growth-promoting-plant-systemic*" (4) and "*ecotoxicity-carbon-amendment-black*" (4), suggesting an alignment of interest concerning plant microbiology and environmental impact. Crop-specific research is there in "*quinoa-breeding-irrigation-temperature*" (3) and advanced materials in "*carbon-nanosphere-nanomaterials-vulgaris*" (3). More general problem-solving meditations, such as "strategies-tolerance-synergistic-emerging" (2), and meta-analyses such as "*bibliometric-research-hotspots-analysis*" (2) are indicative of more specific or emerging topics. The data as a whole provide clear evidence that the research focus between 2013 and 2026 was clearly on the usage of biochar in sustainable agriculture and environmental resilience, whereas other associated scientific areas were not equally emphasized.

Figure 15. The BERT trending topic modeling reveals the links between biochar and soil management, emphasizing their roles in soil science and agriculture.

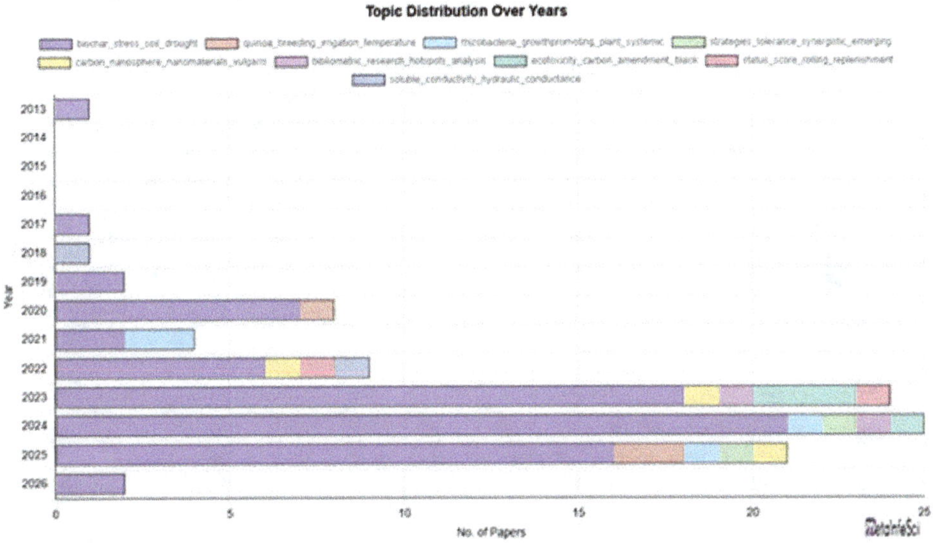

(B) Network Approaches

Country Network Analysis

The country collaboration world map Visualization in scientific querying is a valuable tool to illustrate the worldwide picture of international cooperation in one specific field (Fig 16) (Akrami et al., 2023). The map visualizes complex data with graphic detail: countries are expressed as nodes which often have color or size coding to illustrate a country's output and volume of work in the field (Aria et al., 2017).

Figure 16. The map illustrating international collaboration highlights the connections between biochar in drought and salinity, underscoring their significance in the fields of soil and agriculture.

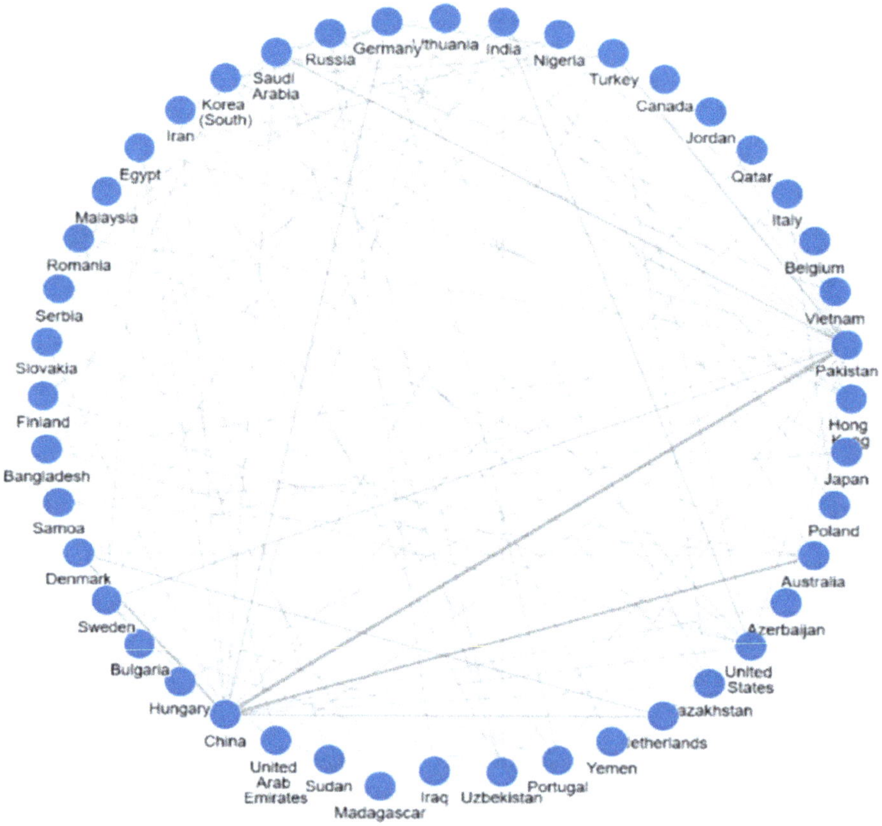

4. CONCLUSION

The bibliometric profile reveals a pronounced tendency of expansion in terms of scientific production of the topic on biochar application for salinity and drought stress mitigation during the last decade, and remarkable diversity regarding publication types and research foci. Biochar has been of increasing interest for sustainable agriculture; improvement of soil and crop stress tolerance coupled with the increased utilization of nanotechnology. Geographical distribution of research contributions occurs primarily in Asia and the Middle East, funded by main funding bodies and publishing companies, despite limited international collaboration. Open access journals are gaining popularity in this digital age due to the ease of making articles public. In conclusion, the review identifies trends and major contributors but also discrepancies in understanding of biochar science as a roadmap for future research towards realizing effective biochar applications for sustainable soil–plant management under environmental stresses.

ACKNOWLEDGMENT

Higher Education and Science Committee of the MESCS Republic of Armenia provided a 25AA-4C051 grant to support SS. KG has received support from the Higher Education and Science Committee of the MESCS Republic of Armenia through grant number 24FP-4C013.

REFERENCES

Abbasi, G. H., Akhtar, J., Ahmad, R., Jamil, M., Anwar-ul-Haq, M., Ali, S., & Ijaz, M. (2015). Potassium application mitigates salt stress differentially at different growth stages in tolerant and sensitive maize hybrids. *Plant Growth Regulation, 76*(1), 111–125. DOI: 10.1007/s10725-015-0050-1

Akbari, M., Khodayari, M., Danesh, M., Davari, A., & Padash, H. (2020). A bibliometric study of sustainable technology research. *Cogent Business & Management, 7*(1), 1751906. Advance online publication. DOI: 10.1080/23311975.2020.1751906

Anwar, T., Munwwar, F., Qureshi, H., Siddiqi, E. H., Hanif, A., Anwaar, S., Gul, S., Waheed, A., Alwahibi, M. S., & Kamal, A. (2023). Synergistic effect of biochar-based compounds from vegetable wastes and gibberellic acid on wheat growth under salinity stress. *Scientific Reports, 13*(1), 1–18. DOI: 10.1038/s41598-023-46487-0

Anwar, T., Shehzadi, A., Qureshi, H., Shah, M. N., Danish, S., Salmen, S. H., & Ansari, M. J. (2023). Alleviation of cadmium and drought stress in wheat by improving growth and chlorophyll contents amended with GA3 enriched deashed biochar. *Scientific Reports, 13*(1), 18503-. DOI: 10.1038/s41598-023-45670-7

Aria, M., & Cuccurullo, C. (2017). bibliometrix: An R-tool for comprehensive science mapping analysis. *Journal of Informetrics, 11*(4), 959–975. DOI: 10.1016/j.joi.2017.08.007

Bargaz, A., Nassar, R. M. A., Rady, M. M., Gaballah, M. S., Thompson, S. M., Brestic, M., Schmidhalter, U., & Abdelhamid, M. T. (2016). Improved Salinity Tolerance by Phosphorus Fertilizer in Two Phaseolus vulgaris Recombinant Inbred Lines Contrasting in Their P-Efficiency. *Journal Agronomy & Crop Science, 202*(6), 497–507. DOI: 10.1111/jac.12181

Bodner, G., Nakhforoosh, A., & Kaul, H. P. (2015). Management of crop water under drought: a review. *Agronomy for Sustainable Development, 35*(2), 401–442. DOI: 10.1007/s13593-015-0283-4

Burke, E. J., Brown, S. J., & Christidis, N. (2006). Modeling the Recent Evolution of Global Drought and Projections for the Twenty-First Century with the Hadley Centre Climate Model. *Journal of Hydrometeorology, 7*(5), 1113–1125. DOI: 10.1175/JHM544.1

Chandran, R., & Chandran MC, S. (2024). Green finance and sustainability: mapping research development through bibliometric analysis. *Discover Sustainability, 5*(1), 339-. DOI: 10.1007/s43621-024-00549-z

Corsi, S., Ruggeri, G., Zamboni, A., Bhakti, P., Espen, L., Ferrante, A., Noseda, M., Varanini, Z., & Scarafoni, A. (2022). A Bibliometric Analysis of the Scientific Literature on Biostimulants. *Agronomy, 12*(6), 1257. DOI: 10.3390/agronomy12061257

Cruz De Carvalho, M. H. (2008). Drought stress and reactive oxygen species: Production, scavenging and signaling. *Plant Signaling & Behavior, 3*(3), 156–165. DOI: 10.4161/psb.3.3.5536 PMID: 19513210

Dehaan, R. L., & Taylor, G. R. (2002). Field-derived spectra of salinized soils and vegetation as indicators of irrigation-induced soil salinization. *Remote Sensing of Environment, 80*(3), 406–417. DOI: 10.1016/S0034-4257(01)00321-2

Dengxiao, Z., Hongbin, J., Wenjing, Z., Qingsong, Y., Zhihang, M., Haizhong, W., Wei, R., Shiliang, L., & Daichang, W. (2024). Combined biochar and water-retaining agent application increased soil water retention capacity and maize seedling drought resistance in Fluvisols. *The Science of the Total Environment, 907*, 167885. DOI: 10.1016/j.scitotenv.2023.167885 PMID: 37863232

Duan, M., Yan, R., Wang, Q., Zhou, B., Zhu, H., Liu, G., Guo, X., & Zhang, Z. (2023). Integrated microbiological and metabolomics analyses to understand the mechanism that allows modified biochar to affect the alkalinity of saline soil and winter wheat growth. *The Science of the Total Environment, 866*, 161330. DOI: 10.1016/j.scitotenv.2022.161330 PMID: 36603639

El Akrami, N., Hanine, M., Flores, E. S., Aray, D. G., & Ashraf, I. (2023). Unleashing the Potential of Blockchain and Machine Learning: Insights and Emerging Trends From Bibliometric Analysis. *IEEE Access : Practical Innovations, Open Solutions, 11*, 78879–78903. DOI: 10.1109/ACCESS.2023.3298371

Gąsior, D., & Tic, W. J. (2017). Application of the Biochar-Based Technologies as the Way of Realization of the Sustainable Development Strategy. *Economic and Environmental Studies, 17*(3 (43)), 597–611. DOI: 10.25167/ees.2017.43.9

Gran, M., Carrera, J., Olivella, S., & Saaltink, M. W. (2011). Modeling evaporation processes in a saline soil from saturation to oven dry conditions. *Hydrology and Earth System Sciences*, *15*(7), 2077–2089. DOI: 10.5194/hess-15-2077-2011

Hazman, M., Fawzy, S., Hamdy, A., Khaled, A., Mahmoud, A., Khalid, E., Ibrahim, H. M., Gamal, M., Abo Elyazeed, N., Saber, N., Ehab, M., & Kabil, F. (2023). Enhancing rice resilience to drought by applying biochar–compost mixture in low-fertile sandy soil. *Beni-Suef University Journal of Basic and Applied Sciences, 12*(1), 74-. DOI: 10.1186/s43088-023-00411-7

Irfan, M., Dawar, K., Fahad, S., Mehmood, I., Alamri, S., Siddiqui, M. H., Saud, S., Khattak, J. Z. K., Ali, S., Hassan, S., Nawaz, T., Hammad, H. M., Banout, J., & Nasim, W. (2022). Exploring the potential effect of Achnatherum splendens L.–derived biochar treated with phosphoric acid on bioavailability of cadmium and wheat growth in contaminated soil. *Environmental Science and Pollution Research, 29*(25), 37676–37684. DOI: 10.1007/s11356-021-17950-0

Ismail, A., Munsi, H., Yusuf, A. M., & Hijjang, P. (2025). Mapping One Decade of Identity Studies: A Comprehensive Bibliometric Analysis of Global Trends and Scholarly Impact. *Social Sciences, 14*(2), 92. DOI: 10.3390/socsci14020092

Kirby, A. (2023). Exploratory Bibliometrics: Using VOSviewer as a Preliminary Research Tool. *Publications, 11*(1), 10. DOI: 10.3390/publications11010010

Lehmann, J. (2007). A handful of carbon. *Nature, 447*(7141), 143–144. DOI: 10.1038/447143a

Luo, X., Li, J., Guo, S., Yu, H., Zeng, X., Zhou, Z., Shangguan, Y., He, M., Ouyang, Y., Chen, K., Chen, Z., & Qin, Y. (2025). Analysis of research status and trends on nano-agricultural application: A bibliometric study. *Frontiers in Plant Science*, *16*, 1530629. DOI: 10.3389/fpls.2025.1530629 PMID: 40034150

Maas, E. V., & Grattan, S. R. (1999). *Crop Yields as Affected by Salinity*. DOI: 10.2134/agronmonogr38.c3

Marris, E. (2006). Putting the carbon back: Black is the new green. *Nature*, *442*(7103), 624–626. DOI: 10.1038/442624a PMID: 16900176

Murtaza, G., Ahmed, Z., Eldin, S. M., Ali, B., Bawazeer, S., Usman, M., Iqbal, R., Neupane, D., Ullah, A., Khan, A., Hassan, M. U., Ali, I., & Tariq, A. (2023). Biochar-Soil-Plant interactions: A cross talk for sustainable agriculture under changing climate. *Frontiers in Environmental Science*, *11*, 1059449. DOI: 10.3389/fenvs.2023.1059449

Noman, A., Ali, S., Naheed, F., Ali, Q., Farid, M., Rizwan, M., & Irshad, M. K. (2015). Foliar application of ascorbate enhances the physiological and biochemical attributes of maize (Zea mays L.) cultivars under drought stress. *Archives of Agronomy and Soil Science, 61*(12), 1659–1672. DOI: 10.1080/03650340.2015.1028379

Okebalama, C. B., & Marschner, B. (2023). Reapplication of biochar, sewage waste water, and NPK fertilizers affects soil fertility, aggregate stability, and carbon and nitrogen in dry-stable aggregates of semi-arid soil. *The Science of the Total Environment, 866*, 161203. DOI: 10.1016/j.scitotenv.2022.161203 PMID: 36581285

Park, J. H., Yun, J. J., Kim, S. H., Park, J. H., Acharya, B. S., Cho, J. S., & Kang, S. W. (2023). Biochar improves soil properties and corn productivity under drought conditions in South Korea. *Biochar, 5*(1), 66-. DOI: 10.1007/S42773-023-00267-1/TABLES/3

Qadir, M., Quillérou, E., Nangia, V., Murtaza, G., Singh, M., Thomas, R. J., Drechsel, P., & Noble, A. D. (2014). Economics of salt-induced land degradation and restoration. *Natural Resources Forum, 38*(4), 282–295. DOI: 10.1111/1477-8947.12054

Quintarelli, V., Ben Hassine, M., Radicetti, E., Stazi, S. R., Bratti, A., Allevato, E., Mancinelli, R., Jamal, A., Ahsan, M., Mirzaei, M., & Borgatti, D. (2024). Advances in Nanotechnology for Sustainable Agriculture: A Review of Climate Change Mitigation. *Sustainability, 16*(21), 9280. DOI: 10.3390/su16219280

Ran, C., Gulaqa, A., Zhu, J., Wang, X., Zhang, S., Geng, Y., Guo, L., Jin, F., & Shao, X. (2020). Benefits of Biochar for Improving Ion Contents, Cell Membrane Permeability, Leaf Water Status and Yield of Rice Under Saline–Sodic Paddy Field Condition. *Journal of Plant Growth Regulation, 39*(1), 370–377. DOI: 10.1007/s00344-019-09988-9

Raza, M. A. S., Ibrahim, M. A., Ditta, A., Iqbal, R., Aslam, M. U., Muhammad, F., Ali, S., Çiğ, F., Ali, B., Muhammad Ikram, R., Muzamil, M. N., Rahman, M. H. ur, Alwahibi, M. S., & Elshikh, M. S. (2023). Exploring the recuperative potential of brassinosteroids and nano-biochar on growth, physiology, and yield of wheat under drought stress. *Scientific Reports, 13*(1), 15015-. DOI: 10.1038/s41598-023-42007-2

Sharmaa, K., Khurana, P., & Uddina, Z. (2025). *MetaInfoSci: An Integrated Web Tool for Scholarly Data Analysis*. Retrieved from https://arxiv.org/pdf/2506.09056

Sheer, A., Fahad Sardar, M., Younas, F., Zhu, P., Noreen, S., Mehmood, T., Ur Rahman Farooqi, Z., Fatima, S., & Guo, W. (2024). Trends and social aspects in the management and conversion of agricultural residues into valuable resources: A comprehensive approach to counter environmental degradation, food security, and climate change. *Bioresource Technology, 394*, 130258. DOI: 10.1016/j.biortech.2023.130258 PMID: 38151206

Siddiqui, M. H., Al-Khaishany, M. Y., Al-Qutami, M. A., Al-Whaibi, M. H., Grover, A., Ali, H. M., Al-Wahibi, M. S., & Bukhari, N. A. (2015). Response of Different Genotypes of Faba Bean Plant to Drought Stress. *International Journal of Molecular Sciences, 16*(5), 10214–10227. DOI: 10.3390/ijms160510214

Singh, A., Rajput, V. D., Singh, R. K., Elshikh, M. S., El-Ramady, H., Prokisch, J., Liu, C., Rajput, P., Minkina, T., Sousa, J. R., Alexiou, A., Singh, S., Varagyan, V., & Ghazaryan, K. (2025). Emerging Frontiers in Nanotechnology for Salinity Stress Management: A Global Bibliometric Analysis. *Magallat Diyalá Li-l-'ulum al-Zira'iyyat, 17*(2), 46–69. DOI: 10.52951/dasj.25170204

Stopar, K. (2016). Presence of nanotechnology in agriculture: Bibliometric approach. *Acta Agriculturae Slovenica, 107*(2), 497–507. DOI: 10.14720/aas.2016.107.2.20

Suzuki, N., Rivero, R. M., Shulaev, V., Blumwald, E., & Mittler, R. (2014). Abiotic and biotic stress combinations. *The New Phytologist, 203*(1), 32–43. DOI: 10.1111/nph.12797 PMID: 24720847

Trenberth, K. E., Dai, A., Van Der Schrier, G., Jones, P. D., Barichivich, J., Briffa, K. R., & Sheffield, J. (2013). Global warming and changes in drought. *Nature Climate Change, 4*(1), 17–22. DOI: 10.1038/nclimate2067

Wang, J., Shi, D., Huang, C., Zhai, B., & Feng, S. (2023). Effects of Common Biochar and Acid-Modified Biochar on Growth and Quality of Spinach in Coastal Saline Soils. *Plants, 12*(18), 3232. DOI: 10.3390/plants12183232

Wang, L., Chen, W., & Zhou, W. (2014). Assessment of future drought in Southwest China based on CMIP5 multimodel projections. *Advances in Atmospheric Sciences, 31*(5), 1035–1050. DOI: 10.1007/s00376-014-3223-3

Wong, V. N. L., Dalal, R. C., & Greene, R. S. B. (2009). Carbon dynamics of sodic and saline soils following gypsum and organic material additions: A laboratory incubation. *Applied Soil Ecology, 41*(1), 29–40. DOI: 10.1016/j.apsoil.2008.08.006

Xia, S., Wang, X., Su, G., & Shi, G. (2015). Effects of drought on cadmium accumulation in peanuts grown in a contaminated calcareous soil. *Environmental Science and Pollution Research, 22*(23), 18707–18717. DOI: 10.1007/s11356-015-5063-9

Yang, A., Akhtar, S. S., Iqbal, S., Amjad, M., Naveed, M., Zahir, Z. A., & Jacobsen, S. E. (2016). Enhancing salt tolerance in quinoa by halotolerant bacterial inoculation. *Functional Plant Biology, 43*(7), 632–642. DOI: 10.1071/FP15265 PMID: 32480492

Zhang, W., Wei, J., Guo, L., Fang, H., Liu, X., Liang, K., Niu, W., Liu, F., & Siddique, K. H. M. (2023). Effects of Two Biochar Types on Mitigating Drought and Salt Stress in Tomato Seedlings. *Agronomy, 13*(4), 1039. DOI: 10.3390/agronomy13041039

Zhu, J. K. (2016). Abiotic Stress Signaling and Responses in Plants. *Cell, 167*(2), 313–324. DOI: 10.1016/j.cell.2016.08.029 PMID: 27716505

Chapter 17
Prospects of Algae-Based Biofuel Production for the Development of a Carbon-Neutral Economy

Maria Vetrova
http://orcid.org/0000-0001-6860-727X
Saint Petersburg State University, Russia

ABSTRACT

In the context of aggravation of the climate agenda and continuing instability in the global energy market, the diversification of the energy structure in favour of innovative and low-carbon technologies and resources, among which bioenergy occupies a special place, is becoming an urgent area of the countries' strategies. According to the Global Bioenergy Statistics Report 2024, biofuels, including bioethanol, biodiesel, biochar, and biogas, account for about 9% of global energy production, which is characterised by low greenhouse gas emissions and high energy efficiency. Taking into account the current trends in the development of bioenergy technologies and its potential to address the climate agenda, this study analyses the possibilities of implementing biocarbon projects and provides an economic assessment of alternative options for biomass production with a focus on the production of third-generation biofuels based on seaweed. The object of special attention is the systematisation of technologies and infrastructural conditions for scaling up third-generation bioenergy, including for diversification of the global energy balance and development of a carbon-neutral economy. As one of the key results of the study, recommendations on the formation of institutional and infrastructural environment for achieving commercial feasibility and effective scaling of biocarbon projects have been developed.

DOI: 10.4018/979-8-3373-1524-9.ch017

INTRODUCTION

Taking into account population growth, global economic development and large-scale digitalisation, energy consumption may increase by 40-50% over the next 15 years (Asselt, 2021). Fossil fuels currently comprise approximately 80% of the global energy mix. Maintaining this composition will inevitably lead to a rise in greenhouse gas (GHG) emissions (UN, 2024). The growth of GHG emissions in 50 years could lead to a climate catastrophe that would leave 3.3 billion people in the zone of annual natural disasters and 14% of biological species at risk of extinction (IPCC, 2023). To mitigate dependence on fossil fuels and associated climate risks, nations are developing and implementing a comprehensive suite of measures and policy instruments. Particular focus is placed on carbon-neutral technologies, renewable energy sources, carbon capture, utilization, and storage (CCUS), and enhancing carbon sequestration by natural ecosystems (Vetrova et al., 2024).

Biomass, utilized for the production of biofuels, constitutes one of the renewable energy sources. Its application for developing low-carbon energy systems gained significant momentum in the early 20th century (Nedbaylov, 2023). The development of bioenergy has passed through a number of stages, each of which is based on a different type of biomass. While the first generation biofuel production was based on sugarcane and corn, which was in conflict with the food sector, raising food prices, and the second generation was based on wood, which is directly associated with environmental risks due to deforestation as a key source of raw materials, the third generation of biofuels as a source of biomass is based on seaweed, which has a number of advantages compared to previous types of biomass: rapid growth, low operating costs, essentially The main attention of researchers and business structures is directed towards the third generation of biofuels based on micro and macro-algae, as evidenced by the growth of projects with a total investment of more than several billion dollars, as well as scientific publications, which numbered more than 500 in the last decade (Pugazhendi et al., 2021). While the biofuels market size was estimated at 1.89 million barrels of oil equivalent per day in 2024, it is projected to grow by 30% by 2030 (Mordor Intelligence, 2024).

At the same time, today there are no climate projects on CO2 sequestration by algae with simultaneous production of biofuels from them, so the purpose of this study is to assess the economic efficiency and infrastructural conditions for the implementation of bioclimatic projects based on marine ecosystems. To achieve the set goal, the study set and solved the following tasks: firstly, prioritisation of algae-based biofuel production technologies; secondly, investment modelling of the project on biofuel production from algae and their storm emissions; thirdly, sensitivity analysis of the bioclimatic project and determination of critical values to maintain its commercial efficiency.

MATERIALS AND METHODS

Data Sources and Literature Analysis

The search for relevant literature was carried out as of 15 January 2025 in Web of Science and Scopus databases as the most rigorous in terms of quality and large databases in terms of scientific publications (Baas et al., 2020), also the Russian database of scientific publications E-Library was used for the analysis of research conducted in the territory of the Russian Federation. The first stage of literature selection is based on the search of sources by keywords in various combinations using the logical operators 'AND' and 'OR', among the keywords in Russian and English were selected 'biofuel', "macroalgae", 'third generation of biofuel', 'assessment of the economic efficiency of the project', 'climate project based on marine ecosystems', 'carbon sequestration by algae', 'storm emissions', 'storm emissions recycling', 'storm emissions forecasting', 'macroalgae biomass growth forecasting', 'biofuel production technologies', 'algae cultivation technologies'. This resulted in 510 studies published between 2010 and 2025. The second stage of selection of relevant studies was performed by analysing the titles of articles and abstracts, leaving a total of 55 key sources. The analysis of scientific studies was complemented by statistical and practical data on biofuel production and carbon sequestration projects by algae published on STATISTA, CORDIS, CARBONPLATFORM, National Academy of Sciences (Ocean-Based Carbon Dioxide Removal), IEA platforms.

Numerous studies have analysed the potential of using algae as biomass for fuel production and climate projects (Godvin Sharmila et al., 2021; Konda et al., 2015; Vincevica-Gaile et al., 2022), technological and infrastructural capabilities of bioclimatic projects (Firman et al., 2023), and ecosystem effects in case of storm emissions collection (Rudovica et al., 2021; Zheng et al., 2019). At the same time, the literature analysis has confirmed conflicting results on the commercial feasibility of projects using algae as a biofuel base and CO2 sequestration (Hemavathy et al., 2025). The analysis of the literature revealed the absence of empirical studies on biofuel production projects from algal biomass and their storm emissions with simultaneous implementation of climate projects of carbon sequestration by algae with assessment of their economic efficiency for the BRICS countries.

Assessment of Economic Efficiency of Climate Project Implementation

Climate projects on carbon sequestration by marine ecosystems, as well as biofuel production projects are among investment projects that may have significant ecosystem effects. Therefore, the cost-benefit method is used to evaluate and prioritise among alternative project options. If additional social or environmental effects arise during project implementation, the cost-benefit method is supplemented by a qualitative analysis, including to justify the attraction of government funding. Assessment of economic efficiency of alternative investment projects is carried out by the classical method of financial modelling of net present value (NPV), which is estimated by the formula:

$$NPV = -I + \sum_{t=1}^{N} \frac{CF_t}{(1+i)^t}$$

I - investment investments (capital costs);

t - project implementation period;

CF - operating cash flows, including liquidation costs;

i - discount rate (cost of capital).

Biofuel and carbon management projects utilizing cultivated algae or beach-cast macroalgae are subject to significant uncertainty. This stems largely from the inherent difficulty in forecasting algal biomass production and shoreline deposition volumes, processes highly sensitive to environmental variables such as marine heat-waves, ultraviolet radiation fluctuations, ocean acidification, and deoxygenation, etc. (Leandro et al., 2020). The efficiency of project implementation is also affected by market conditions, which include changes in the price of biofuels and carbon units (Statista, 2024). All this necessitates the application of sensitivity analysis to identify the most significant variables affecting the commercial viability of project implementation, and to assess the critical values of these variables.

RESULTS AND DISCUSSION

Algal biomass processing can be carried out by different technological methods, resulting in different types of biofuels, heat and energy. All technological processes can be categorised into two groups: biochemical and thermochemical, with some processes requiring pre-drying of algae, which increases operational costs (Table 1).

Table 1. Types of biofuels and their production technologies

Group	Production technology	Description of technology	Biofuel type	Limitations
Thermochemical processes	Transesterification	The process of displacing an alcohol from an ester of another alcohol using strong catalysts: alkali or acid. The process uses dry biomass.	Biodiesel	Primary fuel contains methanol, glycerides and other harmful substances, which requires additional purification process
	Liquefaction	Conversion of wet biomass into liquid fuels using heat and a catalyst. The method involves exposure to low temperatures and high pressure, resulting in a mixture of volatile organic acids and hydrocarbons.	Bio-oil	Undesirable by-products CO2, CH4
	Hydrothermal conversion	To convert wet (dry) biomass into biofuels or biochemicals, water at high pressure (2 to 25 MPa) and high temperature (180 to 375 °C) is used, the process takes place in a special reactor and involves hydrothermal carbonisation (HTC), hydrothermal liquefaction (HTL) and hydrothermal gasification (HTG).	Bio-oil, hydro-coal and synthesis gas (bio-hydrogen)	Selection of catalyst is important to increase the yield of useful product
	Gasification	The process of heating biomass under conditions of limited oxygen access to convert dry algae into gaseous fuel suitable for further use in energy production by breaking down the carbon, hydrogen and oxygen contained in the biomass into basic volatile compounds	Biogas	High thermal power required, undesirable by-products CO2, CH4, tar, coke
	Pyrolysis	Thermal decomposition of dry algal biomass into various pyrolytic products. Dry biomass is separated into liquid and gaseous fractions in a special furnace by converting long-chain molecules into short-chain molecules.	Bio-oil, bio-hydrogen, synthesis gas	High thermal power and two-stage pre-drying required
	Direct combustion	Use of furnaces, boilers, steam turbines to burn the pre-dried biomass and convert the chemical energy into heat, mechanical energy or electricity.	Biogas	Emissions of carbon dioxide and nitrogen oxide, particulate matter and ash are reduced.

continued on following page

Table 1. Continued

Group	Production technology	Description of technology	Biofuel type	Limitations
Biochemical processes	Dark fermentation	Enzymatic conversion of wet biomass to biohydrogen through the application of different groups of bacteria, and a series of biochemical reactions	Bio-hydrogen	Uneconomical technology with difficult to predict yield of useful substance
	Anaerobic digestion	The process by which bacteria break down organic matter without access to oxygen from wet biomass.	Biogas, biomethane, digestate, biological waste	Proteins in biomass release ammonia during decomposition, high levels of which can slow down the rate of chemical reactions.
	Fermentation	The process of producing bioethanol from wet biomass by breaking down saccharides, converting the sugars into ethanol and purifying the by-products.	Bioethanol	Sugar- and starch-rich biomass is needed

Source: compiled by the author on the basis of: Chubarenko et al. (2021) and Vikash Babu et al. (2014).

Biochemical processes are associated with risks of instability of processes of obtaining the maximum yield of useful substances, respectively it is possible to reduce economic efficiency in case of deviation of actual values from the target values of biofuel yield from algae. Practically all thermochemical processes require pre-drying of algal biomass, with the exception of hydrothermal processing. Preliminary analysis shows that hydrothermal digestion is the most promising and priority technology for algae-based biofuel production (El-Araby, 2024). At the same time, the products obtained as a result of hydrothermal processing can have different states, due to the possibility of using different approaches in the production process, which in turn can cover the needs of different groups of consumers of finished products. Thus, hydrothermal carbonisation at temperatures from 180 to 260 °C and pressures of 2-6 MPa produces solid biochar, hydrothermal liquefaction at temperatures from 280 to 370 °C and pressures of 10-25 MPa produces bio-oil, and hydrothermal gasification at temperatures of 400-700 °C and pressures of about 25 MPa produces bio-hydrogen. Thus, for further economic analysis we will consider this technology as a priority. As for the product obtained, the production of hydrocoal and bio-oil is the most feasible, due to the technological readiness to scale production and available storage and transport infrastructure.

Obtaining biomass for biofuel production is possible in different ways, which have their own limitations, and not every method can be attributed to the technology of the climate project (Table 2).

Table 2. Methods to obtain biomass for biofuel production

Biomass production method for project implementation	Additional climate effect	Limitations and risks
Stormwater harvesting for conversion into biofuel	Avoidance of GHG emissions from decomposition of storm emissions	Unstable biomass production;
Wild harvesting of algae	-	Possible low quality biomass for biofuel production.
Cultivation and harvesting of algae	Uptake of GHGs by growing algae during photosynthesis	Possible low quality of biomass for biofuel production

Source: Compiled by the author

There is a change in marine ecosystems under the influence of climate, which in turn affects algal growth and migration (Vigouroux & Destouni, 2022). The average volume of storm emissions on coastal zones worldwide varies significantly depending on currents, waves, tides: from 0.1 to 325 kg per m-1 (Hyndes et al., 2021). Russian studies on the volume of storm emissions are presented in several papers for the areas of the Azov, White, Barents, Baltic Seas, Peter the Great and Aniva Bays (Table 3).

Table 3. Storm emissions on the territory of the Russian Federation

Territory	Emissions	Data source
Sakhalin, Aniva Bay	8.5 thousand tonnes - minimum estimate	Zavyalov R., Latkovskaya E., Repina M., 2022
Vladivostok, Peter the Great Bay	1-5 thousand tonnes	Zhiltsova L.V., Prikhodchenko A.V., 2019.
Sea of Azov	Up to 100 thousand tonnes at the end of XX century, no measurements in recent years	Erokhin V.E., 2020
Black Sea	110-1096 tonnes	Turkulova V.N., Solodovnikov A.A., Kryuchkov V.G., Bityutskaya, O.E., 2008
White Sea and Barents Sea	100 thousand tonnes, including 324 tonnes (Chupa Bay)	Andreev V.P., Zakrevsky Y.N., Martynova E.S., Plakhotskaya J.V., 2021

Source: Compiled by the author

NPV modelling is carried out taking into account the possibility of combining biomass production methods up to the maximum production capacity, which is set at the level of 10 thousand tonnes of biomass processed per year. The key assumptions for calculating the economic efficiency of the project are presented in Table 4.

Table 4. Key assumptions for investment modelling of the project on biofuel production from algae and their storm emissions

Indicator (Category)	**Value**	**Description / Unit of Measurement**
A. Climate & Carbon Parameters		
CO2 emissions from decomposition of beach-cast biomass	0.27 t CO_2 / t biomass	Avoided emissions: The amount of CO_2 that would be released if 1 tonne of beach-cast algae decomposed naturally and is now prevented by its collection.
CO2 uptake by farmed algae	0.05 t CO_2 / t biomass	Carbon sequestration: The amount of CO_2 absorbed from the atmosphere during the growth phase of 1 tonne of cultivated algae.
Carbon credit price (CO2e)	14 USD / t CO_2	Market price for one tonne of CO_2 equivalent (CO2e) on voluntary carbon markets. Represents potential revenue from certified emission reductions.
B. Process Yields by Feedstock Type		
From Beach-Cast Biomass (Storm Emissions):		
Biochar yield	20%	Mass yield: Percentage of input wet biomass converted into solid biochar via thermochemical processing.
Bio-oil yield	40%	Mass yield: Percentage of input wet biomass converted into liquid bio-oil via thermochemical processing.
From Wild-Harvested Algae:		
Biochar yield	30%	Mass yield: Percentage of input wild algae biomass converted into biochar.
Bio-oil yield	50%	Mass yield: Percentage of input wild algae biomass converted into bio-oil.
From Cultivated Algae:		
Biochar yield	40%	Mass yield: Percentage of input cultivated algae biomass converted into biochar.
Bio-oil yield	80%	Mass yield: Percentage of input cultivated algae biomass converted into bio-oil.
C. Product Prices & Biomass Volumes		
Biochar market price	640 USD / t	Expected selling price per tonne of biochar.
Bio-oil market price	960 USD / t	Expected selling price per tonne of bio-oil.
Available beach-cast biomass (storm emissions)	8.5 thousand t/year	Estimated annual volume of algal biomass available for collection from shorelines.
Available wild algae biomass	1.5 – 10 thousand t/year	Estimated range for annual sustainable harvest of wild macroalgae.
Available cultivated algae biomass	0.5 thousand t/year	Planned annual production volume from aquaculture.
D. Financial & Tax Assumptions		

continued on following page

Table 4. Continued

Indicator (Category)	Value	Description / Unit of Measurement
Discount rate	20%	Rate used to calculate the present value of future cash flows, reflecting project risk and cost of capital.
Income tax rate	25%	Corporate income tax rate applied to project profits.
Property tax rate	2.2%	Annual tax rate on the value of capital assets (plant, equipment).
Value Added Tax (VAT) rate	20%	Tax rate applied to the sale of goods (biofuels, carbon credits).
E. Capital Expenditures (CAPEX)		
Plant construction (10 kt capacity)	153,000 USD	Base cost for building the processing facility.
Plant infrastructure (10 kt capacity)	2 million USD	Cost for supporting infrastructure (grid connection, water, etc.).
Hydrothermal processing equipment (10 kt)	1.3 million USD	Cost of core technology unit (e.g., HTL reactor) for biofuel conversion.
Facility design & engineering (10 kt)	346,000 USD	Costs for project design, planning, and engineering works.
Cultivation & harvesting system (500 t)	640,000 USD	Capital cost for algae farms and harvesting equipment for cultivated biomass.
Wild algae collection & pre-processing (10 kt)	1.3 million USD	Capital cost for vessels and equipment to harvest and handle wild algae.
Beach-cast biomass collection equipment	226 USD / t biomass	Specific capital cost per tonne of annual capacity for collecting shoreline biomass.
F. Operational Expenditures (OPEX)		
Zoospores for cultivation	23 USD / t biomass	Cost of algal seed material (spores) per tonne of cultivated algae produced.
Energy for biochar production	300 kWh / t biomass	Electricity consumption per tonne of biomass processed into biochar.
Energy for bio-oil production	600 kWh / t biomass	Electricity consumption per tonne of biomass processed into bio-oil.
Electricity price	0.09 USD / kWh	Cost of purchased electricity per kilowatt-hour.
Material costs	12.8 USD / t biomass	Cost of consumables (chemicals, catalysts, etc.) per tonne of processed biomass.
Labor for beach-cast biomass collection	7 USD / t biomass	Labor cost, including payroll taxes, per tonne of collected shoreline biomass.
Labor for algae cultivation & wild harvest	17 USD / t biomass	Labor cost, including payroll taxes, per tonne of cultivated or wild-harvested algae.
Labor for biofuel production	100 USD / t biomass	Labor cost, including payroll taxes, per tonne of biomass processed into biofuels.
Other operating costs	5%	Percentage of other operational expenses (maintenance, administration) relative to main OPEX.
G. Climate Project Costs		

continued on following page

Table 4. Continued

Indicator (Category)	Value	Description / Unit of Measurement
Climate project validation	6,400 USD	One-time upfront cost for an independent third-party assessment of the project design against a carbon standard.
Climate project verification	15 – 150 thousand USD	Recurring cost for periodic audits to certify the actual amount of emissions reduced/sequestered. Range depends on project scale and verifier.
H. Project Scope		
Project lifetime	10 years	The economic modeling horizon for the NPV calculation.
Max. processing capacity	10 thousand t/year	The designed maximum annual throughput of the integrated biorefinery for all feedstock types combined.

Source: compiled by the author on the basis of Association of CIS NPCs (2025), Federal Service for Intellectual Property (2011), Lucas et al. (2019), Novosti (2025), Procurement (2024), Regional Energy Commission of Sakhalinsk Oblast (2023).

The most prioritised method from an economic point of view is the collection of wild algae (Laminaria) and its production into bio-oil, as the NPV for this type of project is $9.47 million with a payback period of 3 years. In this case, the municipalities will continue to bear the costs of stormwater collection to clean up recreational areas, but if these funds are used to pay off the operating and investment costs of stormwater collection for further biofuel production, and the production costs of bio-oil are left to the biofuel producer, the projects become roughly equivalent in terms of economic viability. At the same time, the project on collection and processing of storm emissions will have a significant ecosystem effect.

Due to the fact that the project has a high variability of variables, it is particularly important to analyse the sensitivity of NPV as a key parameter of economic feasibility of the project, as well as to determine the critical values of variables, beyond which the project implementation will not be economically viable (Table 5).

Table 5. Sensitivity analysis of the project 'Storm water discharges and wild fisheries bio-oil'

Indicator	Sensitivity, %	Critical value
Price of UU, USD.	0,05%	385 USD
Biofuel price, USD.	6,51%	790 USD
Annual volume of storm emissions, tonnes	1,75%	6 850 tonnes
Bio-oil yield from processing storm emissions, % of t biomass	1,36%	35%
Increase in operating costs, %	-3,12%	29%
Investment, USD million.	-2,70%	8,3 USD
Discount rate, %	-1,92%	35%

Source: Compiled by the author

The most significant variable affecting the feasibility of the project is the price of biofuel. However, the price of biofuel has a steady upward trend due to the fourth energy transition (Statista, 2025). Other variables are less significant, but at the same time their control is necessary to ensure the feasibility of the project and its payback. This project, given the economic evaluation, significant environmental and social effects, can be recommended for implementation, with government co-financing for storm emissions collection to ensure improved economic efficiency.

CONCLUSION

Algae hold great potential as a sustainable feedstock for energy and other products due to their rapid growth and versatility. However, technological and economic barriers, such as low product yields and high production costs, hinder their commercial viability. Currently, using algae and beach-cast biomass for climate projects is not economically feasible on its own, as carbon credit prices do not cover costs. Such projects can only be viable as a complementary activity to primary production, like biofuel processing. To scale up, supportive policies, public-private partnerships, and further research to reduce costs and improve technology are essential. Public co-financing for processing beach-cast algae could enhance cost-effectiveness while leveraging existing municipal budgets for beach cleaning, thereby improving the feasibility of related climate initiatives.

REFERENCES

Andreev, V.P., Zakrevsky, Y.N., Martynova, E.S., & Plakhotskaya, J.V. (2021). Heavy metals and arsenic in algae of storm emissions from the coastal zone of Severny Island of the Novaya Zemlya Archipelago. *Medico-Ecological and Medico-Social Issues of Health Protection*, *4*(7), 40-48. DOI: 10.22328/2413-5747-2021-7-4-40-48

Asselt, H. (2021). Governing fossil fuel production in the age of climate disruption: Towards an international law of 'leaving it in the ground'. *Earth System Governance*, *9*, 100118. DOI: 10.1016/j.esg.2021.100118

Association of CIS NPCs. (2025). *Tariffs for validation and verification of climate projects*. https://vv.ciscenter.org/

Baas, J., Schotten, M., Plume, A. M., Cote, G., & Karimi, R. (2020). Scopus as a curated, high-quality bibliometric data source for academic research in quantitative science studies. *Quantitative Science Studies*, *1*(1), 377–386. DOI: 10.1162/qss_a_00019

Chubarenko, B., Schubert, H., & Woelfel, J. (2021). Case studies for innovative solutions of beach wrack use. *Rostocker Meeresbiologische Beiträge Heft*, *31*, 124–202.

El-Araby, R. (2024). Biofuel production: Exploring renewable energy solutions for a greener future. *Biotechnology for Biofuels and Bioproducts*, *17*(1), 129. DOI: 10.1186/s13068-024-02571-9

Erokhin, V.E. (2020). On the possibility of using marine plants of the Azov-Black Sea basin for organic farming. *Voprosy sovremennoi algologii*, *1*(22), 1-7. DOI: 10.33624/2311-0147-2020-1(22)-1-7

Federal Service for Intellectual Property. (2011). *Proposal for survival of stormwater discharges from sea habitats*. https://patentimages.storage.googleapis.com/e4/46/bf/43701254a1f51f/RU2473204C1.pdf

Firman, Z., Tri, H., Jeverson, R., Hafiizh, P., Muhammad, S., & Hada, P. (2023). Economic feasibility analysis of macroalgae farming-based carbon dioxide removal. *Journal of Sustainability Science and Management*, *8*(18), 137–145. DOI: 10.46754/jssm.2023.08.008

Godvin Sharmila, V., Dinesh Kumar, M., Pugazhendi, A., Bajhaiya, A., Gugulothu, P., & Rajesh Banu, J. (2021). Biofuel Production from Macroalgae: Present Scenario and Future Scope. *Bioengineered*, *12*(2), 9216–9238. DOI: 10.1080/21655979.2021.1996019

Hemavathy, R. V., Ragini, Y. P., Shruthi, S., Ranjani, S., Subhashini, S., & Thamarai, P. (2025). Biofuel production from marine macroalgae: Pathways, technologies, and sustainable energy solutions. *Industrial Crops and Products, 224*, 120282. Advance online publication. DOI: 10.1016/j.indcrop.2024.120282

Hyndes, G., Berdan, E., Duarte, C., Dugan, J., Emery, K., Hambäck, P., Henderson, C., Hubbard, D., Lastra, M., Mateo, M., Olds, A., & Schlacher, T. (2021). The role of inputs of marine wrack and carrion in sandy-beach ecosystems: A global review. *Biological Reviews of the Cambridge Philosophical Society, 97*(6), 2127–2161. DOI: 10.1111/brv.12886

IPCC. (2023). *AR6 Synthesis Report: Climate Change 2023.* https://www.ipcc.ch/report/sixth-assessment-report-cycle/

Konda, N. V. S. N. M., Singh, S., Simmons, B. A., & Klein-Marcuschamer, D. (2015). An Investigation on the Economic Feasibility of Macroalgae as a Potential Feedstock for Biorefineries. *BioEnergy Research, 8*(3), 1046–1056. DOI: 10.1007/s12155-015-9594-1

Leandro, A., Pereira, L., & Gonçalves, A. M. M. (2020). Diverse Applications of Marine Macroalgae. *Marine Drugs, 18*(1), 17. DOI: 10.3390/md18010017

Lucas, K., Capron, M., & Stewart, J. (2019) DE-AR0000916 AdjustaDepth Ph1 FinalReportTeam. ARPA DOI:DOI: 10.13140/RG.2.2.21140.32643

Mordor Intelligence. (2024). *Analysing Biofuels Market Size and Share - Growth Trends and Forecasts (2024-2029).* https://www.mordorintelligence.com/ru/industry-reports/biofuels-market

Nedbaylov, M. S. (2023). Bioenergy: Main market development trends and prospects for Russia. *Creative Economy., 3*(17), 869–882. DOI: 10.18334/ce.17.3.117454

Novosti, R. I. A. (2025). *A seaweed processing plant will open in Karelia this year.* https://ria.ru/20250117/zavod-1994211751.html?ysclid=m7jjboxk8p817378950

Procurement, E. I. S. (2024). *Construction of a production complex for primary and (or) subsequent (industrial) processing, storage of algae products in the Republic of Karelia: Installation of on-site electrical networks and outdoor lighting system, owned by NPO BIT Ltd.* https://zakupki.gov.ru/epz/order/notice/ea20/view/documents.html?regNumber=2300700043924000008

Pugazhendi, A., Bajhaiya, A. K., & Gugulothu, P. (2021). Biofuel production from Macroalgae: Present scenario and future scope. *Bioengineered, 12*(2), 9216–9238. DOI: 10.1080/21655979.2021.1996019

Regional Energy Commission of Sakhalinsk Oblast. (2023). *On Setting Tariffs of Limited Liability Company Sakhalin East.* https://docs.cntd.ru/document/406920211?ysclid=m7jk2grrwl353827782

Rudovica, V., Rotter, A., Gaudêncio, S. P., Novoveská, L., Akgül, F., Akslen-Hoel, L. K., Alexandrino, D. A. M., Anne, O., Arbidans, L., Atanassova, M., Bełdowska, M., Bełdowski, J., Bhatnagar, A., Bikovens, O., Bisters, V., Carvalho, M. F., Catalá, T. S., Dubnika, A., Erdoğan, A., & Burlakovs, J. (2021). Valorization of Marine Waste: Use of Industrial By-Products and Beach Wrack Towards the Production of High Added-Value Products. *Frontiers in Marine Science*, *8*, 723333. DOI: 10.3389/fmars.2021.723333

Statista. (2024). *Total bioenergy capacity in Russia from 2010 to 2023.* https://www.statista.com/statistics/1154535/total-bioenergy-capacity-in-russia/

Statista. (2025). *Market value of biofuels production worldwide from 2021 to 2023, with a forecast until 2030.* https://translated.turbopages.org/proxy_u/en-ru.ru.147d6c5e-67c0a0f2-657961a2-74722d776562/https/www.statista.com/statistics/217179/global-biofuels-market-size/

Turkulova, V. N., Solodovnikov, A. A., Kryuchkov, V. G., & Bityutskaya, O. E. (2008). Results and Prospects of YugNIRO Research in Mariculture. *Main results of integrated research in the Azov-Black Sea basin and the World Ocean.*

UN. (2024). *Production Gap Report 2023.* https://www.unep.org/ru/resources/production-gap-report-2023

Vetrova, M., Pakhomova, N., & Richter, K. (2024). Strategies for the development of the Russian energy sector in the context of climate challenges and geopolitical instability. *Vestnik of St. Petersburg University. Economics*, *39*(4), 439–469. DOI: 10.21638/spbu05.2023.401

Vigouroux, G., & Destouni, G. (2022). Gap identification in coastal eutrophication research – Scoping review for the Baltic system case. *The Science of the Total Environment*, *839*, 156240. Advance online publication. DOI: 10.1016/j.scitotenv.2022.156240

Vikash Babu, P. D., Ashish Tapliyal, P. D., & Giriyesh Kumar Patel, P. D. (2014). *Renewable Energy: Biofuel Production.* Scrivener Publishing LLC.

Vincevica-Gaile, Z., Sachpazidou, V., Bisters, V., Klavins, M., Anne, O., Grinfelde, I., Hanc, E., Hogland, W., Ibrahim, M. A., Jani, Y., Kriipsalu, M., Pal, D., Pehme, K.-M., Shanskiy, M., Saaremäe, E., Pilecka-Ulcugaceva, J., Celms, A., Rudovica, V., Hendroko Setyobudi, R., & Burlakovs, J. (2022). Applying Macroalgal Biomass as an Energy Source: Utility of the Baltic Sea Beach Wrack for Thermochemical Conversion. *Sustainability (Basel), 14*(21), 13712. DOI: 10.3390/su142113712

Zavyalov, R., Latkovskaya, E., & Repina, M. (2022) Estimation of possible storm discharge volumes of macrophytes on the coast of Aniva Bay (Sea of Okhotsk). Aniva Bay (Sea of Okhotsk). *Issues of modern scientific research: collection of articles of XVII International Scientific and Practical Conference,* 880-889.

Zheng, Y., Jin, R., Zhang, X., Wang, Q., & Wu, J. (2019). The considerable environmental benefits of seaweed aquaculture in China. *Stochastic Environmental Research and Risk Assessment, 33*(4-6), 1203–1221. DOI: 10.1007/s00477-019-01685-z

Zhiltsova, L. V., & Prikhodchenko, A. V. (2023) Rational approach to the collection of storm emissions of Anfeltia in Primorye. *Actual problems of biodiversity and nature management: proceedings of the 2nd National Scientific and Practical Conference, dedicated to the 20th anniversary of the Department of Marine Ecology FGBOU VO 'KSMTU'*, 300-305.

Compilation of References

Abbasi, G. H., Akhtar, J., Ahmad, R., Jamil, M., Anwar-ul-Haq, M., Ali, S., & Ijaz, M. (2015). Potassium application mitigates salt stress differentially at different growth stages in tolerant and sensitive maize hybrids. *Plant Growth Regulation, 76*(1), 111–125. DOI: 10.1007/s10725-015-0050-1

Abdeen, S. A., Rizk, A. H., & El-Hamamsy, S. M. (2015). Effect of salinity stress on growth and macronutrients uptake of some Egyptian barley landraces ((Hordeum vulgare L.). *J.Soil Sci. and Agric. Eng. Mansoura Univ*, *6*(11), 1355–1364.

Abedin, M. R., & Barua, S. (2021). Isolation and purification of glycoglycerolipids to induce apoptosis in breast cancer cells. *Scientific Reports*, *11*(1), 1–13. DOI: 10.1038/s41598-020-80484-x PMID: 33446783

Abideen, Z., Koyro, H. W., Huchzermeyer, B., Ahmed, M. Z., Zulfiqar, F., Egan, T., & Khan, M. A. (2020). Phragmites karka plants adopt different strategies to regulate photosynthesis and ion flux in saline and water deficit conditions. DOI: 10.1080/11263504.2020.1762783

Aboudala, S. I., Minkilabe, D., & Thiou, T. K. T. (2018). Impacts socio-économiques de l'exploitation minière sur les populations riveraines. Cas du gisement de calcaires de Tabligbo dans la préfecture de Yoto au Togo. *Revue ANYASA, 8,* 152-159. https://hal.science/hal-02265636v1

Abou-Zeid, H. M., & Moustafa, Y. (2014). *Physiological and Cytogenetic Responses of Wheat and Barley to Silver Nanopriming Treatment.*

Abrar, M. M., Saqib, M., Abbas, G., Atiq-Urrahman, M., Mustafa, A., Shah, S. A. A., Mehmood, K., Maitlo, A. A., ul-Hassan, M., Sun, N., & Xu, M. (2020). Evaluating the Contribution of Growth, Physiological, and Ionic Components Towards Salinity and Drought Stress Tolerance in Jatropha curcas. *Plants*, *9*(11), 1–18. DOI: 10.3390/plants9111574 PMID: 33203052

Abreu, R., Semedo-Lemsaddek, T., Cunha, E., Tavares, L., & Oliveira, M. (2023). Antimicrobial Drug Resistance in Poultry Production: Current Status and Innovative Strategies for Bacterial Control. *Microorganisms, 11*(4), 953. DOI: 10.3390/microorganisms11040953 PMID: 37110376

Acosta-Motos, J. R., Ortuño, M. F., Bernal-Vicente, A., Diaz-Vivancos, P., Sanchez-Blanco, M. J., & Hernandez, J. A. (2017). Plant Responses to Salt Stress: Adaptive Mechanisms. *Agronomy 2017, Vol. 7, Page 18, 7*(1), 18. DOI: 10.3390/agronomy7010018

Adele, Mariateresa, Giulio, & Mariateresa. (2024). Oxidative Stress: The Role of Antioxidant Phytochemicals in the Prevention and Treatment of Diseases. *Int J Mol Sci., 25*(6), 3264. DOI: 10.3390/ijms25063264

Adewumi, A. J., & Ogundele, O. D. (2024). Hidden hazards in urban soils: A meta-analysis review of global heavy metal contamination (2010-2022), sources and its Ecological and health consequences. *Sustainable Environment, 10*(1), 2293239. Advance online publication. DOI: 10.1080/27658511.2023.2293239

Adhikari, T., Kundu, S., Biswas, A. K., Tarafdar, J. C., & Subba Rao, A. (2015). Characterization of Zinc Oxide Nano Particles and Their Effect on Growth of Maize (*Zea Mays L.*) Plant. *Journal of Plant Nutrition, 38*(10), 1505–1515. DOI: 10.1080/01904167.2014.992536

Adhikary, S., Biswas, B., Chakraborty, D., Timsina, J., Pal, S., Chandra Tarafdar, J., Banerjee, S., Hossain, A., & Roy, S. (2022). Seed priming with selenium and zinc nanoparticles modifies germination, growth, and yield of direct-seeded rice (Oryza sativa L.). *Scientific Reports, 12*(1), 1–14. DOI: 10.1038/s41598-022-11307-4

Adolf, V. I., Jacobsen, S., & Shabala, S. (2012). Salt tolerance mechanisms in quinoa (Chenopodium quinoa Willd.). *Environmental and Experimental Botany, 92*, 43–54. DOI: 10.1016/j.envexpbot.2012.07.004

African Development Bank Group. (2019). Projet d'électrification rurale CIZO – Résumé CPR. African Development Bank. https://golnk.ru/ZB2E4

Agarie, S., Shimoda, T., Shimizu, Y., Baumann, K., Sunagawa, H., Kondo, A., Ueno, O., Nakahara, T., Nose, A., & Cushman, J. C. (2007). Salt tolerance, salt accumulation, and ionic homeostasis in an epidermal bladder-cell-less mutant of the common ice plant Mesembryanthemum crystallinum. *Journal of Experimental Botany, 58*(8), 1957–1967. DOI: 10.1093/jxb/erm057 PMID: 17452753

Agnieszka & Asakawa. (2020). *Terpenoids and Aromatic Compounds from Bryophytes and their Central Nervous System Activity. Current Organic Chemistry, 24*(1), 113 – 128. DOI: 10.2174/1385272824666200120143558

Ahmadi, J., Pour-Aboughadareh, A., Fabriki Ourang, S., Khalili, P., & Poczai, P. (2020). Unraveling salinity stress responses in ancestral and neglected wheat species at early growth stage: A baseline for utilization in future wheat improvement programs. *Physiology and Molecular Biology of Plants : an International Journal of Functional Plant Biology, 26*(3), 537–549. DOI: 10.1007/s12298-020-00768-4 PMID: 32205929

Akbari, M., Khodayari, M., Danesh, M., Davari, A., & Padash, H. (2020). A bibliometric study of sustainable technology research. *Cogent Business & Management, 7*(1), 1751906. Advance online publication. DOI: 10.1080/23311975.2020.1751906

Akramowskii, N. N. (1976) Fauna of the Armenian SSR. Molluscs. Academy of Sciences of the Armenian SSR.

Alalaf, A., & Fekry, W. (2020). Effect of saline stress on growth of fruit plants (review article). *Mağalla Zirā a Al-rāfidayn, 48*(4), 51–59. DOI: 10.33899/magrj.2020.128155.1070

Alekseev, A. N., & Dubinina, H. V. (2000). Abiotic parameters and diel and seasonal activity of Borrelia infected and uninfected *Ixodes persulcatus* (Acarina, Ixodidae). *Journal of Medical Entomology, 37*(1), 9–15. DOI: 10.1603/0022-2585-37.1.9 PMID: 15218900

Ali, R. M., & Abbas, H. M. (2003). Response of salt stressed barley seedlings to phenylurea. *Https://Pse.Agriculturejournals.* DOI: 10.17221/4107-PSE

Ali, A., Rashid, M. A., Huang, Q. Y., & Lei, C. L. (2016). Effect of UV-A radiation as an environmental stress on the development, longevity, and reproduction of the oriental armyworm, Mythimna separata (Lepidoptera: Noctuidae). *Environmental Science and Pollution Research International, 23*(17), 17002–17007. DOI: 10.1007/s11356-016-6865-0 PMID: 27206749

Ali, D., Ali, H., Alifiri, S., Alkahtani, S., Alkahtane, A. A., & Huasain, S. A. (2018). Detection of oxidative stress and DNA damage in freshwater snail Lymnea leuteola exposed to profenofos. *Frontiers of Environmental Science & Engineering, 12*(5), 1. DOI: 10.1007/s11783-018-1039-6

Ali, M. N., Ghosh, B., Gantait, S., & Chakraborty, S. (2014). Selection of rice genotypes for salinity tolerance through Morpho-Biochemical Assessment. *Rice Science, 21*(5), 288–298. DOI: 10.1016/S1672-6308(13)60189-4

Ali, M., Elhagwa, A., Elfaki, J., & Sulieman, M. (2017). Influence of the artisanal gold mining on soil contamination with heavy metals: A case study from Dar-Mali locality, North of Atbara, River Nile State, Sudan. *Eurasian Journal of Soil Science, 6*(1), 28–36. DOI: 10.18393/ejss.284261

Ali, Q., & Ashraf, M. (2011). Induction of Drought Tolerance in Maize (Zea mays L.) due to Exogenous Application of Trehalose: Growth, Photosynthesis, Water Relations and Oxidative Defence Mechanism. *Journal Agronomy & Crop Science, 197*(4), 258–271. DOI: 10.1111/j.1439-037X.2010.00463.x

Al-Maskri, A., Al-Kharusi, L., Al-Miqbali, H., & Mumtaz Khan, M. (2010). Effects of Salinity Stress on Growth of Lettuce (Lactuca sativa) under Closed-recycle Nutrient Film Technique. *Int. J. Agric. Biol, 12*, 377–380.

Al-Mushhin, A., Qari, S., Fakhr, M., Alnusairi, G., Alnusaire, T., ALrashidi, A., Latef, A., Ali, O., Khan, A., & Soliman, M. (2021). Exogenous Myo-Inositol Alleviates Salt Stress by Enhancing Antioxidants and Membrane Stability via the Upregulation of Stress Responsive Genes in Chenopodium quinoa L. *Plants, 10*(11), 2416. DOI: 10.3390/plants10112416 PMID: 34834781

Al-Salama, Y. (2022). Effect of Seed Priming with ZnO Nanoparticles and Saline Irrigation Water in Yield and Nutrients Uptake by Wheat Plants. *The 2nd International Laayoune Forum on Biosaline Agriculture*, 37. DOI: 10.3390/environsciproc2022016037

Alyousef, A.A. (2021) Antifungal Activity and Mechanism of Action of Different Parts of Myrtus communis Growing in Saudi Arabia against Candida Spp. *Hindawi Journal of Nanomaterials*. DOI: 10.1155/2021/3484125

An, D., Li, Q., Wang, X., Yang, H., & Guo, L. (2014). Characterization on hydrogen production performance of a newly isolated Clostridium beijerinckii YA001 using xylose. *International Journal of Hydrogen Energy, 39*(35), 19928–19936. DOI: 10.1016/j.ijhydene.2014.10.014

Andreev, V.P., Zakrevsky, Y.N., Martynova, E.S., & Plakhotskaya, J.V. (2021). Heavy metals and arsenic in algae of storm emissions from the coastal zone of Severny Island of the Novaya Zemlya Archipelago. *Medico-Ecological and Medico-Social Issues of Health Protection, 4*(7), 40-48. DOI: 10.22328/2413-5747-2021-7-4-40-48

Andrews, S. C., Berks, B. C., McClay, J., Ambler, A., Quail, M. A., Golby, P., & Guest, J. R. (1997). A 12-cistron *Escherichia coli* operon (*hyf*) encoding a putative proton-translocating formate hydrogenlyase system. *Microbiology, 143*(11), 3633–3647. DOI: 10.1099/00221287-143-11-3633 PMID: 9387241

Andriolo, J. L., Da Luz, G. L., Witter, M. H., Godoi, R. D. S., Barros, G. T., & Bortolotto, O. C. (2005). Growth and yield of lettuce plants under salinity. *Horticultura Brasileira, 23*(4), 931–934. DOI: 10.1590/S0102-05362005000400014

Angon, P. B., Islam, M. S., Kc, S., Das, A., Anjum, N., Poudel, A., & Suchi, S. A. (2024a). Sources, effects and present perspectives of heavy metals contamination: Soil, plants and human food chain. *Heliyon, 10*(7), e28357. DOI: 10.1016/j.heliyon.2024.e28357 PMID: 38590838

Anwar, T., Munwwar, F., Qureshi, H., Siddiqi, E. H., Hanif, A., Anwaar, S., Gul, S., Waheed, A., Alwahibi, M. S., & Kamal, A. (2023). Synergistic effect of biochar-based compounds from vegetable wastes and gibberellic acid on wheat growth under salinity stress. *Scientific Reports, 13*(1), 1–18. DOI: 10.1038/s41598-023-46487-0

Anwar, T., Shehzadi, A., Qureshi, H., Shah, M. N., Danish, S., Salmen, S. H., & Ansari, M. J. (2023). Alleviation of cadmium and drought stress in wheat by improving growth and chlorophyll contents amended with GA3 enriched deashed biochar. *Scientific Reports, 13*(1), 18503-. DOI: 10.1038/s41598-023-45670-7

Aria, M., & Cuccurullo, C. (2017). bibliometrix: An R-tool for comprehensive science mapping analysis. *Journal of Informetrics, 11*(4), 959–975. DOI: 10.1016/j.joi.2017.08.007

Arif, Y., Singh, P., Siddiqui, H., Bajguz, A., & Hayat, S. (2020). Salinity induced physiological and biochemical changes in plants: An omic approach towards salt stress tolerance. *Plant Physiology and Biochemistry, 156*, 64–77. DOI: 10.1016/j.plaphy.2020.08.042 PMID: 32906023

Arora, N. K., & Mishra, I. (2019). United Nations Sustainable Development Goals 2030 and environmental sustainability: race against time. *Environmental Sustainability, 2,* 339 342. DOI: 10.1007/s42398-019-00092-y

Asgaria, H. R., Cornelisb, W., & Dammec, P. V. (2012). *Salt stress effect on wheat (Triticum aestivum L.) growth and leaf ion concentrations.*

Ashraf, M., & Akram, N. A. (2009). Improving salinity tolerance of plants through conventional breeding and genetic engineering: An analytical comparison. *Biotechnology Advances, 27*(6), 744–752. DOI: 10.1016/j.biotechadv.2009.05.026 PMID: 19500659

Assad, R., Reshi, Z. A., Jan, S., & Rashid, I. (2017). Biology of amaranths. *Botanical Review, 83*(4), 382–436. DOI: 10.1007/s12229-017-9194-1

Asselt, H. (2021). Governing fossil fuel production in the age of climate disruption: Towards an international law of 'leaving it in the ground'. *Earth System Governance*, *9*, 100118. DOI: 10.1016/j.esg.2021.100118

Association of CIS NPCs. (2025). *Tariffs for validation and verification of climate projects*. https://vv.ciscenter.org/

Awogbemi, O., & Von Kallon, D. V. (2022). Pretreatment techniques for agricultural waste. *Case Studies in Chemical and Environmental Engineering*, *6*, 100229. DOI: 10.1016/j.cscee.2022.100229

Baas, J., Schotten, M., Plume, A. M., Cote, G., & Karimi, R. (2020). Scopus as a curated, high-quality bibliometric data source for academic research in quantitative science studies. *Quantitative Science Studies*, *1*(1), 377–386. DOI: 10.1162/qss_a_00019

Babayan, A., Vassilian, A., Poladyan, A., & Trchounian, K. (2024). Role of the *Escherichia coli* FocA and FocB formate channels in controlling proton/potassium fluxes and hydrogen production during osmotic stress in energy-limited, stationary phase fermenting cells. *Biochimie*, *221*, 91–98. DOI: 10.1016/j.biochi.2024.01.017 PMID: 38307245

Babenko, L. V., & Leonovich, S. A. (1985). Ecology. Behavior upon contact with the host and attachment sites. In Filippova, N. A. (Ed.), *Taiga tick Ixodes persulcatus Schulze (Acarina, Ixodidae). Morphology, systematics, ecology, medical significance* (pp. 249–251). Nauka.

Bajji, M., Lutts, S., & Kinet, J. (2001). Water deficit effects on solute contribution to osmotic adjustment as a function of leaf ageing in three durum wheat (Triticum durum Desf.) cultivars performing differently in arid conditions. *Plant Science*, *160*(4), 669–681. DOI: 10.1016/S0168-9452(00)00443-X PMID: 11448742

Baňas, M., Šofranková, L., Kurimský, J., Pavlík, M., Pikalík, M., Majláthová, V., Cimbala, R., Pipová, N., Wurfl, L., & Majláth, I. (2023). Interspecific differences in the behavioral response of ticks exposed to radiofrequency electromagnetic radiation. *Experimental & Applied Acarology*, *91*(3), 477–485. DOI: 10.1007/s10493-023-00847-7 PMID: 37819593

Bandaru, N., Patil, Y. P., Ekghara, S. D., & Bonthu, M. G. (2025). Exploring marine-derived compounds as potential anti-cancer agents: Mechanisms and therapeutic implications. *Cancer Pathogenesis and Therapy*. Advance online publication. DOI: 10.1016/j.cpt.2025.08.004 PMID: 41732213

Bargaz, A., Nassar, R. M. A., Rady, M. M., Gaballah, M. S., Thompson, S. M., Brestic, M., Schmidhalter, U., & Abdelhamid, M. T. (2016). Improved Salinity Tolerance by Phosphorus Fertilizer in Two Phaseolus vulgaris Recombinant Inbred Lines Contrasting in Their P-Efficiency. *Journal Agronomy & Crop Science*, *202*(6), 497–507. DOI: 10.1111/jac.12181

Bartels, D., & Sunkar, R. (2005). Drought and salt tolerance in plants. *Critical Reviews in Plant Sciences*, *24*(1), 23–58. DOI: 10.1080/07352680590910410

Barzegargolchini, B., Movafeghi, A., Dehestani, A., & Mehrabanjoubani, P. (2017). Morphological and anatomical changes in stems of Aeluropus littoralis under salt stress. *Journal of Plant Molecular Breeding*, *5*(1), 40–48. DOI: 10.22058/jpmb.2017.63945.1133

BBC News. (1999). Massacre Sierra Leone (Afrique). BBC News. http://news.bbc.co.uk/2/hi/africa/325227.stm

Bebenek, K., & Kunkel, T. A. (2004). DNA repair and replication. *Advances in Protein Chemistry*, *69*, 137–165. DOI: 10.1016/S0065-3233(04)69005-X PMID: 15588842

Bedouhene, A., Kelanemer, R., Medrouh, B., Kernif, T., Saidi, F., Tail, G., & Ziam, H. (2022). Seasonal Dynamics and Predilection Sites of Ticks (Acari: Ixodidae) Feeding in Cows in the Western Parts of the Djurdjura, Algeria. *Frontiers in Tropical Diseases*, *3*, 856179. DOI: 10.3389/fitd.2022.856179

Belozerov, V. N. (1981). Ecological rhythms in Ixodid ticks and their regulation. *Parassitologia*, *30*, 22–35.

Benning, C. (1998). Biosynthesis and function of the sulfolipid sulfoquinovosyl diacylglycerol. *Annual Review of Plant Biology*, *49*(1), 53–75. DOI: 10.1146/annurev.arplant.49.1.53 PMID: 15012227

Benning, C. (2008). A role for lipid trafficking in chloroplast biogenesis. *Progress in Lipid Research*, *47*(5), 381–389. DOI: 10.1016/j.plipres.2008.04.001 PMID: 18440317

Beyer, L., Doberenz, C., Falke, D., Hunger, D., Suppmann, B., & Sawers, R. G. (2013). Coordinating FocA and pyruvate formate-lyase synthesis in *Escherichia coli*: Preferential translocation of formate over other mixed-acid fermentation products. *Journal of Bacteriology*, *195*(7), 1428–1435. DOI: 10.1128/JB.02166-12 PMID: 23335413

Bhandari, K. P., Sapkota, D. R., Jamarkattel, M. K., Stillion, Q., & Collins, R. W. (2023). Zinc Oxide Nanoparticles—Solution-Based Synthesis and Characterizations. *Nanomaterials (Basel, Switzerland), 13*(11), 1795. DOI: 10.3390/nano13111795 PMID: 37299698

Bhattacharya, M., & Basak, T. (2017). A comprehensive analysis on the effect of shape on the microwave heating dynamics of food materials. *Innovative Food Science & Emerging Technologies*, *39*, 247–266. DOI: 10.1016/j.ifset.2016.12.002

BioLegend. (2024). *CFSE Cell Division Tracker Kit: Protocol* (Cat. No. 9396). https://www.biolegend.com/en-us/products/cfse-cell-division-tracker-kit-9396

BioLegend. (2024). *ELISA MAX™ Standard Set Human IL-2: Data sheet and protocol* (Cat. No. 431801). https://www.biolegend.com/Files/Images/media_assets/pro_detail/datasheets/431801_V02_JYH.pdf

Bisht, B., Kumar, V., Gururani, P., Tomar, M. S., Nanda, M., Vlaskin, M. S., Kumar, S., & Kurbatova, A. (2021). The potential of nuclear magnetic resonance (NMR) in metabolomics and lipidomics of microalgae-a review. *Archives of Biochemistry and Biophysics*, *710*, 108987. DOI: 10.1016/j.abb.2021.108987 PMID: 34260946

Bishwakarma, H., Kumar, N., Anand, M., Das, A. K., Singh, P. K., & Mishra, S. K. (2022). Magnetic Assisted Synthesis of ZnO Nanoparticle by Electrochemical Discharge Method. *Materials Today: Proceedings*, *56*, 857–861. DOI: 10.1016/j.matpr.2022.02.511

Blair, M. F., Kokabian, B., & Gude, V. G. (2014). Light and growth medium effect on Chlorella vulgaris biomass production. *Journal of Environmental Chemical Engineering*, *2*(1), 665–674. DOI: 10.1016/j.jece.2013.11.005

Blair, M. W., Riggins, C., & De La Rosa, A. P. B. (2023). Editorial: Amaranthus: naturally stress-resistant resources for improved agriculture and human health, volume II. *Frontiers in Plant Science*, *14*, 1329377. Advance online publication. DOI: 10.3389/fpls.2023.1329377 PMID: 38152149

Bligh, E. G., & Dyer, W. J. (1959). A rapid method of total lipid extraction and purification. *Canadian Journal of Biochemistry and Physiology*, *37*(8), 911–917. DOI: 10.1139/o59-099 PMID: 13671378

Bodner, G., Nakhforoosh, A., & Kaul, H. P. (2015). Management of crop water under drought: a review. *Agronomy for Sustainable Development, 35*(2), 401–442. DOI: 10.1007/s13593-015-0283-4

Bok, E., Jo, M., Lee, S., Lee, B. R., Kim, J., & Kim, H. J. (2019). Dietary Restriction and Neuroinflammation: A Potential Mechanistic Link. *International Journal of Molecular Sciences*, *20*(3), 464. DOI: 10.3390/ijms20030464 PMID: 30678217

Bolognesi, C. (2003). Genotoxicity of pesticides: A review of human biomonitoring studies. *Mutation Research/Reviews in Mutation Research*, *543*(3), 251–272. DOI: 10.1016/S1383-5742(03)00015-2 PMID: 12787816

Borsai, O., Hassan, M. A., Negru ier, C., Raigón, M. D., Boscaiu, M., Sestra , R. E., & Vicente, O. (2020). Responses to Salt Stress in Portulaca: Insight into Its Tolerance Mechanisms. *Plants*, *9*(12), 1660. DOI: 10.3390/plants9121660 PMID: 33260911

Borsani, O., Valpuesta, V., & Botella, M. A. (2003). Developing salt tolerant plants in a new century: A molecular biology approach. *Plant Cell, Tissue and Organ Culture*, *73*(2), 101–115. DOI: 10.1023/A:1022849200433

Bose, J., Rodrigo-Moreno, A., & Shabala, S. (2014). ROS homeostasis in halophytes in the context of salinity stress tolerance. *Journal of Experimental Botany*, *65*(5), 1241–1257. DOI: 10.1093/jxb/ert430 PMID: 24368505

Bourioug, M., Ezzaza, K., Bouabid, R., Alaoui-Mhamdi, M., Bungau, S., Bourgeade, P., Alaoui-Sossé, L., Alaoui-Sossé, B., & Aleya, L. (2020). Influence of Hydro- and Osmo-Priming on Sunflower Seeds to Break Dormancy and Improve Crop Performance under Water Stress. *Environmental Science and Pollution Research International*, *27*(12), 13215–13226. DOI: 10.1007/s11356-020-07893-3 PMID: 32016860

Bowman, E. J., Siebers, A., & Altendorf, K. (1988). Inhibitors of the ATP-dependent proton pump in yeast plasma membranes. *Biochemistry*, *27*(9), 3089–3095.

Bøyum, A. (1976). Isolation of lymphocytes, granulocytes and macrophages. *Scandinavian Journal of Immunology*, *5*(s5, Suppl. 5), 9–15. DOI: 10.1111/j.1365-3083.1976.tb03851.x PMID: 1052391

Brevik, E. C., Slaughter, L., Singh, B. R., Steffan, J. J., Collier, D., Barnhart, P., & Pereira, P. (2020). Soil and Human Health: Current Status and Future Needs. *Air, Soil and Water Research*, *13*. Advance online publication. DOI: 10.1177/1178622120934441

Briassoulis, D. (2023). Agricultural plastics as a potential threat to food security, health, and environment through soil pollution by microplastics: Problem definition. *The Science of the Total Environment*, *892*, 164533. DOI: 10.1016/j.scitotenv.2023.164533 PMID: 37285997

Bromham, L., & Bennett, T. H. (2014). Salt tolerance evolves more frequently in C4 grass lineages. *Journal of Evolutionary Biology*, *27*(3), 653–659. DOI: 10.1111/jeb.12320 PMID: 24494637

Buchanan, D., Martindale, W., Romeih, E., & Hebishy, E. (2023). Recent advances in whey processing and valorisation: Technological and environmental perspectives. *International Journal of Dairy Technology*, *76*(2), 291–312. DOI: 10.1111/1471-0307.12935

Buch, K., Peters, T., Nawroth, T., Sänger, M., Schmidberger, H., & Langguth, P. (2012). Determination of cell survival after irradiation via clonogenic assay versus multiple MTT Assay-A comparative study. *Radiation Oncology (London, England)*, *7*(1), 1. DOI: 10.1186/1748-717X-7-1 PMID: 22214341

Burke, E. J., Brown, S. J., & Christidis, N. (2006). Modeling the Recent Evolution of Global Drought and Projections for the Twenty-First Century with the Hadley Centre Climate Model. *Journal of Hydrometeorology*, *7*(5), 1113–1125. DOI: 10.1175/JHM544.1

Butter, A., Aliyev, K., Hillebrandt, K., Raschzok, N., Kluge, M., Seiffert, N., Tang, P., Napierala, H., Muhamma, A. I., Reutzel-Selke, A., Andreou, A., Pratschke, J., Sauer, I. M., & Struecker, B. (2017). Evolution of graft morphology and function after recellularization of decellularized rat livers. *Journal of Tissue Engineering and Regenerative Medicine*, *12*(2), e807–e816. DOI: 10.1002/term.2383 PMID: 27957815

Capozzoli, A., Curcio, C., D'Agostino, F., & Liseno, A. (2024). A Review of the Antenna Field Regions. *Electronics (Basel)*, *13*(11), 2194. DOI: 10.3390/electronics13112194

Carbone, M. C., Tatone, C., Monache, S. D., Marci, R., Caserta, D., Colonna, R., & Amicarelli, F. (2003). Antioxidant enzymatic defences in human follicular fluid: Characterization and age-dependent changes. *Molecular Human Reproduction*, *9*(11), 639–643. DOI: 10.1093/molehr/gag090 PMID: 14561807

Carter, J. L., Colmer, T. D., & Veneklaas, E. J. (2005). Variable tolerance of wetland tree species to combined salinity and waterlogging is related to regulation of ion uptake and production of organic solutes. *The New Phytologist*, *169*(1), 123–134. DOI: 10.1111/j.1469-8137.2005.01552.x PMID: 16390424

Casadaban, M. J. (1976). Transposition and fusion of the *lac* genes to selected promoters in *Escherichia coli* using bacteriophage lambda and Mu. *Journal of Molecular Biology*, *104*(3), 541–555. DOI: 10.1016/0022-2836(76)90119-4 PMID: 781293

Ceaser, R., Montané, D., Constantí, M., & Medina, F. (2024). Current progress on lignocellulosic bioethanol including a technological and economical perspective. *Environment, Development and Sustainability*, *27*(11), 25993–26038. Advance online publication. DOI: 10.1007/s10668-024-04792-2

Chandran, R., & Chandran MC, S. (2024). Green finance and sustainability: mapping research development through bibliometric analysis. *Discover Sustainability, 5*(1), 339-. DOI: 10.1007/s43621-024-00549-z

Chaudhuri, K., & Choudhuri, M. A. (1997). Effects of short-term NaCl stress on water relations and gas exchange of two jute species. DOI: 10.1023/A:1001013913773

Chaves, M. M., Flexas, J., & Pinheiro, C. (2008). Photosynthesis under drought and salt stress: Regulation mechanisms from whole plant to cell. *Annals of Botany, 103*(4), 551–560. DOI: 10.1093/aob/mcn125 PMID: 18662937

Chen, C., Wang, Y., Qian, Y., Zhao, X., & Wang, Q. (2015). The synergistic toxicity of the multiple chemical mixtures: Implications for risk assessment in the terrestrial environment. *Environment International, 77*, 95–105. DOI: 10.1016/j.envint.2015.01.014 PMID: 25667058

Chen, Z., Newman, I., Zhou, M., Mendham, N., Zhang, G., & Shabala, S. (2005). Screening plants for salt tolerance by measuring K+ flux: A case study for barley. *Plant, Cell & Environment, 28*(10), 1230–1246. DOI: 10.1111/j.1365-3040.2005.01364.x

Cherepanov, P. P., & Wackernagel, W. (1995). Gene disruption in *Escherichia coli*: TcR and KmR cassettes with the option of Flp-catalyzed excision of the antibiotic-resistance determinant. *Gene, 158*(1), 9–14. DOI: 10.1016/0378-1119(95)00193-A PMID: 7789817

Chhabra, A., Song, H. G., Grzelak, K. A., Polacheck, W. J., Fleming, H. E., Chen, C. S., & Bhatia, S. N. (2022). A vascularized model of the human liver mimics regenerative responses. *Proceedings of the National Academy of Sciences of the United States of America, 119*(28), e2115867119. DOI: 10.1073/pnas.2115867119 PMID: 35763565

Chi, N. T. L., Duc, P. A., Mathimani, T., & Pugazhendhi, A. (2019). Evaluating the potential of green alga Chlorella sp. for high biomass and lipid production in biodiesel viewpoint. *Biocatalysis and Agricultural Biotechnology, 17*, 184–188. DOI: 10.1016/j.bcab.2018.11.011

Choi, I. S., Lee, Y. G., Khanal, S. K., Park, B. J., & Bae, H.-J. (2015). A low-energy, cost-effective approach to fruit and citrus peel waste processing for bioethanol production. *Applied Energy, 140*, 65–74. DOI: 10.1016/j.apenergy.2014.11.070

Chourasia, K. N., Lal, M. K., Tiwari, R. K., Dev, D., Kardile, H. B., Patil, V. U., Kumar, A., Vanishree, G., Kumar, D., Bhardwaj, V., Meena, J. K., Mangal, V., Shelake, R. M., Kim, J., & Pramanik, D. (2021). Salinity stress in potato: Understanding physiological, biochemical and molecular responses. *Life (Chicago, Ill.)*, *11*(6), 545. DOI: 10.3390/life11060545 PMID: 34200706

Christophe, B. G., Hermann, P., Séraphin, A. Z., Agapit, D. W., Stanley, L., David, H. M., Françoise, A. K., & Armel, C. G. M. (2018). Effects of salinity stress on growth in relation to gas exchanges parameters and water status in amaranth (Amaranthus cruentus). *International Journal of Plant Physiology and Biochemistry*, *10*(3), 19–27. DOI: 10.5897/IJPPB2018.0280

Chubarenko, B., Schubert, H., & Woelfel, J. (2021). Case studies for innovative solutions of beach wrack use. *Rostocker Meeresbiologische Beiträge Heft*, *31*, 124–202.

Cianciullo, P., Maresca, V., Sorbo, S., & Basile, A. (2022). Antioxidant and Antibacterial Properties of Exstracts and Bioactive Compounds in Bryophytes. *Applied Sciences (Basel, Switzerland)*, *12*(1), 160. DOI: 10.3390/app12010160

Ciarkowska, K., Gargiulo, L., & Mele, G. (2016). Natural restoration of soils on mine heaps with similar technogenic parent material: A case study of long-term soil evolution in Silesian-Krakow Upland Poland. *Geoderma*, *261*, 141–150. DOI: 10.1016/j.geoderma.2015.07.018

Ciftel, S., Suleyman, B., Mammadov, R., Coskun, R., Coban, T. A., Mokhtare, B., Suleyman, H., Cerrah, S., Cicek, B., & Suleyman, Z. (2024). Adverse effects of metamizole on heart, lung, liver, kidney, and stomach in rats. *BMC Pharmacology & Toxicology*, *25*(1), 55. Advance online publication. DOI: 10.1186/s40360-024-00780-4 PMID: 39175070

Commisso, M., Guarino, F., Marchi, L., Muto, A., Piro, A., & Degola, F. (2021). Bryo-Activities: A Review on How Bryophytes Are Contributing to the Arsenal of Natural Bioactive Compounds against Fungi. *Plants*, *10*(2), 203. DOI: 10.3390/plants10020203 PMID: 33494524

Comşa, Ş., Cimpean, A. M., & Raica, M. (2015). The story of MCF-7 breast cancer cell line: 40 years of experience in research. *Anticancer Research*, *35*(6), 3147–3154. PMID: 26026074

Convention on Biological Diversity of RA. (2014). 5th National Report.

Cope, J. E., Norton, G. J., George, T. S., & Newton, A. C. (2022). Evaluating Variation in Germination and Growth of Landraces of Barley (Hordeum vulgare L.) Under Salinity Stress. *Frontiers in Plant Science, 13*, 863069. DOI: 10.3389/fpls.2022.863069 PMID: 35783948

Cordiano, R., Di Gioacchino, M., Mangifesta, R., Panzera, C., Gangemi, S., & Minciullo, P. L. (2023). Malondialdehyde as a Potential Oxidative Stress Marker for Allergy-Oriented Diseases: An Update. *Molecules (Basel, Switzerland), 28*(16), 5979. DOI: 10.3390/molecules28165979 PMID: 37630231

Corsi, S., Ruggeri, G., Zamboni, A., Bhakti, P., Espen, L., Ferrante, A., Noseda, M., Varanini, Z., & Scarafoni, A. (2022). A Bibliometric Analysis of the Scientific Literature on Biostimulants. *Agronomy, 12*(6), 1257. DOI: 10.3390/agronomy12061257

Costanza, R., d'Arge, R., de Groot, R., Farber, S., Grasso, M., Hannon, B., Limburg, K., Naeem, S., O'Neill, R. V., Paruelo, J., Raskin, R. G., Sutton, P., & van den Belt, M. (1997). the value of the world's ecosystem services and natural capital. *Nature, 387*(6630), 253–260. DOI: 10.1038/387253a0

Council of Europe. (1986). *European convention for the protection of vertebrate animals used for experimental and other scientific purposes* (ETS No. 123). https://rm.coe.int/168007a67b

Cruz De Carvalho, M. H. (2008). Drought stress and reactive oxygen species: Production, scavenging and signaling. *Plant Signaling & Behavior, 3*(3), 156–165. DOI: 10.4161/psb.3.3.5536 PMID: 19513210

Cruz, J. L., Coelho, E. F., Filho, M. A. C., & dos Santos, A. A. (2018). Salinity reduces nutrients absorption and efficiency of their utilization in cassava plants. *Ciência Rural, 48*(11), e20180351. DOI: 10.1590/0103-8478cr20180351

Cui, Z., He, F., Li, X., Li, Y., Huo, C., Wang, H., Qi, Y., Tian, G., Zong, W., & Liu, R. (2023). Response pathways of superoxide dismutase and catalase under the regulation of triclocarban-triggered oxidative stress in Eisenia foetida: Comprehensive mechanism analysis based on citotoxicity and binding model. *The Science of the Total Environment, 854*, 158821. DOI: 10.1016/j.scitotenv.2022.158821 PMID: 36116645

Ćwieląg-Drabek, M., Piekut, A., Gut, K., & Grabowski, M. (2020). Risk of cadmium, lead and zinc exposure from consumption of vegetables produced in areas with mining and smelting past. *Scientific Reports, 10*(1), 3363. DOI: 10.1038/s41598-020-60386-8 PMID: 32099081

D, S. T., & MA, S. (2012). Calculating Integrated Pollution Indices for Heavy Metals in Ecological Geochemistry Assessment Near Sugar Mill. *Journal of Research in Biology, 2*(5), 489–498. https://ojs.jresearchbiology.com/index.php/jrb/article/view/241

Dailianis, S., Piperakis, S. M., & Kaloyianni, M. (2005). Cadmium effects on ROS production and DNA damage via adrenergic receptors stimulation: Role of Na+/H+ exchanger and PKC. *Free Radical Research, 39*(10), 1059–1070. DOI: 10.1080/10715760500243765 PMID: 16298731

Dai, Q., Jiang, W., Huang, F., Song, F., Zhang, J., & Zhao, H. (2022). Recent Advances in Liver Engineering With Decellularized Scaffold. *Frontiers in Bioengineering and Biotechnology, 10*, 831477. DOI: 10.3389/fbioe.2022.831477 PMID: 35223793

Dasgan, H., Aktas, H., Abak, K., & Cakmak, I. (2002). Determination of screening techniques to salinity tolerance in tomatoes and investigation of genotype responses. *Plant Science, 163*(4), 695–703. DOI: 10.1016/S0168-9452(02)00091-2

de Jesus, S. S., Ferreira, G. F., Moreira, L. S., Maciel, M. R. W., & Maciel Filho, R. (2019). Comparison of several methods for effective lipid extraction from wet microalgae using green solvents. *Renewable Energy, 143*, 130–141. DOI: 10.1016/j.renene.2019.04.168

Dehaan, R. L., & Taylor, G. R. (2002). Field-derived spectra of salinized soils and vegetation as indicators of irrigation-induced soil salinization. *Remote Sensing of Environment, 80*(3), 406–417. DOI: 10.1016/S0034-4257(01)00321-2

Deinlein, U., Stephan, A. B., Horie, T., Luo, W., Xu, G., & Schroeder, J. I. (2014). Plant salt-tolerance mechanisms. *Trends in Plant Science, 19*(6), 371–379. DOI: 10.1016/j.tplants.2014.02.001 PMID: 24630845

Dengxiao, Z., Hongbin, J., Wenjing, Z., Qingsong, Y., Zhihang, M., Haizhong, W., Wei, R., Shiliang, L., & Daichang, W. (2024). Combined biochar and water-retaining agent application increased soil water retention capacity and maize seedling drought resistance in Fluvisols. *The Science of the Total Environment, 907*, 167885. DOI: 10.1016/j.scitotenv.2023.167885 PMID: 37863232

Deng, Z., Xia, A., Zhang, Z., Zhu, X., Huang, Y., Zhu, X., & Liao, Q. (2025). A continuous hydrothermal pretreatment system for efficient sugar production from wheat straw. *Chemical Engineering Journal, 508*, 160867. DOI: 10.1016/j.cej.2025.160867

Devenish, R.J., Prescott, M., & Rodgers, A.J.W. (2008). The structure and function of mitochondrial F1F0-ATP synthase. *Int Rev Cell Mol Biol,* 267, 1–58. DOI: 10.1016/S1937-6448(08)00601-1

Devi, R., Kaur, T., Kour, D., Yadav, A., Yadav, A.N., Suman, A., Ahluwalia, A.S., & Saxena, A.K. (2022). Minerals solubilizing and mobilizing microbiomes: A sustainable approach for managing minerals' deficiency in agricultural soil. *J Appl Microbiol., 133*(3), 1245-1272. DOI: 10.1111/jam.15627

Donadio, G., Mensitieri, F., Santoro, V., Parisi, V., Bellone, M. L., De Tommasi, N., Izzo, V., & Dal Piaz, F. (2021). Interactions with Microbial Proteins Driving the Antibacterial Activity of Flavonoids. *Pharmaceutics, 13*(5), 660. DOI: 10.3390/pharmaceutics13050660 PMID: 34062983

Druart, C., Millet, M., Scheifler, R., Delhomme, O., Raeppel, C., & de Vaufleury, A. (2011). Snails as indicators of pesticide drift, deposit, transfer and effects in the vineyard. *The Science of the Total Environment, 409*(20), 4280–4288. DOI: 10.1016/j.scitotenv.2011.07.006 PMID: 21784506

Duan, M., Yan, R., Wang, Q., Zhou, B., Zhu, H., Liu, G., Guo, X., & Zhang, Z. (2023). Integrated microbiological and metabolomics analyses to understand the mechanism that allows modified biochar to affect the alkalinity of saline soil and winter wheat growth. *The Science of the Total Environment, 866*, 161330. DOI: 10.1016/j.scitotenv.2022.161330 PMID: 36603639

Dubko, A. D., Yurkevich, M. Y., Labai, M. V., Svirskaya, A. V., Yukhanov, N. G., Savitskaya, T. V., Nizheharodava, D. B., & Zafranskaya, M. M. (2021). Engineering a whole rat liver scaffold. *Doklady Natsional'noi Akademii Nauk Belarusi, 65*(4), 448–455. DOI: 10.29235/1561-8323-2021-65-4-448-455

Dudley, N., Attwood, S. J., Goulson, D., Jarvis, D., Bharucha, Z. P., & Pretty, J. (2017). How should conservationists respond to pesticides as a driver of biodiversity loss in agroecosystems? *Biological Conservation, 209*, 449–453. DOI: 10.1016/j.biocon.2017.03.012

Dunuweera, A. N., Nikagolla, D. N., & Ranganathan, K. (2021). Fruit Waste Substrates to Produce Single-Cell Proteins as Alternative Human Food Supplements and Animal Feeds Using Baker's Yeast (Saccharomyces cerevisiae). *Journal of Food Quality, 2021*, 1–6. DOI: 10.1155/2021/9932762

Duse, L., Pinnapireddy, S. R., Strehlow, B., Jedelská, J., & Bakowsky, U. (2018). Low level LED photodynamic therapy using curcumin loaded tetraether liposomes. *European Journal of Pharmaceutics and Biopharmaceutics, 126*, 233–241. DOI: 10.1016/j.ejpb.2017.10.005 PMID: 29017954

Duvigneau, S., Kettner, A., Carius, L., Griehl, C., Findeisen, R., & Kienle, A. (2021). Fast, inexpensive, and reliable HPLC method to determine monomer fractions in poly(3-hydroxybutyrate-co-3-hydroxyvalerate). *Applied Microbiology and Biotechnology, 105*(11), 4743–4749. DOI: 10.1007/s00253-021-11265-3 PMID: 34014345

Dziwak, M., Wróblewska, K., Szumny, A., & Galek, R. (2022). Modern Use of Bryophytes as a Source of Secondary Metabolites. *Agronomy (Basel), 12*(6), 1456. DOI: 10.3390/agronomy12061456

Dzurendova, S., Zimmermann, B., Kohler, A., Reitzel, K., Nielsen, U. G., Dupuy—Galet, B. X., Leivers, S., Horn, S. J., & Shapaval, V. (2021). Calcium affects polyphosphate and lipid accumulation in Mucoromycota fungi. *Journal of Fungi (Basel, Switzerland), 7*(4), 300. DOI: 10.3390/jof7040300 PMID: 33920847

Ehlui, K. S., Atakpama, W., von Wehrden, H., Bah, A., Kola, E., Anthony-Krueger, C., Egbelou, H., Kokou, K. B., & Boukpessi, T. (2024). Anthropogenic Threats to Degraded Forest Land in the Savannahs' Region of Togo from 1984 to 2020, West Africa. *Journal of Geoscience and Environment Protection, 12*(1), 164–179. DOI: 10.4236/gep.2024.121011

El Akrami, N., Hanine, M., Flores, E. S., Aray, D. G., & Ashraf, I. (2023). Unleashing the Potential of Blockchain and Machine Learning: Insights and Emerging Trends From Bibliometric Analysis. *IEEE Access : Practical Innovations, Open Solutions, 11*, 78879–78903. DOI: 10.1109/ACCESS.2023.3298371

El-Araby, R. (2024). Biofuel production: Exploring renewable energy solutions for a greener future. *Biotechnology for Biofuels and Bioproducts, 17*(1), 129. DOI: 10.1186/s13068-024-02571-9

Erokhin, V.E. (2020). On the possibility of using marine plants of the Azov-Black Sea basin for organic farming. *Voprosy sovremennoi algologii, 1*(22), 1-7. DOI: 10.33624/2311-0147-2020-1(22)-1-7

Estrada, Y., Fernández-Ojeda, A., Morales, B., Egea-Fernández, J. M., Flores, F. B., Bolarín, M. C., & Egea, I. (2021). Unraveling the strategies used by the under-exploited amaranth species to confront salt stress: Similarities and differences with quinoa species. *Frontiers in Plant Science, 12*, 604481. Advance online publication. DOI: 10.3389/fpls.2021.604481 PMID: 33643343

European Convention for the Protection of Vertebrate Animals used for Experimental and Other Scientific Purposes. (1987). In *Medical entomology and zoology.*

Evelin, H., Kapoor, R., & Giri, B. (2009). Arbuscular mycorrhizal fungi in alleviation of salt stress: A review. *Annals of Botany*, *104*(7), 1263–1280. DOI: 10.1093/aob/mcp251 PMID: 19815570

Faccenda, D., & Campanella, M. (2012). Molecular regulation of the mitochondrial F1F0-ATP synthase: Physiological and pathological significance of the inhibitory factor 1 (IF1*). International Journal of Cell Biology*, *1*, 1–12. DOI: 10.1155/2012/367934 PMID: 22966230

Falke, D., Schulz, K., Doberenz, C., Beyer, L., Lilie, H., Thiemer, B., & Sawers, R. G. (2010). Unexpected oligomeric structure of the FocA formate channel of *Escherichia coli*: A paradigm for the formate-nitrite transporter family of integral membrane proteins. *FEMS Microbiology Letters*, *303*(1), 69–75. DOI: 10.1111/j.1574-6968.2009.01862.x PMID: 20041954

Favaro, L., Basaglia, M., & Casella, S. (2012). Processing wheat bran into ethanol using mild treatments and highly fermentative yeasts. *Biomass and Bioenergy*, *46*, 605–617. DOI: 10.1016/j.biombioe.2012.07.001

Fayvush, G. (2023). *Biodiversity of Armenia*. Springer International Publishing. DOI: 10.1007/978-3-031-34332-2

Federal Service for Intellectual Property. (2011). *Proposal for survival of stormwater discharges from sea habitats*. https://patentimages.storage.googleapis.com/e4/46/bf/43701254a1f51f/RU2473204C1.pdf

Fernández-Blanco, C., Robles-Iglesias, R., Naveira-Pazos, C., Veiga, M. C., & Kennes, C. (2023). Production of biofuels from <scp> C $_1$ </scp> -gases with *Clostridium* and related bacteria—Recent advances. *Microbial Biotechnology*, *16*(4), 726–741. DOI: 10.1111/1751-7915.14220 PMID: 36661185

Fernando, V. C. D. (2020). Major transcription factor families involved in salinity stress tolerance in plants. *Transcription Factors for Abiotic Stress Tolerance in Plants*, 99–109. DOI: 10.1016/B978-0-12-819334-1.00007-1

Firman, Z., Tri, H., Jeverson, R., Hafiizh, P., Muhammad, S., & Hada, P. (2023). Economic feasibility analysis of macroalgae farming-based carbon dioxide removal. *Journal of Sustainability Science and Management*, *8*(18), 137–145. DOI: 10.46754/jssm.2023.08.008

Flowers, T. J. (2004). Improving crop salt tolerance. *Journal of Experimental Botany*, *55*(396), 307–319. DOI: 10.1093/jxb/erh003 PMID: 14718494

Flowers, T. J., & Colmer, T. D. (2015). Plant salt tolerance: Adaptations in halophytes. *Annals of Botany*, *115*(3), 327–331. DOI: 10.1093/aob/mcu267 PMID: 25844430

Foseid, L., Devle, H., Naess-Andresen, C. F., & Ekeberg, D. (2022). Laminaria hyperborea as a Source of Valuable Glyceroglycolipids—A Characterization of Galactosyldiacilglycerols in Stipe and Blade by HPLC-MS/MS. *AppliedChem*, *2*(4), 185–198. DOI: 10.3390/appliedchem2040013

Fratini, F., Cilia, G., Mancini, S., & Felicioli, A. (2016). Royal Jelly: An ancient remedy with remarkable antibacterial properties. *Microbiological Research*, *192*, 130–141. DOI: 10.1016/j.micres.2016.06.007 PMID: 27664731

Ganju, S., & Gogate, P. R. (2017). A review on approaches for efficient recovery of whey proteins from dairy industry effluents. *Journal of Food Engineering*, *215*, 84–96. DOI: 10.1016/j.jfoodeng.2017.07.021

Gao, Y., Feng, J., Lv, J., Liu, Q., Nan, F., Liu, X., & Xie, S. (2019). Physiological changes of Parachlorella Kessleri TY02 in lipid accumulation under nitrogen stress. *International Journal of Environmental Research and Public Health*, *16*(7), 1188. DOI: 10.3390/ijerph16071188 PMID: 30987041

Gąsior, D., & Tic, W. J. (2017). Application of the Biochar-Based Technologies as the Way of Realization of the Sustainable Development Strategy. *Economic and Environmental Studies*, *17*(3 (43)), 597–611. DOI: 10.25167/ees.2017.43.9

Ge, L., Ali, M. M., Osman, A. I., Elgarahy, A. M., Samer, M., Xu, Y., & Liu, Z. (2025). A critical review on conversion technology for liquid biofuel production from lignocellulosic biomass. *Renewable & Sustainable Energy Reviews*, *217*, 115726. DOI: 10.1016/j.rser.2025.115726

Gemünde, A., Gail, J., & Holtmann, D. (2024). Redox mediator interaction with Cupriavidus necator – spectroelectrochemical online analysis. *Electrochemistry Communications*, *162*, 107705. DOI: 10.1016/j.elecom.2024.107705

Genwali, G. R., Acharya, P. P., & Rajbhandari, M. (2013). Isolation of gallic acid and estimation of total phenolic content in some medicinal plants and their anti-oxidant activity. *Nepal Journal of Science and Technology*, *14*(1), 95–102. DOI: 10.3126/njst.v14i1.8928

George, D., & Mallery, P. (2024). *IBM SPSS statistics 29 step by step: A simple guide and reference*. Routledge. DOI: 10.4324/9781032622156

Gevorgyan, G. A., Movsesyan, H. S., Grigoryan, K. V., & Ghazaryan, K. A. (2015). Environmental risks of heavy metal pollution of the soils around Kajaran Town, Armenia. *Proceedings of the YSU B: Chemical and Biological Sciences*, *49*(2 (237)), 50–55. DOI: 10.46991/PYSU:B/2015.49.2.050

Gevorgyan, H., Grigoryan, G., Atoyan, H., Rukhkyan, M., & Hakobyan, A. (2019). Evidence of Crimean - Congo Haemorrhagic Fever Virus Occurrence in Ixodidae Ticks of Armenia.

Gevorgyan, H., Seitzer, U., & Beer, D. (2014). Implementation or molecular epidemiological approaches as a tool for characterization of tick-borne diseases in Armenia. *Proceedings of the International Conference "Biological Diversity and Conservation Problems of the Fauna of the Caucasus - 2". Yerevan,* 135-138.

Ghaleb, W., Ahmed, L. Q., Wagner, M.-H., Eprinchard-Ciesla, A., Olivares-Rodríguez, W. E., Perrot, C., Chenu, K., Norton, M., & Escobar-Gutiérrez, A. J. (2022). The Concepts of Seed Germination Rate and Germinability: A Re-Evaluation for Cool-Season Grasses. *Agronomy (Basel), 12*(6), 1291. DOI: 10.3390/agronomy12061291

Ghazaryan, K. A., Margaryan, G. H., Minkina, T. M., Rajput, V. D., & Movsesyan, H. S. (2024). Chenopodium album L.: A promising NaCl accumulating and tolerant plant for mitigation of salinity stress. *Revista Brasileira de Botânica. Brazilian Journal of Botany, 47*(3), 837–856. DOI: 10.1007/s40415-023-00974-w

Ghazaryan, K., Movsesyan, H., Gevorgyan, A., Minkina, T., Sushkova, S., Rajput, V., & Mandzhieva, S. (2020). Comparative hydrochemical assessment of groundwater quality from different aquifers for irrigation purposes using IWQI: A case-study from Masis province in Armenia. *Groundwater for Sustainable Development, 11*, 100459. DOI: 10.1016/j.gsd.2020.100459

Ghorbani, A., Omran, V. O. G., Razavi, S. M., Pirdashti, H., & Ranjbar, M. (2019). Piriformospora indica confers salinity tolerance on tomato (Lycopersicon esculentum Mill.) through amelioration of nutrient accumulation, K+/Na+ homeostasis and water status. *Plant Cell Reports, 38*(9), 1151–1163. DOI: 10.1007/s00299-019-02434-w PMID: 31152194

Ghorbani, F., Ekhtiari, M., Chaghervand, B. M., Moradi, L., Mohammadi, B., & Kajbafzadeh, A. (2022). Detection of the residual concentration of sodium dodecyl sulfate in the decellularized whole rabbit kidney extracellular matrix. *Cell and Tissue Banking, 23*(1), 119–128. DOI: 10.1007/s10561-021-09921-z PMID: 33909237

Gigen, S., & Harutyunyan, E. S. (1991). An attempt to detect *Borrelia burgdorferi* in Armenian Ixodid ticks. *Biological Journal of Armenia, 44*(4), 304–307.

Gilbert, T., Sellaro, T., & Badylak, S. (2006). Decellularization of tissues and organs. *Biomaterials, 27*(19), 3675–3683. DOI: 10.1016/j.biomaterials.2006.02.014 PMID: 16519932

Gill, P. K., Dev Sharma, A., Singh, P., & Singh Bhullar, S. (2001). Effect of various abiotic stresses on the growth. *Soluble Sugars and Water Relations of Sorghum Seedlings Grown in Light and Darkness*, *27*(2), 72–84.

Ginovyan, M., Hovhannisyan, S., Javrushyan, H., Sevoyan, G., Karabekian, Z., Zakaryan, N., Sahakyan, N., & Avtandilyan, N. (2022). Screening revealed the strong cytotoxic activity of *Alchemilla smirnovii* and *Hypericum alpestre* ethanol extracts on different cancer cell lines. *AIMS Biophysics*, *10*(1), 12–22. DOI: 10.3934/biophy.2023002

Ginovyan, M., Javrushyan, H., Petrosyan, G., Kusznierewicz, B., Koss-Mikołajczyk, I., Koziara, Z., Kuczyńska, M., Jakubek, P., Karapetyan, A., Sahakyan, N., Maloyan, A., Bartoszek, A., & Avtandilyan, N. (2023). Anti-cancer effect of Rumex obtusifolius in combination with arginase/nitric oxide synthase inhibitors via downregulation of oxidative stress, inflammation, and polyamine synthesis. *The International Journal of Biochemistry & Cell Biology*, *158*, 158. DOI: 10.1016/j.biocel.2023.106396 PMID: 36918141

Ginovyan, M., Tadevosyan, S., Shirvanyan, A., Babayan, A., Kusznierewicz, B., Koss-Mikołajczyk, I., Mróz, M., Bartoszek, A., & Sahakyan, N. (2025). The potential of blackcurrant, fig, and grape leaf extracts in the development of new preparations for overcoming antibiotic resistance and enhancing the efficacy of chemotherapeutic agents. *BMC Complementary Medicine and Therapies*, *25*(1), 148. DOI: 10.1186/s12906-025-04859-1 PMID: 40269785

Ginovyan, M., & Trchounian, A. (2019). Novel approach to combat antibiotic resistance : Evaluation of some Armenian herb crude extracts for their antibiotic modulatory and antiviral properties. *Journal of Applied Microbiology*, *127*(2), 472–480. DOI: 10.1111/jam.14335 PMID: 31136046

Global Forest Watch. (2025). Togo Dashboard. https://golnk.ru/gr325

Godoy, P., Hewitt, N. J., Albrecht, U., Andersen, M. E., Ansari, N., Bhattacharya, S., Bode, J. G., Bolleyn, J., Borner, C., Böttger, J., Braeuning, A., Budinsky, R. A., Burkhardt, B., Cameron, N. R., Camussi, G., Cho, C., Choi, Y., Rowlands, J. C., Dahmen, U., & Hengstler, J. G. (2013). Recent advances in 2D and 3D in vitro systems using primary hepatocytes, alternative hepatocyte sources and non-parenchymal liver cells and their use in investigating mechanisms of hepatotoxicity, cell signaling and ADME. *Archives of Toxicology*, *87*(8), 1315–1530. DOI: 10.1007/s00204-013-1078-5 PMID: 23974980

Godvin Sharmila, V., Dinesh Kumar, M., Pugazhendi, A., Bajhaiya, A., Gugulothu, P., & Rajesh Banu, J. (2021). Biofuel Production from Macroalgae: Present Scenario and Future Scope. *Bioengineered*, *12*(2), 9216–9238. DOI: 10.1080/21655979.2021.1996019

Gran, M., Carrera, J., Olivella, S., & Saaltink, M. W. (2011). Modeling evaporation processes in a saline soil from saturation to oven dry conditions. *Hydrology and Earth System Sciences*, *15*(7), 2077–2089. DOI: 10.5194/hess-15-2077-2011

Gratão, P. L., Monteiro, C. C., Rossi, M. L., Martinelli, A. P., Peres, L. E. P., Medici, L. O., Lea, P. J., & Azevedo, R. A. (2009). Differential ultrastructural changes in tomato hormonal mutants exposed to cadmium. *Environmental and Experimental Botany*, *67*(2), 387–394. DOI: 10.1016/j.envexpbot.2009.06.017

Grattan, S. R., & Grieve, C. M. (1998). Salinity–mineral nutrient relations in horticultural crops. *Scientia Horticulturae*, *78*(1–4), 127–157. DOI: 10.1016/S0304-4238(98)00192-7

Gray, J. S., Kahl, O., Lane, R. S., Levin, M. L., & Tsao, J. I. (2016). Diapause in ticks of the medically important *Ixodes ricinus* species complex. *Ticks and Tick-Borne Diseases*, *7*(5), 992–1003. DOI: 10.1016/j.ttbdis.2016.05.006 PMID: 27263092

Grech-Angelini, S., Stachurski, F., Lancelot, R., Boissier, J., Allienne, J.-F., Marco, S., Maestrini, O., & Uilenberg, G. (2016). Ticks (Acari: Ixodidae) infesting cattle and some other domestic and wild hosts on the French Mediterranean island of Corsica. *Parasites & Vectors*, *582*(9), 1–11. DOI: 10.1186/s13071-016-1876-8 PMID: 27842608

Grigoryan, L., Babayan, A., Vassilian, A., Poladyan, A., Sawers, G., & Trchounian, K. (2025). *Escherichia coli* FocA/B-dependent H^+ and K^+ fluxes: Influence of exogenous versus endogenous formate. *Biophysical Reports*, *5*(3), 100225. DOI: 10.1016/j.bpr.2025.100225 PMID: 40784583

Grote, U., Fasse, A., Nguyen, T. T., & Erenstein, O. (2021). Food Security and the Dynamics of Wheat and Maize Value Chains in Africa and Asia. *Frontiers in Sustainable Food Systems*, *4*(February), 617009. DOI: 10.3389/fsufs.2020.617009

Grotkopp, E., Rejmánek, M., & Rost, T. L. (2002). Toward a causal explanation of plant invasiveness: Seedling Growth and Life-History Strategies of 29 pine (Pinus) species. *American Naturalist*, *159*(4), 396–419. DOI: 10.1086/338995 PMID: 18707424

Guérin, B., Labbe, P., & Solmo, P. (1979). Preparation of Yeast Mitochondria (*Saccharomyces cerevisiae*) with good P/O and respiratory control Ratios. *Methods in Enzymology, 55*, 149–159. DOI: 10.1016/0076-6879(79)55021-6 PMID: 379498

Gunstone, F. D., & Harwood, J. L. (2007). *The lipid handbook with CD-ROM*. CRC press., DOI: 10.1201/9781420009675

Gupta, B., & Huang, B. (2014). Mechanism of salinity tolerance in plants: Physiological, biochemical, and molecular characterization. *International Journal of Genomics, 2014*, 1–18. Advance online publication. DOI: 10.1155/2014/701596 PMID: 24804192

Güray, U., Sibilay, K., Nazife, A., Gülden, V., & Mihrican, B. (2025). Investigation of the Antimicrobial and Antifungal Activities of Some Moss Species (*Scleropodium touretii* (Brid.) L.F.Koch, *Hypnum cupressiforme* Hedw. var. *lacunosum* Brid., *Brachythecium glareosum* (Bruch ex Spruce) Schimp.). *Anatolian Bryology., 11*(1), 72–82. DOI: 10.26672/anatolianbryology.1718549

Gustafson, K. R. (2015). Mining the extensive chemical diversity of the NCI natural products repository for new agents that can target HIV. *Planta Medica, 81*(11), IL1. DOI: 10.1055/s-0035-1556098

Haft, R. J. F., Keating, D. H., Schwaegler, T., Schwalbach, M. S., Vinokur, J., Tremaine, M., Peters, J. M., Kotlajich, M. V., Pohlmann, E. L., Ong, I. M., Grass, J. A., Kiley, P. J., & Landick, R. (2014). Correcting direct effects of ethanol on translation and transcription machinery confers ethanol tolerance in bacteria. *Proceedings of the National Academy of Sciences of the United States of America, 111*(25). Advance online publication. DOI: 10.1073/pnas.1401853111 PMID: 24927582

Haghighizadeh, A., Rajabi, O., Nezarat, A., Hajyani, Z., Haghmohammadi, M., Hedayatikhah, S., Asl, S. D., & Aghababai Beni, A. (2024a). Comprehensive analysis of heavy metal soil contamination in mining Environments: Impacts, monitoring Techniques, and remediation strategies. *Arabian Journal of Chemistry, 17*(6), 105777. DOI: 10.1016/j.arabjc.2024.105777

Hakanson, L. (1980). An ecological risk index for aquatic pollution control.a sedimentological approach. *Water Research, 14*(8), 975–1001. DOI: 10.1016/0043-1354(80)90143-8

Hakobyan, V. J., Boyakhchyan, G. H., Shcherbakov, O. V., & Vardanyan, M. V. (2004). Fauna of pasture ticks of Armavir Marz, Armenia. *Proceedings for Conference dedicated to the 100th Anniversary of S.M. Yablokov-Khndzoryan,* 17-18.

Hakobyan, V. J., Boyakhchyan, G. H., Vardanyan, M. V., & Shcherbakov, O. V. (2003). Some aspects of epidemiology of the bovine piroplasmosis and its control. *Proceedings for Conference "Research and Conservation of Fauna in Southern Caucasus"*, 24-26.

Hakobyan, B., Pinske, C., Sawers, G., Trchounian, A., & Trchounian, K. (2018). pH and a mixed carbon-substrate spectrum influence FocA- and FocB-dependent, formate-driven H_2 production in *Escherichia coli. FEMS Microbiology Letters, 365*, 21. DOI: 10.1093/femsle/fny233 PMID: 30247527

Hammer, E. C., Nasr, H., Pallon, J., Olsson, P. A., & Wallander, H. (2010). Elemental composition of arbuscular mycorrhizal fungi at high salinity. *Mycorrhiza, 21*(2), 117–129. DOI: 10.1007/s00572-010-0316-4 PMID: 20499112

Harutyunyan, B., Mühlstädt, G., Seibel, L., Hilpmann, G., Sayadyan, A., Aghabekyan, A., & Goginyan, V. (2025). Algal biomass as a food and feed supplement: Process engineering analysis. *Functional Foods in Health and Disease, 15*(7), 380-395. DOI: 10.31989/ffhd.v15i7.1662

Hasanaklou, N. T., Mohagheghi, V., Hasanaklou, H. T., Ma'mani, L., Malekmohammadi, M., Moradi, F., & Dalvand, Y. (2023). Seed Nano-Priming Using Silica Nanoparticles: Effects in Seed Germination and Physiological Properties of Stevia Rebaudiana Bertoni. *Chemical and Biological Technologies in Agriculture, 10*(1), 96. Advance online publication. DOI: 10.1186/s40538-023-00445-0

Hasana, R., & Miyake, H. (2017). Salinity Stress Alters Nutrient Uptake and Causes the Damage of Root and Leaf Anatomy in Maize. *KnE Life Sciences, 3*(4), 219. DOI: 10.18502/kls.v3i4.708

Hasanuzzaman, M., Bhuyan, M. H. M. B., Zulfiqar, F., Raza, A., Mohsin, S. M., Al Mahmud, J., Fujita, M., & Fotopoulos, V. (2020). Reactive Oxygen Species and Antioxidant Defense in Plants under Abiotic Stress: Revisiting the Crucial Role of a Universal Defense Regulator. *Antioxidants, 9*(8), 681. DOI: 10.3390/antiox9080681

Hasanuzzaman, M., Davies, N. W., Shabala, L., Zhou, M., Brodribb, T. J., & Shabala, S. (2017). Residual transpiration as a component of salinity stress tolerance mechanism: A case study for barley. *BMC Plant Biology, 17*(1), 1–12. DOI: 10.1186/s12870-017-1054-y PMID: 28629324

Hassan, M. A., Pacurar, A., López-Gresa, M. P., Donat-Torres, M. P., Llinares, J. V., Boscaiu, M., & Vicente, O. (2016). Effects of salt stress on three ecologically distinct plantago species. *PLoS One, 11*(8), e0160236. DOI: 10.1371/journal.pone.0160236 PMID: 27490924

Hayat, K., Menhas, S., & Hayat, S. (2022). Stress-Tolerant Species of Medicinal Plants and Phytoremediation Potential. In *Environmental Challenges and Medicinal Plants*. Springer International Publishing. DOI: 10.1007/978-3-030-92050-0_18

Hayat, M. T., Nauman, M., Nazir, N., Ali, S., & Bangash, N. (2019). Environmental Hazards of Cadmium: Past, Present, and Future. *Cadmium Toxicity and Tolerance in Plants: From Physiology to Remediation*, 163–183. DOI: 10.1016/B978-0-12-814864-8.00007-3

Hazman, M., Fawzy, S., Hamdy, A., Khaled, A., Mahmoud, A., Khalid, E., Ibrahim, H. M., Gamal, M., Abo Elyazeed, N., Saber, N., Ehab, M., & Kabil, F. (2023). Enhancing rice resilience to drought by applying biochar–compost mixture in low-fertile sandy soil. *Beni-Suef University Journal of Basic and Applied Sciences, 12*(1), 74-. DOI: 10.1186/s43088-023-00411-7

Heidari, N. M. (2011). Effects of salinity stress on growth, chlorophyll content and osmotic components of two basil (Ocimum basilicum L.) genotypes. *African Journal of Biotechnology*, *11*(2). Advance online publication. DOI: 10.5897/AJB11.2572

Hemavathy, R. V., Ragini, Y. P., Shruthi, S., Ranjani, S., Subhashini, S., & Thamarai, P. (2025). Biofuel production from marine macroalgae: Pathways, technologies, and sustainable energy solutions. *Industrial Crops and Products*, *224*, 120282. Advance online publication. DOI: 10.1016/j.indcrop.2024.120282

Herre, B., & Arriagada, P. (2023). The Human Development Index and related indices. *Published online at OurWorldinData.org*. https://golnk.ru/MvWEZ

Higbie, S. M., Wang, F., Stewart, J. McD., Sterling, T. M., Lindemann, W. C., Hughs, E., & Zhang, J. (2010). Physiological Response to Salt (NaCl) Stress in Selected Cultivated Tetraploid Cottons. *International Journal of Agronomy*, *2010*, 1–12. DOI: 10.1155/2010/643475

Hnilickova, H., Kraus, K., Vachova, P., & Hnilicka, F. (2021). Salinity Stress Affects Photosynthesis, Malondialdehyde Formation, and Proline Content in Portulaca oleracea L. *Plants*, *10*(5), 845. DOI: 10.3390/plants10050845 PMID: 33922210

Hofstede, G., Hofstede, G. J., & Minkov, M. (2010). Table data (Russia, Armenia, Togo, Sierra Leone and East Africa). Cultures and Organizations: Software of the Mind (3rd ed.). McGraw-Hill. https://golnk.ru/851O5

Hölzl, G., & Dörmann, P. (2007). Structure and function of glycoglycerolipids in plants and bacteria. *Progress in Lipid Research*, *46*(5), 225–243. DOI: 10.1016/j.plipres.2007.05.001 PMID: 17599463

Hong, C. C., Kume, T., & Peterson, R. T. (2008). Role of crosstalk between phosphatidylinositol 3-kinase and extracellular signal-regulated kinase/mitogen-activated protein kinase pathways in artery–vein specification. *Circulation Research, 103*(6), 573–579. DOI: 10.1161/CIRCRESAHA.108.180745 PMID: 18796644

Hong, S., & Pederson, P. L. (2008). ATP Synthase and the Actions of Inhibitors Utilized To Study Its Roles in Human Health, Disease, and Other Scientific Areas. *Microbiology and Molecular Biology Reviews : MMBR*, *72*(4), 590–641. DOI: 10.1128/MMBR.00016-08 PMID: 19052322

Hopper, S., & Böck, A. (1995). Effector-mediated stimulation of ATPase activity by the sigma 54-dependent transcriptional activator FHLA from *Escherichia coli. Journal of Bacteriology*, *177*(10), 2798–2803. DOI: 10.1128/jb.177.10.2798-2803.1995 PMID: 7751289

Hossain, M. N., Sarker, U., Raihan, M. S., Al-Huqail, A. A., Siddiqui, M. H., & Oba, S. (2022). Influence of Salinity Stress on Color Parameters, Leaf Pigmentation, Polyphenol and Flavonoid Contents, and Antioxidant Activity of Amaranthus lividus Leafy Vegetables. *Molecules (Basel, Switzerland)*, *27*(6), 1821. DOI: 10.3390/molecules27061821 PMID: 35335185

Hou, C.-C., Chen, Y.-P., Wu, J.-H., Huang, C.-C., Wang, S.-Y., Yang, N.-S., & Shyur, L.-F. (2007). A galactolipid possesses novel cancer chemopreventive effects by suppressing inflammatory mediators and mouse B16 melanoma. *Cancer Research*, *67*(14), 6907–6915. DOI: 10.1158/0008-5472.CAN-07-0158 PMID: 17638902

Hovnanyan, K., Marutyan, S., Marutyan, S., Hovnanyan, M., Navasardyan, L., & Trchounian, A. (2020). Ultrastructural investigation of acidocalcisomes and ATPase activity in yeast Candida guilliermondii NP-4 as 'complementary'stress-targets. *Letters in Applied Microbiology, 71*(4), 413-419. DOI: 10.1016/j.jff.2024.106505

Hovnanyan, K., Marutyan, S., Marutyan, S. A., Hovnanyan, M., Navasardyan, L., & Trchounian, A. (2020). Ultrastructural investigation of acidocalcisomes and ATPase activity in yeast *Candida guilliermondii* NP-4 as "complementary" stress-targets. *Letters in Applied Microbiology*, *71*, 413–419. DOI: 10.1111/lam.13348 PMID: 32623751

Huang, Z., Long, X., Wang, L., Kang, J., Zhang, Z., Zed, R., & Liu, Z. (2012). Growth, photosynthesis and H+-ATPase activity in two Jerusalem artichoke varieties under NaCl-induced stress. *Process Biochemistry (Barking, London, England)*, *47*(4), 591–596. DOI: 10.1016/j.procbio.2011.12.016

Hübscher, U., Maga, G., & Spadari, S. (2002). Eukaryotic DNA polymerases. *Annual Review of Biochemistry*, *71*(1), 133–163. DOI: 10.1146/annurev.biochem.71.090501.150041 PMID: 12045093

Hunger, D., Doberenz, C., & Sawers, R. G. (2014). Identification of key residues in the formate channel FocA that control import and export of formate. *Biological Chemistry*, *395*(7-8), 813–825. DOI: 10.1515/hsz-2014-0154 PMID: 24659605

Hunger, D., Röcker, M., Falke, D., Lilie, H., & Sawers, R. G. (2017). The *C*-terminal six amino acids of the FNT channel FocA are required for formate translocation but not homopentamer integrity. *Frontiers in Microbiology*, *8*, 1616. DOI: 10.3389/fmicb.2017.01616 PMID: 28878762

Hussein, K. H., Ahmadzada, B., Correa, J. C., Sultan, A., Wilken, S., Amiot, B., & Nyberg, S. L. (2024). Liver tissue engineering using decellularized scaffolds: Current progress, challenges, and opportunities. *Bioactive Materials*, *40*, 280–305. DOI: 10.1016/j.bioactmat.2024.06.001 PMID: 38973992

Hu, Y., & Schmidhalter, U. (2005). Drought and salinity: A comparison of their effects on mineral nutrition of plants. *Journal of Plant Nutrition and Soil Science*, *168*(4), 541–549. DOI: 10.1002/jpln.200420516

Hyndes, G., Berdan, E., Duarte, C., Dugan, J., Emery, K., Hambäck, P., Henderson, C., Hubbard, D., Lastra, M., Mateo, M., Olds, A., & Schlacher, T. (2021). The role of inputs of marine wrack and carrion in sandy-beach ecosystems: A global review. *Biological Reviews of the Cambridge Philosophical Society*, *97*(6), 2127–2161. DOI: 10.1111/brv.12886

Ibrahim, A. M., Ahmed, A. K., Bakry, F. A., & Abdel-Ghaffar, F. (2018). Hematological, physiological and genotoxicological effects of Match 5% EC insecticide on *Biomphalaria alexandrina* snails. *Ecotoxicology and Environmental Safety*, *147*, 1017–1022. DOI: 10.1016/j.ecoenv.2017.09.059 PMID: 29976004

Ibrahim, A. M., Ahmed, A. K., Bakry, F. A., Rabei, I., & Abdel-Ghaffar, F. (2019). Toxicological impact of butralin, glyphosate-isopropylammonium and pendimethalin herbicides on physiological parameters of Biomphalaria alexandrina snails. *Molluscan Research*, *39*(3), 224–233. DOI: 10.1080/13235818.2019.1592296

Ilanidis, D., Stagge, S., Jönsson, L. J., & Martín, C. (2021). Hydrothermal Pretreatment of Wheat Straw: Effects of Temperature and Acidity on Byproduct Formation and Inhibition of Enzymatic Hydrolysis and Ethanolic Fermentation. *Agronomy (Basel)*, *11*(3), 487. DOI: 10.3390/agronomy11030487

Inesa, Gayane, Henrikh, & Armen. (2020). Antibacterial Properties and Flavonoids Content of Some Mosses Common in Armenia. *Iranian Journal of Pharmaceutical Sciences, 16*(4), 31-42. www.ijps.ir

Interagency Coordination Group on Antimicrobial Resistance. (2019). *No Time to Wait: Securing the Future from Drug-Resistant Infections. Report to the Secretary General of the United Nations.* https://www.who.int/publications/i/item/no-time-to -wait-securing-the-future-from-drugresistant-infe

International Organization for Standardization. (2009). *Biological evaluation of medical devices - Part 5: Tests for in vitro cytotoxicity* (ISO Standard No. 10993-5:2009/(R)2014). https://www.iso.org/standard/36406.html

IPCC. (2023). *AR6 Synthesis Report: Climate Change 2023.* https://www.ipcc.ch/report/sixth-assessment-report-cycle/

Irfan, M., Dawar, K., Fahad, S., Mehmood, I., Alamri, S., Siddiqui, M. H., Saud, S., Khattak, J. Z. K., Ali, S., Hassan, S., Nawaz, T., Hammad, H. M., Banout, J., & Nasim, W. (2022). Exploring the potential effect of Achnatherum splendens L.–derived biochar treated with phosphoric acid on bioavailability of cadmium and wheat growth in contaminated soil. *Environmental Science and Pollution Research, 29*(25), 37676–37684. DOI: 10.1007/s11356-021-17950-0

Isayenkov, S. V., & Maathuis, F. J. M. (2019). Plant salinity stress: Many unanswered questions remain. *Frontiers in Plant Science, 10*, 80. DOI: 10.3389/fpls.2019.00080 PMID: 30828339

Isenring, R. (2010). *Pesticides and the loss of biodiversity.* Pesticide Action Network Europe.

Ishizuka, M., Abe, F., Sano, Y., Takahashi, K., Inoue, K., Nakajima, M., Kohda, T., Komatsu, N., Ogura, S., & Tanaka, T. (2011). Novel development of 5-aminolevurinic acid (ALA) in cancer diagnoses and therapy. *International Immunopharmacology, 11*(3), 358–365. DOI: 10.1016/j.intimp.2010.11.029 PMID: 21144919

Iskandaryan, M., Baghdasaryan, L., Minasyan, E., Trchounian, K., Antranikian, G., & Poladyan, A. (2025). A novel, cost-effective approach for the production of hydrogenase enzymes and molecular hydrogen from recycled whey-based by-products. *International Journal of Hydrogen Energy, 140*, 1191–1202. DOI: 10.1016/j.ijhydene.2024.10.256

Iskandaryan, M., Blbulyan, S., Sahakyan, M., Vassilian, A., Trchounian, K., & Poladyan, A. (2023). L-amino acids affect the hydrogenase activity and growth of Ralstonia eutropha H16. *AMB Express, 13*(1), 33. DOI: 10.1186/s13568-023-01535-w PMID: 36932299

Ismail, A., Munsi, H., Yusuf, A. M., & Hijjang, P. (2025). Mapping One Decade of Identity Studies: A Comprehensive Bibliometric Analysis of Global Trends and Scholarly Impact. *Social Sciences, 14*(2), 92. DOI: 10.3390/socsci14020092

Ismail, S., Rao, N. K., & Dagar, J. C. (2019). Identification, evaluation, and domestication of alternative crops for saline environments. In *Springer eBooks* (pp. 505–536). DOI: 10.1007/978-981-13-5832-6_17

ITA. (2025). Mobile-Driven Africa. https://www.trade.gov/rise-ecommerce-africa

Ito, F., Sono, Y., & Ito, T. (2019). Measurement and Clinical Significance of Lipid Peroxidation as a Biomarker of Oxidative Stress: Oxidative Stress in Diabetes, Atherosclerosis, and Chronic Inflammation. *Antioxidants, 8*(3), 72. DOI: 10.3390/antiox8030072 PMID: 30934586

Jessen, H., & Bui, T. P. N. (2025). Metabolism of inositol derivatives by the gut microbiome. *Gut Microbes, 17*(1), 17. DOI: 10.1080/19490976.2025.2564765 PMID: 41021306

Jhariya, U., Dafale, N. A., Srivastava, S., Bhende, R. S., Kapley, A., & Purohit, H. J. (2021). Understanding Ethanol Tolerance Mechanism in Saccharomyces cerevisiae to Enhance the Bioethanol Production: Current and Future Prospects. *BioEnergy Research, 14*(2), 670–688. DOI: 10.1007/s12155-020-10228-2

Jha, S., Singh, J., Chouhan, C., Singh, O., & Srivastava, R. K. (2022). Evaluation of Multiple Salinity Tolerance Indices for Screening and Comparative Biochemical and Molecular Analysis of Pearl Millet [Pennisetum Glaucum (L.) R. Br.] Genotypes. *Journal of Plant Growth Regulation, 41*(4), 1820–1834. DOI: 10.1007/s00344-021-10424-0

Jiang, M., Hong, K., Mao, Y., Ma, H., Chen, T., & Wang, Z. (2022). Natural 5-aminolevulinic acid: Sources, biosynthesis, detection, and applications. *Frontiers in Bioengineering and Biotechnology, 10*, 841443. DOI: 10.3389/fbioe.2022.841443 PMID: 35284403

Jiang, N., Naz, S., Ma, Y., Ullah, Q., Khan, M. Z., Wang, J., Lu, X., Luosang, D.-Z., Tabassum, S., Chatha, A. M. M., & Basang, W. D. (2023). An overview of comet assay application for detecting DNA damage in aquatic animals. *Agriculture, 13*(3), 623. DOI: 10.3390/agriculture13030623

Ji, W., Shen, Z., & Wen, Y. (2015). Hydrolysis of wheat straw by dilute sulfuric acid in a continuous mode. *Chemical Engineering Journal, 260*, 20–27. DOI: 10.1016/j.cej.2014.08.089

Johanson, J., & Vahlne, J.-E. (2009). The Uppsala internationalization process model revisited. *Journal of International Business Studies, 40*(9), 1411–1431. DOI: 10.1057/jibs.2009.24

Johnson, R., Vishwakarma, K., Hossen, M. S., Kumar, V., Shackira, A. M., Puthur, J. T., Abdi, G., Sarraf, M., & Hasanuzzaman, M. (2022). Potassium in Plants: Growth Regulation, Signaling, and Environmental Stress Tolerance. *Plant Physiology and Biochemistry, 172*(February), 56–69. DOI: 10.1016/j.plaphy.2022.01.001 PMID: 35032888

Jomova, K., Alomar, S. Y., Valko, R., Liska, J., Nepovimova, E., Kuca, K., & Valko, M. (2025). Flavonoids and their role in oxidative stress, inflammation, and human diseases. *Chemico-Biological Interactions, 413*, 111489. DOI: 10.1016/j.cbi.2025.111489 PMID: 40147618

Jönsson, L. J., & Martín, C. (2016). Pretreatment of lignocellulose: Formation of inhibitory by-products and strategies for minimizing their effects. *Bioresource Technology, 199*, 103–112. DOI: 10.1016/j.biortech.2015.10.009 PMID: 26482946

Joseph, J., Boby, S., Mooyottu, S. & Muyyarikkandy, M. S. (2025). Antibiotic potentiators as a promising strategy for combating antibiotic resistance. *NPJ Antimicrobials and Resistance, 3*, 53.

Kachoueiyan, F., Karbassi, A., Nasrabadi, T., Rashidiyan, M., & De-la-Torre, G. E. (2024). Speciation characteristics, ecological risk assessment, and source apportionment of heavy metals in the surface sediments of the Gomishan wetland. *Marine Pollution Bulletin, 198*, 115835. DOI: 10.1016/j.marpolbul.2023.115835 PMID: 38039575

Kalipci, E., Cüce, H., Ustaoğlu, F., Dereli, M. A., & Türkmen, M. (2023). Toxicological health risk analysis of hazardous trace elements accumulation in the edible fish species of the Black Sea in Türkiye using multivariate statistical and spatial assessment. *Environmental Toxicology and Pharmacology, 97*, 104028. DOI: 10.1016/j.etap.2022.104028 PMID: 36455837

Kamakura, Y., Yamanashi, K., & Sato, A. (2012). Freeze-dried royal jelly maintains its developmental and physiological bioactivity in *Drosophila melanogaster. Bioscience, Biotechnology, and Biochemistry, 76*(11), 2107–2111. DOI: 10.1271/bbb.120496 PMID: 23132575

Kammel, M., Erdmann, C., & Sawers, R. G. (2024). The formate-hydrogen axis and its impact on the physiology of enterobacterial fermentation. *Advances in Microbial Physiology, 84*, 51–82. DOI: 10.1016/bs.ampbs.2024.02.002 PMID: 38821634

Kammel, M., Hunger, D., & Sawers, R. G. (2021). The soluble cytoplasmic *N*-terminal domain of the FocA channel gates bidirectional formate translocation. *Molecular Microbiology, 115*(4), 758–773. DOI: 10.1111/mmi.14641 PMID: 33169422

Kammel, M., Pinske, C., & Sawers, R. G. (2022). FocA and its central role in fine-tuning pH homeostasis of enterobacterial formate metabolism. *Microbiology, 168*(10), 10. DOI: 10.1099/mic.0.001253 PMID: 36197793

Kammel, M., & Sawers, R. G. (2022). The FocA channel functions to maintain intracellular formate homeostasis during *Escherichia coli* fermentation. *Microbiology, 168*(4), 4. DOI: 10.1099/mic.0.001168 PMID: 35377837

Kammel, M., Trebbin, O., Pinske, C., & Sawers, R. G. (2022). A single amino acid-exchange converts FocA into a unidirectional efflux channel for formate. *Microbiology, 168*(1), 001132. DOI: 10.1099/mic.0.001132 PMID: 35084298

Kandil, A. A., Sharief, A. E., & Fatma, M. (2019). Evaluation of Some Rice Cultivars to Salt Tolerance under Antioxidant Using Physiological Indices. *International Journal of Environment, Agriculture and Biotechnology, 4*(4), 1069–1079. DOI: 10.22161/ijeab.4427

Kapeleka, J.A., Sauli, E., & Ndakidemi, P.A. (2021) Pesticide exposure and genotoxic effects as measured by DNA damage and human monitoring biomarkers. *Int J Environ Health Res., 31*(7), 805-822. DOI: 10.1080/09603123.2019.1690132

Karapetyan, H., Marutyan, S., Muradyan, A., Badalyan, H., Marutyan, S. V., & Trchounian, K. (2024, April). Changes in ATPase activity, antioxidant enzymes and proline biosynthesis in yeast *Candida guilliermondii* NP-4 under X-irradiation. *Journal of Bioenergetics and Biomembranes, 56*(2), 141–148. DOI: 10.1007/s10863-024-10003-4 PMID: 38308068

Karimi, J., & Mohsenzadeh, S. (2016). Effects of Silicon Oxide Nanoparticles on Growth and Physiology of Wheat Seedlings. *Физиология растений, 63*(1), 126–130. DOI: 10.7868/S0015330316010103

Kasaeian, A., Fereidooni, L., Ahmadbeigi, A., Kahedi, S., Koohshoori, M. S., Ghafarian, S., Tajmousavilangerudi, A., & Salaripoor, H. (2025). Review on Bioethanol Production from Fruit Peels. *Waste and Biomass Valorization, 16*(6), 2669–2691. Advance online publication. DOI: 10.1007/s12649-025-02938-5

Kasravi, M., Ahmadi, A., Babajani, A., Mazloomnejad, R., Hatamnejad, M. R., Shariatzadeh, S., Bahrami, S., & Niknejad, H. (2023). Immunogenicity of decellularized extracellular matrix scaffolds: A bottleneck in tissue engineering and regenerative medicine. *Biomaterials Research, 27*(1), 10. Advance online publication. DOI: 10.1186/s40824-023-00348-z PMID: 36759929

Kator, L., Hosea, Z. Y., & Oche, O. D. (2015). Sclerotium rolfsii: Causative organism of southern blight, stem rot, white mold and sclerotia rot disease. *Annals of Biological Research, 6*(11), 78–89.

Kaya, C., Tuna, A. L., Ashraf, M., & Altunlu, H. (2007). Improved salt tolerance of melon (Cucumis melo L.) by the addition of proline and potassium nitrate. *Environmental and Experimental Botany, 60*(3), 397–403. DOI: 10.1016/j.envexpbot.2006.12.008

Kazimierowicz, J., Dębowski, M., Zieliński, M., Kasiński, S., & Cruz Sanchez, J. (2024). Biotechnological Valorization of Waste Glycerol into Gaseous Biofuels—A Review. *Energies, 17*(2), 338. DOI: 10.3390/en17020338

Kenneth, S. K., Nneka, O. F., Kalu, A. K., & Joseph, A. O. (2023). Radioprotective Potencies of Allium Cepa Extract (ACE) against Radiation-Induced Hepatoxicity in Wistar Rats. *International Journal of Medical Physics, Clinical Engineering and Radiation Oncology, 12*(3), 59–83. DOI: 10.4236/ijmpcero.2023.123007

Khajeh-Hosseini, M., Lomholt, A., & Matthews, S. (2009). Mean Germination Time in the Laboratory Estimates the Relative Vigour and Field Performance of Commercial Seed Lots of Maize (Zea Mays L.). *Seed Science and Technology, 37*(2), 446–456. DOI: 10.15258/sst.2009.37.2.17

Kim, Y. H., Choi, J. S., Hong, J., Yoo, J. S. & Kim, M. S. (1999). Identification of acylated glycoglycerolipids from a cyanobacterium, Syn-echocystis sp., by tandem mass spectrometry. *Lipids, 34*, 847–853. DOI: 10.1007/s11745-999-0432-2

Kim, Y. H., Kim, E.-H., Lee, C., Kim, M.-H. & Rho, J.-R. (2007). Two new monogalactosyl diacylglycerols from brown alga Sargassum thunbergii. *Lipids, 42*, 395–399. DOI: 10.1007/s11745-007-3035-7

Kim, J. H., & Chan, K. L. (2022). Natural salicylaldehyde for fungal and pre-and post-emergent weed control. *Applied Sciences (Basel, Switzerland), 12*(8), 3749.

Kim, Y., Kim, H., & Kim, Y. (2025). Advancing hepatotoxicity assessment: Current advances and future directions. *Toxicological Research, 41*(4), 303–323. DOI: 10.1007/s43188-025-00289-w PMID: 40612519

Kinjo, Y., Tupin, E., Wu, D., Fujio, M., Garcia-Navarro, R., Benhnia, M. R.-E.-I., Zajonc, D. M., Ben-Menachem, G., Ainge, G. D., Painter, G. F., Khurana, A., Hoebe, K., Behar, S. M., Beutler, B., Wilson, I. A., Tsuji, M., Sellati, T. J., Wong, C.-H., & Kronenberg, M. (2006). Natural killer T cells recognize diacylglycerol antigens from pathogenic bacteria. *Nature Immunology*, *7*(9), 978–986. DOI: 10.1038/ni1380 PMID: 16921381

Kirby, A. (2023). Exploratory Bibliometrics: Using VOSviewer as a Preliminary Research Tool. *Publications, 11*(1), 10. DOI: 10.3390/publications11010010

Klerkx, L. (2024, December 18). The evolution of precision agriculture and food safety: A bibliometric review. *Frontiers in Sustainable Food Systems*, *8*, 1475602. DOI: 10.3389/fsufs.2024.1475602

Knappe, J., Neugebauer, F. A., Blaschkowski, H. P., & Gänzler, M. (1984). Post-translational activation introduces a free radical into pyruvate formate-lyase. *Proceedings of the National Academy of Sciences of the United States of America*, *81*(5), 1332–1335. DOI: 10.1073/pnas.81.5.1332 PMID: 6369325

Köhler, H. R., & Triebskorn, R. (2013). Wildlife Ecotoxicology of Pesticides: Can We Track Effects to the Population Level and Beyond? *Science*, *341*(6147), 759–765. DOI: 10.1126/science.1237591 PMID: 23950533

Koller, M. (2019). Switching from petro-plastics to microbial polyhydroxyalkanoates (PHA): The biotechnological escape route of choice out of the plastic predicament? *The EuroBiotech Journal*, *3*(1), 32–44. DOI: 10.2478/ebtj-2019-0004

Kombate, A., Folega, F., Atakpama, W., Dourma, M., Wala, K., & Goïta, K. (2022). Characterization of Land-Cover Changes and Forest-Cover Dynamics in Togo between 1985 and 2020 from Landsat Images Using Google Earth Engine. *Land (Basel)*, *11*(11), 1889. DOI: 10.3390/land11111889

Konda, N. V. S. N. M., Singh, S., Simmons, B. A., & Klein-Marcuschamer, D. (2015). An Investigation on the Economic Feasibility of Macroalgae as a Potential Feedstock for Biorefineries. *BioEnergy Research*, *8*(3), 1046–1056. DOI: 10.1007/s12155-015-9594-1

Kotsinyan, M. E. (1959). Epidemic rickettsiosis in the Armenian SSR. *Tenth Meeting on Parasitological Problems and Natural Focal Diseases,* 87–88.

Kottenhahn, P., Schuchmann, K., & Müller, V. (2018). Efficient whole cell biocatalyst for formate-based hydrogen production. *Biotechnology for Biofuels and Bioproducts*, *11*(1), 93. DOI: 10.1186/s13068-018-1082-3 PMID: 29619089

Koval, E. Z., Rudenko, A. V., & Voloshuk, N. M. (2016). *Penicillium*. NNIRXV.

Kozlov, A. V., Javadov, S., & Sommer, N. (2024). Cellular ROS and Antioxidants: Physiological and Pathological Role. *Antioxidants, 13*(5), 602. DOI: 10.3390/antiox13050602 PMID: 38790707

Krisai, P., Rudin, D., Grünig, D., Scherer, K., Pichler, W., Terracciano, L., & Krähenbühl, S. (2019). Acute liver failure in a patient treated with metamizole. *Frontiers in Pharmacology, 10*, 996. Advance online publication. DOI: 10.3389/fphar.2019.00996 PMID: 31572180

Kubier, A., Wilkin, R. T., & Pichler, T. (2019). Cadmium in soils and groundwater: A review. *Applied Geochemistry, 108*, 104388. DOI: 10.1016/j.apgeochem.2019.104388 PMID: 32280158

Küchlin, R., Huber, T., & Hausmann, R. (2021). A new class of medium-chain-length polyhydroxyalkanoates produced by Pseudomonas citronellolis. *Applied Microbiology and Biotechnology, 105*, 7421–7435. https://doi.org/DOI%2520not%2520found

Kumar, P., & Sharma, P. K. (2020). Soil Salinity and Food Security in India. *Frontiers in Sustainable Food Systems, 4*(October), 533781. Advance online publication. DOI: 10.3389/fsufs.2020.533781

Kumar, R., Thakur, A., Kumar, S., & Hajam, Y. A. (2024). Royal jelly a promising therapeutic intervention and functional food supplement. *A systematic review. Heliyon, 10*(17), e37138. DOI: 10.1016/j.heliyon.2024.e37138 PMID: 39296128

Kumar, S., Li, G., Yang, J., Huang, X., Ji, Q., Liu, Z., Ke, W., & Hou, H. (2021). Effect of Salt Stress on Growth, Physiological Parameters, and Ionic Concentration of Water Dropwort (Oenanthe Javanica) Cultivars. *Frontiers in Plant Science, 12*(June), 660409. Advance online publication. DOI: 10.3389/fpls.2021.660409 PMID: 34234795

Kurakov, A. V., & Kokaeva, L. Yu. (2023). *Methods for studying the diversity of fungi in terrestrial and aquatic ecosystems*. Public National Academy of Mycology.

Lal, R., Bouma, J., Brevik, E., Dawson, L., Field, D. J., Glaser, B., Hatano, R., Hartemink, A. E., Kosaki, T., Lascelles, B., Monger, C., Muggler, C., Ndzana, G. M., Norra, S., Pan, X., Paradelo, R., Reyes-Sánchez, L. B., Sandén, T., Singh, B. R., & Zhang, J. (2021). Soils and sustainable development goals of the United Nations: An International Union of Soil Sciences perspective. *Geoderma Regional, 25*, e00398. DOI: 10.1016/j.geodrs.2021.e00398

Lan, M. M., Liu, C., Liu, S. J., Qiu, R. L., & Tang, Y. T. (2020). Phytostabilization of Cd and Pb in Highly Polluted Farmland Soils Using Ramie and Amendments. *International Journal of Environmental Research and Public Health, 17*(5), 1661. DOI: 10.3390/ijerph17051661

Laudiño, F. A. R., Agtong, R. J. M., Jumawan, J. C., Fukuyama, M., & Elvira, M. V. (2024). Assessment of contamination and potential ecological risks of heavy metals in the bottom sediments of Lake Mainit, Philippines. *Journal of Hazardous Materials Advances, 15*, 100443. DOI: 10.1016/j.hazadv.2024.100443

Le Gall, M., & Tooker, J. F. (2017). Developing ecologically based pest management programs for terrestrial molluscs in field and forage crops. *Journal of Pest Science, 90*(3), 825–838. DOI: 10.1007/s10340-017-0858-8

Leandro, A., Pereira, L., & Gonçalves, A. M. M. (2020). Diverse Applications of Marine Macroalgae. *Marine Drugs, 18*(1), 17. DOI: 10.3390/md18010017

Lecina, M., Nadal, G., Solà, C., Prat, J., & Cairó, J. J. (2016). Optimization of ferric chloride concentration and pH to improve both cell growth and flocculation in Chlorella vulgaris cultures. Application to medium reuse in an integrated continuous culture bioprocess. *Bioresource Technology, 216*, 211–218. DOI: 10.1016/j.biortech.2016.05.063 PMID: 27240237

Lee, S.-J., Kim, K.-Y., Kim, G., Moon, S., Park, Y.-C., Cho, H.-S., & Oh, Y. (2024). Potential Tick Defense Associated with Skin and Hair Characteristics in Korean Water Deer (Hydropotes inermis argyropus). *Animals (Basel), 14*(2), 185–192. DOI: 10.3390/ani14020185 PMID: 38254354

Lehmann, J. (2007). A handful of carbon. *Nature, 447*(7141), 143–144. DOI: 10.1038/447143a

Lema, M., Ali, M. Y., & Retuerto, R. (2019). Domestication influences morphological and physiological responses to salinity in Brassica oleracea seedlings. *AoB Plants, 11*(5), plz046. Advance online publication. DOI: 10.1093/aobpla/plz046 PMID: 31579110

Lenz, O., Lauterbach, L., Frielingsdorf, S., & Friedrich, B. (2015). 4 Oxygen-tolerant hydrogenases and their biotechnological potential. In Rögner, M. (Ed.), *Biohydrogen* (pp. 61–96). GRUYTER. DOI: 10.1515/9783110336733.61

Leonhartsberger, S., Korsa, I., & Böck, A. (2002). The molecular biology of formate metabolism in enterobacteria. *Journal of Molecular Microbiology and Biotechnology, 4*, 269–276. https://pubmed.ncbi.nlm.nih.gov/11931558/ PMID: 11931558

Leonovich, S. A. (2019). Comparative analysis of sexual behavior in ixodid ticks (Ixodidae). *The Journal of Parasitology, 53*(4), 294–306. DOI: 10.1134/S0031184719040033

Liao, Q., Gu, S., Kang, S., Du, T., Tong, L., Wood, J. D., & Ding, R. (2021). Mild water and salt stress improve water use efficiency by decreasing stomatal conductance via osmotic adjustment in field maize. *The Science of the Total Environment, 805*, 150364. DOI: 10.1016/j.scitotenv.2021.150364 PMID: 34818800

Li, J.-L., Wie, L.-L., Chen, C., Liu, D., Gu, Y.-Q., Duan-Mu, J.-X., Chen, G.-T., & Song, Y. (2020). Bioactive Constituents from the Bryophyta Hypnum Plumaeforme. *Chemistry & Biodiversity, 17*(12), e2000552. DOI: 10.1002/cbdv.202000552 PMID: 33098214

Lisnjak, I. O., Kutsenok, V. V., Polyschuk, L. Z., Gorobets, O. B., & Gamaleia, N. F. (2005). Effect of photodynamic therapy on tumor angiogenesis and metastasis in mice bearing Lewis lung carcinoma. *Experimental Oncology, 27*(4), 333–335. DOI: 10.1007/s00018-009-0016-4 PMID: 16404357

Liu, J., Sun, Z., Zou, Y., Li, W., He, F., Huang, X., Lin, C., Cai, Q., Wisniewski, M., & Wu, X. (2022). Pre- and postharvest measures used to control decay and mycotoxigenic fungi in potato (*Solanum tuberosum* L.) during storage. *Critical Reviews in Food Science and Nutrition, 62*(2), 415–428. DOI: 10.1080/10408398.2020.1818688 PMID: 32924541

Liu, L., & Wang, B. (2021). Protection of halophytes and their uses for cultivation of Saline-Alkali soil in China. *Biology (Basel), 10*(5), 353. DOI: 10.3390/biology10050353 PMID: 33922035

Liu, P., Hu, W., Tian, K., Huang, B., Zhao, Y., Wang, X., Zhou, Y., Shi, B., Kwon, B. O., Choi, K., Ryu, J., Chen, Y., Wang, T., & Khim, J. S. (2020). Accumulation and ecological risk of heavy metals in soils along the coastal areas of the Bohai Sea and the Yellow Sea: A comparative study of China and South Korea. *Environment International, 137*, 105519. DOI: 10.1016/j.envint.2020.105519 PMID: 32014790

Liu, S., An, Z., & Lai, Z. (2024). Amaranth's growth and physiological responses to salt stress and the functional analysis of ATRTCP1 gene. *International Journal of Molecular Sciences, 25*(10), 5437. DOI: 10.3390/ijms25105437 PMID: 38791475

Lloyd, C., Tan, K. H., Lim, K. L., Valu, V. G., Fun, S. M. Y., Chye, T. R., Mak, H. M., Sim, W. X., Musa, S. L., Ng, J. J. Q., Bte Nordin, N. S., Bte Md Aidzil, N., Eng, Z. Y. W., Manickavasagam, P., & New, J. Y. (2021). Identification of microalgae cultured in Bold's Basal medium from freshwater samples, from a high-rise city. *Scientific Reports, 11*(1), 4474. DOI: 10.1038/s41598-021-84112-0 PMID: 33627771

Long, N. V., Dolstra, O., Malosetti, M., Kilian, B., Graner, A., Visser, R. G. F., & van der Linden, C. G. (2013). Association mapping of salt tolerance in barley (Hordeum vulgare L.). *Theoretical and Applied Genetics*, *126*(9), 2335–2351. DOI: 10.1007/s00122-013-2139-0 PMID: 23771136

Loudari, A., Benadis, C., Naciri, R., Soulaimani, A., Zeroual, Y., Gharous, M. E., Kalaji, H. M., & Oukarroum, A. (2020). Salt stress affects mineral nutrition in shoots and roots and chlorophyll a fluorescence of tomato plants grown in hydroponic culture. *Journal of Plant Interactions*, *15*(1), 398–405. DOI: 10.1080/17429145.2020.1841842

Louhimies, S. (2002). Directive 86/609/EEC on the protection of animals used for experimental and other scientific purposes. *Alternatives to Laboratory Animals*, *30*(2, Suppl.), 217–219. DOI: 10.1177/026119290203002S36 PMID: 12513679

Lowry, O. H., Rosebrough, H. J., Farr, A. L., & Randall, R. J. (1951). Protein Measurement with the Folin Phenol Reagent. *The Journal of Biological Chemistry*, *193*(1), 265–278. DOI: 10.1016/S0021-9258(19)52451-6 PMID: 14907713

Lucas, K., Capron, M., & Stewart, J. (2019) DE-AR0000916 AdjustaDepth Ph1 FinalReportTeam. ARPA DOI:DOI: 10.13140/RG.2.2.21140.32643

Lungoci, C., Motrescu, I., Filipov, F., Jitareanu, C. D., Teliban, G.-C., Ghitau, C. S., Puiu, I., & Robu, T. (2022). The Impact of Salinity Stress on Antioxidant Response and Bioactive Compounds of Nepeta Cataria L. *Agronomy (Basel)*, *12*(3), 562. DOI: 10.3390/agronomy12030562

Luo, X., Li, J., Guo, S., Yu, H., Zeng, X., Zhou, Z., Shangguan, Y., He, M., Ouyang, Y., Chen, K., Chen, Z., & Qin, Y. (2025). Analysis of research status and trends on nano-agricultural application: A bibliometric study. *Frontiers in Plant Science*, *16*, 1530629. DOI: 10.3389/fpls.2025.1530629 PMID: 40034150

Luo, Y., Li, Z., Li, X., Liu, X., Fan, J., Clark, J. H., & Hu, C. (2019). The production of furfural directly from hemicellulose in lignocellulosic biomass: A review. *Catalysis Today*, *319*, 14–24. DOI: 10.1016/j.cattod.2018.06.042

Lü, W., Du, J., Wacker, T., Gerbig-Smentek, E., Andrade, S. L. A., & Einsle, O. (2011). pH-dependent gating in a FocA formate channel. *Science*, *332*(6027), 352–354. DOI: 10.1126/science.1199098 PMID: 21493860

Luyckx, A., Lutts, S., & Quinet, M. (2023). Comparison of Salt Stress Tolerance among Two Leaf and Six Grain Cultivars of Amaranthus cruentus L. *Plants*, *12*(18), 3310. DOI: 10.3390/plants12183310 PMID: 37765474

Maas, E. V., & Grattan, S. R. (1999). *Crop Yields as Affected by Salinity*. DOI: 10.2134/agronmonogr38.c3

Maeda, N., Kokai, Y., Ohtani, S., Hada, T., Yoshida, H., & Mizushina, Y. (2009). Inhibitory effects of preventive and curative orally administered spinach glycoglycerolipid fraction on the tumor growth of sarcoma and colon in mouse graft models. *Food Chemistry, 112*(1), 205–210. DOI: 10.1016/j.foodchem.2008.05.059 PMID: 23251235

Maeda, N., Matsubara, K., Yoshida, H., & Mizushina, Y. (2011). Anti-cancer effect of spinach glycoglycerolipids as angiogenesis inhibitors based on the selective inhibition of DNA polymerase activity. *Mini-Reviews in Medicinal Chemistry, 11*(1), 32–38. DOI: 10.2174/138955711793564042 PMID: 21034405

Maeda, T., Sanchez-Torres, V., & Wood, T. K. (2008). Metabolic engineering to enhance bacterial hydrogen production. *Microbial Biotechnology, 1*(1), 30–39. DOI: 10.1111/j.1751-7915.2007.00003.x PMID: 21261819

Magalhães, C. P., Alves, J. I., Duber, A., Oleskowicz-Popiel, P., Stams, A. J. M., & Cavaleiro, A. J. (2024). Metabolic versatility of anaerobic sludge towards platform chemical production from waste glycerol. *Applied Microbiology and Biotechnology, 108*(1), 419. DOI: 10.1007/s00253-024-13248-6 PMID: 39012392

Mahawar, L., Živčák, M., Barboricova, M., Kovár, M., Filaček, A., Ferencova, J., Vysoká, D. M., & Brestič, M. (2024). Effect of Copper Oxide and Zinc Oxide Nanoparticles on Photosynthesis and Physiology of Raphanus Sativus L. under Salinity Stress. *Plant Physiology and Biochemistry, 206*(January), 108281. DOI: 10.1016/j.plaphy.2023.108281 PMID: 38157834

Mahdavian, K. (2023). Application of salicylic acid on chlorophyll, carotenoids, and proline in radish under salinity stress. *Proceedings of the National Academy of Sciences. India. Section B, Biological Sciences, 93*(4), 809–818. DOI: 10.1007/s40011-023-01484-1

Mahmoud, H., & Hussein, N. (2016). Data supporting the spectrophotometric method for the estimation of catalase activity. *Data in Brief, 6*, 194–199. DOI: 10.1016/j.dib.2015.12.012 PMID: 26862558

Mangal, M., Rao, C. V., & Banerjee, T. (2023). Bioplastic: An eco-friendly alternative to non-biodegradable plastic. *Polymer International, 72*(11), 984–996. DOI: 10.1002/pi.6555

Marija, V. (2023). Analysis of the Qualitative and Quantitative Content of the Phenolic Compounds of Selected Moss Species under NaCl Stress. *Molecules (Basel, Switzerland), 28*(4), 1794. DOI: 10.3390/molecules28041794 PMID: 36838781

Marris, E. (2006). Putting the carbon back: Black is the new green. *Nature*, *442*(7103), 624–626. DOI: 10.1038/442624a PMID: 16900176

Martínez-Abaigar, J., & Núñez-Olivera, E. (2021). Novel biotechnological substances from bryophytes. In Sinha, R. P., & Häder, D.-P. (Eds.), *Natural Bioactive Compounds* (pp. 233–248). Academic Press., DOI: 10.1016/B978-0-12-820655-3.00011-2

Martins, M., & Costa, P. M. (2015). The comet assay in Environmental Risk Assessment of marine pollutants: Applications, assets and handicaps of surveying genotoxicity in non-model organisms. *Mutagenesis*, *30*(1), 89–106. DOI: 10.1093/mutage/geu037 PMID: 25527732

Marutyan, E. M. (1979). Ticks - carriers as a factor in the preservation of foci of some babesiosis in nature. *All-Union conference on natural foci of diseases. Alma-Ata.*, *1*, 127–128.

Marutyan, S., Marutyan, S., Navasardyan, L., Hovnanyan, K., & Trchounian, A. (2021). Changes in growth kinetic parameters, morphology and mitotic activity of yeast Candida guilliermondii exposed to the low-intensity waves of 51.8-GHz frecuency. *Archives of Microbiology*, *203*(6), 3707–3714. DOI: 10.1007/s00203-021-02336-0 PMID: 33938972

Marutyan, Ye. M. (1964). *Haemosporidiosis situation in Ijevan Region of Armenian SSR*. Proceedings for Armenian Scientific Research Institute of Animal Husbandry and Veterinary.

Marutyan, Ye. M., Markosyan, V. M., & Oganesyan, V. V. (1974). Some biological and ecological features of Ixodid ticks, and their role in transmission of piroplasmidoses in mountain forest area of Armenia. *Proceedings for Armenian Scientific Research Institute of Animal Husbandry and Veterinary.*, *12*, 649–659.

Marutyan, Ye. M., & Oganesyan, V. V. (1974). Epidemiological features of the bovine piroplasmidosis in mountain forest area of Armenian SSR. *Proceedings for Armenian Scientific Research Institute of Animal Husbandry and Veterinary.*, *12*, 639–647.

Ma, T. H. (1999). The international program on plant bioassays and the report of the follow-up study after the hands-on workshop in China. *Mutation Research. Fundamental and Molecular Mechanisms of Mutagenesis*, *426*(2), 103–106. DOI: 10.1016/S0027-5107(99)00049-4 PMID: 10350579

Ma, T. H., Cabrera, G. L., Cebulska-Wasilevska, A., Chen, R., Loarca, A. I., Vandenberg, A. L., & Salamone, M. F. (1994a). Tradescantia stamen hair mutation bioassay. *Mutation Research, 310*(2), 211–220. DOI: 10.1016/0027-5107(94)90114-7 PMID: 7523892

Ma, T. H., Cabrera, G. L., Chen, R., Gill, B. S., Sandhu, S. S., Vandenberg, A. L., & Salamone, M. F. (1994b). Tradescantia micronucleus bioassay. *Mutation Research, 310*(2), 220–230. DOI: 10.1016/0027-5107(94)90115-5 PMID: 7523893

Mathesius, U. (2018). Flavonoid Functions in Plants and Their Interactions with Other Organisms. *Plants, 7*(2), 30. DOI: 10.3390/plants7020030 PMID: 29614017

Mazza, G., Rombouts, K., Rennie Hall, A., Urbani, L., Vinh Luong, T., Al-Akkad, W., Longato, L., Brown, D., Maghsoudlou, P., Dhillon, A. P., Fuller, B., Davidson, B., Moore, K., Dhar, D., De Coppi, P., Malago, M., & Pinzani, M. (2015). Decellularized human liver as a natural 3D-scaffold for liver bioengineering and transplantation. *Scientific Reports, 5*(1), 13079. Advance online publication. DOI: 10.1038/srep13079 PMID: 26248878

McDowall, J. S., Murphy, B. J., Haumann, M., Palmer, T., Armstrong, F. A., & Sargent, F. (2014). Bacterial formate hydrogenlyase complex. *Proceedings of the National Academy of Sciences of the United States of America, 111*(38), E3948–E3956. DOI: 10.1073/pnas.1407927111 PMID: 25157147

Mejía-Barajas, J., Montoya-Perez, R., Salgado-Garciglia, R., Aguilera-Aguirre, L., Cortés-Rojo, C., Mejía-Zepeda, R., Arellano-Plaza, M., & Saavedra-Molina, A. (2017). Oxidative stress and antioxidant response in a thermotolerant yeast. *Brazilian Journal of Microbiology : Publication of the Brazilian Society for Microbiology, 48*(2), 326–332. DOI: 10.1016/j.bjm.2016.11.005 PMID: 28094115

Melek, S. M. (1994). How much will coverage for universal access to comprehensive behavioral healthcare really cost? *Behavioral Healthcare Tomorrow, 3*, 31–38. PMID: 10141167

Mendibil, U., Ruiz-Hernandez, R., Retegi-Carrion, S., Garcia-Urquia, N., Olalde-Graells, B., & Abarrategi, A. (2020). Tissue-specific decellularization methods: Rationale and strategies to achieve regenerative compounds. *International Journal of Molecular Sciences, 21*(15), 5447. Advance online publication. DOI: 10.3390/ijms21155447 PMID: 32751654

Metcalfe, G., Sargent, F., & Hippler, M. (2022). Hydrogen production in the presence of oxygen by *Escherichia coli* K-12. *Microbiology, 168*(3), 001167. DOI: 10.1099/mic.0.001167 PMID: 35343886

Miller, J. (1972). *Experiments in Molecular Genetics*. Cold Spring Harbor Laboratory.

Mirzakulova, A., Sarsembaeva, T., Suleimenova, Z., Kowalski, Ł., Gajdzik, B., Wolniak, R., & Bembenek, M. (2025). Whey: Composition, Processing, Application, and Prospects in Functional and Nutritional Beverages—A Review. *Foods*, *14*(18), 3245. DOI: 10.3390/foods14183245 PMID: 41008217

Miyashita, A., Kizaki, H., Sekimizu, K., & Kaito, C. (2016). Body-enlarging effect of royal jelly in a non-holometabolous insect species. *Gryllus bimaculatus. Biology Open*, *5*(6), 770–776. DOI: 10.1242/bio.019190 PMID: 27185266

Moghrovyan, A., Sahakyan, N., Babayan, A., Chichoyan, N., Petrosyan, M., & Trchounian, A. (2019). Essential Oil and Ethanol Extract of Oregano (Origanum vulgare L.) from Armenian Flora as a Natural Source of Terpenes, Flavonoids and other Phytochemicals with Antiradical, Antioxidant, Metal Chelating, Tyrosinase Inhibitory and Antibacterial Activity. *Current Pharmaceutical Design*, *25*(16), 1809–1816. DOI: 10.2174/1381612825666190702095612 PMID: 31267860

Mora, O., Le Mouël, C., De Lattre-Gasquet, M., Donnars, C., Dumas, P., Réchauchère, O., Brunelle, T., Manceron, S., Marajo-Petitzon, E., Moreau, C., Barzman, M., Forslund, A., & Marty, P. (2020). Exploring the future of land use and food security: A new set of global scenarios. *PLoS One*, *15*(7), e0235597. DOI: 10.1371/journal.pone.0235597 PMID: 32639991

Mordor Intelligence. (2024). *Analysing Biofuels Market Size and Share - Growth Trends and Forecasts (2024-2029)*. https://www.mordorintelligence.com/ru/industry-reports/biofuels-market

Muchate, N. S., Nikalje, G. C., Rajurkar, N. S., Suprasanna, P., & Nikam, T. D. (2016). Plant Salt Stress: Adaptive Responses, Tolerance Mechanism and Bioengineering for Salt Tolerance. *The Botanical Review, 82*(4), 371–406. DOI: 10.1007/s12229-016-9173-y

Munns, R. (2005). Genes and salt tolerance: Bringing them together. *The New Phytologist*, *167*(3), 645–663. DOI: 10.1111/j.1469-8137.2005.01487.x PMID: 16101905

Munns, R., & Termaat, A. (1986). Whole-Plant Responses to Salinity. *Australian Journal of Plant Physiology*, *13*(1), 143–160. DOI: 10.1071/PP9860143

Munns, R., & Tester, M. (2008). Mechanisms of salinity tolerance. *Annual Review of Plant Biology*, *59*(1), 651–681. DOI: 10.1146/annurev.arplant.59.032607.092911 PMID: 18444910

Murad, M. A., Khan, A. L., & Muneer, S. (2020). Silicon in Horticultural Crops: Cross-talk, Signaling, and Tolerance Mechanism under Salinity Stress. *Plants*, *9*(4), 460. DOI: 10.3390/plants9040460 PMID: 32268477

Murakami, C., Kumagai, T., Hada, T., Kanekazu, U., Nakazawa, S., Kamisuki, S., Maeda, N., Xu, X., Yoshida, H., Sugawara, F., Sakaguchi, K., & Mizushina, Y. (2003). Effects of glycolipids from spinach on mammalian DNA polymerases. *Biochemical Pharmacology*, *65*(2), 259–267. DOI: 10.1016/S0006-2952(02)01483-1 PMID: 12504801

Murkute, A. A., Sharma, S., & Singh, S. K. (2006). Studies on salt stress tolerance of citrus rootstock genotypes with arbuscular mycorrhizal fungi. *Horticultural Science (Prague)*, *33*(2), 70–76. DOI: 10.17221/3742-HORTSCI

Murtaza, G., Ahmed, Z., Eldin, S. M., Ali, B., Bawazeer, S., Usman, M., Iqbal, R., Neupane, D., Ullah, A., Khan, A., Hassan, M. U., Ali, I., & Tariq, A. (2023). Biochar-Soil-Plant interactions: A cross talk for sustainable agriculture under changing climate. *Frontiers in Environmental Science*, *11*, 1059449. DOI: 10.3389/fenvs.2023.1059449

Mustafa, G., Chaudhari, S. K., Manzoor, M., Batool, S., Hatami, M., & Hasan, M. (2024). Zinc Oxide Nanoparticles Mediated Salinity Stress Mitigation in Pisum Sativum: A Physio-Biochemical Perspective. *BMC Plant Biology*, *24*(1), 835. Advance online publication. DOI: 10.1186/s12870-024-05554-y PMID: 39243061

Muthmainna, S. K. (2020). A Systematic Review of Slash-and-Burn Agriculture as an Obstacle to Future-Proofing Climate Change. *International Journal of Research in Agricultural Sciences*, *4*(1), 1–8. https://golnk.ru/5DArv

Naghashyan, H. Z., Shcherbakov, O. V., & Movsisyan, L. A. (2013). Hard ticks as vectors of Haemosporidia in Goris Region of Armenia. *Proceedings for International Conference "Fundamental and Applied Aspects for Study of Parasitic Arthropodes in XXI Century"*, 109-110.

Nanyan, N. S. B. M., & Takagi, H. (2020). Proline homeostasis in *Saccharomyces cerevisiae*: How does the stress-responsive transcription. factor Msn2 play a role? *Frontiers in Genetics*, *11*, 438. DOI: 10.3389/fgene.2020.00438 PMID: 32411186

Navasardyan, L., Marutyan, S., Hovnanyan, K., & Trchounyan, A. (2017). Survival and changes in morphology, mitotic and metabolic activity of yeast *Candida guilliermondii* exposed to X-irradiation. *Indian Journal of Biochemistry & Biophysics*, *54*, 273–280. http://nopr.niscair.res.in/handle/123456789/43366

NCBI. (2023). https://www.ncbi.nlm.nih.gov/

Nedbaylov, M. S. (2023). Bioenergy: Main market development trends and prospects for Russia. *Creative Economy.*, *3*(17), 869–882. DOI: 10.18334/ce.17.3.117454

Neto, D., José, F., Carneiro, D. C. D. S., & Putti, F. F. (2024). Physiological Indexes in Seed Germination and Seedling Growth of Rangpur Lime (Citrus Limonia L. Osbeck) under Plant Growth Regulators. *Agronomy (Basel)*, *14*(9), 2066. DOI: 10.3390/agronomy14092066

Ng, H.-S., Chang, C.-Y., Chiu, C.-A., & Lan, J. C.-W. (2025). Metabolic engineering of Ralstonia eutropha for enhanced poly(3-hydroxybutyrate) production from crude glycerol. *Bioresource Technology*, *133531*. Advance online publication. DOI: 10.1016/j.biortech.2025.133531 PMID: 41139026

Nikalje, G. C., Srivastava, A. K., Pandey, G. K., & Suprasanna, P. (2017). Halophytes in biosaline agriculture: Mechanism, utilization, and value addition. *Land Degradation & Development*, *29*(4), 1081–1095. DOI: 10.1002/ldr.2819

Nippi, Inc. (2024). *BioMasher® II: Protocol manual* (Cat. No. BM-02N). https://www.nippi-inc.com/en/products/biomashers/biomasher-ii/

Niste, M. G., Stoian, V., Monica, N., Vidican, R., Rotar, I., Pop, R., & Miclea, R. (2014). *Plant Nutrition Affected by Soil Salinity and Response of Rhizobium Regarding the Nutrients Accumulation Ioan Rotar Plant Nutrition Affected by Soil Salinity and Response of Rhizobium Regarding the Nutrients Accumulation.*

Noman, A., Ali, S., Naheed, F., Ali, Q., Farid, M., Rizwan, M., & Irshad, M. K. (2015). Foliar application of ascorbate enhances the physiological and biochemical attributes of maize (Zea mays L.) cultivars under drought stress. *Archives of Agronomy and Soil Science*, *61*(12), 1659–1672. DOI: 10.1080/03650340.2015.1028379

Novosti, R. I. A. (2025). *A seaweed processing plant will open in Karelia this year.* https://ria.ru/20250117/zavod-1994211751.html?ysclid=m7jjboxk8p817378950

Nsabimana, A. (2022). *Évaluation et stratégie du financement et développement du Togo.* PNUD, Ministère de l'Economie et des Finances (MEF) et du Ministère de la Planification du Développement et de la Coopération (MPDC). https://inff.org/assets/resource/efd_pnud-togo-dfa-final.pdf

Obrador, E., & Montoro, A. (2023). Ionizing Radiation, Antioxidant Response and Oxidative Damage: Radiomodulators. *Antioxidants*, *12*(6), 1219. DOI: 10.3390/antiox12061219 PMID: 37371949

Oehlenschläger, K., Lorenz, M., Schepp, E., Di Nonno, S., Holtmann, D., & Ulber, R. (2025). Integrated co-cultivation and subsequent esterification: Harnessing Saccharomyces cerevisiae and Clostridium tyrobutyricum for streamlined ester production. *Biotechnology for Biofuels and Bioproducts, 18*(1), 98. DOI: 10.1186/s13068-025-02698-3 PMID: 40890840

Oganjanyan, A. M., Harutyunyan, E. S., & Rukhkyan, M. Ya. (1990). Distribution of Ixodid ticks in the Armenian SSR. *Advances of Medical Entomology and Acarology in the USSR, X Congress of the All-Union Entomological Society, Leningrad, Russia, 11-15 September 1989,* 118-119.

Ogden, N.H., Ben Beard, C., Ginsberg, H.S., & Jean I. Tsao, J.I. (2021). Possible Effects of Climate Change on Ixodid Ticks and the Pathogens They Transmit: Predictions and Observations *Journal of Medical Entomology, 58*(4), 1536–1545. DOI: 10.1093/jme/tjaa220

Okebalama, C. B., & Marschner, B. (2023). Reapplication of biochar, sewage waste water, and NPK fertilizers affects soil fertility, aggregate stability, and carbon and nitrogen in dry-stable aggregates of semi-arid soil. *The Science of the Total Environment, 866*, 161203. DOI: 10.1016/j.scitotenv.2022.161203 PMID: 36581285

Olivares-Marin, I. K., González-Hernández, J. C., Regalado-Gonzalez, C., & Madrigal-Perez, L. A. (2018). <em>Saccharomyces cerevisiae</em> Exponential Growth Kinetics in Batch Culture to Analyze Respiratory and Fermentative Metabolism. *Journal of Visualized Experiments, 139*(139). Advance online publication. DOI: 10.3791/58192 PMID: 30320748

Oliveira, A. S., Ferreira, C., Pereira, J. O., Pintado, M. E., & Carvalho, A. P. (2025). Valorisation of protein-rich extracts from spent brewer's yeast (Saccharomyces cerevisiae): An overview. *Biomass Conversion and Biorefinery, 15*(2), 1771–1793. DOI: 10.1007/s13399-022-02636-5

Omamt, E. N., Hammes, P. S., & Robbertse, P. J. (2006). Differences in salinity tolerance for growth and water-use efficiency in some amaranth(Amaranthusspp.) genotypes. *New Zealand Journal of Crop and Horticultural Science, 34*(1), 11–22. DOI: 10.1080/01140671.2006.9514382

Osborne, C. K., Hobbs, K., & Trent, J. M. (1987). Biological differences among MCF-7 human breast cancer cell lines from different laboratories. *Breast Cancer Research and Treatment, 9*(2), 111–121. DOI: 10.1007/BF01807363 PMID: 3620713

Oyebamiji, A. O., Olaolorun, O. A., Popoola, O. J. P., & Zafar, T. (2024). Assessment of heavy metal pollution in soils of Jebba Area, Nigeria: Concentrations, source analysis and implications for ecological and human health risks. *The Science of the Total Environment*, *945*, 173860. DOI: 10.1016/j.scitotenv.2024.173860 PMID: 38871321

Özdemir, F., Bor, M., Demiral, T., & Türkan, I. (2004). Effects of 24-epibrassinolide on seed germination, seedling growth, lipid peroxidation, proline content and antioxidative system of rice (Oryza sativa L.) under salinity stress. *Plant Growth Regulation*, *42*(3), 203–211. DOI: 10.1023/B:GROW.0000026509.25995.13

Padoan, E., Hernandez Kath, A., Vahl, L. C., & Ajmone-Marsan, F. (2020). Potential Release of Zinc and Cadmium From Mine-Affected Soils Under Flooding, a Mesocosm Study. *Archives of Environmental Contamination and Toxicology*, *79*(4), 421–434. DOI: 10.1007/s00244-020-00777-0 PMID: 33175188

Papp, J. C., Ball, M. C., & Terry, N.PAPP. (1983). A comparative study of the effects of NaCl salinity on respiration, photosynthesis, and leaf extension growth in Beta vulgaris L. (sugar beet). *Plant, Cell & Environment*, *6*(8), 675–677. DOI: 10.1111/1365-3040.ep11589273

Parida, A. K., & Das, A. B. (2004). Effects of NaCl stress on nitrogen and phosphorous metabolism in a true mangrove Bruguiera parviflora grown under hydroponic culture. *Journal of Plant Physiology*, *161*(8), 921–928. DOI: 10.1016/j.jplph.2003.11.006 PMID: 15384403

Parida, A. K., & Das, A. B. (2005a). Salt tolerance and salinity effects on plants: A review. *Ecotoxicology and Environmental Safety*, *60*(3), 324–349. DOI: 10.1016/j.ecoenv.2004.06.010 PMID: 15590011

Park, J. H., Yun, J. J., Kim, S. H., Park, J. H., Acharya, B. S., Cho, J. S., & Kang, S. W. (2023). Biochar improves soil properties and corn productivity under drought conditions in South Korea. *Biochar, 5*(1), 66-. DOI: 10.1007/S42773-023-00267-1/TABLES/3

Partheeban, C., Chandrasekhar, C. N., Jeyakumar, P., Ravikesavan, R., & Gnanam, R. (2017). Effect of PEG Induced Drought Stress on Seed Germination and Seedling Characters of Maize (Zea Mays L.) Genotypes. *International Journal of Current Microbiology and Applied Sciences*, *6*(5), 1095–1104. DOI: 10.20546/ijcmas.2017.605.119

Pathak, Mehta, & Thakur. (2023). In Vitro Seed Germination Studies in Barley Exposed to Zinc Oxide Nanoparticles. *Agricultural Reviews*. .DOI: 10.18805/ag.R-2637

Pattanagul, W., & Thitisaksakul, M. (2008). Effect of salinity stress on growth and carbohydrate metabolism in three rice (*Oryza sativa* L.) cultivars differing in salinity tolerance. *IJEB, 46*, 736–742.

Paul, D. (2013). Osmotic stress adaptations in rhizobacteria. *Journal of Basic Microbiology, 53*(2), 101–110. DOI: 10.1002/jobm.201100288 PMID: 22581676

Pedersen, M. B. (1986). DNA sequence polymorphisms in the genus saccharomyces III. Restriction endonuclease fragment patterns of chromosomal regions in brewing and other yeast strains. *Carlsberg Research Communications, 51*(3), 163–183. DOI: 10.1007/BF02907322

Peltonen-Sainio, P., Jauhiainen, L., & Laurila, I. P. (2009). Cereal yield trends in northern European conditions: Changes in yield potential and its realisation. *Field Crops Research, 110*(1), 85–90. DOI: 10.1016/j.fcr.2008.07.007

Pescuma, M., De Valdez, G. F., & Mozzi, F. (2015). Whey-derived valuable products obtained by microbial fermentation. *Applied Microbiology and Biotechnology, 99*(15), 6183–6196. DOI: 10.1007/s00253-015-6766-z PMID: 26124070

Peters, K., & Sargent, F. (2023). Formate hydrogenlyase, formic acid translocation and hydrogen production: Dynamic membrane biology during fermentation. *Biochimica et Biophysica Acta. Bioenergetics, 1864*(1), 148919. DOI: 10.1016/j.bbabio.2022.148919 PMID: 36152681

Petridis, P., Jha, A. N., & Langston, W. J. (2009). Measurements of the genotoxic potential of (xeno-)oestrogens in the bivalve mollusc Scrobicularia plana, using the Comet assay. *Aquatic Toxicology (Amsterdam, Netherlands), 94*(1), 8–15. DOI: 10.1016/j.aquatox.2009.05.010 PMID: 19540004

Pidoplichko, N. M., & Milko, A. A. (1971). *Atlas of Mucoral Fungi*. Naukova Dumka.

Pietrzykowski, M., Antonkiewicz, J., Gruba, P., & Pajak, M. (2018). Content of Zn, Cd and Pb in purple moor-grass in soils heavily contaminated with heavy metals around a zinc and lead ore tailing landfill. *Open Chemistry, 16*(1), 1143–1152. DOI: 10.1515/chem-2018-0129

Pignatelli, P., Umme, S., D'Antonio, D. L., Piattelli, A., & Curia, M. C. (2023). Reactive oxygen species produced by 5-aminolevulinic acid photodynamic therapy in the treatment of cancer. *International Journal of Molecular Sciences, 24*(10), 8964. DOI: 10.3390/ijms24108964 PMID: 37240309

Pirzadi, Z., & Meshkani, F. (2022). From glycerol production to its value-added uses: A critical review. *Fuel, 329*, 125044. DOI: 10.1016/j.fuel.2022.125044

Plaga, W., Frank, R., & Knappe, J. (1988). Catalytic-site mapping of pyruvate formate lyase: Hypophosphite reaction on the acetyl-enzyme intermediate affords carbon-phosphorus bond synthesis (1-hydroxyethylphosphonate). *European Journal of Biochemistry, 178*(2), 445–450. DOI: 10.1111/j.1432-1033.1988.tb14468.x PMID: 3061816

Poladyan, A., Blbulyan, S., Sahakyan, M., Lenz, O., & Trchounian, A. (2019). Growth of the facultative chemolithoautotroph Ralstonia eutropha on organic waste materials: Growth characteristics, redox regulation and hydrogenase activity. *Microbial Cell Factories, 18*(1), 201. DOI: 10.1186/s12934-019-1251-5 PMID: 31739794

Poladyan, A., Trchounian, K., Paloyan, A., Minasyan, E., Aghekyan, H., Iskandaryan, M., Khoyetsyan, L., Aghayan, S., Tsaturyan, A., & Antranikian, G. (2023). Valorization of whey-based side streams for microbial biomass, molecular hydrogen, and hydrogenase production. *Applied Microbiology and Biotechnology, 107*(14), 4683–4696. DOI: 10.1007/s00253-023-12609-x PMID: 37289241

Pour-Aboughadareh, A., Sanjani, S., Nikkhah-Chamanabad, H., Mehrvar, M. R., Asadi, A., & Amini, A. (2021). Identification of salt-tolerant barley genotypes using multiple-traits index and yield performance at the early growth and maturity stages. *Bulletin of the National Research Centre, 45*(1), 1–16. DOI: 10.1186/s42269-021-00576-0

Prasad, S., Singh, A., & Joshi, H. C. (2007). Ethanol as an alternative fuel from agricultural, industrial and urban residues. *Resources, Conservation and Recycling, 50*(1), 1–39. DOI: 10.1016/j.resconrec.2006.05.007

Prasertwasu, S., Khumsupan, D., Komolwanich, T., Chaisuwan, T., Luengnaruemitchai, A., & Wongkasemjit, S. (2014). Efficient process for ethanol production from Thai Mission grass (Pennisetum polystachion). *Bioresource Technology, 163*, 152–159. DOI: 10.1016/j.biortech.2014.04.043 PMID: 24811442

Priyanka, S., Gautam, N., Arora, A., & Ghanghoriya, H. (2024). A Review of Phytochemical Constituents and their Pharmacological Activities. *Frontiers in Health Informatics, 14*(2), 2645-2658. DOI: 10.63682/fhi2726

Procurement, E. I. S. (2024). *Construction of a production complex for primary and (or) subsequent (industrial) processing, storage of algae products in the Republic of Karelia: Installation of on-site electrical networks and outdoor lighting system, owned by NPO BIT Ltd.* https://zakupki.gov.ru/epz/order/notice/ea20/view/documents.html?regNumber=2300700043924000008

Pulvento, C., Sellami, M. H., & Lavini, A. (2021). Yield and quality of Amaranthus hypochondriacus grain amaranth under drought and salinity at various phenological stages in southern Italy. *Journal of the Science of Food and Agriculture, 102*(12), 5022–5033. DOI: 10.1002/jsfa.11088 PMID: 33448400

Puppi, D., Pecorini, G., & Chiellini, F. (2019). Biomedical Processing of Polyhydroxyalkanoates. *Bioengineering (Basel, Switzerland), 6*(4), 108. DOI: 10.3390/bioengineering6040108 PMID: 31795345

Qadir, M., Quillérou, E., Nangia, V., Murtaza, G., Singh, M., Thomas, R. J., Drechsel, P., & Noble, A. D. (2014). Economics of salt-induced land degradation and restoration. *Natural Resources Forum, 38*(4), 282–295. DOI: 10.1111/1477-8947.12054

QIAGEN. (2024). *QIAamp DNA Mini and Blood Mini handbook* (Cat. No. 51304). https://www.qiagen.com/us/resources/resourcedetail?id=bf602afb-cc1e-4032-b762-6e364adf39ff&lang=en

Qin, J., Dong, W. Y., He, K. N., Yu, Y., Tan, G. D., Han, L., Dong, M., Zhang, Y. Y., Zhang, D., Li, A. Z., & Wang, Z. L. (2010). NaCl salinity-induced changes in water status, ion contents and photosynthetic properties of Shepherdia argentea (Pursh) Nutt. seedlings. *Plant, Soil and Environment, 56*(7), 325–332. DOI: 10.17221/209/2009-PSE

Qi, Y., Wei, X., Zhao, M. J., Pan, W., Jiang, C., Wu, J., & Li, W. C. (2022). Heavy metal pollution characteristics and potential ecological risk assessment of soils around three typical antimony mining areas and watersheds in China. *Frontiers in Environmental Science, 10*, 913293. DOI: 10.3389/fenvs.2022.913293

Quah, B. J. C., Warren, H. S., & Parish, C. R. (2007). Monitoring lymphocyte proliferation in vitro and in vivo with the intracellular fluorescent dye carboxyfluorescein diacetate succinimidyl ester. *Nature Protocols, 2*(9), 2049–2056. DOI: 10.1038/nprot.2007.296 PMID: 17853860

Quejada, L. F., Hernandez, A. X., & Chitiva, L. C. (2024) Unmasking Antifungal Might:Anacardium occidentale Leaf ExtractUnleashes Mitochondrial Disruption,Thwarting Candida albicans withHeightened Reactive Oxygen Species. *Preprints*.org. DOI: 10.20944/preprints202405.0579.v1

Quéméneur, M., Hamelin, J., Barakat, A., Steyer, J.-P., Carrère, H., & Trably, E. (2012). Inhibition of fermentative hydrogen production by lignocellulose-derived compounds in mixed cultures. *International Journal of Hydrogen Energy, 37*(4), 3150–3159. DOI: 10.1016/j.ijhydene.2011.11.033

Quintarelli, V., Ben Hassine, M., Radicetti, E., Stazi, S. R., Bratti, A., Allevato, E., Mancinelli, R., Jamal, A., Ahsan, M., Mirzaei, M., & Borgatti, D. (2024). Advances in Nanotechnology for Sustainable Agriculture: A Review of Climate Change Mitigation. *Sustainability, 16*(21), 9280. DOI: 10.3390/su16219280

Radwan, M. A., El-Gendy, K. S., & Gad, A. F. (2020). Biomarker responses in terrestrial gastropods exposed to pollutants: A comprehensive review. *Chemosphere, 257*, 127218. DOI: 10.1016/j.chemosphere.2020.127218 PMID: 32497833

Rahmani, A. M., Tyagi, V. K., Kazmi, A. A., & Ojha, C. S. P. (2023). Hydrothermal and thermal-acid pretreatments of wheat straw: Methane yield, recalcitrant formation, process inhibition, kinetic modeling. *Energy, 283*, 129083. DOI: 10.1016/j.energy.2023.129083

Rajan Selin, A., & Shanmugasundram, S. (2024). Traditional approaches of agriculture soil remediation: boon or bane for global food security. *Bioremediation of Emerging Contaminants from Soils: Soil Health Conservation for Improved Ecology and Food Security*, 43–62. DOI: 10.1016/B978-0-443-13993-2.00003-7

Rajeswari, S., Sood, N., Swarup, T. G., Subramanian, R., Salcedo-Reyes, J. C., Rajeswari, S., Sood, N., Swarup, T. G., Subramanian, R., & Salcedo-Reyes, J. C. (2019). Assessing salt-stress tolerance in barley. *Universitas Scientiarum, 24*(1), 91–109. DOI: 10.11144/Javeriana.SC24-1.asst

Rajput, V. D., Minkina, T., Fedorenko, A., Chernikova, N., Hassan, T., Mandzhieva, S., Sushkova, S., Lysenko, V., Soldatov, M. A., & Burachevskaya, M. (2021). Effects of Zinc Oxide Nanoparticles on Physiological and Anatomical Indices in Spring Barley Tissues. *Nanomaterials, 11*(7), 1722. DOI: 10.3390/nano11071722

Rajput, P., Singh, A., Agrawal, S., Ghazaryan, K., Rajput, V. D., Movsesyan, H., Mandzhieva, S., Minkina, T., & Alexiou, A. (2024). Effects of environmental metal and metalloid pollutants on plants and human health: Exploring nano-remediation approach. *Stress Biology, 4*(1), 1–25. DOI: 10.1007/s44154-024-00156-y PMID: 38777953

Ramachandran, A., & Jaeschke, H. (2019). Acetaminophen hepatotoxicity. *Seminars in Liver Disease, 39*(2), 221–234. DOI: 10.1055/s-0039-1679919 PMID: 30849782

Ramanathan, A. N. K. G., Nair, A. J., & Sugunan, V. S. (2016). A review on Royal Jelly proteins and peptides. *Journal of Functional Foods, 44*, 255–264. DOI: 10.1016/j.jff.2018.03.008

Ramón, S., Kielland-Brandt, M. C., & Fink, G. R. (1986). Yeast plasma membrane ATPase is essential for growth and has homology with ($Na^+ + K^+$), K^+- and Ca^{2+}-ATPases. *Nature, 319*(6055), 689–693. DOI: 10.1038/319689a0 PMID: 3005867

Ran, C., Gulaqa, A., Zhu, J., Wang, X., Zhang, S., Geng, Y., Guo, L., Jin, F., & Shao, X. (2020). Benefits of Biochar for Improving Ion Contents, Cell Membrane Permeability, Leaf Water Status and Yield of Rice Under Saline–Sodic Paddy Field Condition. *Journal of Plant Growth Regulation, 39*(1), 370–377. DOI: 10.1007/s00344-019-09988-9

Rasel, M., Tahjib-Ul-Arif, M., Hossain, M. A., Hassan, L., Farzana, S., & Brestic, M. (2021). Screening of Salt-Tolerant Rice Landraces by Seedling Stage Phenotyping and Dissecting Biochemical Determinants of Tolerance Mechanism. *Journal of Plant Growth Regulation, 40*(5), 1853–1868. DOI: 10.1007/s00344-020-10235-9

Rashid, A., Schutte, B. J., Ulery, A., Deyholos, M. K., Sanogo, S., Lehnhoff, E. A., & Beck, L. (2023). Heavy Metal Contamination in Agricultural Soil: Environmental Pollutants Affecting Crop Health. *Agronomy (Basel), 13*(6), 1521. DOI: 10.3390/agronomy13061521

Ravenscroft, P., Brammer, H., & Richards, K. (2011). *Arsenic pollution: a global synthesis*. John Wiley & Sons.

Raza, M. A. S., Ibrahim, M. A., Ditta, A., Iqbal, R., Aslam, M. U., Muhammad, F., Ali, S., Çiğ, F., Ali, B., Muhammad Ikram, R., Muzamil, M. N., Rahman, M. H. ur, Alwahibi, M. S., & Elshikh, M. S. (2023). Exploring the recuperative potential of brassinosteroids and nano-biochar on growth, physiology, and yield of wheat under drought stress. *Scientific Reports, 13*(1), 15015-. DOI: 10.1038/s41598-023-42007-2

Regional Energy Commission of Sakhalinsk Oblast. (2023). *On Setting Tariffs of Limited Liability Company Sakhalin East*. https://docs.cntd.ru/document/406920211?ysclid=m7jk2grrwl353827782

Rengasamy, P. (2006). World salinization with emphasis on Australia. *Journal of Experimental Botany, 57*(5), 1017–1023. DOI: 10.1093/jxb/erj108 PMID: 16510516

Restiawaty, E., Marwani, E., Steven, S., Mega Rahayu, G., Hanif, F., & Prakoso, T. (2023). Cultivation of Chlorella vulgaris in mediums with varying nitrogen sources and concentrations to induce the lipid yield. *Indian Chemical Engineer, 65*(4), 369–380. DOI: 10.1080/00194506.2022.2164525

Richards, L. A. (1954). Diagnosis and improvement of saline and alkali soils. *Soil Science, 78*(2), 154. DOI: 10.1097/00010694-195408000-00012

Rifai, A., Jamilah, S., Admirasari, R., & Prayitno, J. (2022, June). Alternative Source of Nutrients for Microalgae Cultivation in a Photobioreactor System. In *IOP Conference Series: Earth and Environmental Science* (Vol. 1034, No. 1, p. 012005). IOP Publishing. DOI: 10.1088/1755-1315/1034/1/012005

Rimington, W. R., Pressel, S., Duckett, J. G., Field, K. J., Read, D. J., & Bidartondo, M. I. (2018). Ancient Plants with Ancient Fungi: Liverworts Associate with Early-Diverging Arbuscular Mycorrhizal Fungi. *Proceedings. Biological Sciences*, *285*(1888), 20181600. DOI: 10.1098/rspb.2018.1600 PMID: 30305437

Robertson, M. J., Soibam, B., O'Leary, J. G., Sampaio, L. C., & Taylor, D. A. (2018). Recellularization of rat liver: An in vitro model for assessing human drug metabolism and liver biology. *PLoS One*, *13*(1), e0191892. DOI: 10.1371/journal.pone.0191892 PMID: 29377912

Robitaille, P. (2004). Les diamants en Sierra Leone, de la mine à la mort. https://ceim.uqam.ca/db/spip.php?page=article-cirdis&id_article=2065

Rojas, A., González, I., & Morales, M. A. (2025). Natural products and cancer: The urgent need to bridge the gap between preclinical and clinical research. *World Journal of Gastrointestinal Oncology*, *17*(4). Advance online publication. DOI: 10.4251/wjgo.v17.i4.100484 PMID: 40235887

Ro ca, M., Mihalache, G., & Stoleru, V. (2023). Tomato Responses to Salinity Stress: From Morphological Traits to Genetic Changes. *Frontiers in Plant Science*, *14*(February), 1118383. Advance online publication. DOI: 10.3389/fpls.2023.1118383 PMID: 36909434

Rossmann, R., Sawers, G., & Böck, A. (1991). Mechanism of regulation of the formate-hydrogenlyase pathway by oxygen, nitrate and pH: Definition of the formate regulon. *Molecular Microbiology*, *5*(11), 2807–2814. DOI: 10.1111/j.1365-2958.1991.tb01989.x PMID: 1779767

Rudovica, V., Rotter, A., Gaudêncio, S. P., Novoveská, L., Akgül, F., Akslen-Hoel, L. K., Alexandrino, D. A. M., Anne, O., Arbidans, L., Atanassova, M., Bełdowska, M., Bełdowski, J., Bhatnagar, A., Bikovens, O., Bisters, V., Carvalho, M. F., Catalá, T. S., Dubnika, A., Erdoğan, A., & Burlakovs, J. (2021). Valorization of Marine Waste: Use of Industrial By-Products and Beach Wrack Towards the Production of High Added-Value Products. *Frontiers in Marine Science*, *8*, 723333. DOI: 10.3389/fmars.2021.723333

Rudramurthy, Paul, Chakrabarti, Mouton., & Meis. (2019). Invasive Aspergillosis by *Aspergillus flavus*: Epidemiology, Diagnosis, Antifungal Resistance, and Management. DOI: 10.3390/jof5030055

Rukhkyan, M. Ya. (2012). Olfactometry phyto preferendum for Ixodid ticks in the steppe landscape of Armenia. Proceedings for the XI European Multicolloquium of Parasitology (XI EMOP), 447–448.

Rukhkyan, M. Ya., & Hovhannisyan, R. L. (2012). Evaluation of biodiversity of Armnia ixodids. *Proceedings for XIV International Conference "Biological Diversity of Caucasus"*, 12-15.

Rukhkyan, M.Ya. (1986). Hibernation of *Dermacentor marginatus* under the conditions.

Rukhkyan, M. Ya., & Alekseev, A. N. (1991). One of developmental deadlines or the way of habitation of the pasture tick *Dermacentor marginatus. Journal of Medical Parasitology and Parasitic Diseases.*, *2*, 55–57.

Rukhkyan, M. Ya., Dubinina, E. V., & Alekseev, A. N. (2011). On the possibility of circulation of the causative agent of tick-borne borreliosis in the territory of Armenia. *International Conference "Biological Diversity and Problems of Conservation of the Fauna of the Caucasus"*, 267–271.

Rusănescu, C. O., Ciobanu, M., Rusănescu, M., & Dinculoiu, R. L. (2024). Pretreatments Applied to Wheat Straw to Obtain Bioethanol. *Applied Sciences (Basel, Switzerland)*, *14*(4), 1612. DOI: 10.3390/app14041612

Russell, W. M. S., & Burch, R. L. (1959). *The principles of humane experimental technique*. Methuen.

Rynkowska, A., Stępniak, J., & Karbownik-Lewińska, M. (2020). Fenton Reaction-Induced Oxidative Damage to Membrane Lipids and Protective Effects of 17β-Estradiol in Porcine Ovary and Thyroid Homogenates. *International Journal of Environmental Research and Public Health*, *17*(18), 6841. DOI: 10.3390/ijerph17186841 PMID: 32962175

Sabetkish, S., Kajbafzadeh, A.-M., Sabetkish, N., Khorramirouz, R., Akbarzadeh, A., Seyedian, S. L., Pasalar, P., Orangian, S., Beigi, R. S. H., Aryan, Z., Akbari, H., & Tavangar, S. M. (2015). Whole-organ tissue engineering: Decellularization and recellularization of three-dimensional matrix liver scaffolds. *Journal of Biomedical Materials Research. Part A*, *103*(4), 1498–1508. DOI: 10.1002/jbm.a.35291 PMID: 25045886

Safe, S. (2024). Natural products as anticancer agents and enhancing their efficacy by a mechanism-based precision approach. *Exploration of Drug Science*, *2*(4), 408–427. DOI: 10.37349/eds.2024.00054

Sage, R. F., Sage, T. L., Pearcy, R. W., & Borsch, T. (2007). The taxonomic distribution of C4 photosynthesis in Amaranthaceae sensu stricto. *American Journal of Botany, 94*(12), 1992–2003. DOI: 10.3732/ajb.94.12.1992 PMID: 21636394

Sahakyan, N., Bartoszek, A., Jacob, C., Petrosyan, M., & Trchounian, A. (2020). *Bioavailability of Tannins and Other Oligomeric Polyphenols: a Still to Be Studied Phenomenon. Current Pharmacology Reports* (Vol. 6). DOI: 10.1007/s40495-020-00217-6

Sahakyan, N., Petrosyan, M., Koss-Mikołajczyk, I., Bartoszek, A., Sad, T. G., Nasim, M. J., Vanidze, M., Kalandia, A., Jacob, C., & Trchounian, A. (2019). The Caucasian flora: A still-to-be-discovered rich source of antioxidants. *Free Radical Research, 53*(sup1), 1153–1162. DOI: 10.1080/10715762.2019.1648799 PMID: 31510813

Salam, A., Ali, A., Afridi, M. S., Ali, S., & Ullah, Z. (2022). Agrobiodiversity: Effect of Drought Stress on the Eco-Physiology and Morphology of Wheat. In Öztürk, M., Khan, S. M., Altay, V., Efe, R., Egamberdieva, D., & Khassanov, F. O. (Eds.), *Biodiversity, Conservation and Sustainability in Asia*. Springer International Publishing. DOI: 10.1007/978-3-030-73943-0_33

Salazar, H. G. M., Salem, A. Z. M., Kholif, A. E., Monroy, H., Perez, L. S., Zamora, J. L., & Gutiérez, A. (2016). Yeast: Desciption and structure. In *Yeast additive and animal production, India*. PubMed Central Research Publishing Services. DOI: 10.1016/0014-5793(83)80237-3

Salerno, S., Campana, C., Morelli, S., Drioli, E., & De Bartolo, L. (2011). Human hepatocytes and endothelial cells in organotypic membrane systems. *Biomaterials, 32*(34), 8848–8859. DOI: 10.1016/j.biomaterials.2011.08.004 PMID: 21871658

Salim, B. B., Hikal, M. S., & Osman, H. S. (2019). Ameliorating the deleterious effects of saline water on the antioxidants defense system and yield of eggplant using foliar application of zinc sulphate. *Annals of Agricultural Science, 64*(2), 244–251. DOI: 10.1016/j.aoas.2019.12.008

Samapika & Dey. (2020). Bibenzyls and bisbybenzyls of bryophytic origin as promising source of novel therapeutics: pharmacology, synthesis and structure-activity. *Daru Journal of Pharmaceutical Sciences, 28*(2), 701–734. DOI: 10.1007/s40199-020-00341-0

Sambrook, J., Fritsch, E. F., & Maniatis, T. (1989). *Molecular Cloning: a Laboratory Manual*. Cold Spring Harbor Laboratory.

Sánchez-Bayo, F. (2011). Impacts of agricultural pesticides on terrestrial ecosystems. *Ecological impacts of toxic chemicals,* 63-87. .DOI: 10.2174/978160805121210063

Sargsyan, A., Hovhannisyan, G., Simonyan, A., Arakelyan, M., Arzumanyan, M., & Aroutiounian, R. (2022). Application of land snail Helix lucorum for evaluation of genotoxicity of soil pollution. *Mutation Research/Genetic Toxicology and Environmental Mutagenesis*, *878*, 503500. DOI: 10.1016/j.mrgentox.2022.503500 PMID: 35649673

Satoh, H., Sakamoto, T., Kuroki, Y., Kudo, Y., & Mino, T. (2016). Application of the Alkaline-Digestion-HPLC Method to the Rapid Determination of Polyhydroxyalkanoate in Activated Sludge. *Journal of Water and Environment Technology*, *14*(5), 411–421. DOI: 10.2965/jwet.16-027

Sawers, G., & Böck, A. (1989). Novel transcriptional control of the pyruvate formate-lyase gene: Upstream regulatory sequences and multiple promoters regulate anaerobic expression. *Journal of Bacteriology*, *171*(5), 2485–2498. DOI: 10.1128/jb.171.5.2485-2498.1989 PMID: 2651404

Sawers, R. G. (2025). How FocA facilitates fermentation and respiration of formate by *Escherichia coli*. *Journal of Bacteriology*, *207*(2), e0050224. DOI: 10.1128/jb.00502-24 PMID: 39868885

Sayadi, M. H., & Sayyed, M. R. G. (2011). Comparative assessment of baseline concentration of the heavy metals in the soils of Tehran (Iran) with the comprisable reference data. *Environmental Earth Sciences*, *63*(6), 1179–1188. DOI: 10.1007/s12665-010-0792-z

Sayed, S. a. a. E., Hellal, F. A., El–Rab, N. G., & Zewainy, R. M. (2019). Ameliorative effects of potassium on the salinity stress in plants: a review. *Asian Journal of Soil Science and Plant Nutrition*, 1–15. DOI: 10.9734/ajsspn/2019/v4i230040

Schäfer, C., Friedrich, B., & Lenz, O. (2013). Novel, Oxygen-Insensitive Group 5 [NiFe]-Hydrogenase in Ralstonia eutropha. *Applied and Environmental Microbiology*, *79*(17), 5137–5145. DOI: 10.1128/AEM.01576-13 PMID: 23793632

Schlensog, V., Lutz, S., & Böck, A. (1994). Purification and DNA-binding properties of FHLA, the transcriptional activator of the formate hydrogenlyase system from *Escherichia coli*. *The Journal of Biological Chemistry*, *269*(30), 19590–19596. DOI: 10.1016/S0021-9258(17)32210-X PMID: 8034727

Schneider, C. A., Rasband, W. S., & Eliceiri, K. W. (2012). NIH Image to ImageJ: 25 years of image analysis. *Nature Methods*, *9*(7), 671–675. DOI: 10.1038/nmeth.2089 PMID: 22930834

Schneiter, R., Brügger, B., Sandhoff, R., Zellnig, G., Leber, A., Lampl, M., Athenstaedt, K., Hrastnik, C., Eder, S., Daum, G., Paltauf, F., Wieland, F. T., & Kohlwein, S. D. (1999). Electrospray ionization tandem mass spectrometry (ESI-MS/MS) analysis of the lipid molecular species composition of yeast subcellular membranes reveals acyl chain-based sorting/remodeling of distinct molecular species en route to the plasma membrane. *The Journal of Cell Biology*, *146*(4), 741–754. DOI: 10.1083/jcb.146.4.741 PMID: 10459010

Seda, Karapetyan, Khachatryan, Muradyan, Marutyan, Poladyan, & Trchounian. (2024). The antimicrobial effects of silver nanoparticles obtained through the royal jelly on the yeasts *Candida guilliermondii* NP-4. *Scientific Reports, 14*, 19163. DOI: 10.1038/s41598-024-70197-w

Seker, I., Karlidag, S., Akyol, A., Yil, G., Koseman, A., Uyumlu, A. B., Edrdogan, S., & Yilmaztekin, M. (2025). Volatile organic compounds of royal jellies obtained from honeybee (Apis mellifera L.) colonies fed with different industrial sugars at different locations. *Journal of Food Measurement and Characterization*, *19*(3), 2019–2034. DOI: 10.1007/s11694-025-03097-w

Sen, K. (2023). Recent Advances in the Phytochemistry of Bryophytes: Distribution. *Structures and Biological Activity of Bibenzyl and Bisbibenzyl Compounds Plants*, *12*, 4173. DOI: 10.3390/plants12244173 PMID: 38140499

Seo, H.-B., Kim, H.-J., Lee, O.-K., Ha, J.-H., Lee, H.-Y., & Jung, K.-H. (2009). Measurement of ethanol concentration using solvent extraction and dichromate oxidation and its application to bioethanol production process. *Journal of Industrial Microbiology & Biotechnology*, *36*(2), 285–292. DOI: 10.1007/s10295-008-0497-4 PMID: 18989713

Seregin, I. V., & Kozhevnikova, A. D. (2008). Roles of root and shoot tissues in transport and accumulation of cadmium, lead, nickel, and strontium. *Russian Journal of Plant Physiology: a Comprehensive Russian Journal on Modern Phytophysiology*, *55*(1), 1–22. DOI: 10.1134/S1021443708010019

Shabala, S., Bose, J., & Hedrich, R. (2014). Salt Bladders: Do They Matter? *Trends in Plant Science*, *19*(11), 687–691. DOI: 10.1016/j.tplants.2014.09.001 PMID: 25361704

Shankar, S., & Shanker, U., & Shikha. (2014). Arsenic contamination of groundwater: A review of sources, prevalence, health risks, and strategies for mitigation. *TheScientificWorldJournal*, •••, 304524.

Sharmaa, K., Khurana, P., & Uddina, Z. (2025). *MetaInfoSci: An Integrated Web Tool for Scholarly Data Analysis*. Retrieved from https://arxiv.org/pdf/2506.09056

Sharma, V., Sharma, D., Saini, M., Jain, A., Chaudhary, J., Kaur, N., Sahoo, S., Singh, S. K., Goyal, K., Rekha, A., Ali, H., Gupta, S., Hussain, M. S., & Gupta, G. (2025). Flavonoids as Antimicrobial Agents: A Comprehensive Review of Mechanisms and Therapeutic Potential. *Current Pharmaceutical Biotechnology*, *26*, 26. DOI: 10.2174/0113892010384002250628163849 PMID: 40619652

Sheer, A., Fahad Sardar, M., Younas, F., Zhu, P., Noreen, S., Mehmood, T., Ur Rahman Farooqi, Z., Fatima, S., & Guo, W. (2024). Trends and social aspects in the management and conversion of agricultural residues into valuable resources: A comprehensive approach to counter environmental degradation, food security, and climate change. *Bioresource Technology*, *394*, 130258. DOI: 10.1016/j.biortech.2023.130258 PMID: 38151206

Shewry, P. R. (2009). Wheat. *Journal of Experimental Botany*, *60*(6), 1537–1553. DOI: 10.1093/jxb/erp058 PMID: 19386614

Shin, Y. K., Bhandari, S. R., Jo, J. S., Song, J. W., Cho, M. C., Yang, E. Y., & Lee, J. G. (2020). Response to Salt Stress in Lettuce: Changes in Chlorophyll Fluorescence Parameters, Phytochemical Contents, and Antioxidant Activities. *Agronomy, 10*(11), 1627. DOI: 10.3390/agronomy10111627

Shirvanyan, A. H. (2023). Evaluation of ethanol and biomass production rate by different $saccharomyces~cerevisiae strains depending on external ph and temperature. *Proceedings of the YSU B: Chemical and Biological Sciences, 57*(2 (261)), 141–153. DOI: 10.46991/PYSU:B/2023.57.2.141

Shirvanyan, A., Mirzoyan, S., & Trchounian, K. (2023). Relationship between proton/potassium fluxes and central carbon catabolic pathways in different Saccharomyces cerevisiae strains under osmotic stress conditions. *Process Biochemistry (Barking, London, England)*, *133*, 309–318. DOI: 10.1016/j.procbio.2023.09.015

Shirvanyan, A., & Trchounian, K. (2024). Sodium transport and redox regulation in Saccharomyces cerevisiae under osmotic stress depending on oxygen availability. *Scientific Reports*, *14*(1), 23982. DOI: 10.1038/s41598-024-75108-7 PMID: 39402154

Siddiqui, M. H., Al-Khaishany, M. Y., Al-Qutami, M. A., Al-Whaibi, M. H., Grover, A., Ali, H. M., Al-Wahibi, M. S., & Bukhari, N. A. (2015). Response of Different Genotypes of Faba Bean Plant to Drought Stress. *International Journal of Molecular Sciences, 16*(5), 10214–10227. DOI: 10.3390/ijms160510214

Sidonia, Fuentes, & Carballo. (2022). Antioxidant Activity, Total Phenolic Content and Total Flavonoid Content in Sweet Chestnut (Castanea sativa Mill.) Cultivars Grown in Northwest Spain under Different Environmental Conditions. *Foods, 11*, 3519. DOI: 10.3390/foods11213519

Silva, B. R. S., Batista, B. L., & Lobato, A. K. S. (2020). Anatomical changes in stem and root of soybean plants submitted to salt stress. *Plant Biology*, *23*(1), 57–65. DOI: 10.1111/plb.13176 PMID: 32841475

Silvard, T., Babayan, A., & Sahakyan, N. (2025). Antioxidant and Antibiotic-modulatory Potential of *Teucrium polium* Ethanolic Extract Against Genetically-modified Antibiotic-resistant *E. coli* Strains. *Majallah-i Mikrub/Shinasi-i Pizishki-i Iran*, *19*(2), 107–117. DOI: 10.30699/ijmm.19.2.107

Singh, A., Rajput, V. D., Singh, R. K., Elshikh, M. S., El-Ramady, H., Prokisch, J., Liu, C., Rajput, P., Minkina, T., Sousa, J. R., Alexiou, A., Singh, S., Varagyan, V., & Ghazaryan, K. (2025). Emerging Frontiers in Nanotechnology for Salinity Stress Management: A Global Bibliometric Analysis. *Magallat Diyalá Li-l-'ulum al-Zira'iyyat*, *17*(2), 46–69. DOI: 10.52951/dasj.25170204

Singh, A., Sengar, R. S., Rajput, V. D., Agrawal, S., Ghazaryan, K., Minkina, T., Tawaha, A. R. A., Mahmoud Al Zoubi, O., & Habeeb, T. (2023). Impact of Zinc Oxide Nanoparticles on Seed Germination Characteristics in Rice (*Oryza Sativa* L.) Under Salinity Stress. *Journal of Ecological Engineering*, *24*(10), 142–156. DOI: 10.12911/22998993/169142

Singh, G., & Patidar, S. K. (2018). Microalgae harvesting techniques: A review. *Journal of Environmental Management*, *217*, 499–508. DOI: 10.1016/j.jenvman.2018.04.010 PMID: 29631239

Singh, N., Gupta, V. K., Kumar, A., & Sharma, B. (2017). Synergistic effects of heavy metals and pesticides in living systems. *Frontiers in Chemistry*, *5*, 70. DOI: 10.3389/fchem.2017.00070 PMID: 29075624

Sohan, D., Jasoni, R., & Zajicek, J. (1999). Plant–water relations of NaCl and calcium-treated sunflower plants. *Environmental and Experimental Botany*, *42*(2), 105–111. DOI: 10.1016/S0098-8472(99)00027-1

Sohn, Y. J., Son, J., Jo, S. Y., Park, S. Y., Yoo, J. I., Baritugo, K.-A., Na, J. G., Choi, J., Kim, H. T., Joo, J. C., & Park, S. J. (2021). Chemoautotroph *Cupriavidus necator* as a potential game-changer for global warming and plastic waste problem: A review. *Bioresource Technology*, *340*, 125693. DOI: 10.1016/j.biortech.2021.125693 PMID: 34365298

Soltani, N., Keshavarzi, B., Moore, F., Sorooshian, A., & Ahmadi, M. R. (2017). Distribution of potentially toxic elements (PTEs) in tailings, soils, and plants around Gol-E-Gohar iron mine, a case study in Iran. *Environmental Science and Pollution Research International*, *24*(23), 18798–18816. DOI: 10.1007/s11356-017-9342-5 PMID: 28620857

Sonenshine, D. E., & Roe, R. M. (2014). External and internal anatomy of ticks. In Sonenshine, D. E., & Roe, R. M. (Eds.), *Biology of ticks* (Vol. 2, pp. 74–98). Oxford University Press.

Sonkar, V., Jaswal, V., Chenlak, S., & Nandabalan, Y. K. (2024). Pollution status and health risk assessment of heavy metals in the soil of Sahibzada Ajit Singh (SAS) Nagar district of Punjab, India and its source apportionment. *Journal of Geochemical Exploration*, *261*, 107453. DOI: 10.1016/j.gexplo.2024.107453

Sor, R., Ngor, P. B., Boets, P., Goethals, P. L., Lek, S., Hogan, Z. S., & Park, Y. S. (2020). Patterns of mekong mollusc biodiversity: Identification of emerging threats and importance to management and livelihoods in a region of globally significant biodiversity and endemism. *Water (Basel)*, *12*(9), 2619. DOI: 10.3390/w12092619

Spanidi, E., Athanasopoulou, S., Liakopoulou, A., Chaidou, A., Hatziantoniou, S., & Gardikis, K. (2022). Royal jelly components encapsulation in a controlled release system - skin functionality and biochemical activity for skin applications. *Pharmaceuticals (Basel, Switzerland)*, *15*(8), 907. DOI: 10.3390/ph15080907 PMID: 35893731

Srivastava, A. P., Luo, M., Zhou, W., Symersky, J., Bai, D., Chambers, M. G., Faraldo-Gómez, J. D., Liao, M., & Mueller, D. M. (2018). High-resolution cryo-EM analysis of the yeast ATP synthase in a lipid membrane. *Science*, *360*(6389), 1–8. DOI: 10.1126/science.aas9699 PMID: 29650704

Standard Methods for the Examination of Water and Wastewater, 20th Edition - Google Books. (1998). APHA American Public Health Association. https://books.google.am/books/about/Standard_Methods_for_the_Examination_of.html?id=2BcoYAAACAAJ&redir_esc=y

Statista. (2024). *Total bioenergy capacity in Russia from 2010 to 2023*. https://www.statista.com/statistics/1154535/total-bioenergy-capacity-in-russia/

Statista. (2025). *Market value of biofuels production worldwide from 2021 to 2023, with a forecast until 2030*. https://translated.turbopages.org/proxy_u/en-ru.ru.147d6c5e-67c0a0f2-657961a2-74722d776562/https/www.statista.com/statistics/217179/global-biofuels-market-size/

Steglińska, A., Sulyok, M., Janas, R., Grzesik, M., Liszkowska, W., Kręgiel, D., & Gutarowska, B. (2023). Metabolite formation by fungal pathogens of potatoes (*Solanum tuberosum* L.) in the presence of bioprotective agents. *International Journal of Environmental Research and Public Health*, *20*(6), 5221. DOI: 10.3390/ijerph20065221 PMID: 36982130

Steinhilper, R., Höff, G., Heider, J., & Murphy, B. J. (2022). Structure of the membrane-bound formate hydrogenlyase complex from *Escherichia coli. Nature Communications, 13*(1), 5395. DOI: 10.1038/s41467-022-32831-x PMID: 36104349

Stevens, C. M. B., Bojang, O. A., Amara, E., & Msuya, A. S. (2025). Comprehensive Overview of Deforestation Dynamics and Sustainable Forest Management in Sierra Leone. *European Journal of Ecology. Biology and Agriculture, 2*(2), 18–29. DOI: 10.59324/ejeba.2025.2(2).03

Stolyhwo, A., & Privett, O. S. (1973). Studies on the analysis of lipid classes by gradient elution adsorption chromatography. *Journal of Chromatographic Science, 11*(1), 20–25. DOI: 10.1093/chromsci/11.1.20

Stopar, K. (2016). Presence of nanotechnology in agriculture: Bibliometric approach. *Acta Agriculturae Slovenica, 107*(2), 497–507. DOI: 10.14720/aas.2016.107.2.20

Stummer, W., Pichlmeier, U., Meinel, T., Wiestler, O. D., Zanella, F., & Reulen, H. J.ALA-Glioma Study Group. (2006). Fluorescence-guided surgery with 5-aminolevulinic acid for resection of malignant glioma: A randomised controlled multicentre phase III trial. *The Lancet. Oncology, 7*(5), 392–401. DOI: 10.1016/S1470-2045(06)70665-9 PMID: 16648043

Su, C., Jiang, L., & Zhang, W. (2014). *A review on heavy metal contamination in the soil worldwide: Situation, impact and remediation techniques. The International Academy of Ecology and Environmental Sciences.* IAEES.

Sudhir, P., Pogoryelov, D., Kovacs, L., Garab, G., & Murthy, S. D. (2005). The Effects of Salt Stress on Photosynthetic Electron Transport and Thylakoid Membrane Proteins in the Cyanobacterium Spirulina platensis. *BMB Reports, 38*(4), 481–485. DOI: 10.5483/BMBRep.2005.38.4.481 PMID: 16053716

Sugawara, T., & Miyazawa, T. (1999). Separation and determination of glycolipids from edible plant sources by high-performance liquid chromatography and evaporative light-scattering detection. *Lipids, 34*(11), 1231–1237. DOI: 10.1007/s11745-999-0476-3 PMID: 10606047

Sulieman, M. M., Elfaki, J. T., Adam, M. M., Dafalla, M. S., Ali, S. H., Ahmed, H. M., & Ali, M. E. (2017). Assessment of heavy metals contamination in the Nile River water and adjacent sediments: A case study from Khartoum City and Nile River State, Sudan. *Eurasian Journal of Soil Science, 6*(3), 285–294. DOI: 10.18393/ejss.298949

Suppmann, B., & Sawers, G. (1994). Isolation and characterisation of hypophosphite-resistant mutants of *Escherichia coli*: Identification of the FocA protein, encoded by the *pfl* operon, as a putative formate transporter. *Molecular Microbiology, 11*(5), 965–982. DOI: 10.1111/j.1365-2958.1994.tb00375.x PMID: 8022272

Supriya, J. (2022). Bryophytes as a safeguard of fruits from postharvest fungal diseases: A Review. *Natr Resour Human Health, 2*(3), 327–334. DOI: 10.53365/nrfhh/145476

Suzuki, N. (2015) ROS as Key Players of Abiotic Stress Responses in Plants. *Reactive Oxygen Species and Oxidative Damage in Plants Under Stress,* 57–82.

Suzuki, N., Rivero, R. M., Shulaev, V., Blumwald, E., & Mittler, R. (2014). Abiotic and biotic stress combinations. *The New Phytologist, 203*(1), 32–43. DOI: 10.1111/nph.12797 PMID: 24720847

Sysmex Partec. (2014). *CyFlow® Cube 6 instrument operating manual* (CyView™ software version 1.5; Rev. 05). https://www.wolflabs.co.uk/documents/Sysmex-partec_flow-cytometers_cyflow-Cube-6_manual.pdf

Tadevosyan, S. (2025). B. A. S. N. Antioxidant and Antibiotic-modulatory Potential of Teucrium polium Ethanolic Extract Against Genetically-modified Antibiotic-resistant E. coli Strains. *Iran J Med Microbiol 19*(2).

Tadevosyan, S. K., Shirvanyan, A. H., Markosyan, A. A., Petrosyan, M. T., & Sahakyan, N. Z. (2023). Chemical composition and antibacterial activity of essential oil of $mentha~arvensis$ l. harvested at high altitude Armenian flora. *Proceedings of the YSU B: Chemical and Biological Sciences, 57*, 230–237. DOI: 10.46991/PYSU:B/2023.57.3.230

Tadevosyan, S., & Sahakyan, N. (2024). Influence of menthol on membrane-associated properties of tetracycline-resistant *Escherichia coli. AIMS Biophysics, 11*(3), 329–339. DOI: 10.3934/biophy.2024018

Taffouo, V. D., Kouamou, J. K., Ngalangue, L. M. T., Ndjeudji, B. A. N., & Akoa, A. (2009). Effects of salinity stress on growth, ions partitioning and yield of some cowpea (Vigna unguiculata L. Walp.) Cultivars. *International Journal of Botany, 5*(2), 135–143. DOI: 10.3923/ijb.2009.135.143

Taghavi, M., Bakhshi, K., Zarei, A., Hoseinzadeh, E., & Gholizadeh, A. (2024). Soil pollution indices and health risk assessment of metal(loid)s in the agricultural soil of pistachio orchards. *Scientific Reports, 14*(1), 8971. Advance online publication. DOI: 10.1038/s41598-024-59450-4 PMID: 38637594

Tao, R., Ding, J., Li, C., Zhu, X., Guo, W., & Zhu, M. (2021). Evaluating and screening of Agro-Physiological indices for salinity stress tolerance in wheat at the seedling stage. *Frontiers in Plant Science*, *12*, 646175. Advance online publication. DOI: 10.3389/fpls.2021.646175 PMID: 33868346

Tarabishi, A. A., Mashhoud, J., & Tahan, Z. S. (2024). Quercetin and rutin as a dual approach to antibacterial and anti-biofilm activity via iron chelation mechanism. *Discover Food*, *4*(1), 189. DOI: 10.1007/s44187-024-00265-7

Tarchoun, N., Saadaoui, W., Mezghani, N., Pavli, O. I., Falleh, H., & Petropoulos, S. A. (2022). The Effects of Salt Stress on Germination, Seedling Growth and Biochemical Responses of Tunisian Squash (Cucurbita Maxima Duchesne) Germplasm. *Plants*, *11*(6), 800. DOI: 10.3390/plants11060800 PMID: 35336682

Tarin, T., Nolan, R. H., Medlyn, B. E., Cleverly, J., & Eamus, D. (2019). Water-use efficiency in a semi-arid woodland with high rainfall variability. *Global Change Biology*, *26*(2), 496–508. DOI: 10.1111/gcb.14866 PMID: 31597216

Temel, S., & Keskin, B. (2022). Determination of forage quality properties of plant parts in different amaranth varieties cultivated under irrigated and rainfed conditions. *Atatürk University Journal of Agricultural Faculty*, *53*(2), 122–132. DOI: 10.54614/AUAF.2022.1034402

Tepanosyan, G., Sahakyan, L., Belyaeva, O., & Saghatelyan, A. (2016). Origin identification and potential ecological risk assessment of potentially toxic inorganic elements in the topsoil of the city of Yerevan, Armenia. *Journal of Geochemical Exploration*, *167*, 1–11. DOI: 10.1016/j.gexplo.2016.04.006

Thanapirom, K., Caon, E., Papatheodoridi, M., Frenguelli, L., Al-Akkad, W., Zhenzhen, Z., Vilia, M. G., Pinzani, M., Mazza, G., & Rombouts, K. (2021). Optimization and validation of a novel three-dimensional co-culture system in decellularized human liver scaffold for the study of liver fibrosis and cancer. *Cancers (Basel)*, *13*(19), 4936. Advance online publication. DOI: 10.3390/cancers13194936 PMID: 34638417

Thangjam, U., & Sahoo, U. K. (2016). Effect of Seed Mass on Germination and Seedling Vigour of Parkia Timoriana (DC.) Merr. *Current Agriculture Research Journal*, *4*(2), 171–178. DOI: 10.12944/CARJ.4.2.06

The water-culture method for growing plants without soil by D. R. Hoagland | Open Library. (n.d.). Retrieved from https://openlibrary.org/books/OL25240089M/The_water-culture_method_for_growing_plants_without_soil

Thermo Fisher Scientific. (2023). *Quant-iT™ dsDNA broad-range assay kit: User guide* (Cat. No. Q33130). https://assets.thermofisher.com/TFS-Assets/LSG/manuals/Quant_iT_dsDNA_BR_Assay_UG.pdf

Tian, K., Wu, Q., Liu, P., Hu, W., Huang, B., Shi, B., Zhou, Y., Kwon, B. O., Choi, K., Ryu, J., Seong Khim, J., & Wang, T. (2020). Ecological risk assessment of heavy metals in sediments and water from the coastal areas of the Bohai Sea and the Yellow Sea. *Environment International*, *136*, 105512. DOI: 10.1016/j.envint.2020.105512 PMID: 31999973

Tice, R. R., Agurell, E., Anderson, D., Burlinson, B., Hartmann, A., Kobayashi, H., Miyamae, Y., Rojas, E., Ryu, J. C., & Sasaki, Y. F. (2000). Single cell gel/comet assay: Guidelines for in vitro and in vivo genetic toxicology testing. *Environmental and Molecular Mutagenesis*, *35*(3), 206–221. DOI: 10.1002/(SICI)1098-2280(2000)35:3<206::AID-EM8>3.0.CO;2-J PMID: 10737956

Tıpırdamaz, R., Karakas, S., & Dikilitas, M. (2021). Halophytes and the future of agriculture. In *Springer eBooks* (pp. 2225–2239). DOI: 10.1007/978-3-030-57635-6_91

Togo First. (2024). Près de 75% des emplois dans le secteur privé créés par les TPME. https://golnk.ru/BdewE

Töre, Y., Ustaoğlu, F., Tepe, Y., & Kalipci, E. (2021). Levels of toxic metals in edible fish species of the Tigris River (Turkey); Threat to public health. *Ecological Indicators*, *123*, 107361. DOI: 10.1016/j.ecolind.2021.107361

Trchounian, K., Mirzoyan, S., Poladyan, A., & Trchounian, A. (2017). Hydrogen production by Escherichia coli growing in different nutrient media with glycerol: Effects of formate, pH, production kinetics and hydrogenases involved. *International Journal of Hydrogen Energy*, *42*(38), 24026–24034. DOI: 10.1016/j.ijhydene.2017.08.032

Trchounian, K., Müller, N., Schink, B., & Trchounian, A. (2017). Glycerol and mixture of carbon sources conversion to hydrogen by Clostridium beijerinckii DSM791 and effects of various heavy metals on hydrogenase activity. *International Journal of Hydrogen Energy*, *42*(12), 7875–7882. DOI: 10.1016/j.ijhydene.2017.01.011

Trchounian, K., Poladyan, A., Vassilian, A., & Trchounian, A. (2012). Multiple and reversible hydrogenases for hydrogen production by *Escherichia coli*: Dependence on fermentation substrate, pH and the F_0F_1-ATPase. *Critical Reviews in Biochemistry and Molecular Biology*, *47*(3), 236–249. DOI: 10.3109/10409238.2012.655375 PMID: 22313414

Trenberth, K. E., Dai, A., Van Der Schrier, G., Jones, P. D., Barichivich, J., Briffa, K. R., & Sheffield, J. (2013). Global warming and changes in drought. *Nature Climate Change, 4*(1), 17–22. DOI: 10.1038/nclimate2067

Trompenaars, F. (2012). Culture Xplore. *Cultural Dimensions Tool*. https://golnk.ru/KkyEO

Tse, T. J., Wiens, D. J., & Reaney, M. J. T. (2021). Production of Bioethanol—A Review of Factors Affecting Ethanol Yield. *Fermentation (Basel, Switzerland), 7*(4), 268. DOI: 10.3390/fermentation7040268

Tunccedil, M., Tuuml, Rk, Ruuml, Tunccedil, V., Buuml, Yildirim, N., Ccedil, V., & Iftccedil. (2011). Effect of salinity stress on plant fresh weight and nutrient composition of some Canola (Brassica napus L.) cultivars. *African Journal of Biotechnology, 10*(10), 1827–1832. DOI: 10.5897/ajb10.1618

Turhan, I., Bialka, K. L., Demirci, A., & Karhan, M. (2010). Ethanol production from carob extract by using Saccharomyces cerevisiae. *Bioresource Technology, 101*(14), 5290–5296. DOI: 10.1016/j.biortech.2010.01.146 PMID: 20189805

Turkulova, V. N., Solodovnikov, A. A., Kryuchkov, V. G., & Bityutskaya, O. E. (2008). Results and Prospects of YugNIRO Research in Mariculture. *Main results of integrated research in the Azov-Black Sea basin and the World Ocean.*

Tyagi, A., Bhardwaj, S., Sinha, A. K., & Dhiman, M. (2025). Advances in hydrogen-driven bicarbonate conversion to formic acid and formate salts. *ChemSusChem, 18*(14), e202500483. DOI: 10.1002/cssc.202500483 PMID: 40367116

Ullah, A., Munir, S., Badshah, S. L., Khan, N., Ghani, L., Poulson, B. G., Emwas, A.-H., & Jaremko, M. (2020). Important Flavonoids and Their Role as a Therapeutic Agent. *Molecules (Basel, Switzerland), 25*(22), 5243. DOI: 10.3390/molecules25225243 PMID: 33187049

UN. (2024). *Production Gap Report 2023.* https://www.unep.org/ru/resources/production-gap-report-2023

Usmani, Z., Sharma, M., Gaffey, J., Sharma, M., Dewhurst, R. J., Moreau, B., Newbold, J., Clark, W., Thakur, V. K., & Gupta, V. K. (2022). Valorization of dairy waste and by-products through microbial bioprocesses. *Bioresource Technology, 346*, 126444. DOI: 10.1016/j.biortech.2021.126444 PMID: 34848333

Vácha, R. (2021). Heavy metal pollution and its effects on agriculture. *Agronomy (Basel), 11*(9), 1719. Advance online publication. DOI: 10.3390/agronomy11091719

Van de Loosdrecht, A. A., Beelen, R. H. J., Ossenkoppele, G., Broekhoven, M. G., & Langenhuijsen, M. M. A. C. (1994). A tetrazolium-based colorimetric MTT assay to quantitate human monocyte mediated cytotoxicity against leukemic cells from cell lines and patients with acute myeloid leukemia. *Journal of Immunological Methods, 174*(1-2), 311–320. DOI: 10.1016/0022-1759(94)90034-5 PMID: 8083535

Van Zelm, E., Zhang, Y., & Testerink, C. (2020). Salt Tolerance Mechanisms of Plants. DOI: 10.1146/annurev-arplant-050718-100005

Vanyan, L., Aghekyan, H., Vassilian, A., Poladyan, A., & Trchounian, K. (2025). Biotechnological potential of spent coffee grounds for biohydrogen production by Escherichia coli. *International Journal of Hydrogen Energy, 142*, 1121–1131. DOI: 10.1016/j.ijhydene.2024.12.380

Vanyan, L., Kammel, M., Sawers, R. G., & Trchounian, K. (2024). Evidence for bidirectional formic acid translocation *in vivo* via the *Escherichia coli* formate channel FocA. *Archives of Biochemistry and Biophysics, 752*, 109877. DOI: 10.1016/j.abb.2023.109877 PMID: 38159898

Vardanyan, M. V., Hakobyan, V. J., Shcherbakov, O. V., & Boyakhchyan, G. H. (2005). The course of the bovine piroplasmosis in foothill area of Armenia, and comparative efficaciousness of its control. *Bulletin of Armenian Agricultural Academy., 4*, 53–56.

Vardhan, K. H., Kumar, P. S., & Panda, R. C. (2019). A review on heavy metal pollution, toxicity and remedial measures: Current trends and future perspectives. *Journal of Molecular Liquids, 290*, 111197. DOI: 10.1016/j.molliq.2019.111197

Vasilachi, I. C., Stoleru, V., & Gavrilescu, M. (2023). Analysis of Heavy Metal Impacts on Cereal Crop Growth and Development in Contaminated Soils. *Agriculture, 13*(10), 1983. DOI: 10.3390/agriculture13101983

Vassilian, A., & Trchounian, A. (2009). Environment oxidation-reduction potential and redox sensing by bacteria. In *Bacterial Membranes* (pp. 163–195). Research Signpos.

Vetrova, M., Pakhomova, N., & Richter, K. (2024). Strategies for the development of the Russian energy sector in the context of climate challenges and geopolitical instability. *Vestnik of St. Petersburg University. Economics, 39*(4), 439–469. DOI: 10.21638/spbu05.2023.401

Vigouroux, G., & Destouni, G. (2022). Gap identification in coastal eutrophication research – Scoping review for the Baltic system case. *The Science of the Total Environment, 839*, 156240. Advance online publication. DOI: 10.1016/j.scitotenv.2022.156240

Vikash Babu, P. D., Ashish Tapliyal, P. D., & Giriyesh Kumar Patel, P. D. (2014). *Renewable Energy: Biofuel Production*. Scrivener Publishing LLC.

Vincevica-Gaile, Z., Sachpazidou, V., Bisters, V., Klavins, M., Anne, O., Grinfelde, I., Hanc, E., Hogland, W., Ibrahim, M. A., Jani, Y., Kriipsalu, M., Pal, D., Pehme, K.-M., Shanskiy, M., Saaremäe, E., Pilecka-Ulcugaceva, J., Celms, A., Rudovica, V., Hendroko Setyobudi, R., & Burlakovs, J. (2022). Applying Macroalgal Biomass as an Energy Source: Utility of the Baltic Sea Beach Wrack for Thermochemical Conversion. *Sustainability (Basel), 14*(21), 13712. DOI: 10.3390/su142113712

Wahid, A., & Ghazanfar, A. (2005). Possible involvement of some secondary metabolites in salt tolerance of sugarcane. *Journal of Plant Physiology, 163*(7), 723–730. DOI: 10.1016/j.jplph.2005.07.007 PMID: 16616583

Waight, A. B., Love, J., & Wang, D.-N. (2010). Structure and mechanism of a pentameric formate channel. *Nature Structural & Molecular Biology, 17*(1), 31–37. DOI: 10.1038/nsmb.1740 PMID: 20010838

Walker, A. R., Bouattour, A., Camicas, J.-L., & Estrada-Peña, A. (2003). Ticks of Domestic Animals in the Mediterranean Region. A Guide to Identification of Species. University of Zaragoza.

Wang, J., Shi, D., Huang, C., Zhai, B., & Feng, S. (2023). Effects of Common Biochar and Acid-Modified Biochar on Growth and Quality of Spinach in Coastal Saline Soils. *Plants, 12*(18), 3232. DOI: 10.3390/plants12183232

Wang, L., Chen, W., & Zhou, W. (2014). Assessment of future drought in Southwest China based on CMIP5 multimodel projections. *Advances in Atmospheric Sciences, 31*(5), 1035–1050. DOI: 10.1007/s00376-014-3223-3

Wang, J., & Yin, Y. (2019). Progress in microbiology for fermentative hydrogen production from organic wastes. *Critical Reviews in Environmental Science and Technology, 49*(10), 825–865. DOI: 10.1080/10643389.2018.1487226

Wang, W., Vinocur, B., & Altman, A. (2003). Plant responses to drought, salinity and extreme temperatures: Towards genetic engineering for stress tolerance. *Planta, 218*(1), 1–14. DOI: 10.1007/s00425-003-1105-5 PMID: 14513379

Wang, Y., Chen, X., Li, X., Song, Y., Wang, J., Huang, J., & Xue, R. (2021). Exogenous application of 5-aminolevulinic acid alleviated damage to wheat chloroplast ultrastructure under drought stress by transcriptionally regulating genes correlated with photosynthesis and chlorophyll biosynthesis. *Acta Physiologiae Plantarum, 44*(1), 12. Advance online publication. DOI: 10.1007/s11738-021-03347-6

Wang, Y., Huang, Y., Wang, J., Cheng, C., Huang, W., Lu, P., Xu, Y.-N., Wang, P., Yan, N., & Shi, Y. (2009). Structure of the formate transporter FocA reveals a pentameric aquaporin-like channel. *Nature*, *462*(7272), 467–472. DOI: 10.1038/nature08610 PMID: 19940917

Wang, Z., & Benning, C. (2011). Arabidopsis thaliana polar glycerolipid profiling by thin layer chromatography (TLC) coupled with gas-liquid chromatography (GLC). *Journal of Visualized Experiments*, *e2518*(49). DOI: 10.3791/2518-v PMID: 21445048

Warnasuriya, S. D., Udayanga, D., Manamgoda, D. S., & Biles, C. (2023). Fungi as environmental bioindicators. *The Science of the Total Environment*, *892*, 164583. DOI: 10.1016/j.scitotenv.2023.164583 PMID: 37277042

Watanabe, Y., Ichinomiya, Y., Shimada, D., Saika, A., Abe, H., Taguchi, S., & Tsuge, T. (2012). Development and validation of an HPLC-based screening method to acquire polyhydroxyalkanoate synthase mutants with altered substrate specificity. *Journal of Bioscience and Bioengineering*, *113*(3), 286–292. DOI: 10.1016/j.jbiosc.2011.10.015 PMID: 22088761

Weissmannová, H. D., & Pavlovský, J. (2017). Indices of soil contamination by heavy metals – methodology of calculation for pollution assessment (minireview). *Environmental Monitoring and Assessment*, *189*(12), 1–25. DOI: 10.1007/s10661-017-6340-5 PMID: 29116419

Wei, Y., Zhang, Y. J., Cai, Y., & Xu, M. H. (2015). The role of mitochondria in mTOR-regulated longevity. *Biological Reviews of the Cambridge Philosophical Society*, *90*(1), 167–181. DOI: 10.1111/brv.12103 PMID: 24673778

Wieczorek, J., Baran, A., Urbański, K., Mazurek, R., & Klimowicz-Pawlas, A. (2018). Assessment of the pollution and ecological risk of lead and cadmium in soils. *Environmental Geochemistry and Health*, *40*(6), 2325–2342. DOI: 10.1007/s10653-018-0100-5 PMID: 29589150

Włodarczyk, K., & Smolińska, B. (2022). The Effect of Nano-ZnO on Seeds Germination Parameters of Different Tomatoes (Solanum Lycopersicum L.) Cultivars. *Molecules (Basel, Switzerland)*, *27*(15), 4963. DOI: 10.3390/molecules27154963 PMID: 35956913

Wolf, J., Ross, I. L., Radzun, K. A., Jakob, G., Stephens, E., & Hankamer, B. (2015). High-throughput screen for high performance microalgae strain selection and integrated media design. *Algal Research*, *11*, 313–325. DOI: 10.1016/j.algal.2015.07.005

Wong, V. N. L., Dalal, R. C., & Greene, R. S. B. (2009). Carbon dynamics of sodic and saline soils following gypsum and organic material additions: A laboratory incubation. *Applied Soil Ecology, 41*(1), 29–40. DOI: 10.1016/j.apsoil.2008.08.006

Wouyou, A., Gandonou, C., Komlan, F., Montcho, D., Zanklan, A., Lutts, S., & Gnancadja, S. (2017). Salinity Resistance of Five Amaranth (Amaranthus cruentus) Cultivars at Young Plants Stage. *International Journal of Plant and Soil Science, 14*(3), 1–11. DOI: 10.9734/IJPSS/2017/31611

Wu, Yuwen, Wenjie, & Feng. (2024). Measurement of cellular MDA content through MTBE-extraction based TBA assay by eliminating cellular interferences. *Journal of Pharmaceutical and Biomedical Analysis, 248*, 116332 DOI: 10.1016/j.jpba.2024.116332

Wuana, R. A., & Okieimen, F. E. (2011). Heavy Metals in Contaminated Soils: A Review of Sources, Chemistry, Risks and Best Available Strategies for Remediation. *ISRN Ecology, 2011*(1), 402647. DOI: 10.5402/2011/402647

Wyszkowska, J., Borowik, A., Kucharski, M., & Kucharski, J. (2013). Oddzia£ywanie kadmu, miedzi i cynku na roœliny, drobnoustroje i enzymy glebowe. *Journal of Elementology, 18*(4), 769–796. DOI: 10.5601/jelem.2013.18.4.455

Xia, S., Wang, X., Su, G., & Shi, G. (2015). Effects of drought on cadmium accumulation in peanuts grown in a contaminated calcareous soil. *Environmental Science and Pollution Research, 22*(23), 18707–18717. DOI: 10.1007/s11356-015-5063-9

Xianzhao, L., Chunzhi, W., & Qing, S. (2013). Screening for Salt Tolerance in Eight Halophyte Species from Yellow River Delta at the Two Initial Growth Stages. *ISRN Agronomy, 2013*, 1–8. DOI: 10.1155/2013/592820

Xia, Y., Gao, T., Liu, Y., Qi, M., Zhu, J. M., Tong, H., Lv, Y., & Liu, C. (2025). Cd/Pb behavior during combustion in a coal-fired power plant and their spatiotemporal impacts on soils: New insights from Cd/Pb isotopes. *Journal of Environmental Sciences (China), 150*, 582–593. DOI: 10.1016/j.jes.2024.03.002 PMID: 39306431

Xing, W., Zheng, Y., Scheckel, K. G., Luo, Y., & Li, L. (2019). Spatial Distribution of Smelter Emission Heavy Metals on Farmland Soil. *Environmental Monitoring and Assessment, 191*(2), 115. DOI: 10.1007/s10661-019-7254-1 PMID: 30697651

Yang, A., Akhtar, S. S., Iqbal, S., Amjad, M., Naveed, M., Zahir, Z. A., & Jacobsen, S. E. (2016). Enhancing salt tolerance in quinoa by halotolerant bacterial inoculation. *Functional Plant Biology, 43*(7), 632–642. DOI: 10.1071/FP15265 PMID: 32480492

Yang, F., Hanna, M. A., & Sun, R. (2012). Value-added uses for crude glycerol—A byproduct of biodiesel production. *Biotechnology for Biofuels, 5*(1), 13. DOI: 10.1186/1754-6834-5-13 PMID: 22413907

Yang, W., Tian, Y., Han, M., & Miao, X. (2017). Longevity extension of worker honey bees (Apis mellifera) by royal jelly. Optimal dose and active ingredient. *PeerJ, 5*, e3118. DOI: 10.7717/peerj.3118 PMID: 28367370

Yi, Q., & Huang, C. (2019). Review of heavy metal pollution by mining. *E3S Web of Conferences, 118*, 04028. DOI: 10.1051/e3sconf/201911804028

Zagrodnik, R., & Sobociński, D. (2024). Continuous dark-fermentative H2 production using carbon components of lignocellulose hydrolysates: Insight into the difference between mixed and single substrates. *International Journal of Hydrogen Energy, 52*, 378–388. DOI: 10.1016/j.ijhydene.2023.06.119

Zavyalov, R., Latkovskaya, E., & Repina, M. (2022) Estimation of possible storm discharge volumes of macrophytes on the coast of Aniva Bay (Sea of Okhotsk). Aniva Bay (Sea of Okhotsk). *Issues of modern scientific research: collection of articles of XVII International Scientific and Practical Conference,* 880-889.

Zeeshan, M., Lu, M., Sehar, S., Holford, P., & Wu, F. (2020). Comparison of biochemical, anatomical, morphological, and physiological responses to salinity stress in wheat and barley genotypes deferring in salinity tolerance. *Agronomy (Basel), 10*(1), 127. DOI: 10.3390/agronomy10010127

Zhang, W., Wei, J., Guo, L., Fang, H., Liu, X., Liang, K., Niu, W., Liu, F., & Siddique, K. H. M. (2023). Effects of Two Biochar Types on Mitigating Drought and Salt Stress in Tomato Seedlings. *Agronomy, 13*(4), 1039. DOI: 10.3390/agronomy13041039

Zhang, L., Ma, H., Chen, T., Pen, J., Yu, S., & Zhaoe, X. (2014). Morphological and Physiological Responses of Cotton (Gossypium hirsutum L.) Plants to Salinity. *PLoS One, 9*(11), e112807. DOI: 10.1371/journal.pone.0112807 PMID: 25391141

Zhao, C.-R., You, Z.-L., & Bai, L. (2024). Fungal Plasma Membrane H^+-ATPase: Structure, Mechanism, and Drug Discovery. *Journal of Fungi (Basel, Switzerland), 10*(4), 273. DOI: 10.3390/jof10040273 PMID: 38667944

Zheng, Y., Jin, R., Zhang, X., Wang, Q., & Wu, J. (2019). The considerable environmental benefits of seaweed aquaculture in China. *Stochastic Environmental Research and Risk Assessment, 33*(4-6), 1203–1221. DOI: 10.1007/s00477-019-01685-z

Zheng, Y., Sun, J., Luo, Z., & Huang, Y. (2024). Emerging mechanisms of lipid peroxidation in regulated cell death and its physiological implications. *Cell Death & Disease, 15*(11), 859. DOI: 10.1038/s41419-024-07244-x PMID: 39587094

Zhiltsova, L. V., & Prikhodchenko, A. V. (2023) Rational approach to the collection of storm emissions of Anfeltia in Primorye. *Actual problems of biodiversity and nature management: proceedings of the 2nd National Scientific and Practical Conference, dedicated to the 20th anniversary of the Department of Marine Ecology FGBOU VO 'KSMTU'*, 300-305.

Zhu, J. K. (2003). *Salt and drought stress signal transduction in plants.* DOI: 10.1146/annurev.arplant.53.091401.143329

Zhu, J. K. (2016). Abiotic Stress Signaling and Responses in Plants. *Cell, 167*(2), 313–324. DOI: 10.1016/j.cell.2016.08.029 PMID: 27716505

Zhu, X., Cao, Q., Sun, L., Yang, X., Yang, W., & Zhang, H. (2018). Stomatal conductance and morphology of arbuscular mycorrhizal wheat plants response to elevated co2 and NaCl Stress. *Frontiers in Plant Science*, *9*, 410525. DOI: 10.3389/fpls.2018.01363 PMID: 30283478

Zhu, Z., Hu, Y., Teixeira, P. G., Pereira, R., Chen, Y., Siewers, V., & Nielsen, J. (2020). Multidimensional engineering of Saccharomyces cerevisiae for efficient synthesis of medium-chain fatty acids. *Nature Catalysis*, *3*(1), 64–74. DOI: 10.1038/s41929-019-0409-1

Zikmanis, P., Kolesovs, S., & Semjonovs, P. (2020). Production of biodegradable microbial polymers from whey. *Bioresources and Bioprocessing*, *7*(1), 36. DOI: 10.1186/s40643-020-00326-6

Zilfyan, V. N., & Mnatsakanyan, A. G. (1964). Geographic distribution and some features of natural tularemia foci in the Armenian SSR. *Journal of Microbiology. Epidemiology and Immunology*, *4*(2), 14.

Zulfiqar, F., & Ashraf, M. (2021). Nanoparticles potentially mediate salt stress tolerance in plants. *Plant Physiology and Biochemistry*, *160*, 257–268. DOI: 10.1016/j.plaphy.2021.01.028 PMID: 33529801

Zuroff, T. R., Xiques, S. B., & Curtis, W. R. (2013). Consortia-mediated bioprocessing of cellulose to ethanol with a symbiotic Clostridium phytofermentans/yeast co-culture. *Biotechnology for Biofuels*, *6*(1), 59. DOI: 10.1186/1754-6834-6-59 PMID: 23628342

About the Contributors

Karen Ghazaryan, Dean of the Faculty of Biology at Yerevan State University, boasts over 22 years of experience, excelling in both teaching and research. His work has significantly advanced soil ecology, water resource management, climate change impacts, ecological management, environmental assessment, and biodiversity conservation. With a Scopus H-index of 29 and over 124 publications, Dr. Ghazaryan leads three major projects on salinity, heavy metal, and drought stress, aligning with the Sustainable Development Goals (SDGs). He supervises Master's and PhD students and ranks 36th nationally and 13th at Yerevan State University according to the AD Scientific Index (2025). As a reviewer, he contributes to journals such as Acta Biologica Hungarica, Environmental Advances, Journal of Hydrology, and Science of the Total Environment. He is on the editorial boards of the Egyptian Journal of Soil Science and Discover Agriculture.

Karen Trchounian is a distinguished academic and researcher specializing in microbiology, bioenergetics, and biotechnology. Born on May 14, 1989, in Yerevan, Armenia, he holds Armenian citizenship and is proficient in Armenian, Russian, and English, with additional knowledge of German and Spanish. Currently, Prof. Trchounian serves as a Professor in the Department of Biochemistry, Microbiology, and Biotechnology at the Faculty of Biology, Yerevan State University (YSU), Armenia. He is also the Director of the Research Institute of Biology and leads the Laboratory of Microbiology, Bioenergetics, and Biotechnology at YSU. In addition, he heads the YSU Microbial Biotechnologies and Biofuel Innovation Center, reflecting his dedication to advancing microbial biotechnology and sustainable energy solutions. Prof. Trchounian's research is highly regarded, with a Scopus h-index of 28 and a Google Scholar h-index of 31, supported by nearly 2700 citations. His work focuses on microbial bioenergetics, hydrogen production, and biotechnological applications, contributing significantly to sustainable energy and environmental solutions. His extensive academic output, leadership in innovation, and commitment to scientific excellence make him a leading figure in his field. He is an active contributor to global research communities and is recognized for his impactful work in advancing microbial biotechnology.

Abhishek Singh, a Senior Scientist (Senior Researcher) at Yerevan State University, focuses on crop resilience and productivity under abiotic stresses like salinity, drought, heavy metal toxicity, and biochar applications. His research delves into physio-biochemical responses and anatomical adaptations for crop resilience, and the use of metallic nanoparticles to enhance nutrient uptake, bioaccumulation, and stress tolerance, promoting sustainable agriculture. According to the AD Scientific Index, Dr. Singh holds the 39th position in Armenia and ranks 15th at Yerevan State University. This ranking is based on his impressive academic record, which includes a Scopus H-index of 30 and a portfolio of more than 135 published works. In 2024, he received BioThink - National Academy of Biosciences' "Best Scientist Award." Dr. Singh is an active peer reviewer for journals like Acta Ecologica Sinica, Biocatalysis and Agricultural Biotechnology, Bulletin of Environmental Contamination and Toxicology, and more. He also serves on the editorial boards of Scientific Reports, Discover Sustainability, and Discover Agriculture.

* * *

Christopher Erdmann studied Biology at the University of Rostock, Germany and is now studying his PhD at the Martin Luther University Halle-Wittenberg in the Institute of Microbiology.

Elena Kalita is a researcher at joint Laboratory of Experimental Biology, CANDLE Synchrotron Research Institute, Yerevan, Armenia, and Laboratory of Cell Technologies, Institute of Molecular Biology NAS RA, Yerevan, Armenia. Her work focuses on the biomedical effects of ultrashort pulsed electron beam (UPEB) irradiation at the AREAL accelerator, using 2D and 3D cancer models. In parallel, she has experience in drug safety evaluation and anticancer activity screening, integrating in vitro studies with in silico modeling and bioinformatics approaches. ORCID: 0000-0002-8785-2817.

Anzhela Sargsyan is a PhD in Genetics and a researcher and lecturer at the Faculty of Biology, Yerevan State University (YSU). She is currently an Assistant at the Department of Genetics and Cytology and a Researcher in the Genomic Instability and Molecular Mutagenesis Group at the Research Institute of Biology, YSU. Dr. Sargsyan completed her Bachelor's, Master's, and PhD studies in Genetics and Cytology at Yerevan State University. Her research focuses on genetics, molecular and cellular biology, mutagenesis, genotoxicology, ecogenotoxicology, epigenetics, and molecular cytogenetics. She has extensive experience applying genetic and epigenetic biomarkers to assess DNA damage and genomic instability in human cells and environmental bioindicator organisms. She has participated in multiple international research projects and competitive grants supported by Armenian and international funding agencies, and has undertaken numerous long-term research visits to the Institute of Human Genetics in Jena, Germany. Dr. Sargsyan is the author and co-author of peer-reviewed publications in high-impact international journals and has actively participated in international conferences and scientific workshops. In parallel, she is deeply engaged in teaching undergraduate and graduate courses in genetics, molecular biology, genomics, and epigenetics.

Gary Sawers studied biochemistry at Heriot-Watt University in Edinburgh and completed a PhD in Biochemistry at the University of Dundee. After a post-doctoral stay at the Ludwig Maximilian University at Munich, he became a project leader at the John-Innes Center in Norwich, U.K., before becoming chair of Microbiology at the Martin Luther University Halle-Wittenberg in Germany.

Rupesh Kumar Singh is a passionate researcher with 10 years of post-PhD experience in diverse area of Agronomy, plant molecular biology, Vitiviniculture, plant physiology, Plant pathology, genetic engineering and modern genome editing technology, tissue culture, biochemistry, nanotechnology, biocontrol agent development and formulations, transcriptomics, proteomics, metabolomics. The research profile includes 3212 citations, H index-24 and 92 peer reviewed published scientific communications indexed in Scopus database including 1 granted patent and 2 provisional submissions recently. He gained academic and industrial research experience while working in national and international institutes and industries of good repute. He did his first postdoc project (September 2013- January 2016) at department of Biology, University of Minho, Braga, Portugal where he studied the recalcitrance nature of Hypericum perfoliatum and the complex molecular mechanism with its interaction to Agrobacterium tumefaciens. He did his second postdoc (July 2016-March 2020) in Department of Agronomy, Vitiviniculture lab, University of Tras-O-Montes and Alto Douro, Vila Real, Portugal. During this tenure, he worked as a senior member in the lab and management of one research line in the INNOVINE&WINE Innovation platform of Vineyard and Wine (cofinanced by European Regional Development Fund, FEDER, with a total budget of 5 293 984,76) project. He has been leading a team of 7 peoples (supervise 2 MSc) for performing large scale field trials on different vineyards in Douro region. The research was performed in collaboration with academia and industries and the major outcome during this work was chitosan-based nanomaterial preparation and application in vineyards. The findings were promising which was patented nationally and for European Union territories; while several good articles have been published. Additionally; his research has been ranked second for ENOFORUM AWARD in The Spanish edition of Enoforum 2020 among 35 scientific works in the wine sector from 21 research groups of different countries. After 2 postdoc projects, he moved to Ireland to gain experience in industrial research and started working as Principal Investigator in MBIO (Bio-control group, Monaghan Biosciences, Monaghan county, Ireland). Major duties during this tenure were leading the team of scientists, technicians and research trainees, Project management, establishment of new research facilities, Collaborations with universities. He has been also responsible for scale-up process development, validation of biocontrol/bio stimulants/plant growth promoting biological products and product commercialization. Later he joined as PhD researcher (January 2022-December 2022) in InnovPlantProtect Collaborative Laboratory, Department of Protection of Specific Crops, Elvas, Portugal. Currently he is working as Researcher biologist in UTAD. He did his PhD from Plant Molecular Biology and Genetic Engineering Department, Defense Institute of Bioenergy Research, India. He has been invited as an external expert by National Research and Development Agency (ANID), Government of Chile for evaluation of research proposals for national funding. Additionally he has been interviewed by french magazine regarding his recent research in Grapevine as a new and sustainable practice in vitiviniculture. Acknowledgments Authors acknowledge the financial support by National Funds by FCT - Portuguese Foundation for Science and Technology, under the project UIDB/04033/ 2020 (https://doi.org/10.54499/UIDB/04033/2020), ECONUTRI project, the European Union Horizon Europe Innovation program under the Grant Agreement No. 101081858.

João Ricardo Sousa holds a PhD in Soil and Environmental Sciences and is a Professor specializing in soil chemistry, fertilization, and plant nutrition. His research focuses on soil pollution and remediation, nutrient cycling, and soil quality indicators. He has expertise in composting, organic residue assessment, and environmental risk evaluation. Dr. Sousa has participated in more than ten national and international projects on soil–organic matter dynamics, nutrient management, and nitrogen loss mitigation, advancing sustainable soil management.

Index

A

Agro-Industrial Residues 61, 62, 63
Alashkert 40, 41, 53
Ampicillin-Resistant E. coli 286, 289, 291, 294
Antibacterial Activity 134, 285, 286, 287, 289, 296, 298, 300
Antibiotic Modulatory Activity 287
Anticancer Activity 176, 177, 190
Antioxidant Enzymes 131, 135, 137, 138, 142, 264, 265, 278
Ararat Plain 41, 103, 104
Armenian Barley 41
Assessment of Economic Efficiency 349, 350
ATPase 12, 14, 16, 18, 126, 132, 135, 137, 138, 144, 145, 146, 150, 264, 265, 267, 268, 269, 270, 274, 278, 280, 288

B

Bibliometric 215, 315, 316, 319, 320, 321, 335, 336, 338, 340, 341, 342, 344, 358
Biochar 232, 315, 316, 317, 318, 320, 322, 323, 324, 325, 326, 327, 328, 329, 330, 331, 332, 333, 334, 336, 338, 339, 340, 341, 342, 343, 344, 345, 347, 352, 354, 355
Bioenergy 81, 104, 347, 348, 359, 360
Biofuels 17, 23, 36, 38, 62, 63, 74, 79, 81, 82, 84, 347, 348, 350, 351, 355, 358, 359, 360
Biomass 22, 25, 28, 33, 37, 40, 42, 44, 46, 62, 63, 68, 69, 74, 79, 81, 82, 83, 89, 92, 98, 105, 106, 107, 110, 111, 112, 114, 122, 131, 135, 136, 137, 141, 142, 145, 178, 179, 180, 183, 184, 185, 186, 190, 191, 192, 193, 204, 230, 266, 317, 334, 347, 348, 349, 350, 351, 352, 353, 354, 355, 356, 357, 361
Bryophytes 132, 133, 134, 135, 138, 139, 146, 147, 148, 149

C

Carbon-Neutral Economy 347
Catalase 131, 137, 143, 144, 145, 264, 265, 268, 271, 272, 274, 275, 277, 279
Cattle 303, 305, 306, 307, 308, 309, 311
Cell Viability 153, 156, 157, 159, 162, 167, 168, 176, 182, 190
Chlorella vulgaris 177, 178, 184, 185, 186, 187, 189, 190, 191, 194, 195
Clostridium beijerinckii 63, 76, 80, 84
Copper 101, 219, 221, 226, 229, 232
Cupriavidus necator H16 21, 23
Cytoplasmic Membrane 2, 7, 10, 11
Cytotoxicity 151, 154, 156, 157, 159, 162, 171, 181, 182, 189, 190, 196, 229

D

Decellularization 152, 154, 155, 161, 168, 171, 172, 173
Dilijan National Park 304, 305, 309
DNA Damage 242, 243, 244, 246, 254, 255, 256, 257, 259, 260
Drought 54, 57, 60, 86, 101, 102, 103, 105, 106, 110, 116, 123, 124, 128, 129, 315, 316, 317, 318, 320, 322, 323, 324, 325, 326, 327, 328, 329, 330, 331, 332, 333, 334, 336, 338, 340, 341, 342, 343, 344, 345

E

Ecological Resources 197, 198, 202, 210
Environmental Health 218
Environmental Risk 261
Escherichia coli 1, 15, 16, 17, 18, 19, 24, 83, 84, 284, 300
Extracellular Matrix 152, 171, 172

F

Flavonoids 131, 133, 134, 136, 139, 147, 265, 283, 284, 286, 287, 289, 291, 292, 296, 298, 299, 300, 301
Formate Homeostasis 1, 2, 3, 9, 10, 13, 17
Formate Hydrogenlyase 1, 2, 15, 17, 18

Formic Acid 1, 2, 3, 7, 8, 9, 10, 11, 12, 13, 14, 17, 19

G

Genotoxicity 248, 249, 250, 257, 258, 259, 261, 262
Glycerol 22, 23, 24, 25, 28, 29, 30, 31, 32, 33, 34, 36, 37, 38, 70, 72, 83, 84
Glycoglycerolipids 176, 184, 191, 193, 194
Green Energy 213
Growth Parameters 25, 28, 44, 47, 89, 92

H

Heavy Metals 84, 217, 218, 219, 221, 223, 224, 225, 226, 229, 232, 233, 235, 236, 237, 238, 239, 240, 262, 317, 358
Hepatotoxicity 153, 168, 171, 172, 173
Hydrogenase Enzymes 24, 33, 34, 35
Hypophosphite 1, 2, 3, 4, 5, 6, 7, 8, 9, 10, 11, 18

I

Immunogenicity 151, 152, 154, 157, 163, 167, 172
Ixodidae 303, 304, 310, 311

K

Kanamycin-Resistant E. coli 286, 289, 290, 292, 293

L

Localization 308

M

MCF-7 Cells 182
Methods for Assessing Agricultural Soil Pollution 257
Microalgae 24, 176, 177, 178, 179, 189, 190, 191, 192, 194, 195, 196
Millimeter Waves 141, 263, 264, 270, 272, 274
Monogalactosyldiacylglycerol 175, 176, 177, 187

N

Na+ 37, 40, 41, 43, 49, 50, 51, 52, 108, 113, 117, 118, 122, 125, 240, 246, 259, 270, 280
Neochloris oleoabundans 177, 178, 184, 185, 186, 187, 190

P

pH Homeostasis 16
Phytodesalination 49, 51
Polyhydroxyalkanoate 21, 33, 37, 38
Proliferation 152, 158, 159, 164, 165, 167, 168, 172, 176
Proton Fluxes 287, 292, 294, 295, 296

R

Recellularization 151, 153, 154, 159, 160, 161, 165, 167, 168, 170, 173
Royal Jelly 149, 264, 265, 274, 277, 278, 279, 280

S

Saccharomyces cerevisiae 63, 78, 81, 82, 83, 84, 148, 277, 279
Salinity 22, 39, 40, 41, 44, 45, 46, 47, 48, 49, 50, 51, 52, 53, 54, 55, 56, 57, 58, 59, 60, 85, 86, 87, 89, 90, 91, 93, 94, 95, 96, 97, 98, 100, 101, 102, 103, 104, 105, 106, 107, 108, 109, 110, 111, 112, 113, 114, 115, 116, 117, 118, 119, 120, 121, 122, 123, 124, 125, 126, 127, 128, 129, 315, 316, 317, 318, 320, 322, 323, 324, 325, 326, 327, 328, 329, 330, 331, 332, 333, 334, 338, 340, 342, 344
Salinity Stress 39, 40, 44, 45, 46, 47, 49, 51, 53, 54, 55, 56, 57, 58, 59, 85, 89, 95, 101, 102, 103, 110, 111, 112, 113, 117, 118, 121, 124, 125, 126, 127, 128, 129, 322, 323, 324, 325, 326,

334, 340, 344
Salt-Tolerant Crops 103
Scopus 319, 320, 349, 358
SDGs 198, 199, 201, 212, 213, 218
Sheep 303, 305, 306, 307, 308, 309
SOD 137, 143, 144, 145, 264, 265, 268, 271, 272, 274, 275
Soil Contamination 220, 221, 233, 235, 236, 240
Soil Salinization 89, 104, 106, 341
Sustainable Agriculture 39, 40, 41, 53, 86, 106, 122, 316, 322, 323, 327, 329, 330, 333, 334, 336, 338, 342, 343
Sustainable Development 125, 197, 198, 201, 207, 208, 209, 210, 212, 214, 237, 325, 340, 341

T

Tissue Engineering 151, 152, 170, 171, 172, 173

V

Viability 34, 44, 153, 156, 157, 159, 162, 163, 165, 167, 168, 176, 182, 190, 255, 257, 287, 289, 350, 356, 357

W

Whey 22, 23, 25, 26, 28, 29, 30, 31, 32, 33, 34, 35, 36, 37, 38

Y

Yeast 64, 66, 68, 69, 78, 79, 81, 82, 84, 131, 134, 136, 137, 138, 139, 140, 141, 142, 143, 144, 145, 148, 195, 263, 264, 265, 266, 267, 268, 269, 270, 272, 273, 274, 275, 276, 277, 278, 279, 280

www.ingramcontent.com/pod-product-compliance
Lightning Source LLC
LaVergne TN
LVHW062328300926
849161LV00007B/1617